D0764270

Feeds and Feeding

Fifth Edition

TILDEN WAYNE PERRY
Emeritus Professor of Animal Nutrition
Purdue University

ARTHUR E. CULLISON
Late Professor of Animal Science
University of Georgia

ROBERT S. LOWREY
Professor of Animal Science
University of Georgia

PRENTICE HALL
Upper Saddle River, New Jersey 07458

0399309048

Library of Congress Cataloging-in-Publication Data

Perry, Tilden Wayne.

  Feeds and feeding / Tilden Wayne Perry, Arthur E. Cullison,
Robert S. Lowrey. — 5th ed.

     p.   cm.

  Rev. ed. of: Feeds and feeding / Arthur E. Cullison, 4th ed. 1987.

  Includes index.

  ISBN 0-13-319294-6

  1. Feeds.   2. Animal nutrition.   3. Animal feeding.   I. Lowrey,
Robert S., 1934- . II. Cullison, Arthur Edison, 1914- Feeds and
feeding.   III. Title.

SF95.P435   1999

636.08'4—dc21   99-408885

                                                         CIP

Publisher: *Charles Stewart*
Editorial Assistant: *Jennifer Stagman*
Production Editor: *Lori Harvey/Carlisle Publishers Services*
Production Liaison: *Eileen M. O'Sullivan*
Director of Manufacturing & Production: *Bruce Johnson*
Managing Editor: *Mary Carnis*
Production Manager: *Marc Bove*
Marketing Manager: *Melissa Bruner*
Cover Designer: *Liz Nemeth*
Formatting/page make-up: *Carlisle Communications, Ltd.*
Printer: *R. R. Donnelley Harrisonburg*

10 9 8 7 6 5 4 3 2 1

ISBN 0-13-319294-6

Prentice-Hall International (UK) Limited, *London*
Prentice-Hall of Australia Pty. Limited, *Sydney*
Prentice-Hall Canada Inc., *Toronto*
Prentice-Hall Hispanoamericana, S.A., *Mexico*
Prentice-Hall of India Private Limited, *New Delhi*
Prentice-Hall of Japan, Inc., *Tokyo*
Simon & Schuster Asia Pte. Ltd., *Singapore*
Editora Prentice-Hall do Brasil, Ltda., *Rio de Janeiro*

Contents

Preface

Feeds and Feeding, fifth edition, has been prepared by the senior author in order to keep current the text that was first written by Dr. Arthur E. Cullison and published in 1975. Following the traditional design and intent of Dr. Cullison—and also Dr. Robert S. Lowrey in the fourth edition—the current revision has been prepared as a text for an undergraduate course in animal nutrition and feeding. Therefore, only information pertinent to such a course has been included. While this book answers most basic questions pertaining to animal feeding and nutrition, it does not deal with more unusual feeds and is not intended as an all-purpose reference on such matters.

The overall importance of poultry meat—especially broiler chicken and turkeys—as sources of human food, has increased very rapidly, whereas meat from beef, pork, lamb, and veal has tended to either remain fairly constant or to decline. Charts have been prepared to demonstrate changes in consumption of animal and poultry meat per person, in the United States, between about 1960 and 1997, and presented in connection with the subject matter relevant to each of the species. Therefore, the fifth edition was designed to contain sections on the nutrition of laying hens, chicks for the production of broiler meat, and turkeys for the production of turkey meat. This represents a change in the format of the book.

This edition, following the tradition of the first four editions, is not meant as a review of original research literature. Most complete literature citations are found in each of National Research Council (NRC) Bulletins covering each of the species of animals for which nutrition and feeding programs have been included. Since each of the NRC Bulletins has been prepared by a committee of outstanding researchers and teachers, such bulletins have served as the ultimate authority in preparing nutrient and feed recommendations presented in this book. However, there are several other textbooks that contain excellent reviews of literature that deserve to be perused when the student wishes to go into greater depth on a subject. Although the last edition of Morrison's *Feeds and Feeding* was published nearly forty years ago, it still has a great deal of pertinent information that was useful to the senior author in preparing the fifth edition. *Feeds and Nutrition,* by Ensminger, Oldfield, and Heinemann, containing more than 1500 pages, is probably one of the most nearly complete books on the subject. In addition, more basic nutrition books such as Maynard's *Animal Nutrition* might be helpful reference books.

The fifth edition provides the latest information available on feed composition for use in ration formulation. However, because NRC Bulletins set up nutrient requirements quite differently for monogastric and ruminant animals—and even for species within a category—it is necessary to provide two sets of feed tables, one for the ruminant animals and one for the monogastric animals. Naturally, this necessitates additional pages in the text but such changes also become necessary as research findings and changes in types of animals develop. Such feed tables have been developed from those presented in the respective NRC Bulletins. The student should be made aware that different specie committees may differ in the manner in which feed table date are presented. In other words, the fact that some feed data are presented on a 100% dry matter basis, whereas other feed data are presented on an air-dry basis is a reflection of the specie committee preparing that bulletin. Either method of calculating formulations is quite effective as long as the student takes any such effects into account.

The authors wish to express appreciation to the National Academy of Sciences for permission to use the data that have been generated by the respective specie committees and presented in the respective specie bulletins. The following NRC Bulletins are cited in this fifth revision:

Nutrient Requirements of Beef Cattle, Seventh Rev. Ed., 1996.
Nutrient Requirements of Dairy Cattle, Sixth Rev. Ed., 1989.
Nutrient Requirements of Horses, Fifth Rev. Ed., 1989.
Nutrient Requirements of Poultry, Ninth Rev. Ed., 1994.
Nutrient Requirements of Sheep, Sixth Rev. Ed., 1985.
Nutrient Requirements of Swine, Tenth Rev. Ed., 1998.

The senior author, who prepared the fifth edition, would like to thank the following people for their contributions: Nancy Perry for preparation of the meat consumption per capita charts; James Herndon for the many photographic illustrations added to this revision; many people for personal communications concerning many subjects; many persons at several universities for counsel and assistance; and his wife Ena Perry for her patience while the text was being revised. However, no dedication of this edition is being made because it is a continuation of a project begun by my colleague, Dr. Arthur E. Cullison more than a quarter of a century ago.

Tilden Wayne Perry,
Senior Author, Fifth Edition

Introduction

It appears that people have been aware of how dependent they are upon all of the life of the earth, the sky, and the sea, and have attempted to aid in the propagation and nourishment of such life. Naturally, there have been exceptions wherein poor judgment in tilling the soil and management of the sources of the seas have resulted in decline in potential value of such resources. In more recent years, people are attempting to realign their vision and thinking in management of these very valuable resources.

This text was meant to be of assistance in feeding several species of animals on which people are so dependent for their daily meat, milk, and eggs. A table is presented in this section that depicts how dependent people are upon the animal life for their daily food. These U.S. Department of Agriculture data demonstrate that we consume more than one-half pound (240 g) of livestock meat, poultry, and fish/shellfish per day, plus one egg every two days. Through the sciences of genetics, environmental control, nutrition, and physiology, the efficiency of meat and egg animal production has been improved. It is the hope of the authors of this text that *Feeds and Feeding* will assist the student and producer in continuing to increase the production of abundant supplies of wholesome and healthful meat animal products. The graph presented shows that people continue to utilize livestock and poultry as a source of food. In fact, the graph shows that our appetite for animal products has increased 29% (140.1 vs 180.0 lb per capita/year) from 1960 projected through 1998.

LIVESTOCK MEAT, POULTRY, AND FISH/SHELLFISH CONSUMPTION IN BONELESS EQUIVALENT WEIGHT, POUNDS PER PERSON, 1960–98[1]

	Beef	Pork	Lamb and Mutton	Veal	Live-stock Meat	Young Chicken	Turkey	Total Poultry	Fish and Shellfish	Total Fish & Meat	Eggs[2]
1960	59.8	48.9	3.1	4.2	116.0		4.9	24.1	10.3	150.4	—
1961	61.8	47.1	3.3	4.0	116.2		5.9	26.6	10.7	153.5	—
1962	62.5	47.7	3.4	3.8	117.4		5.6	26.2	10.6	154.2	—
1963	66.3	48.9	3.2	3.5	121.9	18.7	5.5	26.8	10.7	159.4	—
1964	70.5	49.2	2.7	3.7	126.1	19.1	5.8	27.4	10.5	164.0	—
1965	69.5	43.8	2.4	3.7	119.4	20.4	5.9	28.9	10.8	159.1	—
1966	73.8	43.1	2.6	3.2	122.7	22.0	6.3	30.8	10.9	164.4	—
1967	75.5	47.4	2.5	2.8	128.2	22.3	6.8	31.9	10.6	170.7	—
1968	77.5	48.7	2.4	2.6	131.2	22.5	6.4	31.7	11.0	173.9	—
1969	78.0	47.5	2.3	2.3	130.1	23.9	6.6	33.0	11.2	174.3	—
1970	79.8	48.6	2.1	2.0	132.5	25.2	6.4	34.1	11.7	178.3	311
1971	79.2	53.0	2.1	1.9	136.2	25.1	6.6	34.3	11.5	182.0	314
1972	80.8	48.1	2.2	1.6	132.7	26.2	7.1	35.8	12.5	181.0	308
1973	76.8	43.4	1.8	1.3	123.1	25.5	6.7	34.5	12.7	170.3	294
1974	80.7	47.1	1.5	1.6	130.9	25.5	7.0	34.9	12.1	172.1	288
1975	83.2	38.5	1.3	2.8	125.8	25.2	6.7	34.2	12.1	177.9	277
1976	89.1	41.0	1.2	2.7	134.0	27.4	7.2	36.6	12.9	183.5	270
1977	86.4	42.6	1.1	2.7	132.8	28.1	7.2	37.4	12.6	182.8	287
1978	82.4	42.8	1.0	2.0	128.2	29.2	7.2	39.2	13.4	180.8	272
1979	73.7	49.1	1.0	1.4	125.2	31.6	7.8	41.4	13.0	179.6	278
1980	72.2	52.6	1.0	1.3	125.5	31.5	8.3	41.9	12.5	180.6	273
1981	72.8	45.3	1.1	1.4	120.4	32.2	8.5	42.8	12.7	180.7	265
1982	72.6	45.3	1.1	1.4	120.4	32.4	8.5	43.1	12.5	175.4	264
1983	73.9	47.7	1.1	1.4	124.1	32.7	8.9	43.7	13.4	180.9	261
1984	73.8	47.5	1.1	1.5	123.9	34.1	9.0	44.9	14.2	182.6	261
1985	74.7	48.1	1.1	1.5	125.4	35.2	9.2	45.7	15.1	186.2	255
1986	74.5	45.6	1.0	1.6	122.7	36.0	10.2	47.5	15.5	185.7	252
1987	69.7	46.0	1.0	1.3	117.9	38.1	11.6	51.1	16.2	185.2	249
1988	68.7	49.2	1.0	1.2	120.1	38.3	12.4	52.1	15.2	187.4	244
1989	65.6	48.8	1.0	1.0	116.5	39.8	13.1	54.1	15.6	186.1	237
1990	64.1	46.8	1.0	0.9	112.8	41.1	13.9	56.4	15.0	184.2	235
1991	63.3	47.3	1.0	0.8	112.4	43.3	14.2	60.6	14.8	190.0	234
1992	63.0	49.9	1.0	0.8	114.7	45.4	14.1	62.6	14.9	192.2	235
1993	61.6	49.2	0.8	1.0	112.5	46.4	14.1	62.6	14.9	192.9	239
1994	63.9	49.9	0.9	0.8	115.4	47.4	14.2	62.2	15.1	192.9	236
1995	64.0	49.1	0.9	0.8	114.7	48.5	14.1	63.5	14.9	193.1	234.6
1996	64.2	45.9	0.8	1.0	111.9	49.8	14.5	64.3	14.7	190.9	237.2
1997[3]	63.8	44.9	0.8	0.8	110.3	51.1	14.2	68.3	14.7	193.9	239.7
1998[3]	62.3	48.3	0.8	0.7	112.1	54.1	14.8	68.9	NA	NA	243.4

[1]Adapted from *Feedstuffs Magazine,* Vol. 69 (No.47):14, whose source was the U.S. Department of Agriculture.
[2]Actual numbers.
[3]1997-98 are projections.

Total Livestock and Poultry Meat Consumption per Capita
per Year (Boneless Equivalent) in the United States, 1960–1998

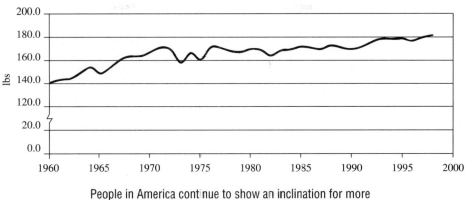

People in America continue to show an inclination for more
meat and poultry. From 1960 through projections for 1998,
there has been a gradual and consistent 29% increase in
consumption (140.1 lb vs 181 lb)
Source: USDA, as presented in *Feedstuffs*, 69(47):14, 1997

1 The Feed Nutrients

The proper feeding of livestock is for the most part a matter of supplying them with the right amounts of those chemical elements and compounds essential for carrying on the different life processes. These elements and compounds are as a group referred to as the feed nutrients. In view of the interchangeability of certain of these nutrients, their number is not exact; there are somewhere around 50 or more. The amount of each required varies but ranges for the different nutrients from less than a microgram per head per day for some to more than several kilograms per head per day for others. Feed materials supply livestock with these nutrients by serving as a source of the nutrients on the one hand and by serving as a carrier of the nutrients in facilitating the feeding operation on the other. It is intended that section 1 serve as a brief review of the chemical nature of the various nutrient materials.

Of the more than 100 known chemical elements, at least 20 enter into the makeup of the various essential feed nutrients. These 20 elements, their symbols, and their atomic weights are as follows:

1

Name	Symbol	Atomic Wt		Name	Symbol	Atomic Wt
Carbon	C	12		Magnesium	Mg	24.3
Hydrogen	H	1		Sodium	Na	23
Oxygen	O	16		Chlorine	Cl	35.5
Phosphorus	P	31		Cobalt	Co	59
Potassium	K	39		Copper	Cu	63.5
Iodine	I	127		Fluorine	F	19
Nitrogen	N	14		Manganese	Mn	55
Sulfur	S	32		Zinc	Zn	65.4
Calcium	Ca	40		Molybdenum	Mo	96
Iron	Fe	55.8		Selenium	Se	79

There is some evidence that chromium, silicon, tin, vanadium, nickel, and possibly others should be included in this group.

II

These elements, either alone or in various combinations, make up what are known as the *feed nutrients*. (The term *feed nutrient* is applied to any feed constituent that may function in the nutritive support of animal life.)

III

Many different feed nutrients are currently recognized, and new ones are still being found. Those currently recognized are as follows:

A. **Carbohydrates.** Carbohydrates contain carbon, hydrogen, and oxygen, with hydrogen and oxygen in the same proportion as in water. They consist largely of hexosans. These are made up of hexose or 6-carbon atom molecules. Pentosans, which are made up of pentose or 5-carbon atom molecules, are sometimes present. Tetrose, triose, and diose compounds are also sometimes present in small amounts but are generally unimportant.

1. **Monosaccharides.** Monosaccharides all have a chemical formula of $C_6H_{12}O_6$. They are formed in plants by the following reaction. This reaction is reversed by animals to release energy.

$$6\,CO_2 + 6\,H_2O + 673\,cal \Rightarrow C_6H_{12}O_6 + 6\,H_2O$$

The more common monosaccharides are:

a. **Glucose.** Glucose (Figure 1-1) is found in corn syrup and also in blood. It is sometimes referred to as *dextrose* because it rotates the plane of polarized light to the right. It is about three-fourths as sweet as cane sugar.

b. **Fructose.** Fructose is found principally in ripe fruits and honey. It is the sweetest of all sugars.

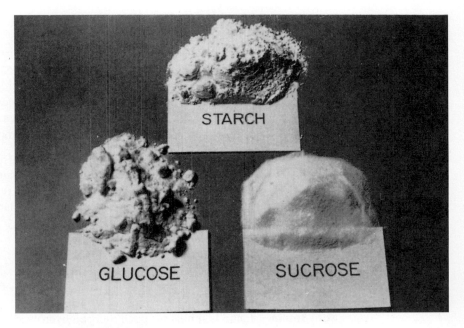

FIGURE 1-1
Essentially pure forms of a monosaccharide (glucose), a dis-
accharide (sucrose), and a polysaccharide (starch).
(Courtesy of the University of Georgia College of Agriculture
Experiment Stations)

   c.   **Galactose.**   Galactose is obtained along with glucose upon the hydrolysis of lactose or milk sugar, a disaccharide listed below.

2.   **Disaccharides.**   Disaccharides all have a chemical formula of $C_{12}H_{22}O_{11}$.
They are formed from two monosaccharide molecules with the loss of one
molecule of water. The more common disaccharides are:

   a.   **Sucrose.**   Sucrose (Figure 1-1) is the same as *cane* and *beet sugar,* commonly used as food sweeteners. It is hydrolyzed by the sucrase enzyme to
glucose and fructose.

$$1 \text{ sucrose} + 1 \text{ H}_2\text{O} \rightarrow 1 \text{ glucose} + 1 \text{ fructose}$$

   b.   **Maltose.**   Maltose is the same as *malt sugar.* It is obtained from the hydrolysis of starch. Maltose is one-fourth as sweet as sucrose. It hydrolyzes
entirely to glucose, by the enzyme maltase.

$$1 \text{ maltose} + 1 \text{ H}_2\text{O} \rightarrow 2 \text{ glucoses}$$

   c.   **Lactose.**   Lactose is commonly referred to as *milk sugar.* It is found principally in milk. Lactose is one-sixth as sweet as sucrose. It is hydrolyzed

by the enzyme lactase to glucose and galactose. (Lactose intolerance in humans is a lactase deficiency).

$$1 \text{ lactose} + 1 \text{ } H_2O \rightarrow 1 \text{ glucose} + 1 \text{ galactose}$$

3. **Polysaccharides.** Polysaccharides all have a chemical formula of $(C_6H_{10}O_5)_n$. They are formed by the combination of unknown numbers of hexose molecules. Those polysaccharides usually regarded as important in animal nutrition are:

    a. **Starch.** Many plants store energy in the form of starch (Figure 1-1). Starch is a major component of most livestock rations (especially fattening rations) and is highly digestible. Hence, it is a primary energy source for livestock. Starch hydrolyzes as follows:

    $$\text{Starch} + H_2O \rightarrow \text{Dextrin}$$
    $$\text{Dextrin} + H_2O \rightarrow \text{Maltose}$$
    $$\text{Maltose} + H_2O \rightarrow \text{Glucose}$$

    b. **Inulin.** Inulin is similar to starch except it hydrolyzes to fructose rather than glucose. It is not very prevalent. Inulin is found especially in Jerusalem artichokes.

    c. **Glycogen.** Glycogen is sometimes referred to as *animal starch*. Found only in the animal body, it is produced in the liver and is the primary carbohydrate reserve in the animal. It hydrolyzes entirely to glucose.

    d. **Hemicellulose.** Hemicellulose is a term used to denote a group of substances that lie chemically between sugars and starch on the one hand and cellulose on the other. Most of such substances are more digestible than cellulose but less digestible than sugars and starch. However, the extent of their presence is not reflected by the conventional proximate analysis. Consequently when they are present in a feed material in significant amounts, additional determinations are required. Hemicelluloses are distributed widely in forage crops and certain other materials frequently used for feeding purposes. They are especially abundant in the extract resulting from certain wood manufacturing processes from which the product known as wood molasses is made.

    e. **Cellulose.** Cellulose (Figure 1-2) is a principal constituent of the cell wall of plants. It is most abundant in the more fibrous feeds. It is generally low in digestibility. Also, it may reduce the digestibility of other nutrients. Cattle, sheep, and horses digest cellulose fairly effectively; it is only slightly digested by swine. Cellulose can be hydrolyzed by special processes to glucose.

    f. **Lignin.** Lignin is not a true carbohydrate. It contains too much carbon, the hydrogen and oxygen are not in the right proportion, and some nitro-

FIGURE 1-2
Two essentially pure forms of cellulose. (Courtesy of the
University of Georgia College of Agriculture Experiment
Stations)

gen usually is present. However, lignin is usually considered with the
polysaccharides. It is found largely in overmature hays, straws, and hulls.
It is essentially indigestible by all livestock. Also, it may reduce the digestibility of other nutrients, especially cellulose. Lignin is of no known
nutritive value except as a bulk factor. In plants, it serves as a structural
material.

B. **Fats.** Fats contain carbon, hydrogen, and oxygen with more carbon and hydrogen
in proportion to the oxygen than do carbohydrates. Fats contain 2.25 times as much
energy per lb or kg as do carbohydrates. They are formed by the combination of 3
fatty acids with glycerol.

$$\text{Fatty acid} + \text{Glycerol} \rightarrow \text{Fat} + \text{Water}$$

Examples of individual fat-forming reactions are:

$$\text{Stearic acid} + \text{Glycerol} \rightarrow \text{Stearin} + \text{Water}$$
$$3\ C_{17}H_{35}COOH + C_3H_5(OH)_3 \rightarrow C_{57}H_{110}O_6 + 3\ H_2O$$

$$\text{Palmitic acid} + \text{Glycerol} \rightarrow \text{Palmitin} + \text{Water}$$
$$3\ C_{15}H_{31}COOH + C_3H_5(OH)_3 \rightarrow C_{51}H_{98}O_6 + 3\ H_2O$$

$$\text{Oleic acid} + \text{Glycerol} \rightarrow \text{Olein} + \text{Water}$$
$$3\ C_{17}H_{33}COOH + C_3H_5(OH)_3 \rightarrow C_{57}H_{104}O_6 + 3\ H_2O$$

1. Stearic acid and palmitic acid are two of many different saturated fatty acids
(no double bonds) that combine with glycerol to form two of the more common saturated fats (stearin and palmitin).

2. Oleic and certain other fatty acids (linoleic, linolenic, and arachidonic) are un- saturated (have one or more double bonds) and combine with glycerol to form unsaturated fats. Oleic acid has one double bond, linoleic two, linolenic three, and arachidonic four.

3. Iodine Number of a fat is a measure of its degree of unsaturation. Iodine is taken up in proportion to the degree of unsaturation. The Iodine Number denotes the g of iodine absorbed per 100 g of fat.

4. Saponification is the reaction of fats with alkali to produce soap and glycerol, as illustrated below.

$$\text{Fat} + \text{Alkali} \rightarrow \text{Soap} + \text{Glycerol}$$

Example:

$$\text{Stearin} + \text{sodium hydroxide} \rightarrow \text{sodium stearate} + \text{glycerol}$$

$$
\begin{array}{ll}
C_{17}H_{35}COOCH_2 & CH_2OH \\
\quad\quad | & \quad\quad | \\
C_{17}H_{35}COOCH + 3\ NaOH \rightarrow 3\ C_{17}H_{35}COONa + CHOH \\
\quad\quad | & \quad\quad | \\
C_{17}H_{35}COOCH_2 & CH_2OH
\end{array}
$$

Other fats and/or other alkalis will produce other soaps.

5. Oils are actually fats and differ from other fats only in melting point. Oils have a low melting point and tend to be liquid at room temperature. Generally speaking, the shorter the carbon chain in the fatty acid and the greater the degree of unsaturation, the lower the melting point of the respective fat.

6. Most common fats consist of mixtures of pure fats.

C. **Protein.** Proteins always contain carbon, hydrogen, oxygen, and nitrogen, and sometimes iron, phosphorus, and/or sulfur. Protein is the only macronutrient that contains nitrogen except for small amounts in lignin. Feed proteins on the average contain 16% nitrogen. They are formed by various combinations of amino acids of which there are some 25 or more to be found in naturally occurring proteins.

The basic structure of an amino acid molecule is illustrated by the formula for the amino acid alanine, which follows:

$$
\begin{array}{c}
CH_3-CH-COOH \\
| \\
NH_2
\end{array}
$$

The carbon chain part of the molecule (CH_3-CH-) varies in both length and
|
structure depending on the amino acid under consideration. An amino acid molecule always contains one or two carboxyl groups ($-COOH$) and usually carries one

or two amino groups (NH_2), although there are some slight variations from the latter generalization.

The reaction of the carboxyl group of one amino acid molecule with the amino group of another amino acid molecule produces a peptide bond or linkage, with the liberation of water. Multiple linkages of different amounts of various amino acid molecules results in the formation of protein. Just as the 26 letters of our alphabet combine in various ways to form countless words, the 25 different naturally-occurring amino acids combine to produce innumerable proteins. By the use of the proper temperature, time, pH, and enzyme(s), proteins may be hydrolyzed into the basic amino acids from which they were formed. (Proteins will be discussed in more detail subsequently.)

D. Minerals. Of the 20 elements that function in animal nutrition, carbon, hydrogen, oxygen, and nitrogen are regarded as the *nonmineral elements*. The other 16 are referred to as the *mineral elements* that function in animal nutrition. Of these, 7 are macro (required in relatively large amounts) and 9 are micro (required in very small or trace amounts). The latter are referred to as the *trace minerals*.

The *macro minerals* are:

Calcium	Sulfur
Phosphorus	Chlorine
Potassium	Magnesium
Sodium	

The *micro* or *trace* minerals are:

Iron	Manganese
Iodine	Zinc
Copper	Molybdenum
Cobalt	Selenium
Fluorine	

(The role of minerals in animal nutrition will be discussed in more detail subsequently.)

E. Vitamins. Vitamins are organic substances required by animals in very small amounts for regulating various body processes toward normal health, growth, production, and reproduction. They are classed as micronutrients. They all contain carbon, hydrogen, and oxygen. In addition, several contain nitrogen. Certain ones also contain one or more of the mineral elements. There are 16 or more that function in animal nutrition. (Vitamins will be discussed in more detail later.)

F. Water. Water contains hydrogen and oxygen. Farm animals will consume from 3 to 8 times as much water as dry matter and will die from lack of water quicker than from lack of any other nutrient. Water is found in all feeds, ranging from about 10% in air-dry feeds to more than 80% in fresh green forages. Besides serving as a nutrient, it has many important implications in feeds and feeding.

1. **Functions of water in the animal body.**

   a. It enters into many biochemical reactions in the body.

   b. It functions in the transport of other nutrients.

   c. It helps to maintain normal body temperature.

   d. It helps to give the body form.

2. **Water is important in feed storage.**

   a. Too much water will cause certain feeds to heat and mold or otherwise lose quality. Some approximate maximum tolerances are as follows:

Ground feeds	11%	Grass hay	20%
Small grains	13%	Molasses	40%
Shelled corn	15%	Silage	75%
Snap corn	18%		

3. **Water is an important factor influencing feed value.**

   a. Water in feed is of no more value to an animal than water from other sources. Hence, appropriate allowances must be made when buying or feeding feeds high in water content.

2 General Functions of Feed Nutrients

I

There are four general functions that nutrients may serve in the animal body. Three of these may be classed as basic functions, the other as an accessory function (Table 2-1).

A. **The three basic functions are:**

1. **As a structural material for building and maintaining the body structure.** Just as boards, blocks, bricks, mortar, etc., are essential for building a house, so are certain nutrients required for the development of the animal body. Just as a house is a physical structure, so is an animal's body—consisting of bones, muscles, skin, organs, connective tissue, teeth, hair, horn, and hoof. Just as a house undergoes constant degeneration and requires more or less constant maintenance, so does an animal's body undergo constant degeneration and require constant maintenance. Essentially the same nutrients are required for body maintenance as for body building. Protein, minerals, fat, and water function in this connection.

2. **As a source of energy for heat production, work, and/or fat deposition.** Just as a house needs sources of energy (electricity, gas, coal, oil, wood, etc.) for heat production and for power to run all the gadgets around a home, so must an animal be supplied with energy for keeping the body warm and for

Table 2-1
SUMMARY OF THE VARIOUS FUNCTIONS THAT
THE DIFFERENT NUTRIENTS MAY SERVE

	Basic Functions			Accessory Function
	As a Structural Material for Body Building and Maintenance	As Energy for Heat Production, Work, and Fat Deposition	As or for the Formation of a Body Regulator	As a Source of Nutrients for Milk (or Egg) Production
Protein	Yes	Yes	Certain amino acids	Yes
Carbohydrates	Only as fat formed from carbohydrates enters into makeup of cellular growth	Yes	Yes	Yes
Fats	Only as fat enters into makeup of cellular growth	Yes	Certain fatty acids	Yes
Minerals	Yes	No	Yes	Yes
Vitamins	No	No	Yes	Yes
Water	Yes	No	Yes	Yes

the capacity to do work—both work of the vital organs and voluntary work. Also, just as coal is sometimes stored in the basement of a home as a reserve supply of stored energy, so is fat sometimes deposited within the tissues of an animal's body as a reserve supply of stored energy for the animal. Carbohydrates, fats, and proteins function here.

3. **For regulating body processes or in the formation of body-produced regulators.** Just as thermostats, switches, faucets, latches, etc., serve to regulate the functioning of various utilities around the home, so are regulators involved in the control of the various functions, processes, and activities in an animal's body. Serving in this capacity are such items as vitamins, enzymes, hormones, minerals, certain amino acids, and certain fatty acids.

B. **The accessory function—milk production.** The production of milk does not actually represent the ultimate use of nutrients but simply the shunting off of a portion of the nutrients, which an animal has consumed and digested, into the product we know as milk. Not until milk is consumed and utilized by another animal do the nutrients therein actually serve ultimate functions. Since milk contains some of almost all of the essential nutrients, essentially all of the various nutrients function in its production. With poultry, egg production would fall into this same category.

3 Proximate Analysis of Feedstuffs

I

A system for approximating the value of a product for feeding purposes, without taking the time and trouble of actually using it in a feeding trial, was developed at the Weende Experiment Station in Germany more than 100 years ago. The system was based on the separation of a sample of the product into nutritive fractions through a series of chemical determinations that in turn would reflect the material's feeding value. The different fractions were:

Water	Crude fiber
Crude protein	Nitrogen-free extract
Crude fat or ether extract	Mineral matter or ash

This system came to be known as the *Proximate Analysis* of feed materials and has served a very basic function over the past century in the development of the science of feeding farm animals. While it has served a most worthwhile purpose in this regard over the years and will continue to do so in the years ahead, certain shortcomings of the system came to light several years ago. These tended to reduce its efficacy for the purpose intended.

The shortcomings have to do especially with the crude fiber and nitrogen-free extract fractions, and in turn the material's roughage and energy value to an animal. These shortcomings are discussed in Section 4 where a new method for the chemical evaluation of feedstuffs, which has been developed and generally accepted by feeding authorities, is presented. However, since certain aspects of the Proximate Analysis procedure are still pertinent in feed evaluation, and since it is important that all students of feeds and feeding be familiar with these determinations, the system of Proximate Analysis is discussed in some detail at this point in the book.

II

In order to make an accurate analysis on a given lot of feed, it is essential that proper procedures be followed in obtaining and preparing a sample. No analysis can be any more accurate than the sample used in making it. Having no information concerning a feed's composition is better than having misinformation. Steps to be followed in obtaining and preparing a sample are as follows:

A. Obtain a small quantity from several locations within a lot of feed.

B. Finely grind or chop the sample and mix until each spoonful is representative of every other spoonful within the sample, as well as of the overall lot of feed.

C. Keep the sample in a tightly closed container. Refrigerate if necessary.

III

The procedures followed in arriving at the amounts of the different fractions ordinarily determined are as follows:

A. Water or moisture.

1. Weigh out a small quantity (usually less than 10 g) of the prepared sample into an appropriate container.

2. Dry in an appropriate oven (Figure 3-1) until there is no further loss in weight.

   a. At 135°C for a short period. (About 2 hours is usually required for air-dry materials.)

   b. At 100°C for a longer period (8–24 hours).

   c. At less than 100°C with forced air or under vacuum.

3. Weigh sample after drying.

4. Calculate the percentage of water or moisture by either of two methods.

Method A

$$\frac{\text{Loss of wt during drying}}{\text{Wt of sample before drying}} \times 100 = \% \text{ water}$$

Method B

$$\frac{\text{Wt of sample after drying}}{\text{Wt of sample before drying}} \times 100 = \% \text{ DM (dry-matter)}$$

$$100 - \% \text{ DM} = \% \text{ water}$$

FIGURE 3-1
Aluminum dishes containing feed samples being placed in a
drying oven for determination of moisture content.
(Courtesy of Karen Anschutz, University of Arkansas, Animal
Nutrition Laboratory, photo by James Herndon)

Sometimes other methods are used for determining moisture content:

a. Volumetric distillation using oil or toluene

b. Electronic—based on conductivity

c. Freeze drying

B. Crude protein. This determination is based on the fact that most nitrogen-containing macromaterials in most feeds are proteins, and proteins on the average are approximately 16% nitrogen. Actually, individual proteins will range from about 15% to more than 18% nitrogen.

1. Weigh out a small quantity of the prepared sample (usually from 1–5 g, depending on its nitrogen content) onto a piece of nitrogen-free filter paper.

2. Proceed to determine the amount of ammoniacal nitrogen (Figure 3-2) in the sample as follows:

a. Digest in concentrated sulfuric acid in the presence of potassium or sodium sulfate plus a catalyst to convert all ammoniacal nitrogen to ammonium sulfate—$(NH_4)_2SO_4$.

b. Add an excess of concentrated sodium hydroxide to make the solution strongly alkaline, causing all ammoniacal nitrogen to form ammonium hydroxide—NH_4OH.

FIGURE 3-2
A Kjeldahl rack for making nitrogen determinations in full operation. At the bottom a set of samples is being subjected to concentrated sulfuric acid digestion while at the top the ammonia is being distilled from an already digested set into beakers containing standard acid just below. (Courtesy of Russell Research Center, Athens, Georgia)

 c. Add water and distill ammonia into a known quantity of a standard acid solution, and determine by titration with standard alkali the amount of acid neutralized by the ammonia formed from the nitrogen in the feed. From this the amount of nitrogen can be calculated.

3. Calculate the amount of protein in the sample by multiplying the amount of nitrogen by 6.25 (since protein is 16% nitrogen, 6.25 times the amount of nitrogen in the sample would equal the total amount or 100% of the protein in the sample).

4. Calculate the percentage of crude protein in the feed as follows:

$$\frac{\text{Amount of protein in sample}}{\text{Wt of sample}} \times 100 = \% \text{ crude protein}$$

Protein so determined is referred to as *crude protein* as contrasted to *true protein,* which is determined by more involved procedures. It is designated as *crude* since it may contain amounts of certain ammoniacal nitrogen-containing materials that are not true proteins such as amino acids, enzymes, certain vitamins, urea, biuret, ammonia, etc. Such materials are referred to as *nonproteins.* They are usually present in natural feedstuffs, however only in very small amounts and so do not ordinarily involve a sizeable error. This is especially true since such materials are usually most prevalent in ruminant feeds, and ruminants can make effective use of nonprotein materials for meeting their protein needs. When the word *protein* is used in this book, it will be used to mean *crude protein* or "N × 6.25" unless otherwise indicated.

C. Amino acids. Especially in swine and poultry nutrition, the balancing of the amino acids (basic building blocks of protein) is most critical. These monogastric animals have a specific requirement for 10 or 11 "essential" amino acids that for the most part must be present, preformed, in the diet. Except for small amounts of nonspecific nitrogenous products from which they can synthesize "nonessential" amino acids, the monogastric animals really do not have a protein requirement per se; rather, they have specific amino acid requirements. More recently, sophisticated equipment that analyzes for amino acids has become available. However, because of the high cost of such equipment, many institutions are not able to afford it. Figure 3-3 shows a liquid chromatograph (HPLC) with a Pickering post column reaction module. In this instrument, 100 mg of prepared sample is digested in 6 N HC1; subsequently, a norleucine standard is added. Pickering brand reagents are added for the amino acid analysis, which separates the amino acids on a column.

D. Crude fat. Includes all of that portion of a feed soluble in ether. Hence, crude fat is commonly referred to as *ether extract* or *EE*. While the crude fat in most feeds is usually mostly true fats, it may also include varying amounts of other ether-soluble materials such as the fat-soluble vitamins, carotene, chlorophyll, sterols, phospholipids, waxes, etc.—hence, the designation "crude" fat. The

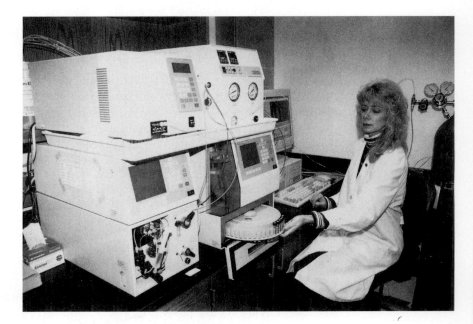

FIGURE 3-3
Waters high performance liquid chromatograph (HPLC) with a
Pickering post column reaction module. This equipment is used
for amino acid analyses. (Courtesy of Dr. Kelly Beers, University of
Arkansas Poultry Science Laboratory, photo by James Herndon)

amounts of ether-soluble materials in a feed that are not true fats, however, usually represent only a very small percentage of the overall feed. Consequently, no sizeable error is ordinarily involved in assuming that the ether-soluble fraction of a feed is mostly true fat. The steps involved in making this determination (Figure 3-4) are as follows:

1. Weigh out a small quantity of the prepared sample (usually less than 5 g) into an extraction thimble.

2. Remove water from the sample by placing it in a drying oven. This is essential to permit thorough penetration of the sample by ether.

3. Extract the sample with ether in a Soxhlet extractor or some other suitable extractor for several hours.

4. Evaporate the ether from the extract and weigh what remains. This is crude fat.

5. Calculate the percentage of crude fat:

$$\frac{\text{Wt of crude fat}}{\text{Wt of sample used}} \times 100 = \% \text{ crude fat (ether extract).}$$

E. **Fatty acids.** Fatty acid profiles and pesticide recognition are important in animal feeding. The Hewlett Packard 6890 gas chromatograph (GC), which is equipped with ionization detection and electron capture detection, (Figure 3-5) can accomplish these tasks. Extracted crude fat from a sample is treated successively with methanolic potassium hydroxide and boron trifluoride/methanol. It is extracted with hexane and eventually subjected to column separation for evaluation.

FIGURE 3-4
Feed samples in the process of being analyzed for ether extract. (Courtesy of the University of Georgia College of Agriculture Experiment Stations)

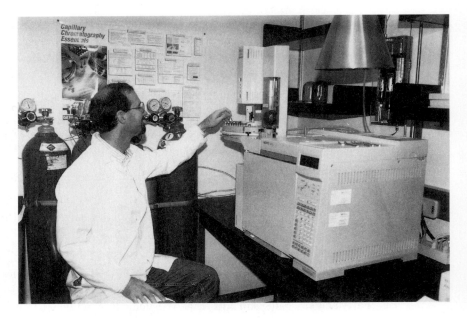

FIGURE 3-5
Hewlett Packard 6890 gas chromatograph (GC) equipped with
flame ionization detection (FID) and electron capture detection
(ECD). This equipment is used in fatty acid profile determina-
tion. (Courtesy of Dr. Kelly Beers, University of Arkansas
Poultry Science Laboratory, photo by James Herndon)

F. **Fat-soluble and water-soluble vitamins.**   Feed formulators tend to discount the
    value of vitamins contained in feedstuffs because of their variability and because it
    is perhaps more practical to purchase a premix package that contains most of the
    critical vitamins. At times, however, it is critical to be able to analyze for exact feed-
    stuff content. Figure 3-6 shows a Waters 2690 HPLC equipped with UV, fluores-
    cence, and electrochemical detection. This equipment can be used for determina-
    tion of fat- and water-soluble vitamins, as well as antioxidants, neurotransmitters,
    biogenic amines, and cholesterol. After sample treatment, the above entities sepa-
    rate out on a column for evaluation.

G. **Crude fiber.**   This fraction was originally designed to include the materials in a
    feed that were of low digestibility. Included here were cellulose, certain hemicel-
    luloses, and lignin, if present. The steps involved in this determination were as fol-
    lows:

    1. Weigh out a small quantity of the prepared sample (usually less than 5 g).

    2. Remove water from the sample by placing it in a drying oven.

    3. Extract the sample with ether to remove crude fat. The same sample used for
        crude fat determination may be used for crude fiber.

FIGURE 3-6
A Waters 2690 HPLC equipped with UV, fluorescence, and electrochemical detection. Used for determination of fat- and water-soluble vitamins, cholesterol, antioxidants, neuro-transmitters, and biogenic amines. (Courtesy of Dr. Kelly Beers, University of Arkansas Poultry Science Laboratory, photo by James Herndon)

4. Boil the remainder of the sample in dilute sulfuric acid (1.25%) for 30 minutes, and filter; then boil in dilute sodium hydroxide (1.25%) for 30 minutes, and filter (Figure 3-7). These extractions remove the protein, sugars, and starches, and the more soluble hemicelluloses and minerals, and also possibly some of the lignin, if present.

5. Dry the residue and weigh. The residue consists of the crude fiber and the more insoluble mineral matter of the feed sample.

6. Ash the residue to oxidize off the crude fiber and weigh the ash.

7. Calculate the amount of crude fiber in the sample by subtracting the weight of the ash in step 6 from the weight of the residue in step 5.

8. Calculate the percentage of crude fiber as follows:

$$\frac{\text{Wt of crude fiber}}{\text{Wt of original sample}} \times 100 = \% \text{ crude fiber}$$

FIGURE 3-7
A feed sample boiled in dilute acid for 30 minutes is being
removed for filtering in a crude fiber determination.
(Courtesy of the University of Georgia College of Agriculture
Experiment Stations)

9. Crude fiber as determined by the Proximate Analysis procedure includes only
a part of the hemicellulose and lignin present in a feed, with a portion of these
substances being included in the nitrogen-free extract fraction. Thus, the use
of crude fiber and nitrogen-free extract as determined over the years for use in
predicting feed value is now recognized as having some definite limitations,
especially with some of the more fibrous feed materials.

H. **Mineral matter or ash.** This fraction includes, for the most part, the inorganic
or mineral components of a feed. It is determined as follows:

1. Weigh out a small quantity of the prepared sample into a small crucible.

2. Ash in a furnace at red heat (600°C) for several hours (Figure 3-8).

3. Weigh the ash, which includes most of the minerals of the feed in the oxide,
chloride, or sulfate forms.

4. Calculate the percentage of ash or mineral matter as follows:

$$\frac{\text{Wt of ash}}{\text{Wt of original sample}} \times 100 = \% \text{ ash or mineral matter}$$

I. **Nitrogen-free extract.** This is commonly referred to as *NFE*. It includes mostly
sugars and starches, and also some of the more soluble hemicelluloses and some of

FIGURE 3-8
A muffle furnace used in determination of ash content.
(Courtesy of Karen Anschutz, University of Arkansas Animal
Nutrition Laboratory, photograph by James Herndon)

the more soluble lignin. Since this fraction was designed to include the more digestible carbohydrates, any lignin that may come out here will tend to distort the meaningfulness of the NFE figure as lignin is essentially indigestible. Nitrogen-free extract is determined by difference—that is, all those fractions discussed above are added together and subtracted from 100, as follows:

$$
\begin{array}{l}
\% \text{ water} \\
\% \text{ crude protein} \\
\% \text{ crude fat} \\
\% \text{ crude fiber} \\
\underline{\% \text{ mineral matter}} \\
100\text{–Total} = \% \text{ nitrogen-free extract}
\end{array}
$$

Over the years nitrogen-free extract has been regarded as consisting mainly of sugars and starches, both of which are highly digestible and similar in feeding value. However, since in some of the more fibrous feeds the NFE fraction may contain significant amounts of indigestible lignin as well as sizable quantities of hemicelluloses of variable feeding value, the use of NFE in estimating the energy value of such feeds is no longer recommended. Nitrogen-free extract is discussed here simply to familiarize the student with what it is and how it has been used over the years in predicting feeding value and why it is no longer used for this purpose, especially with feeds high in fiber.

IV

Vitamins. The amount of total vitamins or of the different individual vitamins is not determined in the routine proximate analysis, although sophisticated equipment is capable of making determinations (Figure 3-6) for the individual vitamins when this seems warranted. Vitamins appear as other nutrients in a routine proximate analysis—that is, as crude protein, crude fat, or nitrogen-free extract. However, they are quantitatively unimportant insofar as they affect the determination of the amounts of these fractions. The total amount of vitamins occurring in any one of these three fractions will usually amount to less than one-tenth of 1.0%. The amount of choline in the crude protein fraction is a frequent exception to this generalization, but choline has protein value for most animals.

V

There are different ways of expressing compositions, as outlined below:

A. **In percent (%).** This simply says that a feed contains so many parts (pounds, grams, milligrams, micrograms, etc.) of a particular feed component per 100 parts of the overall feed.

B. **In parts per million (PPM).** This simply says that a feed contains so many parts (lb, g, mg, mcg, etc.) of a particular feed component per 1,000,000 parts of the overall feed. PPM differs from % only in the location of the decimal point. Since one million is $10,000 \times 100$, to change % to PPM, simply multiply by 10,000 or, in other words, move the decimal point four places to the right. To change PPM to %, simply divide by 10,000 or move the decimal point four places to the left.

C. **In milligrams per kilogram (mg/kg).** This says a feed contains so many mg of some component per kg of the overall feed. Since a kilogram is equal to 1,000,000 mg, then "mg per kg" is the same as "mg per million mg" or "parts per million."

D. **In milligrams per pound (mg/lb).** This says that a feed contains so many mg of some feed component per lb of the overall feed. Since a pound is equal to 453,600 milligrams, "mg per lb" is the same as "mg per 453,600 mg." Hence, since 453,600 will go into one million 2.205 times, to change mg per lb to PPM, simply multiply by 2.205. To change mg per lb to %, multiply by 2.205 and divide by 10,000. To change PPM to mg/lb, divide by 2.205. To change % to mg/lb, multiply by 10,000 and divide by 2.205.

VI

The composition of feeds may be expressed on any one or more of three dry matter bases.

A. **As fed.** Sometimes referred to as the *wet* or *fresh basis*. On this basis dry matter of different feeds may range from near 0% to 100%.

B. **Air-dry.** May be actual or an "assumed dry matter content" basis. The latter is usually 90%. This basis is useful for comparing the composition of feeds having different moisture contents. Most feeds, but not all, are fed in an air-dry state.

C. **Oven-dry.**   Based on a moisture-free or 100% DM state. Also useful for comparing feeds of different moisture contents.

The different bases may be illustrated as follows:

	As fed	*Air-dry*	*Oven-dry*
% water	May be any %	Usually 10%	0%
% crude protein	This is dry		
% crude fat	matter—it		
% crude fiber	is always	Usually 90%	100%
% NFE	100% minus		
% ash	the % water		

Composition figures expressed on one basis may be converted to another basis by the use of a simple ratio, as follows:

$$\frac{\text{\% of any component in a feed on any basis}}{\text{\% DM in that feed on the same basis}} = \frac{\text{\% of any component in the feed on another basis}}{\text{\% of DM in the feed on the same basis}}$$

For example:

If a feed contains 4.0% crude protein on a fresh basis and 75.0% water, the percentage of crude protein on an air-dry basis would be calculated as follows:

100%–75% = 25% DM in the fresh material

$$\frac{4}{25} = \frac{x}{90}$$

$$25x = 360$$

$$x = 14.4 \text{ (\% crude protein in feed on an air-dry basis)}$$

Table 3-1 includes the feeds and feed materials from the appendix on feed composition in the back of the book. The feeds are listed and grouped according to their dry matter content on an as-fed basis. To include such a table of dry matter information at this point in the book is not to suggest that the student should memorize the exact dry matter percentages for the different feeds and products. Neither is it intended that the table serve as a source of information on the dry matter content of an individual feed. Such information can be obtained more readily from the appendix. On the other hand, it is desirable for the student to become familiar with the range of dry matter percentages found in the different dry matter groups and in which group a particular feed would be found. Such information automatically provides similar information about a feed's water content.

Table 3-1
SELECTED FEEDS LISTED AND GROUPED ACCORDING TO THEIR PERCENTAGE OF DRY MATTER CONTENT– AS-FED BASIS

Group A: **Mainly Artificially Dried Feeds Along with Several Feeds High in Fat, Fiber, and/or Minerals**

Corn oil	100.0%	Poultry by-product meal	93.0
Defluorinated phosphate	100.0	Sesame oil meal	93.0
Ground limestone	100.0	Sunflower meal	93.0
Hydrolyzed animal fat	99.0	Tankage with bone	93.0
Oyster shell flour	99.0	Blood meal	92.0
Soybean oil	99.0	Cottonseed	92.0
Urea, 45% N	99.0	36% cottonseed meal	92.0
Dicalcium phosphate	97.0	Dehydrated alfalfa meal	92.0
Dried whole milk	96.0	Dried bakery product	92.0
Steamed bonemeal	95.7	Dried beet pulp	
Meadow hay	95.0	with molasses	92.0
Sericea hay	95.0	Dried brewers grains	92.0
Peanut kernels	94.8	Fish meal	92.0
Dried corn distillers grains	94.0	Oat hulls	92.0
Dried skimmed milk	94.0	Oat straw	92.0
Meat meal	94.0	Peanut meal	92.0
Redtop hay	94.0	Prairie hay	92.0
Dried whey	93.0	Rice hulls	92.0
Hydrolyzed feather meal	93.0	Safflower oil meal	92.0
41% mech-extd cottonseed meal	93.0	Soybean seed	92.0
Meat and bonemeal	93.0	Tankage	92.0

Group B: **Primarily Field Cured Grains, Hays, and Straws Along with Several Artificially Dried Products**

Bahiagrass hay	91.0	Peanut hay	91.0
Barley straw	91.0	Peanut hulls	91.0
Bermudagrass hay	91.0	Potato meal	91.0
Coastal bermudagrass hay, fert.	91.0	Rapeseed meal	91.0
Copra meal	91.0	Reed canarygrass hay	91.0
Corn gluten meal	91.0	Rice bran	91.0
Cottonseed hulls	91.0	41% solv-extd cottonseed meal	91.0
Dehydrated casein	91.0	Soybean hulls	91.0
Dried beet pulp	91.0	Sudangrass hay	91.0
Dried citrus pulp	91.0	Alfalfa stem meal	90.9
Light oats grain	91.0	Dried citrus pulp with molasses	90.3
Linseed meal	91.0	Sweet potato meal	90.2
Oat hay	91.0	Alfalfa hay	90.0
Orchardgrass hay	91.0	Bromegrass hay	90.0
Pangolagrass hay	91.0	Coastal bermudagrass hay	90.0

continued

Table 3-1 (Continued)

Corn gluten feed	90.0	44% soybean meal	89.0
Cowpea hay	90.0	Timothy hay	89.0
Grain sorghum grain	90.0	Wheat bran	89.0
Ground corn cobs	90.0	Wheat grain	89.0
Hominy feed	90.0	Wheat middlings	89.0
Ky 31 fescue hay	90.0	Wheat straw	89.0
Oat groats	90.0	Alsike clover hay	88.0
49% soybean meal	90.0	Barley grain	88.0
Triticale grain	90.0	Rye grain	88.0
Corn grain	89.0	Soybean hay	88.0
Dried alfalfa leaves	89.0	Soybean straw	88.0
Johnsongrass hay	89.0	Wheat shorts	88.0
Kentucky bluegrass hay	89.0	Corn and cob meal	87.0
Lespedeza hay	89.0	Dehy coastal bermudagrass meal	87.0
Oats grain	89.0	Grain sorghum grain 8–10% prot.	87.0
Poultry manure	89.0	Millet hay	87.0
Red clover hay	89.0	Sweetclover hay	87.0

Group C: Molasses

Beet molasses	78.0	Blackstrap molasses	75.0
Wood molasses	76.0	Citrus molasses	68.0

Group D: Mostly the Silages, the Fresh Forages, Wet By-Products, the Root Crops, and Fresh Whole and Skimmed Milk

Corn ear silage	44.0	Fresh orchardgrass	23.0
Fresh Kentucky bluegrass	35.0	Fresh potatoes	23.0
Corn stover silage	31.0	Fresh sudangrass	23.0
Fresh sweet potatoes	30.6	Fresh alfalfa	21.0
Corn silage	30.0	Fresh pangolagrass	21.0
Fresh bahiagrass	30.0	Wet brewers grains	21.0
Fresh bromegrass	30.0	Fresh ladino clover	19.0
Grain sorghum silage	30.0	Fresh citrus pulp	18.3
Fresh coastal		Fresh crimson clover	18.0
bermudagrass	29.0	Fresh cow's milk	12.0
Fresh Ky 31 fescue	29.0	Fresh carrots	12.0
Oat silage	28.0	Wet beet pulp	11.0
Fresh dallisgrass	25.0	Fresh skimmed milk	10.0
Fresh ryegrass	25.0	Fresh cabbage	9.6
Wheat silage	25.0	Fresh turnip roots	9.0
Fresh rye forage	24.0	Fresh whey	7.0

In Table 3-2 the various feedstuffs from the appendix on composition are listed and grouped on an as-fed basis according to their content of ether extract. As with the dry matter data in the preceding table, the student is not expected to memorize the exact percentage of ether extract for each feed nor is this table to serve as a source of information on the ether extract content of an individual feed material. It is intended that this table simply bring out the range of ether extract levels to be found in different types of feeds and to help students bring together in their thinking those feeds of similar ether extract levels. This is especially important since fat has 2¼ times the energy value of other organic nutrients and so has an important bearing, for reasons to be discussed subsequently, on a feed's energy value.

Table 3-2
SELECTED FEEDS LISTED AND GROUPED ACCORDING TO THEIR PERCENTAGE OF ETHER EXTRACT— AS-FED BASIS

Group A: The Pure Fats, Dried Whole Milk, and the Oil Seeds

Corn oil	100.0	Peanut kernels	47.7
Hydrolyzed animal		Dried whole milk	26.6
fat	98.7	Cotton seed	21.3
Soybean oil	95.4	Soybean seed	17.2

Group B: Primarily the Mechanically Extracted Protein Feeds, Other By-Product Concentrates, Some of the Feed Grains, and Certain Other Miscellaneous Products

Rice bran	13.7	41% mech-extd cottonseed meal	4.6
Poultry by-product meal	13.1	Wheat shorts	4.6
Tankage with bone	12.8	Light oats grain	4.5
Dried bakery product	11.7	Wheat middlings	4.3
Meat and bone meal	9.7	36% cottonseed meal	4.2
Fish meal	9.6	Wheat bran	3.9
Dried corn distillers grains	9.2	Corn grain	3.8
Meat meal	9.1	Ky 31 fescue hay	3.8
Tankage	8.9	Dried citrus pulp	
Hominy feed	6.9	with molasses	3.6
Sesame oil meal	6.9	Fresh cow's milk	3.6
Dried brewers grains	6.6	Soybean hay	3.6
Oat groats	6.2	Copra meal	3.5
Linseed meal	5.4	Dried citrus pulp	3.4
Sericea hay	5.1	Dehy coastal bermudagrass meal	3.3
Oats grain	4.8	Corn and cob meal	3.2

Group C: Predominantly the Hays, the Hulls, and the Straws; Some of the Grains; and Certain Solvent-Extracted Materials Along with Several Other Miscellaneous Products

Kentucky bluegrass hay	3.1	Grain sorghum grain 8–10% prot.	2.9
Orchardgrass hay	3.1	Hydrolyzed feather meal	2.9
Peanut hay	3.1	Sunflower meal	2.9

continued

Table 3-2 (Continued)

Cow pea hay	2.8	Steamed bonemeal	1.9
Grain sorghum grain	2.8	Pangolagrass hay	1.8
Reed canarygrass hay	2.8	Peanut hulls	1.8
Alsike clover hay	2.7	Wheat grain	1.8
Dehydrated alfalfa meal	2.7	Alfalfa stem meal	1.7
Dried alfalfa leaves	2.7	Barley straw	1.7
Poultry manure	2.7	Corn ear silage	1.7
Timothy hay	2.7	Rapeseed meal	1.7
Millet hay	2.6	Sweetclover hay	1.7
Redtop hay	2.5	Fresh Kentucky bluegrass	1.6
Meadow hay	2.4	Fresh Ky 31 fescue	1.6
Oat hay	2.4	Oat hulls	1.6
Red clover hay	2.4	Sudangrass hay	1.6
Alfalfa hay	2.3	Wheat straw	1.6
Bromegrass hay	2.3	Coastal bermudagrass hay, fert.	1.5
Corn gluten feed	2.2	Cottonseed hulls	1.5
Corn gluten meal	2.2	Rye grain	1.5
Prairie hay	2.2	Triticale grain	1.5
Coastal bermudagrass hay	2.1	41% solv-extd cottonseed meal	1.4
Johnsongrass hay	2.1	44% soybean meal	1.4
Oat straw	2.1	Wet brewers grains	1.4
Bahiagrass hay	1.9	Blood meal	1.3
Barley grain	1.9	Peanut meal	1.3
Lespedeza hay	1.9	Safflower oil meal	1.3
Soybean hulls	1.9	Soybean straw	1.3

Group D: **Largely High-Moisture Feeds, Mineral Feeds, and Solvent-Extracted Feeds Along with Several Other Miscellaneous Products**

Fresh bromegrass	1.2	Dried beet pulp	
Fresh coastal bermudagrass	1.1	with molasses	0.6
Fresh orchardgrass	1.1	Fresh alfalfa	0.6
Corn silage	1.0	Fresh citrus pulp	0.6
Fresh rye forage	0.9	Fresh crimson clover	0.6
Grain sorghum silage	0.9	Fresh dallisgrass	0.6
Oat silage	0.9	Dried beet pulp	0.5
49% soybean meal	0.9	Fresh bahiagrass	0.5
Sweet potato meal	0.9	Fresh ladino cover	0.5
Dried skimmed milk	0.8	Fresh pangolagrass	0.5
Fresh ryegrass	0.8	Potato meal	0.5
Wheat silage	0.8	Fresh sudangrass	0.4
Corn stover silage	0.7	Fresh sweet potatoes	0.4
Dried whey	0.7	Fresh whey	0.3
Ground corn cobs	0.7	Wood molasses	0.3
Rice hulls	0.7	Beet molasses	0.2
Dehydrated casein	0.6	Citrus molasses	0.2
Fresh cabbage	0.2	Fresh skimmed milk	0.1

continued

Table 3-2 (Continued)

Fresh carrots	0.2	Defluorinated phosphate	***
Fresh turnip roots	0.2	Dicalcium phosphate	***
Wet beet pulp	0.2	Ground limestone	***
Blackstrap molasses	0.1	Oyster shell flour	***
Fresh potatoes	0.1	Urea, 45% N	***

***very little, if any

In Table 3-3 the feeds from the appendix are listed and grouped on an as-fed basis according to their content of crude fiber. While crude fiber during recent years has come to be a less meaningful figure than it once was, it still is the only fibrous fraction on which data are available for many feeds. Also, it continues to be used as a measure of feed fibrousness where more definitive information is not yet available for a feed material. Again it is not intended for the student to memorize the exact crude fiber percentages for the different feeds or to use the table as a source of crude fiber information on individual feeds. However, through careful perusal of the table the student can become acquainted with the relative fibrousness of different types of feeds and group in which a particular feed would appear.

Table 3-3
SELECTED FEEDS LISTED AND GROUPED ACCORDING TO THEIR PERCENTAGE OF CRUDE FIBER— AS-FED BASIS

Group A: **Extremely Low-Quality Dry Roughages Consisting Primarily of Hulls and Straws**

Peanut hulls	57.3	Barley straw	38.3
Cottonseed hulls	43.3	Oat straw	37.3
Rice hulls	39.6	Wheat straw	36.9
Soybean straw	38.9	Soybean hulls	36.4

Group B: **Made Up Almost Entirely of the Hays and Hay Meals**

Alfalfa stem meal	33.9	Reed canarygrass hay	30.1
Orchardgrass hay	33.6	Johnsongrass hay	29.9
Pangolagrass hay	32.8	Coastal bermudagrass hay, fert.	29.6
Sudangrass hay	32.8	Ky 31 fescue hay	29.3
Ground corn cobs	32.7	Sweetclover hay	29.3
Prairie hay	31.1	Redtop hay	29.0
Oat hulls	30.9	Bromegrass hay	28.8
Meadow hay	30.7	Timothy hay	28.4
Peanut hay	30.2	Bermudagrass hay	27.8
Oat hay	27.8	Soybean hay	25.0

continued

Table 3-3 (Continued)

Coastal bermudagrass hay	27.7	Dehy coastal bermudagrass meal	24.3
Kentucky bluegrass hay	27.6	Cowpea hay	24.0
Lespedeza hay	27.4	Dehydrated alfalfa meal	24.0
Alsike clover hay	26.5	Alfalfa hay	23.4
Millet hay	25.8	Sericea hay	22.3
Red clover hay	25.0	Bahiagrass hay	21.2

Group C: **Consists Largely of a Group of Feeds Generally Referred to as the High-Fiber Concentrates**

Cottonseed	19.1	41% solv-extd cottonseed meal	12.1
Dried beet pulp	18.0	Rapeseed meal	12.0
Dried alfalfa leaves	15.8	41% mech-extd cottonseed meal	11.9
Dried beet pulp with molasses	15.1	Rice bran	11.7
Light oats grain	14.4	Dried citrus pulp	11.6
Poultry manure	14.4	Sunflower meal	11.4
36% cottonseed meal	14.3	Dried corn distillers grains	11.3
Copra meal	14.0	Oats grain	10.8
Safflower oil meal	13.5	Dried citrus pulp with molasses	10.3
Dried brewers grains	13.2		

Group D: **Predominantly a Combination of Low-Fiber, Air-Dry Concentrates and High-Moisture Feeds**

Wheat bran	10.0	Fresh orchardgrass	5.8
Peanut meal	9.9	Fresh ryegrass	5.8
Corn stover silage	9.6	Sesame oil meal	5.7
Oat silage	9.4	Soybean seed	5.3
Fresh bahiagrass	9.0	Corn ear silage	5.1
Linseed meal	8.8	Barley grain	5.0
Corn gluten feed	8.7	Fresh crimson clover	5.0
Fresh coastal bermudagrass	8.3	Fresh alfalfa	4.9
Corn and cob meal	8.2	Corn gluten meal	4.4
Grain sorghum silage	8.2	Triticale grain	4.0
Fresh Kentucky bluegrass	8.1	49% soybean meal	3.4
Wheat silage	7.8	Sweet potato meal	3.3
Corn silage	7.5	Wet brewers grains	3.2
Wheat middlings	7.3	Wet beet pulp	3.1
Fresh dallisgrass	7.2	Peanut kernels	2.8
Fresh Ky 31 fescue	7.1	Fresh ladino clover	2.7
Fresh rye forage	6.8	Meat meal	2.7
Fresh sudangrass	6.8	Corn grain	2.6
Wheat shorts	6.8	Wheat grain	2.6
Fresh bromegrass	6.7	Oat groats	2.5
Fresh pangolagrass	6.4	Grain sorghum grain	2.4
44% soybean meal	6.2	Fresh citrus pulp	2.3
Hominy feed	6.0	Grain sorghum grain 8–10% prot.	2.3

continued

Table 3-3 (Continued)

Poultry by-product meal	2.3	Steamed bonemeal	1.9
Meat and bonemeal	2.2	Hydrolyzed feather meal	1.4
Rye grain	2.2	Fresh sweet potatoes	1.3
Tankage with bone	2.2	Dried bakery product	1.2
Potato meal	2.1	Fresh carrots	1.2
Tankage	2.0	Fresh turnip roots	1.1

Group E: The Mineral Feeds, the Pure Fats, the Molasses, the Milk Products, and Certain Animal By-Products

Blood meal	1.0	Corn oil	0.0
Fresh cabbage	1.0	Soybean oil	0.0
Fish meal	0.9	Hydrolyzed animal fat	0.0
Wood molasses	0.8	Citrus molasses	***
Fresh potatoes	0.6	Defluorinated phosphate	***
Dehydrated casein	0.2	Dicalcium phosphate	***
Dried skimmed milk	0.2	Fresh cow's milk	***
Dried whey	0.2	Fresh skimmed milk	***
Dried whole milk	0.2	Fresh whey	***
Beet molasses	0.0	Ground limestone	***
Blackstrap molasses	0.0	Oyster shell flour	***
Urea, 45% N	0.0		

***very little, if any

4 Use of the Van Soest Analysis In Feed Evaluation

I

While the Weende system of feed analysis has served for many years a very important role in predicting the nutritive value of feed materials, certain shortcomings of the procedure have received the attention of nutritionists over the past several years. These have to do primarily with the crude fiber and nitrogen-free extract determinations.

A. Crude fiber, as determined by the Weende procedure, is not a chemically uniform substance but a variable mixture, the major components of which are cellulose, hemicellulose, and lignin. While cellulose and hemicellulose are similar in nutritive value, they have much greater feeding values for ruminants than for nonruminants, whereas lignin is essentially indigestible by all livestock. Further complicating the situation is the fact that only a part of the hemicellulose and lignin comes out in the crude fiber fraction, with the remaining portions showing up as NFE, which is ordinarily thought of as consisting largely of highly digestible sugars and starches. Consequently, to the extent that hemicellulose—at best low in digestibility—and lignin—essentially indigestible—appear in the NFE fraction, this fraction would be larger and would have a lower average digestibility than would be true if it included only sugars and starches. At the same time, the crude fiber value would not reflect all of the more indigestible portion of the feed.

B.   Fortunately, the above situation has been a major problem only with the more fibrous feeds from the standpoint of making effective use of proximate analysis figures in predicting nutritive value. However, with fibrous feeds, problems of considerable magnitude frequently have been experienced in this connection. As a result, numerous workers over the past several years have tested alternative procedures that might provide a more definitive separation of feed carbohydrates than does the Weende system of proximate analysis. This has been especially true from the standpoint of forage crop evaluation.

II

A procedure that has received essentially universal acceptance by feeding authorities for approximating a feed's fiber content and energy value has been developed by Van Soest and associates[1-5] working at the USDA's Agricultural Research Service (ARS) research laboratory in Beltsville, MD, and subsequently at Cornell University. This procedure involves the separation of feed dry matter into two fractions—one of high digestibility and the other of low digestibility—by boiling a sample of the feed in a neutral detergent solution for a period of one hour and then filtering. The more soluble, and accordingly more digestible, nutrients come out in the filtrate while the residue consists of the more fibrous and less digestible fractions of the feed.

A.   The neutral detergent solubles (NDS) consist for the most part of the cell contents. They are composed primarily of lipids, sugars, starches, and protein and are all high in digestibility. Their digestibility does not seem to be materially influenced by the amount of neutral detergent insolubles present.

The neutral detergent insolubles are usually referred to as *neutral detergent fiber* (NDF). For the most part, they embrace the plant cell wall and are sometimes referred to as the *cell wall components* or *cell wall constituents.* They consist primarily of hemicellulose, cellulose, lignin, silica, and some protein.

In the Van Soest procedure, essentially all of the lignin and hemicellulose are included in the NDF fraction, whereas with the Weende method, variable amounts of these two components are lost from the crude fiber to the NFE. As a result, NDF as determined by the Van Soest procedure is considerably higher than the conventional crude fiber values for some feeds.

The different NDF (cell wall) components are at best low in digestibility and are entirely dependent on the microorganisms of the digestive tract for any digestion

---

[1]Van Soest, P. J. and L. A. Moore. "New Chemical Methods for Analysis of Forages for the Purpose of Predicting Nutritive Value." *Proceedings of the Ninth International Grassland Congress, Sao Paulo, Brazil.* Sao Paulo, Brazil: International Grassland Conference, 1966.

[2]Van Soest, P. J. "Non-Nutritive Residues: A System of Analysis for the Replacement of Crude Fiber." *Journal of the Association of Official Agricultural Chemists* 49 (1966): 546.

[3]Van Soest, P. J. "Development of a Comprehensive System of Feed Analyses and Its Application to Forages." *Journal of Animal Science* 26 (1967): 119.

[4]Goering, H. K. and P. J. Van Soest. "Forage Fiber Analyses." *USDA, ARS Agriculture Handbook No. 379.* Washington, D.C.: Government Printing Office, 1970.

[5]Robertson, J. B. and P. J. Van Soest. The Detergent System of Analysis and Its Application to Human Foods, Chapter 8. *The Analysis of Dietary Fiber in Food (1981),* W. P. T. James and Olof Theander, eds., Marcell Dekker, Inc., New York and Basel.

they do undergo. Lignin and silica are essentially indigestible even by microorganisms. Also, lignin has a curvilinear negative influence on cellulose and hemicellulose digestibility.

While, according to Van Soest, NDF corresponds more closely than does conventional crude fiber to the total fiber fraction of a forage feed, it is not a uniform chemical entity but a variable mixture of cell wall components whose overall nutritive availability is influenced to a considerable degree by the proportion of lignin present. Accordingly, Van Soest has proposed that the amount of lignin be determined and appropriate allowance be made for the amount of this component present in attempting to predict NDF digestibility.

B. For the purpose of determining the lignin in a forage sample, Van Soest has proposed the use of what has come to be known as the *acid detergent lignin procedure*.[6] In this connection the acid detergent fiber procedure is used as a preparatory step. This involves the boiling of a sample of air-dry material in an acid detergent solution for one hour and then filtering. The insolubles, or residue, makes up what is known as *acid detergent fiber* (ADF) and consists primarily of cellulose, lignin, and variable amounts of silica. ADF differs from NDF in that NDF contains most of the feed hemicellulose and a limited amount of protein not present in ADF. The difference in the amount of NDF and ADF is an estimate of the hemicellulose in the feed.

In order to determine the amount of lignin present, the ADF is then digested in 72% H_2SO_4 at 15°C for 3 hours and filtered. The residue remaining after washing and drying is weighed and ashed. The ash remaining approximates the silica present, while the loss in weight during ashing approximates the lignin and is referred to as *acid detergent lignin* (ADL) or more specifically as *acid insoluble lignin*.

C. An alternative method for determining lignin, which has advantages for certain materials, involves the oxidation of the lignin of ADF with an excess of acetic acid-buffered potassium permanganate solution. Lignin so determined is referred to as *permanganate lignin*. A variation of this method may be used to allow for the cutin present in many seed hulls, which otherwise would be measured as lignin.

D. Forage processing temperatures higher than 50°C tend to increase lignin yields with either of the above methods largely by the production of artifact lignin via the nonenzymic browning reaction. The nitrogen content of the ADF is considered to be a sensitive measure of the extent of such damage and serves as a basis for estimating artifact lignin.

E. Once NDS, NDF, ADF, and ADL have been determined for a feed material, one or more of these values might be used either alone or in combination with each other, possibly with certain other composition figures, to evaluate the material for feeding purposes.

One of the most immediate uses of such data has been for establishing a feed material's roughage value. This has been especially helpful in the feeding of high-producing dairy animals that must be provided with a certain minimum level of

---

[6]Ibid.

fibrous material in their ration in order to produce at their maximum from both a quality and quantity of milk standpoint. On the other hand, more roughage or fiber than is necessary in a cow's ration will reduce the energy concentration of the ration below that which will support maximum production. It has been found that NDF is a more effective measure for evaluating a feed material from a roughage standpoint than is either ADF[7] or conventional crude fiber as determined over the years. The proper level of NDF will be discussed in the section, "Balancing a Ration for a Dairy Cow."

An even more rewarding objective from the Van Soest analysis is the use of such data for the prediction of a feed material's energy value to an animal. This would mean being able to calculate a feed's TDN content and in turn its net energy (NE) value using data from a Van Soest analysis on the feed either alone or along with other information available on the feed.

Various prediction equations and procedures have been proposed[8] for this purpose, but feeding authorities have not yet decided on the best route to take for accomplishing this goal. Certainly if reliable TDN and NE values can be established for feed materials on which such information is not already available without having to conduct expensive, time-consuming digestion trials and metabolism studies, a most valuable contribution will have been made to the science of livestock feeding.

---

[7]Van Soest, P. J. and D. R. Mertens. The Use of Neutral Detergent Fiber versus Acid Detergent Fiber in Balancing Dairy Rations. *Monsanto Technical Symposium. 1984.* Fresno, CA.

[8]Van Soest, P.J., D. G. Fox, D. R. Mertens, and C. J. Sniffen. Discounts for Net Energy and Protein. *Proceedings 1984 Cornell Nutrition Conference.* Cornell University, Ithaca, NY.

5 The Digestive Tract

The digestive tract, sometimes referred to as the *alimentary tract,* is the passage from the mouth to the anus through which feed passes following consumption as it is subjected to various digestive processes. It consists of the following:

A. **Mouth and pharynx.**

B. **Esophagus.**

C. **Stomach.**

    1. In animals such as the hog and horse this is a single-compartment organ.

    2. In animals such as the cow and sheep this is a multiple-compartment structure consisting of the following:

        a. **Rumen or paunch.**

        b. **Reticulum or honeycomb or water bag.**

        c. **Omasum or manyplies.**

        d. **Abomasum or true stomach.**

    D. **Small intestine** which is divided into three sections:

        1. **Duodenum**—upper section.

        2. **Jejunum**—middle section.

        3. **Ileum**—lower section.

    E. **Cecum or caecum.**

    F. **Large intestine.**

    G. **Anus.**

    H. **Associated glands and organs,** including:

        1. **Salivary glands.**

        2. **Liver.**

        3. **Gall bladder.**

        4. **Pancreas.**

II

Farm animals are classified, according to the nature of their digestive tracts, into two general categories:

    A. **Nonruminants** are animals such as the hog and horse that have a single-compartment stomach (Figure 5-1) and do not chew a cud. Most nonruminants make very poor use of high fiber feeds. However, the horse by means of the microbial fermentation processes of the cecum and large intestine is able to utilize such feeds effectively, and as a result horses are frequently fed rations containing considerable fiber. Hogs, however, are ordinarily fed rations relatively low in fiber.

    B. **Ruminants** are animals such as the cow and sheep which ruminate or, in other words, chew a cud. An animal's cud consists of boluses of feed eaten earlier. Ruminants have a multicompartment stomach (Figure 5-2) as outlined above. By means of the microbial fermentation processes of the rumen, such animals are able to make effective use of high fiber feeds and as a result are frequently fed rations containing high levels of fibrous feeds.

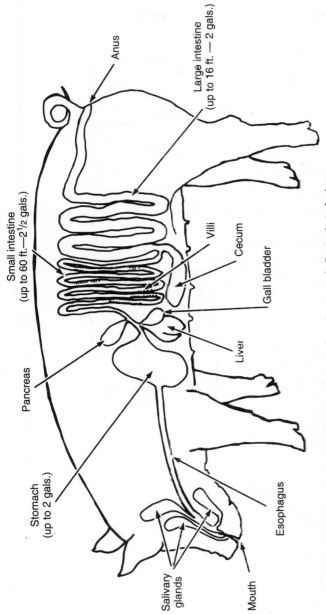

Anus

Large intestine
(up to 16 ft. — 2 gals.)

Small intestine
(up to 60 ft.—2½ gals.)

Villi

Cecum

Gall bladder

Pancreas

Liver

Stomach
(up to 2 gals.)

Salivary
glands

Esophagus

Mouth

Note—The horse is also a nonruminant and has a digestive tract similar to that of a hog,
except:
1. It is proportionately larger.
2. The horse has no gall bladder.
3. The horse has a greatly enlarged cecum (up to 3½ ft. —9 gal.) and large intestine
   (up to 21 ft.—25 gal.)

FIGURE 5-1

The digestive tract of the hog—a nonruminant.

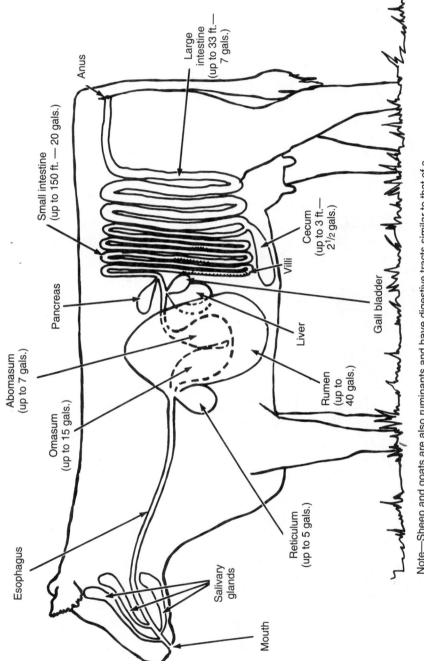

Esophagus

Mouth

Salivary glands

Reticulum (up to 5 gals.)

Rumen (up to 40 gals.)

Abomasum (up to 7 gals.)

Omasum (up to 15 gals.)

Pancreas

Small intestine (up to 150 ft. — 20 gals.)

Anus

Large intestine (up to 33 ft. — 7 gals.)

Cecum (up to 3 ft. — 2½ gals.)

Villi

Liver

Gall bladder

Note—Sheep and goats are also ruminants and have digestive tracts similar to that of a cow, except they are proportionately smaller.

FIGURE 5-2

The digestive tract of the cow—a ruminant.

6 Nutrient Digestion, Absorption, and Transport

I. **Nutrient digestion** includes the processes involved in the conversion of various feed nutrients into forms (called *end products*) that can be absorbed from the digestive tract. Most of these processes are accomplished through the action of various enzymes that are found in the different digestive juices secreted into the digestive tract. Details of these digestive processes are summarized in Tables 6-1 and 6-2.

A. **End products of digestion** are the forms into which a nutrient must be converted in order for it to be absorbed from the digestive tract.

Nutrient	End Product
Protein	Amino acids
Starch	Glucose
Sucrose	Glucose and fructose (fructose converted to glucose upon/during absorption)
Lactose	Glucose and galactose (galactose converted to glucose upon/during absorption)
Cellulose	Organic acids and salts of organic acids
Fats	Primarily fatty acids and glycerol; some soap; micronized globules
Minerals	Any soluble form
Vitamins	Any soluble form

Table 6-1
SUMMARY OF DIGESTION

Nutrient	Mouth	Rumen	Nonruminant Stomach and Ruminant Abomasum	Small Intestine	Cecum and Large Intestine
Protein	None	Some breakdown of protein to amino acids by microbial fermentation. Also some microbial synthesis of essential amino acids.	Enzyme rennin of gastric juice secreted by wall of stomach curdles milk. Enzyme pepsin of gastric juice acts on protein to form intermediate protein breakdown products (IPBP), such as proteoses, polypeptides, and peptides.	Trypsin and certain other enzymes of the pancreatic juice secreted by the pancreas act on protein and IPBP to produce other IPBP and amino acids. Intestinal peptidases (formerly the enzyme erepsin) of the intestinal juices secreted by the intestinal wall act on IPBP to produce amino acids.	Continued action of trypsin, etc., and the intestinal peptidases.
Starch	In nonruminants, enzyme salivary amylase of the saliva secreted by the salivary glands acts on starch to produce maltose. Enzyme salivary maltase acts on maltose to produce glucose.	Some starch undergoes microbial (bacterial and protozoal) fermentation to form primarily acetic, propionic, butyric, and certain other volatile fatty acids (VFA), methane (CH_4), CO_2, and heat.	None	Enzyme pancreatic amylase of the pancreatic juice secreted by the pancreas acts on starch to produce maltose. Enzyme intestinal maltase of the intestinal juices secreted by the intestinal wall acts on maltose to form glucose.	Continued action of pancreatic amylase and intestinal maltase. Much undigested starch is converted to undesirable gases.

Nutrient	Mouth	Rumen	Nunruminant Stomach and Ruminant Abomasum	Small Intestine	Cecum and Large Intestine
Sucrose	None	Same as for starch above.	None	Enzyme intestinal sucrase of the intestinal juices secreted by the intestinal wall acts on sucrose to form glucose and fructose.	Continued action of intestinal sucrase if any sucrose is still present.
Lactose	None	Lactose ordinarily not found in a functioning rumen.	None	Enzyme intestinal lactase of the intestinal juices secreted by the wall of the intestine acts on lactose to form glucose and galactose.	Continued action of intestinal lactase if any lactose is present.
Cellulose	None	Same as for starch above.	None	None	Similar to that in rumen but much less extensive, except in the horse.
Fat	None	Some microbial fermentation of fats to form fatty acids and glycerol. Some fermentation of glycerol to propionic acid.	Enzyme gastric lipase of the gastric juice secreted by wall of stomach acts on some fat to form fatty acids and glycerol.	Certain compounds of bile secreted by the liver react with some fat to form soap and glycerol. Enzyme pancreatic lipase of the pancreatic juice secreted by the pancreas acts on fat to form fatty acids, glycerol, and monoglycerides.	Continued action of pancreatic lipase.

Table 6-2

ENZYMES AND DIGESTIVE JUICES

Enzyme	Enzyme Found in Which Digestive Juice?	Digestive Juice Secreted By?	Digestive Juice Secreted into What Part of Digestive Tract?	Enzyme Acts on What?	Enzyme Active in Acid or Alkaline Medium?	Action Produces?
Salivary amylase	Saliva	Salivary glands	Mouth	Starch	Neutral to slightly alkaline	Maltose
Salivary maltase	Saliva	Salivary glands	Mouth	Maltose	Neutral to slightly alkaline	Glucose
Renin	Gastric juice	Wall of stomach	Stomach or abomasum	Milk protein	Acid	Curd
Pepsin	Gastric juice	Wall of stomach	Stomach or abomasum	Proteins	Acid	Proteoses, polypeptides, peptides
Gastric lipase	Gastric juice	Wall of stomach	Stomach or abomasum	Fats	Acid	Fatty acids, glycerol
Pancreatic amylase	Pancreatic juice	Pancreas	Upper small intestine	Starch	Slightly alkaline	Maltose
Trypsin, etc.	Pancreatic juice	Pancreas	Upper small intestine	Proteins, proteoses, polypeptides, peptides	Slightly alkaline	Intermediate protein break-down products, amino acids
Pancreatic lipase	Pancreatic juice	Pancreas	Upper small intestine	Fats	Slightly alkaline	Fatty acids, glycerol, mono-glycerides
Intestinal peptidases (erepsin)	Intestinal juice	Wall of small intestine	Small intestine	Intermediate protein breakdown products	Slightly alkaline	Amino acids
Intestinal maltase	Intestinal juice	Wall of small intestine	Small intestine	Maltose	Slightly alkaline	Glucose
Sucrase	Intestinal juice	Wall of small intestine	Small intestine	Sucrose	Slightly alkaline	Glucose, fructose
Lactase	Intestinal juice	Wall of small intestine	Small intestine	Lactose	Slightly alkaline	Glucose galactose
	Bile	Liver	Upper small intestine	Bile reacts with fats	Slightly alkaline	Soap, glycerol

B. **Digestive juices** are fluid materials secreted into the digestive tract by glands or tissues along the digestive tract. Those usually considered include the following:

1. **Saliva.**
2. **Gastric juice.**
3. **Bile.**
4. **Pancreatic juice.**
5. **Intestinal juice.**

C. **Enzymes.** Enzymes are organic catalysts. Most are protein in makeup. They are usually specific with respect to substrate and medium requirements. Most enzyme names end in "ase." A few early named enzymes end in "in." Today, enzymes are usually named according to the substrate on which they act, and possibly the action they effect, with an "ase" ending. The following enzymes are usually recognized in connection with feed digestion:

1. **Salivary amylase.**
2. **Salivary maltase.**
3. **Rennin.**
4. **Pepsin.**
5. **Gastric lipase.**
6. **Pancreatic amylase.**
7. **Trypsin.**
8. **Pancreatic lipase.**
9. **Intestinal peptidases (erepsin).**
10. **Intestinal maltase.**
11. **Sucrase.**
12. **Lactase.**

II

Nutrient absorption is the movement of the end products of digestion from the digestive tract into the blood and/or lymph system and is accomplished by diffusion through the semipermeable membranes that line much of the digestive tract. In nonruminants, most nutrient absorption takes place in the small and large intestines. The villi of the small intestine especially facilitate absorption at this location. A large amount of absorption also takes place through the rumen wall in ruminants. All of the end products of digestion, except those of fat digestion and possibly the fat soluble vitamins, are absorbed directly into the blood stream. The end products of fat digestion are absorbed into the lymph system through the lacteals of the villi in the form of chyle. The latter subsequently enters the blood through the thoracic duct and undergoes certain metabolic changes in the liver before being used in the tissues.

III

Nutrient transport is the movement of nutrients from the point of absorption to the point of utilization. Blood is the primary basis for the transport of nutrients and other materials in the animal body. However, the lymph serves as the final link between the blood capillaries and the individual cells. All nutrients are transported in solution in the water soluble form. The various nutrients are absorbed from the lymph into the individual cells again by diffusion.

7 Apparent Digestibility

I Different feeds and nutrients vary greatly in their digestibility. In the evaluation of different feeds for feeding purposes, it is helpful to have information on the digestibility of feeds and the nutrients they contain.

II Information on the digestibility of feeds and nutrients is obtained by carrying out a digestion trial. To carry out a digestion trial, one simply feeds an animal a known amount of feed of a known composition, collects the feces resulting from this feed, and then determines the composition of the feces (Figure 7-1). The difference between the nutrients consumed and the nutrients excreted in the feces is considered to be the amount of nutrients digested. Such a procedure gives one *apparent digestibility* rather than *true digestibility*. Apparent digestibility differs from true digestibility in that:

A. Apparent digestibility considers nutrients lost as methane as having been digested and absorbed by the animal. This is especially a factor with ruminants.

B. Apparent digestibility regards all nutrients remaining in the feces as not digestible, which is not necessarily so. Some may have been digested but not absorbed. Others possibly were not digested but would have been had they remained in the digestive tract for a longer period or were fed again.

FIGURE 7-1
A metabolism stall such as this might be used for conducting digestion trials with calves. Note the pan at the rear for collecting feces. (Courtesy of the University of Georgia College of Agriculture Experiment Stations, Dairy Nutrition Laboratory)

C. Apparent digestibility regards all nutrients in the feces as undigested feed when actually considerable amounts of other materials—especially intestinal mucosa and bacteria—are present in the feces.

III

For all practical purposes, however, apparent digestibility data provide invaluable information for the evaluation of feeds for feeding purposes, and such data are used extensively in carrying out a scientific feeding program.

In order to fully understand and appreciate the meaning and significance of *digestibility* data, it behooves a student of feeds and feeding to be familiar with the basic steps of carrying out a digestion trial and the various calculations that are ordinarily made in that connection.

A. In carrying out a digestion trial it is very difficult, especially in ruminants, to know just which feces resulted from which feed. Consequently, preliminary to carrying out a digestion trial an animal is brought to a constant level of daily feed intake and, hopefully, of daily feces excretion, usually over a period of 7 to 10 days. Even so, the amount of feces excretion will vary from day to day. Hence, fecal collections are usually made over a period of 7 to 10 days and an average obtained.

B. Some of the basic calculations made in connection with carrying out a digestion trial are as follows:

1. Amount of a nutrient in daily feed–Amount of that nutrient in daily feces = Amount of that nutrient digested daily.

    a. Amount of a nutrient in daily feed = (Amount of feed eaten daily × % of nutrient in feed)/100.

    b. Amount of a nutrient in daily feces = (Average amount of feces excreted daily × % of nutrient in feces)/100.

2. Coefficient of digestibility of any nutrient = (Amount of that nutrient digested daily / Amount of the nutrient eaten daily) × 100.

3. % of digestible nutrient in a feed = (% of that nutrient in the feed × Coefficient of digestibility of that nutrient)/100.

    or

    % of digestible nutrient in a feed = (Amount of that nutrient digested daily / Amount of total feed eaten daily) × 100.

The application of the above calculations is illustrated in Table 7-1.

C. Sometimes in determining the digestibility of a feed, rather than using the total fecal collection method described above, digestibility figures are obtained by simply analyzing samples of the feed and the feces for some indigestible reference material such as chromic oxide, which has been added to the feed, or for some essentially indigestible feed component naturally present in the feed such as lignin or acid insoluble ash. The apparent digestibility of a particular nutrient is then calculated from the change in the ratio of the amount of reference material to the amount of the nutrient in the feed compared to the ratio of these same materials in the feces. The use of reference materials for determining digestibility is usually quicker, easier, and cheaper, and frequently is more suitable for a particular research endeavor than the total collection method. However, their use is still being researched and has not as yet received general acceptance for digestibility studies.

Table 7-1
DIGESTION TRIAL CALCULATIONS ILLUSTRATED

Line (L.)	Energy	Dry Matter	Crude Protein	Crude Fat	Crude Fiber	N-Free Extract	Mineral Matter*
1. Amount of feed eaten daily—9.11 kg							
2. Composition of feed (as determined by lab analysis)	3839 kcal/kg	91.50%	8.70%	2.01%	27.89%	48.22%	4.68%
3. Amount of each nutrient in daily feed—(L.1 × L.2)/100*	34,973 kcal	8.34kg	0.79kg	0.18kg	2.54kg	4.39kg	0.43kg
4. Average daily feces excretion—20.61 kg							
5. Composition of feces (as determined by lab analysis)	818.9 kcal/kg	20.03%	1.92%	0.41%	8.09%	8.10%	1.51%
6. Amount of each nutrient in daily feces—(L.4 × L.5)/100**	16,877 kcal	4.13kg	0.40kg	0.08kg	1.67kg	1.67kg	0.31kg
7. Amount of each nutrient digested daily—(L.3–L.6)	18,096 kcal	4.21kg	0.39kg	0.10kg	0.87kg	2.72kg	0.12kg
8. Coefficient of digestibility of each nutrient—(L.7/L.3) × 100	51.7%	50.5%	49.4%	55.5%	34.3%	62.0%	27.9%
9. kcal/kg or % of digestible nutrient in feed—(L.8 × L.2)/100** or (L.7/L.1) × 100*	1985 kcal/kg	46.2%	4.28%	1.10%	9.55%	29.90%	1.31%

*The digestibility of mineral matter usually is not calculated. It is shown here simply to help accomplish certain instructional objectives.
**Does not apply to the Energy column.

8 Total Digestible Nutrients or TDN

TDN is the abbreviation for *total digestible nutrients.* TDN is simply a figure that indicates the relative energy value of a feed to an animal. It is ordinarily expressed in pounds or kilograms or in percent (lb or kg of TDN per 100 lb or kg of feed). It is arrived at by adding together the following:

		As Calculated for Feed from Table 7.1 (See Line 9)
Digestible crude protein	=	4.28%
Digestible crude fiber	=	9.55
Digestible N-free extract	=	29.90
Digestible crude fat $\times$ 2.25 (1.10 $\times$ 2.25)	=	2.48
TOTAL		46.21% TDN

Total digestible nutrients (TDN) is not an actual total of the digestible nutrients in a feed. In the first place, it does not include the digestible mineral matter. Secondly, the digestible fat is multiplied by 2.25 before being included in the TDN figure. The latter step is necessary to allow for the extra energy value of fats compared to carbohydrates and protein. As a result of this step, feeds high in fat will sometimes exceed 100 in percentage of TDN.

II

Factors affecting the TDN value of a feed.

A. **The percentage of dry matter.** Water can in no way contribute in a positive way to the TDN value of a feed. The more water present in a feed, the less there is of other nutrients, and, other things being equal, the lower the TDN value. For example, while milk on a dry basis is quite nutritious, on a fresh basis it is quite low (16%) in TDN because of its high (88%) water content. Silage is low in TDN compared to hay mainly because of a difference in water content. Many other such examples can be cited.

B. **The digestibility of the dry matter.** Unless the dry matter of a feed is digestible, it can have no TDN value. Only digestible dry matter can contribute TDN. For example, mineral oil has a high gross energy value, but it cannot be digested by the animal and so has no digestible energy or TDN value. Lignin would fall into a similar category. Feeds high in fiber are, in general, low in digestibility and relatively low in TDN. Sand would be another form of dry matter that is indigestible and so would have a 0.0 TDN value.

C. **The amount of mineral matter in the digestible dry matter.** Mineral compounds in an animal's ration may be digestible, but they contribute no energy to the animal and so have no TDN value. Such materials as salt, limestone, and defluorinated phosphate are all, in effect, digestible by the animal but would have 0.0 TDN values. The more mineral matter a feed contains, other things being equal, the lower will be its TDN value.

D. **The amount of fat in the digestible dry matter.** As mentioned previously, in calculating TDN the digestible fat is multiplied by 2.25 since fat contributes 2.25 times as much energy per unit of weight as do carbohydrates and protein. Consequently, the more digestible fat a feed contains, other things being equal, the greater will be the TDN value. With feeds exceptionally high in digestible fat, such as peanut kernels or dried whole milk, TDN values may even exceed 100%. In fact, a pure fat that had a coefficient of digestibility of 100% would theoretically have a TDN value of 225% ($100\% \times 2.25$).

III

The following diagram may be helpful in understanding the factors affecting the TDN value of feed.

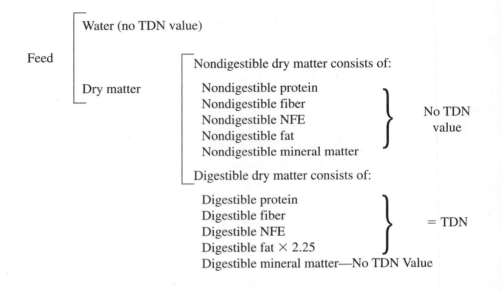

In Table 8-1 the feeds from the appendix tables have been listed and grouped on an as-fed basis according to their content of TDN. Again it is not intended for the student to memorize the exact percentages of TDN for the different feeds or to use the table as a source of information on TDN content for an individual feed. It simply is hoped that the listing and grouping of the feeds on the basis of TDN content will help the student get in mind the range of TDN values found in feed materials and in which type or group a particular feed would be found. Since an economical source of energy represents one of the more critical aspects of livestock feeding, and since TDN reflects a feeds energy value, it is important that the student of feeds and feeding becomes familiar with the relative TDN value of the different feeds as soon as possible.

Table 8-1
FEEDS LISTED AND GROUPED ACCORDING TO THEIR PERCENTAGE OF TDN FOR CATTLE—AS-FED BASIS

Group A: Primarily the Pure Fats and Other High-Fat Feeds, Plus Certain Other Feeds of Generally High Digestibility

Hydrolyzed animal fat	223.0	Cottonseed	88.0
Soybean oil	194.0	Hominy feed	85.0
Corn oil	172.8	Oat groats	85.0
Peanut kernels	131.0	Soybean seed	83.0
Dried whole milk	114.0	Dried bakery product	82.0

continued

Table 8-1 (Continued)

<u>Group B:</u> **Mostly Low-Fiber Grains and Grain By-Products and Low-Fiber and Low–Bone-Containing Protein Feeds, Plus Dried Citrus Pulp**

Dehydrated casein	81.0	Dried whey	75.0
Dried brewers grains	81.0	44% soybean meal	75.0
Dried corn distillers grains	80.0	Barley grain	74.0
Grain sorghum grain 8–10% prot.	80.0	Linseed meal	74.0
Dried skimmed milk	79.0	Potato meal	74.0
Corn gluten meal	78.0	Poultry by-product meal	74.0
Grain sorghum grain	78.0	Rye grain	73.0
49% soybean meal	78.0	Sweet potato meal	73.0
Wheat grain	78.0	Corn and cob meal	72.0
Corn grain	77.0	41% mech-extd cottonseed meal	72.0
Dried citrus pulp with molasses	77.0	Peanut meal	71.0
Triticale grain	76.0	Sesame oil meal	71.0
Corn gluten feed	75.0	Dried beet pulp with molasses	70.0
Dried citrus pulp	75.0	41% solv-extd cottonseed meal	70.0

<u>Group C:</u> **Largely the More Fibrous Grains and By-Product Concentrates, and the High–Bone-Containing Protein Feeds, Plus Beet Molasses**

Soybean hulls	70.0	Wheat shorts	65.0
Copra meal	68.0	Dried alfalfa	64.0
Oats grain	68.0	leaves	
36% cottonseed meal	67.0	Rapeseed meal	63.0
Dried beet pulp	67.0	Rice bran	63.0
Fish meal	67.0	Tankage with bone	63.0
Meat meal	67.0	Wheat bran	63.0
Safflower oil meal	67.0	Beet molasses	61.0
Tankage	67.0	Blood meal	61.0
Meat and bonemeal	66.0	Wheat middlings	61.0
Hydrolyzed feather meal	65.0	Sunflower meal	60.0

<u>Group D:</u> **Predominantly the Hays, the Hay Meals, the Straws, and the Hulls (Except Peanut and Rice Hulls), Plus Corn Cobs, Corn Stover, Citrus, Cane and Wood Molasses**

Light oats grain	59.0	Redtop hay	54.0
Poultry manure	59.0	Cowpea hay	53.0
Oat hay	56.0	Alfalfa hay	52.0
Dehydrated alfalfa		Timothy hay	52.0
meal	55.0	Alsike clover hay	51.0
Meadow hay	55.0	Bromegrass hay	51.0
Dehy coastal bermudagrass		Citrus molasses	51.0
meal	54.6	Millet hay	51.0
Blackstrap molasses	54.0	Sudangrass hay	51.0

continued

Table 8-1 (Continued)

Kentucky bluegrass hay	54.0	Ky 31 fescue hay	50.0
Peanut hay	50.0	Bahiagrass hay	46.0
Reed canarygrass hay	50.0	Wood molasses	46.0
Coastal bermudagrass hay, fertilized	49.1	Barley straw	45.0
		Ground corn cobs	45.0
Coastal bermudagrass hay	49.0	Soybean hay	45.0
Orchardgrass hay	49.0	Bermudagrass hay	42.0
Red clover hay	49.0	Lespedeza hay	42.0
Alfalfa stem meal	48.2	Cottonseed hulls	41.0
Johnsongrass hay	48.0	Pangolagrass hay	41.0
Oat straw	47.0	Sericea hay	40.0
Prairie hay	47.0	Wheat straw	39.0
Sweetclover hay	47.0	Soybean straw	39.0

Group E: For the Most Part High-Moisture and High-Mineral Feeds, Plus Peanut and Rice Hulls

Corn ear silage	32.0	Fresh ryegrass	15.0
Oat hulls	32.0	Wheat silage	15.0
Fresh sweet potatoes	25.0	Fresh sudangrass	14.0
Fresh Kentucky bluegrass	23.0	Fresh alfalfa	13.0
Fresh bromegrass	22.0	Fresh ladino clover	13.0
Corn silage	21.0	Fresh pangolagrass	12.0
Peanut hulls	20.0	Fresh crimson clover	11.0
Fresh coastal bermudagrass	19.0	Rice hulls	11.0
Fresh Ky 31 fescue	19.0	Fresh carrots	10.0
Fresh potatoes	19.0	Fresh skimmed milk	9.0
Wet brewers grains	18.5	Wet beet pulp	9.0
Grain sorghum silage	18.0	Fresh turnip roots	8.8
Corn stover silage	17.0	Fresh cabbage	8.3
Fresh orchardgrass	17.0	Fresh whey	7.0
Fresh rye forage	17.0	Urea, 45%N	0.0
Oat silage	16.8	Defluorinated phosphate	***
Fresh bahiagrass	16.0	Dicalcium phosphate	***
Fresh cow's milk	16.0	Ground limestone	***
Fresh citrus pulp	15.0	Oyster shell flour	***
Fresh dallisgrass	15.0	Steamed bonemeal	***

***Very little, if any

9 Energy Utilization

I

Definitions of basic energy terms:

A. **Calorie (cal).** The amount of energy as heat required to raise 1 gram of water 1°C (precisely from 14.5°C to 15.5°C). Formerly referred to as a small *calorie* and was so designated by being spelled with a lower case *c*.

B. **Kilocalorie (kcal).** The amount of energy as heat required to raise 1 kilogram of water 1°C (from 14.5°C to 15.5°C). Equivalent to 1,000 calories. Formerly referred to as a large *Calorie* and was so designated by being spelled with a capital *C*.

C. **Megacalorie (Mcal).** Equivalent to 1,000 kilocalories or 1,000,000 calories. Formerly referred to as a *Therm*.

D. **British Thermal Unit (BTU).** The amount of energy as heat required to raise 1 lb of water 1°F. Equal to 252 calories. Seldom used in animal nutrition.

E. **Bomb calorimeter.** An instrument used for determining the gross energy content of a material (Figure 9-1). It consists of an insulated water container equipped with a combustion bomb, a thermometer, and certain other accessories.

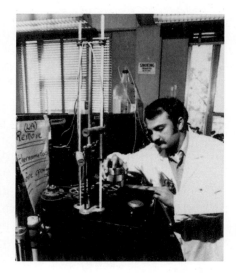

FIGURE 9-1
The different parts of a bomb calorimeter are explained by a
lab supervisor. (Courtesy of the University of Georgia
College of Agriculture Experiment Stations)

II

Energy disposition in the animal may be diagrammed as follows:

Gross energy
{
Energy of feces

Digestible energy
{
Energy of the urine
Energy of methane

Metabolizable energy
{
Heat increment
Heat of fermentation
Heat of nutrient metabolism
Net energy
}
}
}

A.   **Gross energy (GE).**   This is the total heat of combustion of a material as determined with a bomb calorimeter—ordinarily expressed as kilocalories per kilogram
of feed or kcal/kg. The gross energy value of a feed has no relationship to the feed's
digestible, metabolizable, or net energy values, except that the latter can never exceed the GE. Certain products such as coal, mineral oil, and lignin have high gross

energy values but, because of their indigestibility, are of no energy value to the animal. Roughages have gross energy values comparable to those of concentrates, but the two differ greatly in digestible, metabolizable, and net energy values. Fats, because of their greater proportion of carbon and hydrogen, yield 2.25 times more gross energy per kg than do carbohydrates.

B.   **Digestible energy (DE).**   This is the portion of the energy of a feed that does not appear in the feces. It includes metabolizable energy as well as the energy of the urine and methane. DE differs from TDN in that TDN, as calculated, does not include the energy of the urine—at least that from protein metabolism, which is most of it.

C.   **Metabolizable energy (ME).**   This is the portion of the gross energy consumed that is utilized by the animal for accomplishing work, growth, fattening, fetal development, milk production, and/or heat production. It is the portion of the gross energy not appearing in the feces, urine, and gases of fermentation (principally CH_4). It is digestible energy minus the energy of the urine and methane. It is comparable to the energy of TDN minus the energy of the fermentation gases.

D.   **Net energy (NE).**   This is the portion of metabolizable energy that may be used as needed by the animal for work, growth, fattening, fetal development, milk production, and/or heat production. It differs from metabolizable energy in that net energy does not include the heat of fermentation and nutrient metabolism (the heat increment). No net energy is used for heat production unless heat over and above that from other sources is required to keep the animal warm.

E.   **Heat increment (HI).**   This is the difference between ME and NE. It represents the heat unavoidably produced by an animal incidental with nutrient digestion and metabolism. It has been referred to also as *work of digestion, specific dynamic effect,* and *thermogenic effect.* This heat is useful only for keeping an animal warm during very cold weather. At other times the energy represented by this heat is not only a complete loss but also may actually interfere with production by causing the animal to be too warm.

III

Comparison of TDN, DE, ME, and NE as measures of a feed's energy value to an animal.   Total digestible nutrients (TDN), digestible energy (DE), metabolizable energy (ME), and net energy (NE) have all been used over the years for expressing the energy value of different feeds and rations for feeding purposes. Each has served and continues to serve a valuable function in ration formulation.

A.   Of these, DE is probably the least precise measure of a feed's energy value to an animal. It embraces all of the energy of a feed that does not appear in the feces but makes no allowances for other energy losses during digestion and utilization.

B.   TDN is superior to DE in this regard since in attributing the same value to digestible protein as is attributed to digestible carbohydrates in calculating TDN, an approximate correction is effected for that part of the protein energy that is excreted in the urine. Whereas digestible carbohydrates have a DE value of approximately 4.15 kcal per gram, digestible protein has a DE value of approximately 5.65 kcal per gram. However, the two are of similar energy value to the animal and are so considered by the procedure followed in calculating TDN.

C. ME figures, on the other hand, represent a more precise measure of a feed's energy value to an animal than do either DE or TDN figures. In determining the ME of a feed, allowances are made not only for the energy losses in the urine but also for those of the fermentation gases. While the above-mentioned procedure followed in calculating TDN effects an approximate correction for the energy losses of the urine, both TDN and DE values fail to take into account the energy losses of the fermentation gases. ME figures, on the other hand, allow for both. This is especially important from the standpoint of balancing rations for ruminants since such animals normally have significant energy losses of both types, and different feeds and rations vary greatly in the amount of these losses.

D. In the final analysis, however, to the extent that reliable NE values are available, they represent the most precise measure of an animal's energy needs and the capacity of different feeds to meet these needs. Not only do NE values allow for the energy losses of the urine and fermentation gases, but they also take into consideration the energy losses from the heat of nutrient utilization or the heat increment. Actual NE values, however, have been determined for only a limited number of feeds. Most available NE values are estimates that have been calculated for various feeds based on their composition and digestibility in relation to the composition and digestibility of the feeds on which actual NE values have been determined.

10 Study Questions and Problems

I. Give the correct figure for each of the following:

1. Number of chemical elements that function in animal nutrition.
2. Number of mineral elements that function in animal nutrition.
3. Number of nonmineral elements that function in animal nutrition.
4. Number of micro mineral elements that function in animal nutrition.
5. Number of micro elements that function in animal nutrition.
6. Number of chemical elements found in carbohydrates.
7. Minimum number of chemical elements found in a protein.
8. Maximum number of chemical elements found in a protein.
9. Minimum number of chemical elements found in a true fat.
10. Maximum number of chemical elements found in a true fat.
11. Number of carbon atoms in a fructose molecule.
12. Number of hydrogen atoms in a galactose molecule.
13. Number of oxygen atoms in a glucose molecule.
14. Number of carbon atoms in a lactose molecule.
15. Number of hydrogen atoms in a sucrose molecule.
16. Number of oxygen atoms in a maltose molecule.
17. Number of hexose molecules in a disaccharide.
18. Number of hexose molecules in a polysaccharide.

19. Number of hexose molecules in cellulose.
20. Number of glucose molecules in sucrose.
21. Number of glucose molecules in maltose.
22. Number of fructose molecules in lactose.
23. Number of galactose molecules in lactose.
24. Number of galactose molecules in sucrose.
25. Number of carbon atoms in a pentose molecule.
26. Number of known naturally occurring amino acids.
27. Number of naturally occurring proteins.
28. Percentage of nitrogen in crude protein.
29. What times nitrogen equals crude protein?
30. Fats contain how many times as many cal/lb as does starch?
31. What is the percentage of dry matter in air-dry feed?
32. What times % equals parts per million?
33. What times mg/lb equals parts per million?
34. To convert PPM to % divide by what number?
35. To convert PPM to mg/lb divide by what number?
36. What times % equals mg/lb?
37. Number of pounds in a kilogram.

II. **Match (using numbers) the items in the right-hand column with the proper chemical formula in the left-hand column:**

1. $C_6H_{12}O_6$.   1. Sucrose.
2. $C_{12}H_{22}O_{11}$.   2. A pentosan.
3. $(C_6H_{10}O_5)_n$.   3. Oleic acid.
4. $(C_5H_8O_4)_n$.   4. Stearin.
5. $C_{17}H_{35}COOH$.   5. Glucose.
6. $C_3H_5(OH)_3$.   6. An amino acid.
7. $C_{57}H_{110}O_6$.   7. A soap.
8. $C_{17}H_{33}COOH$.   8. Stearic acid.
9. $C_{17}H_{35}COONa$.   9. Glycerol.
10. $C_4H_8 (NH_2) COOH$.   10. Starch.

III. **Indicate by number into which of the following categories each of the items below would fall:**

1. A micro mineral element that functions in animal nutrition.
2. A macro mineral element that functions in animal nutrition.
3. A nonmineral element that functions in animal nutrition.
4. A true fat.
5. A carbohydrate.
6. An enzyme that functions in feed digestion.
7. A part of the digestive tract.
8. None of the above.

1.	Carbon.	*18.*	Trypsin.	*35.*	Hydrogen.
2.	Molybdenum.	*19.*	Sodium.	*36.*	Lactose.
3.	Glucose.	*20.*	Peptidase.	*37.*	Amino acid.
4.	Phosphorus.	*21.*	Galactose.	*38.*	Starch.
5.	Olein.	*22.*	Cadmium.	*39.*	Chlorine.
6.	Lactase.	*23.*	Calcium.	*40.*	Stearin.
7.	Iodine.	*24.*	Pepsin.	*41.*	Abomasum.
8.	Sucrose.	*25.*	Iron.	*42.*	Cobalt.
9.	Fructose.	*26.*	Rumen.	*43.*	Zinc.
10.	Fluorine.	*27.*	Cellulose.	*44.*	Bile.
11.	Rennin.	*28.*	Magnesium.	*45.*	Maltose.
12.	Palmitin.	*29.*	Villi.	*46.*	Reticulum.
13.	$C_{51}H_{98}O_6$.	*30.*	Selenium.	*47.*	Manganese.
14.	Oxygen.	*31.*	Omasum.	*48.*	Sucrase.
15.	Cecum.	*32.*	Potassium.	*49.*	Dextrin.
16.	Nitrogen.	*33.*	Glycogen.	*50.*	Inulin.
17.	Sulfur.	*34.*	Boron.		

IV. **Indicate by number into which of the following categories each of the items below would fall:**

> *1.* A monosaccharide.
> *2.* A disaccharide.
> *3.* A polysaccharide.
> *4.* Other.

1.	Glucose.	*11.*	Glycogen.	*21.*	Beet sugar.
2.	Methane.	*12.*	Insulin.	*22.*	Urea.
3.	Sucrose.	*13.*	Fruit sugar.	*23.*	Lignin.
4.	Glycerol.	*14.*	Maltase.	*24.*	Dextrose.
5.	$C_{12}H_{22}O_{11}$.	*15.*	Milk sugar.	*25.*	Crude fiber.
6.	Cane sugar.	*16.*	Fructose.	*26.*	Starch.
7.	Corn sugar.	*17.*	Galactose.	*27.*	Lactose.
8.	Cellulose.	*18.*	Rennin.	*28.*	Mannose.
9.	Stearin.	*19.*	Maltose.	*29.*	Cotton.
10.	$(C_6H_{10}O_5)_N$.	*20.*	Inulin.	*30.*	$C_6H_{12}O_6$.

V. **Correlate by number the various items in the first column with the most appropriate item in the second column:**

1.	Dry roughage.	*1.*	Olein.
2.	Fructose.	*2.*	Palmitin.
3.	Low in fiber—high in TDN.	*3.*	Fruit sugar.
4.	Sucrose.	*4.*	Ether.

5. Nonmineral element.
6. Glycogen.
7. Enzyme.
8. Lactose.
9. Glycerol.
10. Starch and sugar.
11. Fiber + NFE.
12. Amide.
13. A trace mineral.
14. A fat solvent.
15. mg/kg.
16. $C_{15}H_{31}COONa$.
17. mg/lb.
18. Ash.
19. CH_4.
20. Crude fat.
21. Alkaline digestive juice.
22. Percent.
23. End product of starch digestion.
24. $C_{51}H_{98}O_6$.
25. Ruminant.
26. Rumen.
27. Dry matter.
28. Crude protein.
29. Oven-dry.
30. Manyplies.
31. Reticulum.
32. Unsaturated fat.
33. Abomasum.
34. Amino acid.
35. Almost pure cellulose.
36. $C_{17}H_{35}COOH$.
37. Proteolytic enzyme.

5. Mineral matter.
6. Omasum.
7. Stearic acid.
8. Cotton.
9. Honeycomb.
10. PPM.
11. True stomach.
12. Glucose.
13. Cane sugar.
14. Bile.
15. Milk sugar.
16. Cobalt.
17. End product of protein digestion.
18. (mg/kg)/2.2.
19. Trypsin.
20. Animal starch.
21. Methane.
22. Nitrogen-free extract.
23. Carbon.
24. Nonprotein.
25. Organic catalyst.
26. Paunch.
27. $C_3H_5(OH)_3$.
28. Ether extract.
29. PPM/10,000.
30. Soap.
31. Carbohydrate.
32. Nitrogen $\times$ 6¼.
33. High in fiber—low in TDN.
34. 100% dry matter.
35. Concentrate.
36. Cud-chewing animal.
37. 100 minus % H_2O.

VI. A mixture of the following materials is subjected to a proximate analysis. Indicate by number in which of the following fractions each would appear:

1. Crude protein.
2. Crude fiber.
3. Crude fat.
4. Nitrogen-free extract.
5. Mineral matter.
6. None of the above.

1.	True protein.	13.	Glycogen.	25.	Ether.
2.	Cholesterol.	14.	Fatty acids.	26.	Sand.
3.	Cotton fiber.	15.	Wood ashes.	27.	Glucose.
4.	Amides.	16.	Dextrin.	28.	Hair.
5.	Inulin.	17.	$C_5H_{10}O_5$.	29.	Amino acids.
6.	Vitamin D.	18.	Starch.	30.	Wool fiber.
7.	Maltose.	19.	Disaccharides.	31.	$C_{12}H_{22}O_{11}$.
8.	Salt.	20.	Sucrose.	32.	Soybean oil.
9.	Calcium oxide.	21.	$C_{51}H_{98}O_6N_{11}$.	33.	Sucrase.
10.	Stearin.	22.	$C_{51}H_{98}O_6$.	34.	Lignin.
11.	Mineral oil.	23.	Cellulose.		
12.	Chlorophyll.	24.	Carotene.		

VII. **Indicate by number which of the following materials on an as-fed basis would be among the:**

1. 8 highest in % water.
2. 19 highest in % TDN.
3. 17 highest in % fiber.
4. 14 highest in % fat.

1.	Alfalfa hay.	20.	Steamed bonemeal.
2.	Dehydrated alfalfa meal.	21.	Soybean seed.
3.	Timothy hay.	22.	Dried skimmed milk.
4.	Grain sorghum grain.	23.	Corn silage.
5.	Defluorinated phosphate.	24.	Feather meal.
6.	Wheat grain.	25.	Crimson clover pasture.
7.	Alfalfa pasture.	26.	Barley grain.
8.	Oat straw.	27.	Sorgo silage.
9.	Peanut kernels.	28.	Fish meal.
10.	Corn oil.	29.	Cottonseed.
11.	Wheat bran.	30.	Yellow shelled corn.
12.	Corn starch.	31.	Tankage with bone.
13.	White potatoes.	32.	Dried citrus pulp.
14.	Ground corn cob.	33.	36% cottonseed meal.
15.	Red clover hay.	34.	44% soybean meal.
16.	Cane molasses.	35.	Fresh whole milk.
17.	Lespedeza hay.	36.	Bermudagrass pasture.
18.	Sweet potatoes.	37.	Ground limestone.
19.	Cane sugar.		

VIII. In using the tables on composition of feed, for what does each of the following abbreviations stand:

1.	dehy	6.	Solv-extd	11.	mg/kg	16.	NE lac
2.	IU	7.	mech-extd	12.	mcal/kg	17.	mg/lb
3.	kcal	8.	prot equiv	13.	kcal/kg	18.	mcal/lb
4.	IU/kg	9.	DE	14.	NE main	19.	TDN
5.	lb/bu	10.	ME	15.	NE gain	20.	kcal/lb

IX. Indicate whether each statement is true or false. In order for a statement to be true, it must be completely true.

1. Bile is produced in the liver and functions in fat digestion.
2. Starch is a carbohydrate and is also a form of nitrogen-free extract.
3. The cell walls of plants and animals are chiefly cellulose.
4. Apparently dry feeds may contain more than 10% water.
5. Galactose is a hexose and is a component of milk sugar.
6. The omasum of a cow has a capacity of up to 40 to 50 gals.
7. Maltase acts on maltose to yield fructose and glucose.
8. Most good hays contain more than 60% TDN.
9. Proteins are, on the average, about 6.25% nitrogen.
10. Fat contains a greater proportion of carbon and hydrogen than does starch.
11. The end products of fat digestion are amino acids, glycerol, and soap.
12. Carbohydrates contain carbon, hydrogen, and oxygen.
13. Proteins are soluble in ether and also in dilute H_2SO_4.
14. Another name for crude protein is nitrogen-free extract.
15. Plants get their energy from the soil.
16. Pancreatic lipase is an enzyme that works on fats in the small intestine.
17. Ether extract is another name for crude fat.
18. Fructose is a monosaccharide found in ripe fruits and honey.
19. Fats, on the average, contain 2½ times as much energy per lb as do sugars and starch.
20. The cell walls of animals are high in lignin.
21. Lactose is a disaccharide found in milk.
22. A megacalorie is equal to 1,000 calories.
23. Trypsin is an enzyme that works on protein in the stomach.
24. Sucrose is another name for beet sugar as well as cane sugar.
25. Proteins sometimes contain sulfur, potassium, and/or iron.
26. A pig has no rumen but does have an abomasum.
27. Cane sugar consists of one molecule of glucose and one molecule of fructose.
28. The gross energy value of a feed is a good indication of its TDN value.

29. Pepsin is an enzyme found in the gastric juice, and it acts on protein.
30. Of all farm animals, the horse has the largest cecum.
31. The horse is a ruminant but does not chew a cud.
32. The pig is classified as a nonruminant.
33. The coefficient of digestibility of a nutrient is the percentage of the total amount of that nutrient present that is digestible.
34. The percentage of digestible protein in a feed may be greater than the percentage of crude protein in the feed.
35. The percentage of crude fiber in a feed may exceed 100% under certain conditions.
36. The percentage of crude fat in a feed may exceed 100% under certain conditions.
37. The percentage of crude protein in a feed may exceed 100% under certain conditions.
38. The percentage of nitrogen-free extract in a feed may exceed 100% under certain conditions.
39. The percentage of TDN in a feed may exceed 100% under certain conditions.
40. The percentage of TDN plus the percentage of mineral matter in a feed should equal the percentage of dry matter in the feed.
41. 100% minus the percentage of water in a feed should equal the percentage of dry matter.
42. The percentage of crude protein plus the percentage of crude fat plus the percentage of crude fiber plus the percentage of nitrogen-free extract should equal the percentage of dry matter in the feed.
43. Metabolizable energy minus the heat increment equals net energy.
44. TDN minus (energy of the urine + energy of methane) would be comparable to metabolizable energy.
45. A small calorie, a gram-calorie, and a megacalorie are all the same.
46. A large calorie, kilocalorie, and a therm are all the same.
47. Proteins and carbohydrates are about equal in TDN value per pound.
48. The sum of digestible protein, digestible carbohydrates, and (digestible fat × 2¼) equals TDN.
49. TDN may be expressed in lb or kg or percent.
50. TDN includes all of the digestible nutrients in a feed.
51. TDN indicates the relative digestible energy value of a feed.
52. TDN minus the energy of the fermentation gases would be comparable to metabolizable energy.
53. The heat increment of a feed serves no worthwhile function in livestock production.
54. Starch is a common component of many feeds.
55. Net energy plus the heat increment equals metabolizable energy.
56. Commercially mixed feeds frequently will contain excessive amounts of moisture.
57. Starch and sugar would be comparable in TDN value per pound.

58. Cellulose is the primary constituent of cotton.
59. Reputable feed manufacturing companies welcome rigid feed-mixing controls.
60. A feed component that is highly digestible could have a low TDN value.
61. Crude fiber is one of the chief constituents of wool.
62. Sometimes lignin comes out in the NFE fraction of a feed.
63. Glycogen is a common component of many feeds.
64. The percentage of crude protein plus the percentage of crude fat plus the percentage of crude fiber plus the percentage of nitrogen-free extract should equal the percentage of dry matter in the feed.
65. A feed component that is high in crude fat will always have a high TDN value.
66. A megacalorie and a therm are the same.
67. Inulin is a common component of many feeds.
68. Digestible protein and digestible nitrogen-free extract are about equal in TDN value per pound.

X. A feed has the following composition on an air-dry basis. Calculate its composition on 1. An oven-dry basis; 2. An 80% moisture basis:

Ash	4.7%
Crude fiber	27.6
Ether extract	1.9
N-free extract	49.4
Crude protein	6.4
TOTAL DM	90.0%

SOLUTIONS:

1. Composition on an oven-dry basis.

Ash:

$$\frac{4.7}{90} = \frac{x}{100}$$

$$90x = 470$$

$$x = 5.22 \text{ (\% ash, oven-dry basis)}$$

Calculate the crude fiber, etc.

2. Composition on an 80% moisture basis.

Ash:

$$\frac{4.7}{90} = \frac{x}{100-80}$$

$$90x = 94$$

$$x = 1.04 \ (\% \ \text{ash}, \ 80\% \ \text{moisture basis})$$

Calculate the crude fiber, etc.

XI. A feed has the following composition on a fresh basis:

Ash	2.21%
Crude fiber	6.70%
Ether extract	0.92%
N-free extract	10.00%
Crude protein	4.57%
Calcium	0.40%
Phosphorus	0.06%
Carotene	28.30 mg/lb

1. What is the % dry matter in the fresh feed?
2. What is the % crude fiber in the feed on an air-dry basis?
3. What is the % crude fiber in the feed on an oven-dry basis?
4. What is the % carotene in the fresh feed?
5. How many PPM of phosphorus in the fresh feed?

SOLUTIONS:

1. The % DM in the fresh feed.

$$2.21 + 6.70 + 0.92 + 10.00 + 4.57 = 24.40$$

2. The % crude fiber in the feed on an air-dry basis.

$$\frac{6.70}{24.40} = \frac{x}{90.0}$$

$$24.40x = 603$$

$$x = 24.71$$

3. The % crude fiber in the feed on an oven-dry basis.

$$\frac{6.70}{24.40} = \frac{x}{100} \quad or \quad \frac{24.71}{90} = \frac{x}{100}$$

$$24.40x = 670 \qquad\qquad 90x = 24.71$$

$$x = 27.46 \qquad\qquad x = 27.46$$

4. The % carotene in the fresh feed.

$$\frac{28.3 \times 2.2}{10,000} = 0.00623$$

or

$$\frac{28.3}{453,600} \times 100 = 0.00624$$

(mg in a lb)

5. PPM of phosphorus in the fresh feed.

$$0.06 \times 10,000 = 600$$

XII. A feed has the following composition on a wet basis:

Ash	3.11%
Crude fiber	8.32%
Ether extract	1.10%
N-free extract	12.44%
Crude protein	5.61%
Calcium	0.32%
Phosphorus	0.16%
Carotene	110.0 mg/kg

1. What is the % water in the feed on the wet basis?
2. What is the % NFE in the feed on an air-dry basis?
3. What is the % NFE in the feed on an oven-dry basis?
4. How many mg/lb of phosphorus are in the feed on a wet basis?
5. How many PPM of carotene are in the feed on a wet basis?

ANSWERS:

1. 69.42%.
2. 36.61%.
3. 40.68%.
4. 725.6 mg/lb.
5. 110.0 PPM.

XIII. A feed contains 1.47% calcium on an air-dry basis. Calculate the following:

1. % calcium on an oven-dry basis.
2. % calcium on a 75% water basis.
3. Calcium content of air-dry feed in PPM.
4. Calcium content of air-dry feed in mg/lb.
5. Calcium content of air-dry feed in lb per 100 lb (Cwt) feed.

ANSWERS:

1. 1.63%.
2. 0.408%.
3. 14,700 PPM.
4. 6681.8 mg/lb.
5. 1.47 lb per Cwt.

XIV. A steer consumes 20 lb of feed daily and excretes 40 lb of feces. The feed and feces had the following compositions:

	Feed	Feces
Ash	8.0%	1.33%
Crude fiber	28.6%	9.91%
Ether extract	1.9%	0.11%
N-free extract	36.7%	6.20%
Crude protein	15.3%	2.10%

1. What is the % dry matter in the feed?
2. What is the % water in the feces?
3. What is the % digestible DM in the feed, as-fed basis?
4. What is the % TDN in the feed, as-fed basis?
5. What is the coefficient of digestibility of the DM in the feed?
6. What is the coefficient of digestibility of the NFE in the feed?
7. What is the % of digestible NFE in the feed, as-fed basis?
8. What is the % of digestible NFE in the feed, oven-dry basis?

SOLUTIONS:

1. The % DM in the feed.

$$8.0 + 28.6 + 1.9 + 36.7 + 15.3 = 90.5$$

2. The % water in the feces.

$$100 - (1.33 + 9.91 + 0.11 + 6.20 + 2.10) = 80.35$$

3. The % of digestible DM in the feed, as fed.

	Ash	Fiber	EE	NFE	Protein
Nutrients per lb feed (lb)	0.080	0.286	0.019	0.367	0.153
× 20 lb feed	20	20	20	20	20
= Nutrients consumed (lb)	1.60	5.72	0.38	7.34	3.06

	Ash	Fiber	EE	NFE	Protein
Nutrients per lb feces (lb)	0.0133	0.0991	0.0011	0.0620	0.0210
× 40 lb feces	40	40	40	40	40
= Nutrients excreted (lb)	0.532	3.964	0.044	2.480	0.840
Nutrients digested (lb)	1.068	1.756	0.336	4.860	2.220

$$\text{DM digested per 20 lb feed} = 1.068 + 1.756 + 0.336 + 4.860 + 2.220$$

$$= 10.24 \text{ lb}$$

$$\% \text{ digestible DM in feed} = \frac{10.24}{20} \times 100 = 51.2$$

4. The % TDN in the feed, as fed.

$$\frac{1.756 + (0.336 \times 2.25) + 4.860 + 2.220}{20} \times 100 = 47.96$$

5. The coefficient of digestibility of the DM in the feed.

$$\frac{51.2}{90.5} \times 100 = 56.57\%$$

6. The coefficient of digestibility of the NFE in the feed.

$$\frac{4.86}{7.34} \times 100 = 66.21\%$$

7. The % of digestible NFE in the feed, as fed.

$$\frac{4.86}{20} \times 100 \; or \; \frac{66.21 \times 36.7}{100} = 24.3$$

8. The % of digestible NFE in the feed, oven-dry basis.

$$\frac{24.3}{90.5} = \frac{x}{100}$$

$$90.5x = 2430$$

$$x = 26.85$$

XV. **A steer consumes daily 9.07 kg of a feed containing 11.0% crude protein and excretes daily an average of 14.97 kg of feces containing 2.00% crude protein.**

    1.   What is the coefficient of digestibility of the crude protein in the feed?

    2.   What is the % of digestible crude protein in the feed as fed?

ANSWERS:

    1.   69.99%.

    2.   7.70%.

XVI. **A steer consumes 9.07 kg of feed daily and excretes an average of 6.11 kg of feces. The composition of the feed as fed and the coefficients of digestibility of the respective nutrients are as follows:**

	Composition	Coefficient of Digestibility
Ash	3.2%	52.0%
Crude fiber	6.2%	55.5%
Ether extract	1.9%	80.0%
N-free extract	66.2%	91.0%
Crude protein	11.8%	77.0%

    1.   What is the % digestible DM in the feed, as-fed basis?

    2.   What is the TDN in the feed, as-fed basis?

    3.   What is the % digestible protein in the feed, as-fed basis?

    4.   What was the % DM in the feces?

SOLUTIONS:

    1.   The % digestible DM in the feed, as fed.

	Ash	*Fiber*	*EE*	*NFE*	*Protein*
Nutrients per kg feed (kg)	0.032	0.062	0.019	0.662	0.118
× 9.07 kg feed	9.07	9.07	9.07	9.07	9.07
= Nutrients eaten daily (kg)	0.290	0.562	0.172	6.004	1.070
× C of D	0.520	0.555	0.800	0.910	0.770
= Nutrients digested daily (kg)	0.151	0.312	0.138	5.464	0.824

$$\frac{0.151 + 0.312 + 0.138 + 5.464 + 0.824}{9.07} \times 100 = 75.95$$

    2.   The % TDN in the feed, as fed.

$$\frac{0.312 + (0.138 \times 2.25) + 5.464 + 0.824}{9.07} \times 100 = 76.20$$

3. The % digestible protein in the feed, as fed.

$$\frac{0.824}{9.07} \times 100 \text{ or } \frac{11.8 \times 77.0}{100} = 9.08$$

4. The % DM in the feces.

	Ash	Fiber	EE	NFE	Protein
Nutrients eaten daily (kg)	0.290	0.562	0.172	6.004	1.070
− Nutrients digested daily (kg)	0.151	0.312	0.138	5.464	0.824
= Nutrients in daily feces (kg)	0.139	0.250	0.034	0.540	0.246

$$\frac{0.139 + 0.250 + 0.034 + 0.540 + 0.246}{6.11} \times 100 = 19.79 \ (answer)$$

XVII. **A cow consumes 10.0 kg of feed daily and excretes an average of 18.61 kg of feces. The feed had the following composition (as-fed basis) and coefficients of digestibility for the respective nutrients:**

	Composition	Coefficient of Digestibility
Ash	4.7%	44.0%
Crude fiber	27.8%	55.0%
Ether extract	2.1%	50.0%
N-free extract	46.7%	60.0%
Crude protein	9.2%	65.0%

1. What is the % digestible DM in the feed, as-fed basis?
2. What is the % TDN in the feed, as-fed basis?
3. What is the % digestible fiber in the feed, as-fed basis?
4. What was the % water in the feces?

ANSWERS:

1. 52.41%.
2. 51.65%.
3. 15.29%.
4. 79.53%.

11 Protein Nutrition

I

Needs of animals for protein. Every living animal has a need for protein. It is the basic structural material from which all body tissues are formed. This includes not only the muscles, nerves, skin, connective tissue, and vital organs but also the blood cells, as well as the animal's hair, hoof, and horn. Even the dry matter of bone is more than ⅓ protein, with protein providing the basic cellular matrix within which the bone mineral matter is deposited. Obviously then, protein is essential for an animal's growth and development as well as for fetal development. Also, since all living tissue is in a dynamic state and is undergoing constant degeneration, protein is necessary for its maintenance. Also, protein is required for wool growth and milk production. Furthermore, most body enzymes and hormones are basically protein in composition. Finally, no other nutrient can replace protein in the ration. In view of an animal's many needs for protein and the irreplaceable nature of this nutrient, there is a certain minimum level of dietary protein recommended for each class of animals. This level varies for animals of different classes, depending on the physiological age and type of production, but will usually be somewhere between 8% and 18%.

II

Amino acids. Proteins are complex organic nitrogenous compounds made up of amino acids. *Amino acids* are organic acids that contain one or more amino groups (NH_2). Some 25 or more different amino acids are present in feed proteins, of which

FIGURE 11-1
An amino acid analyzer—used for determining the amino
acid makeup of proteins. (Courtesy of the University of
Georgia College of Agriculture Experiment Stations)

some 20 or more enter into the makeup of animal tissue (Figure 11-1). Of these, some
10 or 11 are classified as *essential* and the others as *nonessential*.

A. Essential amino acid. An essential amino acid is one that is needed by the animal but cannot be synthesized by the animal in the amounts needed and so must be present in the protein of the feed as such. These are sometimes referred to as the *indispensable amino acids*. Rumen microorganisms can synthesize nearly all amino acid requirements.

B. Nonessential amino acid. A nonessential amino acid is one that is needed by the animal that can be formed by the animal from other amino acids or amine groups. Therefore, it does not have to be present as that particular amino acid in the protein of the feed. These are sometimes referred to as the *dispensable amino acids*.

III

Essentiality of amino acids. Those amino acids that function in animal nutrition are usually classified on the basis of their essentiality as follows:

Essential	*Nonessential*
Arginine	Alanine
Histidine	Aspartic acid
Isoleucine	Citrulline
Leucine	Cystine
Lysine	Glutamic acid

continued

Essential (Cont'd)	*Nonessential(Cont'd)*
Methionine*	Glycine
Phenylalanine**	Hydroxyproline
Threonine	Proline
Tryptophan	Serine
Valine	Tyrosine

*May be replaced in part by cystine.
**May be replaced in part by tyrosine.

IV

Limiting essential amino acid.   The essential amino acids are required by livestock in definite proportions. While the proportion may vary for different functions, it is always definite for any given animal performing any given set of functions. The amino acid that is present in a protein in the least amount in relation to the animal's need for that particular amino acid is referred to as the *limiting amino acid.* Other essential amino acids can be used by the animal toward meeting its essential amino acid requirement only to the extent that the so-called limiting amino acid is present.

The above phenomenon is illustrated by a comparison of the requirements of growing pigs for the different essential amino acids with the levels of these amino acids in shelled corn as presented in Table 11-1. It will be noted that lysine is the limiting essential amino acid of corn with tryptophan running a close second (Figure 11-2).

V

Biological value.   The biological value of a protein is the percentage of the digestible protein of a feed or feed mixture that is usable as protein by the animal. The biological value of a protein will depend on the amount of the limiting amino acid present in relation to the other essential amino acids, based on the animal's need for each. A protein that has a desirable balance of essential amino acids will have a high biological value and is described as being a protein of *good quality.* A protein that is extremely deficient in one or more of the essential amino acids will have a low biological value and is described as being a protein of *low quality.*

VI

Supplementary effect of proteins.   When two proteins have different limiting amino acids and one contains an excess of the amino acid that is limiting in the other, then a *supplementary effect* is realized when the two proteins are mixed together. It is for this reason that more than one source of protein is recommended for nonruminants. Fortunately, the protein of most of the more abundant high-protein supplements has a quite favorable supplementary effect with the protein of most farm grains. This phenomenon is well illustrated by the bar graph in Figure 11-3.

VII

Ruminants vs. nonruminants.   While ruminant animals are thought to have just as rigid a physiological requirement for essential amino acids as nonruminants, they do not

Table 11–1
LIMITING ESSENTIAL AMINO ACID OF SHELLED CORN

	Percentage of Each Essential Amino Acid Required in Rations for 20–35 kg (44–77 lb) Pigs, as Fed	Percentage of Each Essential Amino Acid Present in Ground Shelled Corn, as Fed	Percentage of Each Essential Amino Acid Requirement in Corn
Arginine	0.20%	0.43%	215%
Histidine	0.18	0.26	144
Isoleucine	0.50	0.35	70
Leucine	0.60	1.21	202
Lysine	0.70	0.25	36
Methionine + cystine	0.45*	0.17 + 0.22	87
Phenylalanine + tyrosine	0.70**	0.40 + 0.38 (0.35 usable)	119
Threonine	0.45	0.35	78
Tryptophan	0.12	0.08	67
Valine	0.50	0.44	88

*Methionine can fulfill the total requirement; cystine can meet up to 50 percent of the total requirement.

**Phenylalanine can fulfill the total requirement; tyrosine can meet up to 50 percent of the total requirement.

(a)   (b)

FIGURE 11-2
Lysine deficiency. (a) This pig gained 11.4 kg in 28 days after lysine was added to the basal diet (2.0% DL-lysine). (b) A lysine deficient pig—basal diet only. This pig lost 0.9 kg in 28 days. (Courtesy of W. M. Beeson, Purdue University)

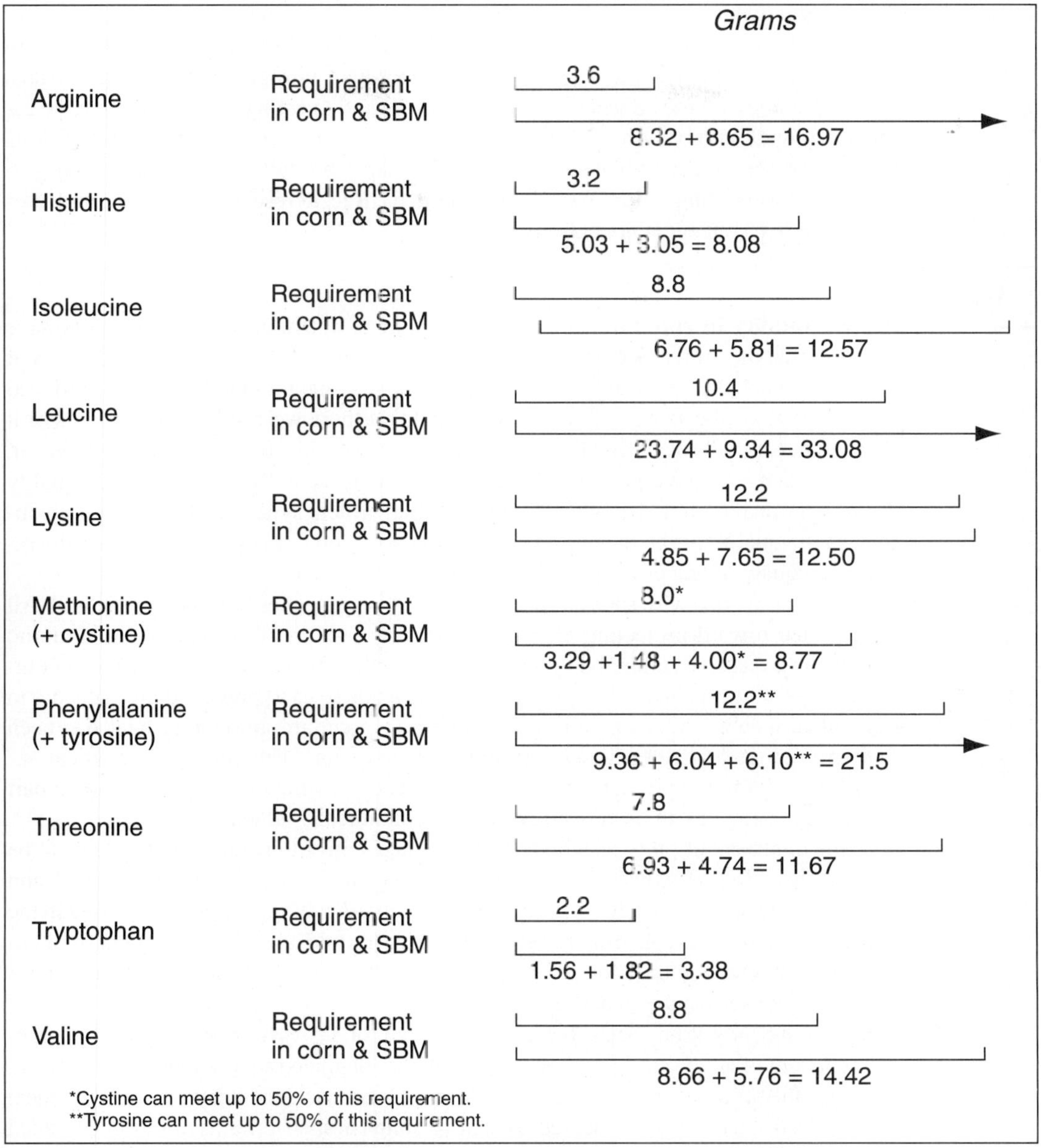

Grams

Arginine
Requirement
in corn & SBM

3.6

8.32 + 8.65 = 16.97

Histidine
Requirement
in corn & SBM

3.2

5.03 + 3.05 = 8.08

Isoleucine
Requirement
in corn & SBM

8.8

6.76 + 5.81 = 12.57

Leucine
Requirement
in corn & SBM

10.4

23.74 + 9.34 = 33.08

Lysine
Requirement
in corn & SBM

12.2

4.85 + 7.65 = 12.50

Methionine
(+ cystine)
Requirement
in corn & SBM

8.0*

3.29 + 1.48 + 4.00* = 8.77

Phenylalanine
(+ tyrosine)
Requirement
in corn & SBM

12.2**

9.36 + 6.04 + 6.10** = 21.5

Threonine
Requirement
in corn & SBM

7.8

6.93 + 4.74 = 11.67

Tryptophan
Requirement
in corn & SBM

2.2

1.56 + 1.82 = 3.38

Valine
Requirement
in corn & SBM

8.8

8.66 + 5.76 = 14.42

*Cystine can meet up to 50% of this requirement.
**Tyrosine can meet up to 50% of this requirement.

FIGURE 11-3
A comparison of the daily requirements of a 47.5 kg growing-finishing pig for the different essential amino acids with the amounts of each of these acids in a daily ration of 1.733 kg corn + .256 kg soybean meal + minerals and vitamins.

have a rigid dietary amino acid requirement. Microbes (bacteria and protozoa) in the rumen synthesize, for their own cellular development, essential amino acids from nonessential amino acids and certain other nonprotein nitrogen-containing materials. These microbes are then digested in the abomasum and intestinal tract to provide the ruminant animal with what has been considered over the years to be most if not all of its needs for essential amino acids. Hence protein quality has not been a matter of consideration in formulating rations for ruminant animals. Recent research, however, indicates that the above concept is not entirely true.

VIII

Protein quality in ruminant nutrition.   Ruminant nutritionists over the years have assumed that the protein leaving the rumen and entering the lower digestive tract was of as good and usually better quality than that fed. It was thought that the protein produced by microbial synthesis was of good quality and that there were sufficient amounts of it to meet the animal's need for all of the essential amino acids. Recent research, however, indicates that a dietary protein of good quality may actually be lowered in quality through ruminal protein degradation and microbial protein synthesis. In fact, in some instances the quality of the dietary protein may be lowered to the point of making the ration inadequate in one or more of the essential amino acids.

In order to cope with such a possibility, several approaches have been proposed. All are directed toward protecting high-quality dietary protein or critical essential amino acids from excessive ruminal degradation. One method that is being studied to accomplish this is to encapsulate such protein or amino acids so as to prevent their exposure to ruminal fermentation. Also, subjecting protein to a certain amount of heat has been found to reduce its tendency toward ruminal degradation. This may be a practical approach to solving the problem, especially with feeds in which heating is already a part of the processing or can be conveniently and economically added.

The treatment of proteins with certain chemicals prior to feeding also appears to be a possibility. Two chemicals that seem promising in this regard are tannic acid and formaldehyde. Other ways for coping with this problem will no doubt be proposed in the future. Protecting proteins from ruminal degradation could have the added benefit of encouraging the ruminal microorganisms to use nonprotein nitrogen rather than protein nitrogen in their fermentation activities.

More recently it has been shown that dietary proteins vary in their "rumen degradability." Some feed proteins, such as skim milk, are degraded to ammonia in the rumen, whereas others, such as hydrolyzed feather meal, blood meal, and dehydrated alfalfa meal are degraded to a much lesser degree. This has resulted in such terms as rumen undegraded protein (RUP) and undegraded intake protein (UIP). The senior author of this textbook demonstrated that a combination of urea (highly degradable protein source) and hydrolyzed feather meal (highly undegraded intake protein) was superior to soybean meal as a source of supplemental protein for cattle. It is standard practice today to take into account the degradability of supplemental protein in calculating dairy and beef cattle rations.

IX

Protein as a source of energy.   Protein fed in excess of an animal's needs or not usable as protein by the animal because of a limiting amino acid is *deaminated* (or *deaminized*) in the liver and used as a source of energy by the animal. In deamination, the NH_2 group is split off the amino acid molecule and excreted as urea in the urine. While high-protein feeds may serve as a source of energy for livestock, they are usually too expensive to be a practical source of feed energy.

X

Nonprotein nitrogen (NPN) utilization by ruminants. As mentioned previously, ruminants through microbial action in the rumen are able to synthesize protein from nonprotein nitrogen. As a result, nonprotein nitrogen may be used as a substitute for protein in ruminant rations. When so used, however, it must be fed along with a readily fermentable carbohydrate source such as corn, grain sorghum, or some type of molasses to supply the energy for the formation of the protein molecule.

XI

Urea as an NPN source. While various products have been and are being studied as nonprotein nitrogen sources, the principal NPN source in use today in livestock feeding is urea. It is produced by combining natural gas with water and air. It has a chemical formula of $CO(NH_2)_2$. Consequently, in pure form it contains 46.67% nitrogen.

Pure crystalline urea is extremely hygroscopic (takes on moisture from the atmosphere) and therefore it is critical that it be coated by a thin film of calcium carbonate. Thus, commercial feed-grade urea contains only 45% nitrogen, giving it a crude protein equivalent of 281% (Figure 11-4). In order for it to be used by rumen microbes, it must first be *hydrolyzed* (chemically combined with water) to form $CO_2 + NH_3$. This hydrolysis is accomplished by the action of an enzyme, urease, secreted by rumen organisms. As NH_3 is liberated in the rumen, it apparently reacts with organic acids of fermentation to form ammonium salts of organic acids, such as ammonium acetate and ammonium propionate. These in turn are metabolized by rumen microbes to form cellular protein.

It should be remembered, however, that some readily fermentable carbohydrate source must be fed along with the urea to provide the energy for the formation of the protein molecule. Shelled corn, grain sorghum, molasses, and citrus pulp are feeds commonly used for this purpose.

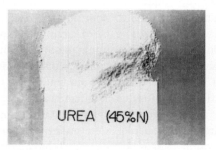

FIGURE 11-4
Much of today's urea is in a beaded form, which helps keep it free flowing for good mixing. (Courtesy of the University of Georgia College of Agriculture Experiment Stations)

Also, urea contains no minerals, and some rations in which it might be included are marginal in their content of sulfur for the adequate synthesis of the sulfur-containing amino acids, cystine and methionine. Rations high in shelled corn and corn silage would fall into this category. Consequently such rations will sometimes give a favorable response to sulfur supplementation when urea is being used as a protein source.

XII

Urea toxicity. Urea as such is a normal component of body fluids, since it is a by-product of protein metabolism in the body. Consequently, in the amounts ordinarily fed, it is not toxic to the animal. However, ammonia that is produced upon urea hydrolysis is quite toxic if liberated in excessive amounts—that is, in amounts that exceed the supply of organic acids of fermentation in the rumen with which the urea NH_3 is supposed to combine. Under such circumstances the NH_3 is absorbed into the blood through the rumen wall to produce alkalosis and, in turn, possibly death of the animal. Such trouble is most likely to occur when animals do not consume sufficient fermentable carbohydrates to maintain a relatively high level of acidity in the rumen. When rumen acidity is low, not only are there insufficient organic acids to react with the NH_3 present, but also urease is most active at a pH of 6 to 8 such as would be found in a rumen low in organic acids of fermentation. Also, the excess NH_3 would tend to form NH_4OH in the rumen to further reduce rumen acidity. In other words, the rumen condition most favorable for urea hydrolysis and NH_3 production is least favorable for effective NH_3 utilization. If urea is to be used effectively for ruminant feeding without toxic results, it must be fed in conjunction with an adequate supply of readily fermentable carbohydrates such as will keep the rumen pH below a level of 6.

XIII

Extent of urea feeding. It has been estimated that urea is presently being used for feeding purposes at the rate of about 500,000 tons per year. This amount of urea would have a nitrogen equivalent of 3.2 million tons of 44% soybean meal, which would be approximately ¼ of the nation's annual use of SBM (12.6 million tons). SBM makes up about ⅔ of all high-protein feeds. While urea in limited amounts is not particularly harmful to nonruminants such as swine, nonruminant animals can make little, if any, effective use of urea since they do not have a rumen. Consequently, urea feeding is restricted to ruminants. Even with ruminants, it is recommended that for effective results it not be used to provide more than ⅓ of the protein nitrogen in the ration. While more than this amount can be used, the hazards of toxicity increase and the efficiency of utilization decreases at higher urea levels. Whether or not urea should be used for ruminant feeding will depend on the prevailing cost of urea and carbohydrate feeds such as corn, on the one hand, and the cost of feeds high in preformed protein such as soybean meal, on the other. The following may be helpful in this connection:

$$1 \text{ lb } 45\% \text{ N urea} + 6.5 \text{ lb grnd sh corn} \leqq 7.5 \text{ lb } 44\% \text{ SBM}$$

If urea is $200 per ton and corn is $2.80 per bu, then the cost of 1 lb urea (10¢) + 6.5 lb corn (32.5¢) would be 42.5¢. If 7.5 lb of SBM would cost more than 42.5¢ (or 5.67¢ per lb or $113.40 per ton), then it would probably be more economical to use urea in the ration.

XIV

Urea Fermentation Potential (UFP) of feeds. Burroughs and Associates at the Iowa Station have published (A.S. Leaflet R190) a set of values they refer to as the Urea Fermentation Potential (UFP) of feeds. These values estimate the amount of urea that different feeds can convert to microbial protein in the rumen. The UFP value of a feed as calculated is based on the amount of fermentable energy in the feed as reflected by its TDN content and the amount of ammonia formed by that feed through protein degradation in the rumen. It is expressed as grams of urea (44.8% N) per kilogram of feed dry matter. While UFP values as published by Burroughs and Associates represent a needed effort to provide better guidance for urea feeding, the authors are of the opinion that these values need further research before being proposed for general usage in ration formulation. This is especially true since the effect that treatment of feeds against ruminal protein degradation will have on this phenomenon needs clarification. Also the Iowa UFP values really do not change anything very much in that those feeds that are low in protein and high in fermentable carbohydrates, such as ground shelled corn, cane molasses, and citrus pulp, continue to top the list as recommended components of urea-based feeds.

XV

Ruminally degraded and ruminally undegraded ingested protein (DIP, UIP). For years, ruminant nutritionists have thought the friendly rumen bacteria and protozoa that inhabit the rumen can utilize nonprotein nitrogen and carbohydrates to synthesize a most nearly perfect balance of amino acid protein—at least as far as the host ruminant is concerned. Therefore, up to a decade or so ago, students were taught that mature ruminating animals did not have any essential amino acid requirements because such animals could manufacture their entire needs. That statement was so nearly correct that its slight inaccuracy went unchallenged for a long time. However, in the mid-1950s, the senior author of this text observed that when a small amount of dehydrated alfalfa meal was included in the diets of feedlot cattle, an unexplained increase in rate of gain occurred. Although many explanations were offered, none seemed to make much sense. (In more recent years, dehydrated alfalfa meal was identified as an excellent source of UIP.) Then, in the late 1970s, the same Purdue University researcher included a small amount of hydrolyzed poultry-feather meal to replace a protein-equivalent of solvent extracted soybean meal, as a source of supplemental protein for finishing cattle. Once again, there was an increase in rate of gain in the cattle. Basic metabolism trials demonstrated that the nitrogen of hydrolyzed poultry-feather meal was not broken down as efficiently, as completely, or as rapidly as that from soybean meal in the cow rumen. The portion of the feather meal protein that was not broken down in the rumen and went on into the small intestine was at first called "by-pass" protein. More recently protein consumed by ruminant animals has been labeled digestible intake protein, (DIP) or undigestible intake protein (UIP). Note the diagram of the rumen (Figure 11-5) to observe the protein pictured near the top of the rumen that is labeled "protein that escapes bacterial breakdown." This represents UIP. Apparently, the rumen does not necessarily synthesize the most desirable mixture of amino acids to meet the host cow's needs, but also needs some preformed amino acids which will go past the rumen unscathed. Ruminant nutritionists place such a high value on taking into account the UIP that the requirements for UIP are listed separately from the total protein requirements—especially in high-producing dairy cows.

XVI

High-protein feeds. High protein feeds are ordinarily named and classified on the basis of their origin and method of processing.

A. On the basis of origin, high-protein feeds are usually classified into two general categories.

1. **Those of animal origin.** Hogs formerly did best with one of these in the ration. This was attributed to a factor known as the *animal protein factor* or *APF*. At the time, this factor was unidentified but has since proved to be what is now known as *vitamin B_{12}*. With the general availability of vitamin B_{12}, high-protein feeds of animal origin are not regarded as essential for swine production and, so, hold little advantage over other high-protein feeds at this time.

2. **Those of plant origin.** These consist, for the most part, of oil-seed by-products. They vary in composition and feeding value depending on the seed from which they are produced, the amount of hull and/or seed coat included, and the method of fat extraction used.

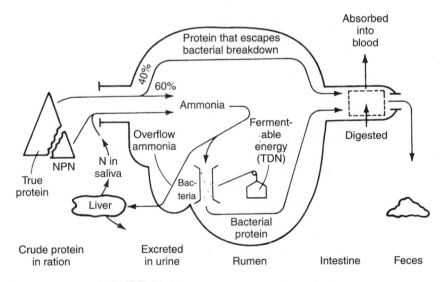

FIGURE 11-5
Schematic flow and breakdown of nitrogen through the rumen. From *Satter, L. D. and R. E. Roffler.* Journal of Dairy Science *58 (1975): 1219.*

B. Three methods of extraction have been used over the past several years for extracting fat from oil seeds.

   1. Hydraulic or old process.

      a. The original process for fat extraction.

      b. Widely used in the past throughout the Cotton Belt.

      c. Based on mechanical extraction using a hydraulic press.

      d. Not a continuous process.

      e. Leaves considerable oil in meal.

      f. Few mills use this process today.

   2. Expeller or new process.

      a. Came in with the soybean industry.

      b. Also based on mechanical extraction using a screw press.

      c. Preferred over the hydraulic process because it permits continuous operation.

      d. Also leaves considerable oil in the meal.

      e. Still extensively used in the Cotton Belt—almost completely replaced in soybean area with solvent process.

   3. Solvent process.

      a. Originally developed in Germany.

      b. Introduced into United States following World War II—late 1940s.

      c. Based on extraction of fat in a manner similar to crude fat determination except on much larger scale—also hexane rather than ether used as solvent.

      d. Brings about almost complete fat extraction—some fat is sometimes added back.

      e. Most new fat extraction plants are this type—have almost completely taken over in soybean section.

      f. Essentially all SBM is solvent process today.

In Table 11-2 the feeds from appendix feed tables have been listed and grouped on an as-fed basis according to their content of crude protein. Once again it is not intended for the student to memorize the exact percentages of crude protein for the different feeds or to use this table as a source of information on the crude protein content of an individual feed. It is hoped that the listing and grouping of the feeds on the basis of their crude protein content will help the student get in mind the range of crude protein values found in feed materials and within which type or group a particular feed would be found.

Since protein, along with energy, represents one of the more critical and costly parts of a ration, it is most important that the student of feeds and feeding become familiar with the relative protein content of the different feed materials as promptly as possible.

Table 11-2
FEEDS LISTED AND GROUPED ACCORDING TO THEIR PERCENTAGE OF CRUDE PROTEIN— AS-FED BASIS

Group A: **Mostly the Meat and Poultry By-Product Meals; Fish Meal, the Oil Seed Meals, and Oil Seeds; and the Dried Milks**

Urea, 45% N	275.8	Corn gluten meal	42.7
Hydrolyzed feather meal	84.9	41% solv-extd cottonseed meal	41.2
Dehydrated casein	84.0	41% mech-extd cottonseed meal	41.0
Blood meal	79.8	Soybean seed	39.2
Fish meal	61.1	36% cottonseed meal	38.6
Tankage	59.4	Rapeseed meal	37.0
Poultry by-product meal	58.7	Linseed meal	34.3
Meat meal	51.4	Dried skimmed milk	33.7
Meat and bonemeal	50.4	Peanut kernels	28.4
49% soybean meal	49.7	Dried corn distillers grains	27.9
Peanut meal	48.1	Dried brewers grains	27.1
Tankage with bone	46.6	Dried whole milk	25.4
Sunflower meal	46.3	Corn gluten feed	23.0
Sesame oil meal	45.0	Cottonseed	22.0
44% soybean meal	44.6	Poultry manure	21.9
Safflower oil meal	43.0	Copra meal	21.3

Group B: **Consists for the Most Part of the Legume Hays and Meals; Barley, Oats, Rye, Wheat, and the Wheat By-Products; and Most of the Grain Sorghums**

Dried alfalfa leaves	20.6	Red clover hay	14.2
Cowpea hay	17.5	Wheat grain	14.2
Dehydrated alfalfa meal	17.3	Sweetclover hay	13.7
Wheat shorts	16.5	Dried whey	13.3
Wheat middlings	16.4	Bromegrass hay	13.2
Sericea hay	16.3	Alsike clover hay	13.1
Oat groats	15.8	Lespedeza hay	12.8
Triticale grain	15.8	Rice bran	12.7
Alfalfa hay	15.3	Alfalfa stem meal	12.5
Wheat bran	15.2	Rye grain	12.1
Dehy coastal bermudagrass meal	15.1	Barley grain	11.9
Soybean hay	14.7	Light oats grain	11.9

continued

Table 11-2 (Continued)

Oats grain	11.8	Hominy feed	10.4
Kentucky bluegrass hay	11.6	Dried bakery product	9.8
Grain sorghum grain	11.1	Peanut hay	9.8
Redtop hay	11.0	Corn grain	9.6
Soybean hulls	11.0		

Group C: **Made Up Primarily of the Nonlegume Hays; Ear and Shelled Corn; and Certain Grain Sorghums**

Reed canarygrass hay	9.4	Potato meal	8.1
Dried beet pulp with molasses	9.3	Corn and cob meal	7.8
Coastal bermudagrass hay, fert.	9.0	Orchardgrass hay	7.6
		Millet hay	7.5
Bermudagrass hay	8.9	Bahiagrass hay	7.4
Dried beet pulp	8.8	Sudangrass hay	.7.3
Grain sorghum grain 8–10% prot.	8.8	Timothy hay	7.2
Johnsongrass hay	8.5	Peanut hulls	7.1
Oat hay	8.5	Steamed bonemeal	7.1
Ky 31 fescue hay	8.3	Beet molasses	6.6
Meadow hay	8.3	Pangolagrass hay	6.5

Group D: **For the Most Part High-Moisture Feeds, Mineral Feeds, Certain Low-Quality Roughages, the Pure Fats, and the Molasses**

Fresh bromegrass	6.3	Fresh cow's milk	3.3
Dried citrus pulp with molasses	6.2	Wheat straw	3.2
Dried citrus pulp	6.1	Fresh crimson clover	3.1
Coastal bermudagrass hay	5.4	Fresh dallisgrass	3.0
Citrus molasses	5.5	Fresh skimmed milk	3.0
Fresh ladino clover	5.3	Rice hulls	3.0
Prairie hay	5.3	Ground corn cobs	2.8
Fresh Kentucky bluegrass	5.2	Oat silage	2.7
Sweet potato meal	4.9	Fresh bahiagrass	2.6
Wet brewers grains	4.9	Corn silage	2.5
Soybean straw	4.6	Fresh potatoes	2.2
Blackstrap molasses	4.4	Grain sorghum silage	2.2
Fresh coastal bermudagrass	4.4	Fresh pangolagrass	2.1
Fresh alfalfa	4.3	Fresh cabbage	2.0
Fresh orchardgrass	4.3	Fresh sudangrass	2.0
Fresh Ky 31 fescue	4.2	Wheat silage	2.0
Oat straw	4.1	Corn stover silage	1.9
Barley straw	4.0	Fresh sweet potatoes	1.7
Corn ear silage	3.9	Soybean oil	1.4
Fresh rye forage	3.8	Fresh carrots	1.2
Cottonseed hulls	3.7	Fresh citrus pulp	1.2
Oat hulls	3.6	Wet beet pulp	1.2
Fresh ryegrass	3.5	Fresh turnip roots	1.1

continued

Table 11-2 (Continued)

Fresh whey	0.9	Dicalcium phosphate	***
Wood molasses	0.6	Ground limestone	***
Corn oil	***	Hydrolyzed animal fat	***
Defluorinated phosphate	***	Oyster shell flour	***

*****very little, if any*

12 Mineral Nutrition— General

Of the 20 or more chemical elements regarded as essential for the proper nutrition of farm animals, all but carbon, hydrogen, oxygen, and nitrogen are referred to as the *inorganic* or *mineral elements*. These are used for a multitude of different functions in the animal body. They are required in greatly different amounts, but the amount required has nothing to do with the essentiality of the function performed. The amounts needed vary from as much as approximately 1.0% calcium in the ration for certain very young, rapidly growing animals to as little as 0.10 ppm (0.000010%) of selenium in rations for most classes of beef cattle. An animal derives its mineral needs from its feed, mineral supplements, drinking water, and/or soil, which it might consume either accidentally or on purpose. From a nutrition standpoint the source is not important as long as the animal receives an adequate supply of each of the required mineral elements. While an animal has a more or less constant metabolic need for each of the required minerals it is able, through the use of reserve supplies within the body on the one hand and the automatic activation of certain mineral conserving processes by the body on the other, to tolerate limited periods of mineral shortages without experiencing irreparable damage. Prolonged shortages of any one of the required minerals, however, can produce dire results. Such incidents should be avoided whenever possible with any progressive program of livestock feeding.

I. **Needs of animals for minerals.**

A. **In tissue growth and repair.**

1. Bones and teeth are high in mineral content.

	Composition of Fresh Bone		
Water	45%	Ca	36%
Ash	25%	P	17%
Protein	20%	Mg	0.8%
Fat	10%		

2. Some in hair, hoofs, and horns.

3. Some in soft tissues.

4. Some in blood cells.

B. **As body regulators or for producing body regulators.**

1. In regulating body processes, minerals function in different forms as listed below and as discussed subsequently.

a. In the ionic form.

b. In the molecular form.

c. As components of vitamins.

d. In the formation of enzymes, hormones, etc.

2. Minerals and mineral-containing enzymes, hormones, and vitamins function in the body to regulate the following:

a. Various metabolic cycles.

b. Molecular concentration—make body fluids physiologically compatible with the tissues.

c. Acid-base balance—help maintain pH of body fluids at about 7.0.

d. Nerve irritability.

e. Muscle stimulation and activity.

C. **In milk production.**

1. Cow's milk contains 5.8% ash or mineral matter on a dry basis.

2. Milk normally contains significant amounts of all of the essential minerals—except iron.

3. The minerals in milk must come either directly or indirectly from the feed and/or water that the animal consumes.

II

Mineral composition of animal body. (Average of analyses of 18 steers of varying ages less content of the digestive tract*)

Calcium	1.33%		
Phosphorus	0.74%		
Potassium	0.19%		
Sodium	0.16%	2.73%	49% Ca
Sulfur	0.15%		27% P
Chlorine	0.11%		24% Other
Magnesium	0.04%		
Iron	0.01%		

Cobalt
Copper
Fluorine
Iodine   Present
Manganese   in
Molybdenum   trace
Selenium   amounts
Zinc   only
Others

Hogan, Albert G. and John L. Nierman. Studies in Animal Nutrition—VI The Distribution of Mineral Elements in the Animal Body as Influenced by Age and Condition. *Missouri Agricultural Experiment Station Research Bulletin 107. 1927.*

13 The Macro Minerals

Of the 16 or more different mineral elements required for the proper nutrition of farm animals, 7, as noted previously, are referred to as the *macrominerals*. These elements, while generally required in small amounts compared with the amounts of carbohydrates, fats, and protein found in balanced rations, are ordinarily present at much higher levels than are the other so-called *microminerals*. The amounts of macrominerals required or present usually are such that they can be worked with conveniently on a percentage basis.

In order that the student of feeds and feeding might properly understand the role of macrominerals in livestock feeding, each of these minerals is considered in the following pages with regard to its functions in animal nutrition, the required amount, the likelihood and nature of a deficiency, and how any necessary supplementation might be best accomplished. While this information is incomplete in many instances, it will guide the livestock producer in coping with most livestock feeding problems related to the macrominerals.

Sodium and chlorine. These are usually considered together because of their close biochemical relationships—and because they are usually provided as common salt (NaCl).

A. **Functions.**

    1. Formation of digestive juices (HCl).

    2. Control of body fluid concentration.

    3. Control of body fluid pH.

    4. Nerve and muscle activity.

B. **Requirements.**

    1. Specific requirements for sodium and chlorine have not been worked out for most livestock classes.

    2. For young pigs weighing 13–35 kg, the requirement would appear to be

        Sodium = 0.08% to 0.10% of the ration.

        Chlorine = 0.12% to 0.13% of the ration.

    3. For livestock in general, the requirement for salt is considered to be

        0.25% to 0.50% of the ration

        or

        0.005% to 0.010% of the body weight daily.

    4. Requirements of different livestock classes will vary depending on

        a. Class of animal.

        b. Type of feed fed.

        c. Activity of animal.

        d. Production of animal.

C. **Deficiency.**

    1. Rations consisting of farm-produced feeds are usually deficient in these two minerals, and animals not receiving supplemental salt will tend to develop sodium and/or chlorine deficiency.

        a. Such a deficiency is slow to develop. Salt is reused or recycled (not excreted) by animals on low salt intake. Development of a deficiency may take several weeks.

        b. Even with the development of a sodium and/or chlorine deficiency, under ordinary feedlot conditions, there are no specific deficiency symptoms— just unthrifty appearance and impaired performance.

    2. With heavily perspiring animals, such as hard-working horses, on low salt intake, an acute salt deficiency may develop resulting in disrupted nerve and muscle function and possible nervous prostration.

D. Supplementation. Farm livestock should be provided with supplemental salt under almost all circumstances. This may be accomplished in any one of several ways.

1. **As block salt.**

   a. **Types of block salt.**

      - *Plain*—contains only NaCl.

      - *Yellow*—contains NaCl plus sulfur.

      - *Red, brown, or purple*—usually contains mostly NaCl plus the critical trace minerals—except selenium—including Co, Cu, Fe, I, Mn, and Zn. Selenium, if added to the ration, must be provided through a specially prepared supplement. Livestock do not normally need supplemental F and Mo.

   b. **Advantages of block salt.**

      - Easy to provide—no protection required.

      - Stimulates salivation.

      - No danger of overconsumption.

   c. **Disadvantages of block salt.**

      - Animal sometimes has difficulty obtaining sufficient salt.

2. **As loose salt.**

   a. **Types of loose salt.**

      - Plain.

      - Trace-mineralized.

   b. **Advantages of loose salt.**

      - Easy for animal to consume.

   c. **Disadvantages of loose salt.**

      - Must have protected mineral box.

      - Must have adequate water available with salt-hungry animals to avoid possible death.

3. **As part of a mineral mix.** From 20% to 50% of either plain or trace-mineralized salt may be included in an overall mineral mix.

   a. **Advantages of a mineral mix.**

      - Easy for animal to consume adequate amount of salt.

      - Induces animal to consume other less palatable minerals.

    b. **Disadvantages of a mineral mix.**

- Must have protected mineral box.

- Must have adequate water available with salt-hungry animals to avoid possible death.

- Forces animals to consume minerals they may not need.

    4. As a component of the overall ration mix.

      a. Usually added at 0.25%–0.5% of ration.

      b. Ensures adequate salt consumption.

      c. It may improve ration palatability.

      d. Free choice salt may or may not be provided in addition.

      e. High levels of salt in the ration may kill an animal if plenty of water is not readily available, but plain salt (NaCl) is not toxic to livestock at any level if they have access to adequate water.

II. Calcium.

A. Functions.

  1. Bone and teeth formation—99% of body calcium in the bones and teeth.

  2. Nerve and muscle function.

  3. Acid-base balance.

  4. Milk production—also egg production.

B. Requirements.

	% of Air-Dry Ration						% of Air-Dry ration (Poultry)	
	Beef Cattle	Dairy Cattle	Horses	Sheep	Swine		Chickens	Turkeys
Growing and fattening	0.54–0.20%	0.31%	0.74–0.26%	0.34–0.24%	0.65–0.50%	0–4 wks	1.0	1.2
Dry pregnant females	0.17	0.31	0.34	0.23	0.75	3–7 wks	0.9	1.0
Lactating females	0.26	0.39–0.48	0.43	0.45	0.75	6–8 wks	0.8	1.0
						10–20 wks	—	0.7
						20–24 wks		0.55
						Layers	3.5–4.0	2.25

C. Deficiency.

1. Calcium may be deficient if a good feed source is not provided. Some good feed sources are

Legume roughages	Meat and bonemeal
Grass roughages from calcium-rich soils	Fish meal
Tankage	Milk
Tankage with bone	Skimmed milk
Poultry by-product meal	Citrus pulp
Meat scrap	Citrus molasses

In Table 13–1 the feeds from the appendix table, Table Feed–3, have been listed and grouped on an as-fed basis according to their content of calcium. As with the other nutritive fractions previously looked at, we do not intend that the student memorize the exact percentages of calcium in the different feeds or use the table as a source of information on the calcium content of an individual feed material. We simply hope that the listing and grouping of the feeds on the basis of their calcium content will help the student get in mind the range of calcium levels in feed materials and within which type or group a particular feed would be found. Since calcium is one of the nutrients for which an animal's ration is balanced, such knowledge of a feed's composition will often facilitate the ration balancing process.

Table 13–1
SELECTED FEEDS LISTED AND GROUPED ACCORDING TO THEIR PERCENTAGE OF CALCIUM—AS-FED BASIS

Group A: The Ca and the Ca + P Supplements

Oyster shell flour	37.62	Defluorinated phosphate	32.00
		Steamed bonemeal	30.92
Ground limestone	34.00	Dicalcium phosphate	26.30

Group B: The Bone-Containing High Protein Feeds and Poultry Manure

Tankage with bone	11.16	Fish meal	5.18
Meat and bonemeal	10.30	Poultry by-product	
Meat meal	8.85	meal	3.51
Tankage	5.86	Poultry manure	2.82

Group C: Primarily Dry Legume Roughages, Dried Milk Products, Dried Citrus Pulp, and Molasses (Except Beet)

Dried alfalfa leaves	2.27	Dried skimmed milk	1.28
Sesame oil meal	2.01	Alfalfa hay	1.27
Dried citrus pulp	1.67	Cowpea hay	1.26
Dried citrus pulp with molasses	1.66	Citrus molasses	1.16

continued

Table 13–1 (Continued)

Dehydrated alfalfa meal	1.40	Soybean hay	1.15
Soybean straw	1.40	Alsike clover hay	1.13
Red clover hay	1.35	Peanut hay	1.12
Wood molasses	1.34	Sweet clover hay	1.11
Lespedeza hay	1.02	Dried whole milk	0.91
Alfalfa stem meal	1.00	Dried whey	0.86
Sericea hay	0.94	Blackstrap molasses	0.75

Group D: **Predominantly the Nonlegume Hays, Fresh Legume Forages, Certain Air-Dry Low-Quality Roughages, and Some of the High-Protein Feeds**

Johnsongrass hay	0.75	Reed canarygrass hay	0.35
Dried beet pulp	0.63	Orchardgrass hay	0.35
Dehydrated casein	0.61	Corn gluten feed	0.33
Rapeseed meal	0.61	Coastal bermudagrass hay	0.31
Redtop hay	0.60	44% soybean meal	0.30
Meadow hay	0.58	Dried brewers grains	0.30
Dried beet pulp with molasses	0.56	Dehy coastal bermudagrass meal	0.29
Sundangrass hay	0.50	Kentucky bluegrass hay	0.29
Fresh alfalfa	0.48	Millet hay	0.29
Bahiagrass hay	0.46	Blood meal	0.29
Soybean hulls	0.45	Peanut meal	0.27
Bermudagrass hay	0.43	Barley straw	0.27
Ky 31 fescue hay	0.43	Hydrolyzed feather meal	0.26
Pangolagrass hay	0.42	49% soybean meal	0.26
Sunflower meal	0.41	Soybean seed	0.25
Linseed meal	0.41	Bromegrass hay	0.25
Fresh citrus pulp	0.40	Peanut hulls	0.25
Coastal bermudagrass hay, fert.	0.39	Fresh crimson clover	0.24
Prairie hay	0.39	Fresh ladino clover	0.24
Timothy hay	0.38	Oat hay	0.22
Safflower oil meal	0.35	Oat straw	0.22

Group E: **A Combination of Fresh Nonlegume Forages, the Silages, Wet By-Products, the Root Crops, the Fresh Milks, the Feed Grains and Grain By-Products, Certain Air-Dry Low-Quality Roughages, the Pure Fats, Urea, and Certain Other Miscellaneous Products**

41% mech-extd cottonseed meal	0.19	Oat hulls	0.14
36% cottonseed meal	0.18	Dried bakery product	0.13
41% solv-extd cottonseed meal	0.17	Beet molasses	0.13
Copra meal	0.17	Fresh orchardgrass	0.13
Fresh bromegrass	0.16	Cottonseed hulls	0.13
Wheat straw	0.16	Fresh skimmed milk	0.13
Corn gluten meal	0.15	Fresh Kentucky bluegrass	0.12
Sweet potato meal	0.15	Fresh cow's milk	0.12

continued

Table 13–1 (Continued)

Fresh Ky 31 fescue	0.15	Corn stover silage	0.12
Fresh dallisgrass	0.14	Wheat middlings	0.11
Fresh coastal bermudagrass	0.14	Wheat bran	0.11
Fresh bahiagrass	0.14	Ground corn cobs	0.11
Cotton seed	0.14	Fresh sudangrass	0.10
Fresh rye forage	0.09	Corn and cob meal	0.06
Rice hulls	0.09	Rye grain	0.06
Corn silage	0.09	Hominy feed	0.05
Oat groats	0.08	Fresh carrots	0.05
Oats grain	0.07	Fresh turnip roots	0.05
Light oats grain	0.07	Triticale grain	0.05
Wheat silage	0.07	Corn ear silage	0.04
Fresh whey	0.07	Barley grain	0.04
Rice bran	0.07	Wheat grain	0.04
Potato meal	0.07	Grain sorghum grain 8–10% prot.	0.04
Wet brewers grains	0.07	Grain sorghum grain	0.03
Fresh cabbage	0.06	Corn grain	0.03
Peanut kernels	0.06	Fresh sweet potatoes	0.03
Wet beet pulp	0.10	Fresh potatoes	0.01
Grain sorghum silage	0.10	Hydrolyzed animal fat	0.00
Oat silage	0.10	Corn oil	0.00
Dried corn distillers grains	0.10	Soybean oil	0.00
Fresh pangolagrass	0.09	Urea, 45% N	0.00
Wheat shorts	0.09		

2. Deficiency symptoms (Figure 13–1) are as follows:

    a. Rickets in young animals. Joints become enlarged. Bones become soft and deformed. Condition may be corrected in early stages with calcium feeding.

    b. Osteomalacia or osteoporosis in older animals. Bones become porous and weak. Condition may be corrected with calcium feeding if bone does not break. Rear-quarter paralysis in swine sometimes due to calcium deficiency resulting in a crushed vertebra and pinched spinal cord.

D. Supplementation.

1. The need for supplementation will depend on the ration. If needed, it may be accomplished by using the following:

    a. Calcium-only supplements.

- Ground limestone.

- Oystershell flour.

- Marble dust.

All three are primarily $CaCO_3$, containing 33–40% calcium. Pure $CaCO_3$ is 40% calcium. All are about equal in nutritive value. Use whichever is cheapest.

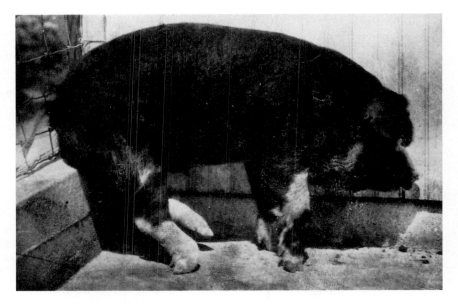

FIGURE 13–1
Calcium deficiency. Note the abnormal bone development
and the rachitic condition in advanced stage of deficiency.
Lack of calcium retards normal skeletal development but
does not usually depress total gain. (Courtesy of N. R. Ellis,
U.S. Department of Agriculture)

The cheapest will probably be the one produced the nearest since transportation is a major factor in determining selling price.

b.   Calcium with phosphorus.

- Steamed bonemeal.

- Defluorinated phosphates.

These usually contain about 30% calcium along with 14% to 20% phosphorus. They are usually considered to be about equal in feeding value per unit of calcium and phosphorus. They are usually more expensive than the calcium-only supplements—hence, they are used as calcium sources only when phosphorus is also needed.

2.   Any of the above calcium sources may either be included in a mineral mix provided free choice or added to the ration at the level required.

3.   While high levels of calcium in the ration are not acutely toxic, they can cause disturbances in the availability and metabolism of certain other elements, especially phosphorus, manganese, magnesium, and zinc. Levels of more than about 1.0% available Ca in the ration DM for extended periods should be avoided.

III   Phosphorus.

A.   Functions.

1. Bone and teeth formation—about 80% of body phosphorus is in the bones and teeth.

2. A component of protein in the soft tissues.

3. Milk production—also egg production.

4. Various metabolic processes.

B.   Requirements.

	% of Air-Dry Ration						% of Air Dry Ration (Poultry), as Non-Phytate Phosphorus	
	Beef Cattle	Dairy Cattle	Horses	Sheep	Swine		Chickens	Turkeys
Growing and fattening	0.39–0.20%	0.23%	0.46–0.22%	0.21–0.15%	0.50–0.40%	0–3 weeks	0.45	—
Dry pregnant females	0.17	0.23	0.26	0.20	0.50	0–4 weeks	—	0.6
Lactating females	0.26	0.30–0.35	0.35	0.33	0.50	3–6 weeks	0.35	—
						4–8 weeks	—	0.5
						6–8 weeks	0.30	—
						8–24 weeks	—	0.35
					Layers	0.30	0.35	

In Table 13–2 the feeds from the appendix tables have been listed and grouped on an as-fed basis according to their content of phosphorus. Once again we do not intend for the student to memorize the exact percentages of phosphorus in the different feeds or to use the table as a source of information on the phosphorus content of an individual feed. We simply hope that the listing and grouping of the feeds on the basis of their phosphorus content will help the student get in mind the range of phosphorus levels in feed materials and within which type or group a particular feed would be found. Since phosphorus is one of those nutrients for which an animal's ration is balanced, such knowledge of a feed's composition will often facilitate the ration balancing process.

Table 13–2
SELECTED FEEDS LISTED AND GROUPED ACCORDING TO THEIR PERCENTAGE OF PHOSPHORUS—AS-FED BASIS

Group A: The Ca + P Supplements

Dicalcium phosphate	18.70	Defluorinated phosphate	18.00
Tankage with bone	5.41	Steamed bonemeal	14.01
Meat and bonemeal	5.10	Tankage	3.07
Meat meal	4.44	Fish meal	2.89
		Poultry by-product meal	1.83

Group B: Mostly Nonbone-Containing High-Protein Feeds and the Wheat By-Products, Along with a Limited Number of Miscellaneous Concentrates

Poultry manure	1.59	Dried whole milk	0.76
Rice bran	1.54	Dried whey	0.76
Sesame oil meal	1.36	Corn gluten feed	0.74
Safflower oil meal	1.29	Cotton seed	0.68
Wheat bran	1.22	Hydrolyzed feather meal	0.67
41% solv-extd cottonseed meal	1.10	49% soybean meal	0.63
41% mech-extd cottonseed meal	1.08	44% soybean meal	0.63
Dried skimmed milk	1.02	Peanut meal	0.62
36% cottonseed meal	0.96	Copra meal	0.60
Rapeseed meal	0.95	Soybean seed	0.60
Sunflower meal	0.91	Hominy feed	0.52
Wheat middlings	0.88	Dried brewers grains	0.51
Linseed meal	0.87	Corn gluten meal	0.45
Dehydrated casein	0.82	Oat groats	0.43
Wheat shorts	0.81	Peanut kernels	0.43

Group C: Predominantly the Hays and the Feed Grains, Along with Just a Few Other Miscellaneous Products

Dried corn distillers grains	0.40	Dehy coastal bermudagrass meal	0.23
Wheat grain	0.37		
Barley grain	0.34	Alsike clover hay	0.23
Oats grain	0.33	Dehydrated alfalfa meal	0.23
Light oats grain	0.33	Reed canary grass hay	0.23
Redtop hay	0.33	Red clover hay	0.22
Orchardgrass hay	0.32	Alfalfa hay	0.22
Rye grain	0.32	Sericea hay	0.22
Cow pea hay	0.31	Kentucky bluegrass hay	0.22
Ky 31 fescue hay	0.31	Sweetclover hay	0.22
Triticale grain	0.30	Pangolagrass hay	0.21
Grain sorghum grain	0.29	Potato meal	0.20
Grain sorghum grain 8–10% prot.	0.29	Oat hay	0.20

continued

Table 13–2 (Continued)

Sudangrass hay	0.28	Bromegrass hay	0.20
Corn grain	0.26	Bahiagrass hay	0.20
Soybean hay	0.25	Lespedeza hay	0.19
Johnsongrass hay	0.25	Soybean hulls	0.19
Corn and cob meal	0.24	Coastal bermudagrass hay	0.18
Dried bakery product	0.24	Timothy hay	0.18
Blood meal	0.24	Millet hay	0.17
Dried alfalfa leaves	0.24	Meadow hay	0.17
Bermudagrass hay	0.16	Prairie hay	0.14
Coastal bermudagrass hay, fert.	0.14	Sweet potato meal	0.14
Peanut hay	0.14	Oat hulls	0.14

Group D: **For the Most Part High-Moisture Feeds, Low-Quality Air-Dry Roughages, the Ca-Only Supplements, the Pure Fats, Plus a Very Limited Number of Other Materials**

Fresh bromegrass	0.13	Oyster shell flour	0.07
Fresh orchardgrass	0.13	Fresh crimson clover	0.06
Corn ear silage	0.13	Fresh bahiagrass	0.06
Wet brewers grains	0.12	Fresh whey	0.06
Fresh Kentucky bluegrass	0.12	Fresh potatoes	0.06
Fresh Ky 31 fescue	0.11	Grain sorghum silage	0.06
Dried citrus pulp	0.11	Peanut hulls	0.06
Dried citrus pulp with molasses	0.11	Oat straw	0.06
Fresh ryegrass	0.10	Soybean straw	0.05
Fresh skimmed milk	0.10	Fresh dallisgrass	0.05
Fresh cow's milk	0.09	Fresh sweet potatoes	0.05
Cottonseed hulls	0.09	Fresh carrots	0.04
Corn stover silage	0.09	Fresh pangolagrass	0.04
Dried beet pulp	0.09	Ground corn cobs	0.04
Dried beet pulp with molasses	0.09	Wheat straw	0.04
Citrus molasses	0.09	Wood molasses	0.04
Blackstrap molasses	0.08	Beet molasses	0.03
Fresh coastal bermudagrass	0.08	Fresh cabbage	0.03
Fresh sudangrass	0.08	Fresh citrus pulp	0.02
Corn silage	0.08	Fresh turnip roots	0.02
Fresh rye forage	0.08	Alfalfa stem meal	0.02
Fresh alfalfa	0.07	Ground limestone	0.02
Fresh ladino clover	0.07	Wet beet pulp	0.01
Oat silage	0.07	Soybean oil	0.00
Wheat silage	0.07	Corn oil	0.00
Barley straw	0.07	Hydrolyzed animal fat	0.00
Rice hulls	0.07	Urea, 45% N	0.00

C. Deficiency.

1. Phosphorus may be deficient if one or more of the following good feed sources of phosphorus are not in the ration:

Wheat bran and middlings   Whole or skimmed milk
All high-protein feeds   Grains in general are fair sources
All bone-containing feeds

2. Symptoms of a phosphorus deficiency (Figures 13–2 and 13–3) are as follows:

     a. Rickets in young animals similar to that of calcium deficiency. May be corrected in the early stage with phosphorus feeding if due to phosphorus shortage. (Figure 13–4.)

     b. Osteomalacia or osteoporosis in older animals similar to calcium deficiency. Seldom results in broken bones. May be corrected with phosphorus feeding if due to phosphorus shortage.

     c. Phosphorus-deficient animals usually show poor appetite, slow gain, lowered milk production, low blood phosphorus, and general unthriftiness. Animals may eat soil and chew on nonfeed objects, but this is not specific for phosphorus deficiency.

D. Supplementation.

1. The need for supplements will depend on the ration but may be accomplished by using either of the following:

     a. Steamed bonemeal.

         • Around 14% phosphorus.

         • Excellent as source of supplemental phosphorus.

         • Frequently in short supply.

     b. Defluorinated phosphates.

         • Are of several types.

         • Vary in phosphorus content from 14% to 20%.

         • Most natural deposits of phosphate contain fluorine at levels that would be toxic if used in their natural state as a primary source of phosphorus. Hence, most must be defluorinated before being fed to livestock.

         • Mineral supplements sold in most states cannot legally contain fluorine in excess of:

             0.30% for cattle   0.45% for swine
             0.35% for sheep   0.60% for poultry

         • American Association of Feed Control Officials (AAFCO) recommend not more than 1 part of fluorine per 100 parts of phosphorus.

         • Most phosphorus supplements offered for sale by reputable companies are satisfactory sources of phosphorus from the standpoint of fluorine content, phosphorus content, and phosphorus availability.

         • Such supplements are usually in adequate supply.

FIGURE 13–2

Phosphorus deficiency. On the left is a typical phosphorus-deficient pig in an advanced stage of deficiency. Leg bones are weak and crooked. The pig on the right received the same ration as the one on the left, except that the ration was adequate in available phosphorus. (Courtesy of M. P. Plumlee and W. M. Beeson, Purdue University)

(a)   (b)

FIGURE 13–3

Phosphorus deficiency in feedlot cattle. (a) This steer was fed a ration consisting of wet beet pulp, alfalfa hay, and beet molasses containing 0.12% phosphorus. (b) This steer received the same ration plus 0.1 lb of steamed bonemeal daily, which brought the phosphorus content up to 0.18% and provided an average of 17 g of phosphorus daily. (Courtesy of W. M. Beeson; taken at Idaho Agricultural Experiment Station)

2. Either of the above phosphorus sources may be either included in a mineral mix for free choice or *ad libitum* feeding or added directly to the ration at the level needed. (Figure 13–5.)

(a)   (b)

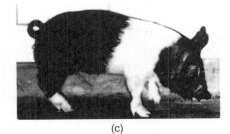

(c)

FIGURE 13–4
(a) Calcium, (b) phosphorus, and (c) vitamin D deficiencies.
Note the similarity of deficiency symptoms in these 5-week-
old pigs. (Courtesy of E. R. Miller and D. E. Ullrey, Michigan
State University)

FIGURE 13–5
An easily cleaned fiberglass two-compartment mineral box
for providing two types of minerals to cattle or sheep. The
box sits on a post and has a wind fin (not shown) on top.
This fin catches the wind and rotates the box away from the
wind and weather.

3. Excessive phosphorus in the forms found in most feed products is not acutely toxic but can cause disturbances in the availability and metabolism of certain other elements, especially calcium and magnesium. Levels of more than about 0.75% in the ration dry matter for extended periods should be avoided. However, feeds and products high in phosphorus are usually too costly to make the use of excessive levels of this element in the ration either necessary or desirable.

IV Magnesium.

A. **Functions.**

1. Necessary for many enzyme systems.

2. Plays a role in carbohydrate metabolism.

3. Necessary for the proper functioning of the nervous system.

B. **Requirements.**

1. Requirements for magnesium under normal conditions seem to be as follows, DM basis:

   a. For lactating cows: .18–.20%.

   b. For nonlactating cattle: .07–.16%.

   c. For range ewes: .06%.

   d. For hogs: <.10%.

   e. For horses: .09%.

   f. For poultry: 500 p.p.m. of diet.

2. Blood serum normally contains about 2.5 mg of magnesium per 100 ml.

3. Most rations under most conditions contain levels of magnesium that are more than adequate to meet the requirements.

C. **Deficiency.**

1. Cattle and sheep grazing on certain growing plants during the late winter and early spring tend to develop a *hypomagnesemia* associated with severe toxic symptoms and frequently death. This is the result of either a low magnesium level in the plant or the presence of some substance in the plant that renders the magnesium unavailable.

2. The condition is commonly referred to as *grass tetany*—also as *grass staggers* and *wheat poisoning.*

3. In such instances, blood magnesium levels usually fall below 1 mg/100 ml causing the development of *hypomagnesemic tetany,* a hyperirritability of the neuromuscular system producing hyperexcitability, incoordination, and frequently death.

4. The condition is most common with older, lactating cattle and sheep grazing on small grains, highly fertilized fescue, and certain other green growing crops in the late winter and early spring.

5. In pigs, a magnesium deficiency is manifested in a peculiar stepping syndrome (Figure 13–6.)

D. **Supplementation.**

1. Most cattle and sheep feeds contain adequate levels of magnesium to meet the animal's requirement under most conditions.

2. Cattle and sheep grazed on small grains and certain other crops during the late winter or early spring probably should be supplied with supplemental sources of magnesium.

3. Under such conditions it is recommended that cattle be provided with 1 + oz of supplemental magnesium per head daily. Presumably proportionately lesser amounts would suffice for sheep.

4. While both $MgSO_4$ and MgO have been used for this purpose, the latter is less purgative and probably is to be preferred for feeding purposes.

5. Consumption of supplemental magnesium may be induced by mixing it with an animal's salt or with a small amount of supplemental feed.

6. A mixture containing 2 parts MgO to 1 part salt used as the only source of salt seems to be a practical method of magnesium administration.

7. A mixture of 30% magnesium oxide, 32% defluorinated phosphate, 30% trace-mineralized salt, and 8% cottonseed meal fed free choice has also been used with good results for minimizing the incidence of grass tetany.

8. While certain magnesium compounds are known to have a laxative effect when provided to animals in relatively large amounts, excessive levels of this

FIGURE 13–6
Magnesium deficiency Five-week-old pig showing stepping syndrome; the pig keeps stepping almost continuously while standing. Weakness of pasterns is apparent. (Courtesy of E. R. Miller and D. E. Ullrey, Michigan State University)

element are not known to be otherwise acutely toxic to farm livestock. However, levels above about 0.6% in the ration dry matter for extended periods probably should be avoided.

V   Potassium.

A. **Functions.**

   1. Potassium is required by livestock for a variety of body functions, such as osmotic relations, acid-base balance, rumen digestion, and the primary intracellular cation in neuromuscular activity.

B. **Requirements.**

   1. The requirements of the different classes of livestock for potassium have been estimated as follows, DM basis:

       a. All classes of dairy cattle: 0.80%.

       b. Beef cattle in general: 0.60–0.80%.

       c. Sheep in general: 0.50%.

       d. Swine in general: 0.20–0.30%.

       e. Mature horses: 0.40%.

       f. Poultry: 0.20–0.30%.

C. **Deficiency.**

   1. A potassium deficiency is most unlikely under ordinary conditions.

   2. Symptoms of a potassium deficiency are rather nonspecific, such as decreased feed consumption, lowered feed efficiency, slow growth, stiffness, and emaciation.

D. **Supplementation.**

   1. Potassium supplementation should not be needed with most practical type livestock rations.

   2. However, grains are generally much lower in potassium than are roughages, and while high-grain rations will usually meet the potassium requirement of hogs and poultry, such rations may be marginal or even deficient in potassium for cattle and sheep. However, no general recommendations have as yet been developed for providing cattle and/or sheep on high- or all-grain rations with supplemental sources of potassium.

   3. Potassium at levels found in ordinary farm feeds and products is not acutely toxic to farm livestock, but some believe that under certain circumstances excessive amounts disturb the normal absorption, metabolism, and/or excretion of certain other elements such as magnesium, phosphorus, sodium, and

chlorine. However, maximum tolerance levels have not yet been established for the different livestock classes.

VI Sulfur.

A. Functions.

1. As a component of the amino acids cystine and methionine and the vitamins biotin and thiamine.

2. In the synthesis of sulfur-containing amino acids in the rumen.

3. In the formation of various body compounds.

B. Requirements.

1. The sulfur requirement of cattle and sheep appears to be around 0.1–0.2% of ration dry matter.

2. In nonruminants, sulfur, for the most part, should be in the form of sulfur-containing proteins.

3. In ruminants and probably horses, it may be supplied as protein, as elemental sulfur, or as sulfate sulfur.

C. Deficiency.

1. Most livestock rations provide more than the required level of sulfur—hence, a deficiency is seldom experienced under ordinary conditions.

2. A deficiency of sulfur will express itself as a protein deficiency—a general unthrifty condition and poor performance.

D. Supplementation.

1. Livestock seldom benefit from sulfur supplementation.

2. Heavy wool-producing sheep being fed on mature grass hay or other low-sulfur feeds may give a positive response to sulfur supplementation.

3. Animals fed urea as a source of nitrogen for rumen protein synthesis may benefit from supplemental sulfur.

4. Supplemental sulfur may be provided as either elemental sulfur or in the sulfate form.

5. Sulfur in either the organic or elemental form is relatively nontoxic to farm livestock, but sulfate sulfur in excess of about 0.05% of the ration dry matter can be toxic and should be avoided.

14 The Micro Minerals

Those minerals necessary for the proper nutrition of farm animals not covered under macrominerals are considered in this section. They are discussed with the same considerations as were the macrominerals. The only difference is that the amounts of the microminerals involved are generally quite small compared to the amounts of macrominerals and, for the most part, are expressed as parts per million (ppm), milligrams per kilogram (mg/kg), or milligrams per pound (mg/lb).

Recent advances in microanalysis methods have greatly facilitated micromineral research in the past several years. However, information on the role of many of the microminerals in animal nutrition is still incomplete and further research is needed.

I   Iron.

A.   Functions.

1.   Necessary for hemoglobin formation.

2.   Essential for the formation of certain enzymes related to oxygen transport and utilization.

3. Enters into the formation of certain compounds that serve as iron stores in the body—especially ferritin, found primarily in the liver and spleen, and hemosiderin, found mainly in the blood.

B. Requirements.

1. The precise minimum requirements for various classes of livestock have not been determined.

2. As little as 80 mg of iron per kg of diet has proved to be more than adequate for most animals, including poultry.

C. Deficiency.

1. Most livestock feeds contain more than 80 mg of iron per kg—hence, most livestock rations are more than adequate in iron content, and an iron deficiency seldom occurs with older animals.

2. Milk is low in iron content—especially sow's milk (20–25 mg iron per kg milk dry matter).

3. Newborn animals normally have sufficient iron reserves in the liver and spleen to carry them through the early nursing period.

4. However, newborn pigs kept on concrete during early life frequently do not possess sufficient iron reserves and will develop an iron deficiency.

5. An iron deficiency in the young pig is characterized by the following:

Low blood hemoglobin	Pale eyelids, ears, and nose
Labored breathing ("the thumps")	Flabby, wrinkled skin
Listlessness	Edema of head and shoulders

6. Iron deficiency in chicks and turkeys causes anemia in which red blood cells are reduced in size and are lower in hemoglobin.

D. Supplementation.

1. Young pigs raised on concrete are usually supplied with supplemental iron in the form of the following:

    a. Concentrated ferrous sulfate or another iron solution administered orally a few drops daily during the first 3–4 weeks or as a weekly drench of ⅓ to 1 teaspoonful.

    b. Injections of iron dextran or comparable compounds: 100 mg of Fe at less than 3 days of age with weaning at 3 weeks of age, or 150 mg of iron at less than 3 days of age with weaning at 5–6 weeks of age.

2. Many livestock producers provide their animals with supplemental iron as a safety measure against a possible deficiency through the use of trace-mineralized salt, which ordinarily contains iron.

3. A level of iron as low as 1000 mg/kg dry feed has produced symptoms of chronic toxicity in farm animals.

II Iodine.

A. Functions.

1. In the production of thyroxine by the thyroid gland.

2. Other apparent iodine functions are probably an outgrowth of thyroxine activity.

B. Requirements.

1. A level of 0.25 mg per kg of air-dry diet is considered to be more than adequate for most classes of livestock.

2. Dairy cows should be provided with a level of 0.5 mg iodine/kg dry matter.

C. Deficiency.

1. The natural feeds and water of most areas of the United States contain sufficient iodine to meet livestock needs.

2. Exceptions to this are the Great Lakes region, the far Northwest, and possibly isolated areas in other sections.

3. In iodine deficient areas, iodine deficiency symptoms are most frequently observed in young animals in the form of the following:

   Goiter at birth or soon thereafter   Hairlessness at birth
   Death or weakness at birth   Infected navels—especially in foals

4. In poultry, reduced thyroid hormone (thyroxine); smaller eggs.

D. Supplementation.

1. Where there is the slightest reason to suspect an iodine deficiency, supplying livestock with supplemental iodine is recommended.

2. Supplementation of livestock with iodine is easily accomplished by feeding iodized salt—salt that contains approximately 0.0076% iodine as potassium iodide, sodium iodide, calcium iodate, pentacalcium orthroperiodate, or some other suitable iodine-containing compound.

3. As little as 4.8 mg iodine/kg dry matter has proven to be toxic for at least certain classes of livestock.

III Cobalt.

A. Functions.

1. As a component of the vitamin B_{12} molecule.

2. In the rumen synthesis of vitamin B_{12}.

B. Requirements.

1. For cattle and sheep, feed that contains from 0.05 to 0.10 mg of cobalt per kg feed prevents any symptoms of a cobalt deficiency.

2. For horses, the requirement appears to be less than that for ruminants.

3. For hogs and poultry, only as vitamin B_{12}.

C. Deficiency.

1. Cobalt deficiencies have been noted in livestock grazing the natural forages of Florida, Massachusetts, New Hampshire, Pennsylvania, New York, Michigan, and Alberta, Canada.

2. The cobalt content in the leaves of catalpa trees is regarded as a good indicator of the adequacy of cobalt in an area.

3. Symptoms of a cobalt deficiency are simply those of general malnutrition—poor appetite, unthriftiness, weakness, anemia, decreased fertility, slow growth, and decreased milk and wool production.

D. Supplementation.

1. Increasing numbers of livestock producers, even outside the cobalt deficient areas, are providing their animals with supplemental cobalt as insurance against a possible cobalt deficiency.

2. Supplemental cobalt is easily provided to livestock by the use of trace-mineralized salt, all brands being fortified with this mineral, usually in the form of cobalt chloride, cobalt sulfate, or cobalt carbonate.

3. A level of 12.5 g of cobalt in any of the above forms per 100 kg of salt would be about right for preparing trace-mineralized salt.

4. While cobalt in the mineral form is not known to be beneficial to hogs, it is still included in mineral mixes for this class of animals.

5. While toxic in large amounts, as much as 30 mg cobalt/kg dry matter have been fed without apparent toxicity.

IV Copper.

A. Functions.

1. In iron absorption.

2. In hemoglobin formation.

3. In synthesis of keratin for hair and wool growth.

4. In various enzyme systems.

B. Requirements.

1. Where diets are not high in molybdenum and/or sulfate, the following levels of copper per kg of diet DM have been found adequate for the following:

  a. Dairy cattle: 10 mg/kg.

  b. Beef cattle and sheep: 4–5 mg/kg.

  c. Swine: 6 mg/kg.

  d. Horses: 5–8 mg/kg.

  e. Poultry: 4–5 mg/kg.

2. High levels of molybdenum and/or sulfate may increase the copper requirement two- to threefold.

C. Deficiency.

1. The copper content of feedstuffs varies considerably but is usually three to four times the requirement—hence, copper deficiencies are seldom experienced in the United States outside of Florida and certain other sections of the Southeast. Copper deficiencies are common in Australia.

2. Copper deficiency symptoms are not specific and may include any of the following:

Low blood and liver copper | Abnormal bone metabolism
Bleaching of hair in cattle | Muscular incoordination
Abnormal wool growth in | Weakness at birth
  sheep | Anemia

D. Supplementation.

1. Supplementation of livestock with copper is essential in copper deficient areas.

2. Supplementation of livestock with copper in other areas as insurance against a possible marginal supply is becoming a common practice.

3. Supplementation is easily accomplished through the use of trace-mineralized salt containing from 0.25% to 0.50% copper sulfate ($CuSO_4 \cdot 5\,H_2O$).

4. Up to about 250 mg copper/kg dry feed are sometimes included in swine rations to improve performance. More than this may be toxic.

5. More than about 100 mg/kg DM may be toxic for cattle while about half this level is toxic to sheep.

V Fluorine.

A. Functions.

1. Reduces incidence of dental caries in humans and possibly other animals.

2. Possibly retards osteoporosis in mature animals.

B. Requirements.

1. About 1 PPM or less in drinking water.

C. Deficiency.

    1. Excesses of fluorine are of greater concern than are deficiencies in livestock production because of its presence at high levels in the drinking water and forages of certain areas; also because of its presence at high (3–4%) levels in most natural phosphate sources.

    2. The only reported symptoms of fluorine deficiency have been noted in children in the form of excessive dental caries.

D. Supplementation.

    1. No need has ever been demonstrated for supplementing livestock with fluorine since most, if not all, livestock rations seem to contain more than an adequate amount of this mineral.

    2. Should fluorine supplementation ever become necessary, it would appear that the addition of 1 PPM to the drinking water should suffice.

    3. No feeding or supplementation practice should be followed that would provide more than 30 mg (20 mg for young stock) of fluorine per kg of diet for breeding cattle (and probably breeding sheep) or more than 100 mg per kg diet for fattening steers and lambs. Swine are somewhat more tolerant of this mineral than are cattle and sheep.

    4. Where the drinking water contains a significant amount of fluorine, the level in the ration should be reduced accordingly, if possible.

VI

Manganese.

A. Functions.

    1. Probably in enzyme systems influencing estrus, ovulation, fetal development, udder development, milk production, growth, and skeletal development.

B. Requirements.

    1. Beef cattle and sheep: 5–20 mg per kg of diet DM with the higher level for breeding animals.

    2. Dairy cattle: 40 mg per kg of diet DM.

    3. Swine: 10–20 mg per kg of diet DM.

    4. Poultry: 40–50 mg per kg of diet DM.

High levels of calcium and phosphorus in the ration may increase the above requirement levels.

C. Deficiency.

    1. Most livestock rations will contain adequate manganese. Most roughages contain 40–140 mg/kg; grains other than corn, 15–45 mg/kg; corn, about 5 mg/kg.

2. Symptoms that have been shown to be the result of a manganese deficiency have been noted in cattle in certain areas.

3. Manganese deficiency symptoms have been produced in swine on a low manganese diet.

4. Manganese deficiency symptoms (Figure 14–1) take the following form:

Delayed estrus	Deformed young
Reduced ovulation	Poor growth
Reduced fertility	Lowered serum alkaline phosphatase
Abortions	Lowered tissue manganese
Resorptions	"Knuckling over" in calves
	Perosis (slipped tendon in poultry)

D. Supplementation.

1. Seldom necessary for beef cattle except on all-concentrate diets based on corn and nonprotein nitrogen supplements—also in certain isolated areas where manganese is unusually low.

2. Conventional swine and poultry rations based on corn and soybean meal may be improved with manganese supplementation.

3. Manganese supplementation of livestock is easily accomplished through the use of trace-mineralized salt containing 0.25% manganese (higher levels used for poultry are all right for livestock).

4. Manganese levels over 1000 mg/kg DM should be suspected in any otherwise unexplained trouble with any class of livestock.

FIGURE 14–1
Manganese deficiency in a newborn calf; legs are weak and deformed. (Courtesy of I. A. Dyer, Washington State University)

VII

Molybdenum.

A. Functions.

1. As a component of the enzyme xanthine oxidase—especially important to poultry for uric acid formation.

2. Stimulates action of rumen organisms.

B. Requirements.

1. Exact needs of livestock for molybdenum are unknown.

2. Probably 1 mg or less per kg of dry diet is adequate for most, if not all, classes of farm livestock.

3. Toxicity symptoms have been noted at levels of 5–10 mg per kg of dry diet. Prevented with copper supplementation.

4. Poultry: 0.15 mg/kg of diet.

C. Deficiency.

1. The essentiality of molybdenum has been demonstrated in poultry.

2. Only evidence of a positive nutritional role of molybdenum with farm livestock has been with lambs, in which it improved growth rate.

D. Supplementation.

1. Since even extremely low levels of molybdenum are sometimes toxic to livestock and since almost all normal rations are adequate in this mineral, it is not recommended that farm livestock be supplemented with molybdenum.

VIII

Selenium.

A. Functions.

1. In vitamin E absorption and utilization.

2. As an essential component of the enzyme glutathione peroxidase, which functions to destroy toxic peroxides in the tissues, thereby having a sparing effect on the vitamin E requirement.

3. Other compounds of selenium seem to work in concert with vitamin E in the maintenance of normal cell functions and membrane health.

B. Requirements.

1. Up to 0.3 mg per kg of dry diet.

2. More than about 5 mg per kg of dry feed may produce toxic symptoms.

C. **Deficiency.**

1. Forages in the following sections of the country are low to very low in selenium and may result in a selenium deficiency:

> Southeastern coastal area
> States adjoining the Great Lakes
> New England states
> Coastal Northwest

2. Corn-SBM based diets are also frequently marginal to low in this mineral.

3. The following selenium deficiency symptoms are, in many respects, similar to those of vitamin E deficiency:

Nutritional muscular dystrophy (White muscle disease) in lambs and calves	Poor growth Low fertility Liver necrosis
Retained placenta in cows	Pancreatic fibrosis in
Heart failure	chicks
Paralysis	

4. In poultry, a selenium deficiency is associated with a vitamin E deficiency, resulting in exudative diathesis, poor growth, muscular dystrophy, and myopathies of the gizzard and heart.

D. **Supplementation.**

1. The use of supplemental selenium in ration formulation is under the regulation of the Food and Drug Administration (FDA) of the federal government.

2. Naturally, additives under regulation can be changed from time to time. Currently, it is legal to include up to 0.3 ppm of selenium in complete feeds, or not to exceed 3 mg/hd/day. In 1996, FDA put out a notice that they would not object to the use of selenium at a level no greater than 0.3 ppm in feeds for any specie, including pet foods, horse feed, and aquaculture.

IX — Zinc.

A. **Functions.**

1. Exact functions not understood.

2. It prevents parakeratosis (Figure 14–2).

3. It promotes general thriftiness and growth.

4. It promotes wound healing.

5. It is related to hair and wool growth, and health (Figure 14–3).

6. Deficiency impairs testicular growth and function.

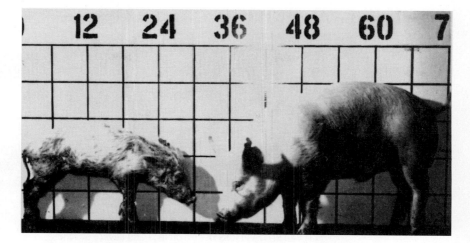

FIGURE 14–2

Zinc deficiency. The pig on the left received 17 PPM of zinc and gained 1.4 kg in 74 days: note the severe dermatosis or parakeratosis. The pig on the right received the same diet as the pig on the left, except that the diet contained 67 PPM of zinc. This pig gained 50.3 kg. (Courtesy of J. H. Conrad and W. M. Beeson, Purdue University)

B. Requirements.

1. Beef cattle: 20–30 mg/kg air-dry feed.

2. Dairy cattle: 40 mg/kg dry matter.

3. Sheep: About the same as for cattle.

4. Horses: About the same as for cattle.

5. Hogs: 50 mg/kg air-dry feed.

6. Poultry: 60 mg/kg air-dry feed.

C. Deficiency.

1. Zinc deficiencies seldom occur in cattle and sheep on normal rations.

2. Zinc deficiencies are frequently experienced in growing and fattening swine being fed on concrete with rations containing recommended levels of calcium. What would otherwise be adequate zinc is sometimes rendered unavailable and inadequate with an adequacy of calcium.

3. Deficiency symptoms include:

Parakeratosis

General unthriftiness

Poor growth

Unhealthy looking hair or wool

Slow wound healing

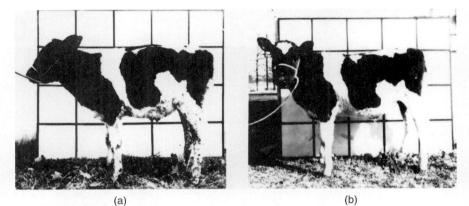

(a)   (b)

FIGURE 14–3
(a) Calf showing loss of hair on legs and severe scaliness, cracking, and thickening of the skin as the result of a zinc deficiency. (b) The same calf after receiving supplemental zinc. (Courtesy of W. J. Miller of the University of Georgia College of Agriculture Experiment Stations)

    4. Zinc deficiency in poultry causes "frayed feathers," retarded growth, and possibly enlarged hock joints.

D. Supplementation.

    1. Modern-day swine and poultry are usually supplemented with about 50 mg of zinc per kg of air-dry feed in the form of $ZnCO_3$ or $ZnSO_4$ added through the supplement or as a component of trace-mineralized salt.

    2. Cattle and sheep rations generally do not require zinc supplementation. However, many are provided with trace-mineralized salt containing zinc as a precautionary measure.

    3. The toxicity threshold of zinc appears to lie between 500 and 1000 mg/kg DM for most classes of livestock, depending on the supply of other nutrients.

15 General Recommendations for Mineral Feeding

I

When livestock are being fed totally or in part on a mixed feed, those minerals that are considered necessary or possibly necessary are usually included in the mixture (Figure 15–1).

A. Salt is included at 0.25–0.50% of the total ration. If the lower level is used, salt may also be provided free choice.

B. Calcium and phosphorus are added as needed to balance the ration—ground limestone or oystershell flour for calcium; bonemeal, defluorinated phosphate, or dicalcium phosphate for phosphorus, or for calcium and phosphorus.

C. If it is thought that any of the trace minerals might be in short supply, trace-mineralized salt will be used. It is cheap insurance, not harmful, and good for the soil.

D. Other minerals will not ordinarily be added unless special circumstances indicate a need for them.

II

When livestock are being fed on unmixed rations or are being run on pasture, minerals may be supplied on a free choice basis in either of two ways:

FIGURE 15–1
Four mineral products widely used in livestock feeding.
(Courtesy of the University of Georgia College of Agriculture
Experiment Stations)

A. **Cafeteria style.** This involves the use of a multi-compartment feeder with a separate compartment for each of several different minerals or mineral combinations. This method is based on the theory that an animal will consume as much of each mineral as it requires on a free-choice basis. This has never been definitely established. Such feeders are more expensive and are more trouble to keep filled with minerals than are feeders for single mixtures.

B. **Single mixtures.** Two general types follow:

1. For livestock primarily on heavy grain feeding:

   1 part salt (usually trace-mineralized)

   1–2 parts defluorinated phosphate, dicalcium phosphate or steamed bonemeal

   1–2 parts ground limestone or oystershell flour ($CaCO_3$)

   Trace-mineralized salt is usually recommended because it is cheap insurance against a possible trace-mineral deficiency (Figure 15–2), no harmful effects have ever been experienced from the proper use of properly formulated trace-

FIGURE 15–2
Livestock frequently exhibit a craving for some nutritive factor by chewing on wooden gates, etc. (Courtesy of Ralston Purina Co., St. Louis, Mo.)

mineralized salt, and the use of trace-mineralized salt will tend to maintain the trace-mineral level of the soil.

2. For livestock primarily on pasture, hay, and/or silage:

   1 part salt (usually trace-mineralized for same reasons as mentioned above)

   1–3 parts defluorinated phosphate, dicalcium phosphate or steamed bonemeal

   Since forages are much more apt to be deficient in phosphorus than in calcium and since the commonly used phosphorus supplements (defluorinated phosphate, dicalcium phosphate and steamed bonemeal) contain calcium as well as phosphorus, no limestone or oystershell flour (sources of calcium only) is needed in this mix.

16 Vitamin Nutrition

I

Definition of vitamin.   Vitamins are organic substances required by animals in very small amounts for regulating various body processes toward normal health, growth, production, and reproduction.

II

Brief history of vitamin nutrition.   The existence of nutritive factors, such as vitamins, was not recognized until about the start of the 20th century. There was no such word as "vitamin" prior to that time. The existence of vitamin-like factors was first recognized in the Orient when prisoners fed on unpolished rice seemed to be freer of beriberi than were nonprisoners consuming polished rice. It was theorized that there was a nutritive factor in rice polishings that prevented this disease. Tests with poultry confirmed this theory. It was subsequently determined that the factor was water soluble and was an amine (nitrogen-containing). Since the factor seemed to be essential for life, and apparently contained nitrogen, it was named *vitamine*.

Soon thereafter, workers in the United States recognized the presence of a factor in milk butterfat that prevented nightblindness in calves. However, the factor seemed to be different from that in rice polishings since the milk fat factor was fat soluble rather than water soluble and also did not contain nitrogen. As a result, it was concluded that there were two factors, both of which were essential for life. However, since only one contained nitrogen, it was agreed that such nutrients should be referred to as *vitamins* rather than *vitamines*—also that the antinightblindness factor should be called vitamin A and the anti-beriberi factor, vitamin B.

Subsequently, the observation was made that there must be a third such factor since there seemed to be something in fresh fruits and vegetables that prevented scurvy in humans. It was believed to be different from either of those discovered earlier since it was water soluble but did not contain nitrogen. Consequently, it was designated as vitamin C.

Soon thereafter, it was proposed that vitamin B, which was initially thought to be a single factor, consisted of two different factors, both of which were water soluble amines. The two factors were differentiated by calling one vitamin B_1 and the other vitamin B_2. As vitamin research was continued, it was determined that instead of consisting of two factors, vitamin B actually consisted of several factors, the total number of which has risen to 11 to date. This overall group has come to be known as the *B complex*. The different members of this group are referred to as vitamins B_1, B_2, etc., or in some instances by their actual chemical names.

A total of 16 different vitamins that function in animal nutrition has been discovered to date. In addition to those already mentioned, there are vitamins D, E, and K. The last vitamin to be discovered was vitamin B_{12}, the existence of which was demonstrated in the late 1940s. It is believed that most, if not all, of the vitamins that function in animal nutrition have been identified as of this time.

III

Vitamin designations.   Initially, in the absence of a more specific identification, the different vitamins were distinguished from each other by using the letters of the alphabet, such as A, B, C, etc. As pointed out previously, it subsequently became necessary to use subscript numbers with these letters in order to distinguish between different vitamins of a particular letter group or to distinguish between different forms of a particular vitamin.

In many instances, vitamins have also been referred to on the basis of the function they performed or the symptoms they prevented. As information concerning the chemical makeup of the different vitamins has been developed, an increasing number of the vitamins are being referred to by their actual chemical names, and today most of the vitamins are referred to in this manner.

IV

Chemical formula.   In dealing with the different vitamins, it is sometimes helpful to have information concerning their elemental composition and molecular structure. In the early stages of the vitamin story such information was quite limited. However, today, as the result of the diligent efforts of chemists, not only is the elemental composition of each vitamin known but also the structural formula of each vitamin molecule.

As may be noted from the empirical formulas for the different vitamins shown in Table 17–1, all of the vitamins contain carbon, hydrogen, and oxygen. In addition, all of the B vitamins except one (inositol) also contain nitrogen. Certain of the B vitamins also contain one or more of the mineral elements in their molecular structure.

V

Solubility.   Most organic materials are soluble either in water or in fats and fat solvents, or else they are not soluble in either. Very few are soluble in both water, on the one hand, and fat and fat solvents, on the other.

All of the B vitamins and vitamin C are soluble in water and so are said to be *water soluble.* Vitamins A, D, E, and K, on the other hand, are soluble in fats and fat solvents and so are said to be *fat soluble.* Many phenomena of vitamin nutrition are related to a vitamin's solubility. Consequently, it behooves every student of animal nutrition to be constantly aware of the solubility differences in vitamins and to make use of such differences whenever possible in developing practices, processes, or procedures.

VI

Color.   Vitamin color does not have much significance from the standpoint of vitamin nutrition. In some instances, however, the presence or absence of a particular vitamin may be indicated by the color of the material in question. Several of the vitamins, however, are colorless and consequently would not lend themselves to such evaluations.

VII

Rumen synthesis.   Under ordinary circumstances most, if not all, of the B vitamins can be synthesized in the rumen in amounts that will more than meet the requirements for these vitamins by ruminant animals. These vitamins are formed in the rumen as metabolic by-products of microbial fermentations. Different strains of microorganisms produce different metabolic by-products. The metabolic by-products of one strain contribute to the nutrition of other strains, and vice versa, in a complicated symbiotic relationship. In turn, the combined metabolic by-products of the various strains of rumen microorganisms represent a significant contribution toward the nutritive needs of the ruminant animal.

VIII

Body reserves.   An animal tends to store reserves of certain vitamins in its body so that a daily intake is not required. This is more true for the fat soluble vitamins than for the water soluble vitamins, except for vitamin B_{12}, which can be stored in the liver. The large amount of water that passes through most animals daily tends to carry out and thereby deplete the water soluble vitamins of the body. Fat soluble vitamins, on the other hand, are more inclined to remain in the body. This is especially true of vitamin A and/or carotene, which may be stored by an animal in its liver and fatty tissue in sufficient quantities to meet its requirements for vitamin A for periods of 6 months or even longer.

IX

Commercial synthesis.   Initially vitamin nutrition was dependent on the feeding of feeds and products that were known to have a high natural content of the respective vitamins. Over the past 40 years, however, industry has developed methods for the laboratory synthesis of the various vitamins. As a result, all of the different vitamins today are obtainable in pure crystalline forms at prices that make them economical for use in the vitamin supplementation of livestock.

X

Basic functions and deficiency symptoms. Each of the vitamins performs one or more basic functions in the regulation of various metabolic processes within the body. As a deficiency of a certain vitamin develops, there usually appears a certain set of deficiency symptoms. Originally it was thought that vitamins functioned simply to prevent these symptoms. However, it is realized today that deficiency symptoms are simply outward manifestations of disturbances of basic metabolic processes that occur as the result of a vitamin shortage. The nature of these functions and the symptoms are covered in Table 17–1.

17 Vitamins in Livestock Feeding

Vitamins in ruminant feeding.

A. While all of the different vitamins are apparently metabolic essentials for all of the various classes of farm livestock, under most conditions only vitamin A needs to be given attention from the standpoint of meeting the dietary needs of ruminants.

Under ordinary circumstances all of the various B vitamins are apparently synthesized in the rumen in sufficient quantities to overcome any dietary shortage of these vitamins in ruminant rations. Also, vitamin C, while apparently a metabolic essential for ruminants, is synthesized within the body tissues in sufficient quantities to meet the animal's needs. Ruminants, certainly under most conditions, will usually receive sufficient exposure to direct sunlight to meet their needs for vitamin D. Most ruminant rations are considered to be more than adequate in vitamin E, assuming that sufficient selenium is present to bring about its effective utilization. Most ruminant rations also are more than adequate in vitamin K. In addition, at least one form of vitamin K (K_2) is synthesized in the rumen. Finally, with the large amounts of pasture, hay, and/or silage included in most ruminant rations and the large intake of carotene that these feeds provide, the vitamin A needs of most ruminants are met without supplementation. This is especially so since reserves of carotene and/or vitamin A can be stored in the liver and body tissues during periods of high intake for use during periods of low intake.

B. Consequently, it would seem that in most instances ruminant animals would have no need for vitamin supplementation. However, it has been observed that under certain feeding conditions there is sometimes a very poor conversion of carotene to vitamin A in the ruminant animal. This condition has especially been noted with feedlot steers, which, in spite of the fact that they are receiving more than the recommended allowance of carotene, frequently develop extremely low blood vitamin A levels and display classical symptoms of vitamin A deficiency. Consequently, it is generally recommended that feedlot steers be supplemented with 20,000–30,000 IU of actual vitamin A per head per day, with the higher amount being fed during hot weather.

1 IU vitamin A = 0.3 microgram vitamin A alcohol
1 IU carotene = 0.6 microgram beta-carotene

(The above holds only for rats and chicks, which convert carotene to vitamin A with an efficiency of 50%. Cattle convert carotene to vitamin A with an efficiency of about 12.0% or less.)

Other classes of ruminants may be supplemented with vitamin A if the ration is low in carotene content or if the animals should exhibit symptoms of a vitamin A deficiency.

Vitamin A deficiency symptoms in cattle include reduced feed intake; slow gains; nightblindness; swollen hocks, knees, and brisket; excess lacrimation; total blindness; diarrhea; muscular incoordination; staggering gait; reduced sexual activity; low fertility in bulls; and poor conception, resorptions, and abortions in cows.

Recent research indicates that with certain feeds and rations under certain conditions, ruminant animals will sometimes respond to B vitamin supplementation. Further study is needed, however, before any general recommendations can be made in the feeding of B vitamins to ruminant animals.

II

Vitamins in swine nutrition. Swine are dependent on their diet for vitamins to a much greater degree than are ruminants. Like ruminants, swine can synthesize vitamin C in their tissues in adequate amounts to meet their needs. Also, swine, like other livestock, can synthesize their own vitamin D, provided they are exposed to sufficient amounts of direct sunlight. Synthesis of varying amounts of other vitamins may also take place in the large intestine of swine through microbial action. The extent of this synthesis, however, is generally quite limited, and this is not regarded as an adequate source of any of the vitamins. On the other hand, conventional swine feeds are good sources of several of the vitamins and so in themselves provide adequate amounts of these vitamins. It is generally recognized at this time that under most conditions of practical swine production, special consideration should be given to the matter of the adequacy of the following vitamins from the standpoint of the need for supplementation.

A. **Vitamin A.**   Rations based primarily on yellow corn, soybean meal, and minerals are at best barely adequate in vitamin A value. Where grain sorghum or barley is substituted for a major portion or all of the corn, vitamin A will definitely be in short supply. Consequently, unless at least 2.5% good quality alfalfa meal is included in such rations or unless such rations are fed on pasture, which is not frequently done in modern-day swine production, then a supplemental source of vitamin A should be used.

B. **Vitamin D.**   Since so many hogs today are fed in confinement where they receive little exposure to direct sunlight, supplementation of modern-day swine rations with vitamin D is generally recommended.

C. **Riboflavin.**   Swine rations that contain little or no alfalfa or milk products are likely to be deficient in riboflavin if not supplemented with this vitamin (Figures 17–1, 17–2). Since milk products are seldom used in swine rations and alfalfa meal is used only in limited amounts, if at all, it is generally recommended that riboflavin be included in present-day swine supplements.

D. **Niacin.**   Since many corn-based rations for swine are marginal in their content of available niacin (without niacin supplementation), it is generally recommended that swine rations be supplemented with this vitamin as a safety measure.

E. **Pantothenic acid.**   Corn is a poor source of pantothenic acid, and other commonly used swine feeds are not sufficiently high in this vitamin (Figure 17–3) to provide adequate levels of it in most ordinary swine rations without supplementation. Consequently, supplementation of swine rations with pantothenic acid is generally recommended.

F. **Vitamin B_{12}.**   Swine rations that do not contain liberal amounts of protein supplements of animal origin will probably be deficient in vitamin B_{12} unless supplemented with this vitamin. Since present-day swine rations for the most part involve the use of soybean meal as the protein source, it is generally desirable to provide a supplemental source of vitamin B_{12}.

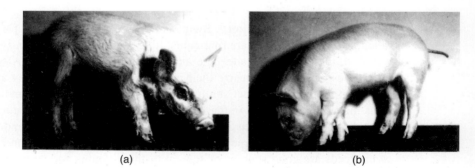

(a)   (b)

FIGURE 17–1
Riboflavin deficiency. (a) Pig that received no riboflavin.
(b) Pig that received adequate riboflavin. (Courtesy of R. W.
Luecke, Michigan Agricultural Experiment Station)

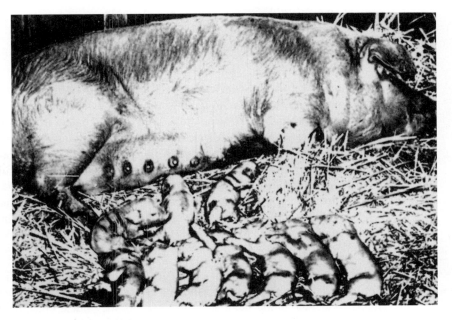

FIGURE 17–2
Seven of the ten pigs farrowed by this sow were born dead.
The sow received a riboflavin deficient ration. (Courtesy of
T. J. Cunha and J. P. Bowland, Washington State University)

FIGURE 17–3
Pantothenic acid deficiency. Locomotor incoordination
(goose-stepping) was produced by feeding a ration (corn-
soybean meal) low in pantothenic acid. (Courtesy of R. W.
Luecke, Michigan Agricultural Experiment Station)

G. Choline. While choline is usually adequate in most swine rations, extra choline has a sparing effect on the methionine requirement. Since choline can be provided more cheaply than methionine, choline is frequently included in supplements for swine.

H. Vitamin E. Most swine rations normally contain adequate levels of vitamin E. However, effective vitamin E utilization seems to be dependent on the presence of adequate selenium, and selenium is sometimes deficient in feeds from certain areas. Before selenium supplementation of swine rations received FDA approval, some swine producers were supplementing their rations with vitamin E in an effort to offset a possible selenium shortage and a resulting poor vitamin E utilization through providing extra amounts of this vitamin. However, with the FDA's recent approval of the use of supplemental selenium in swine rations, the use of supplemental vitamin E in such rations should not be necessary in the future if adequate selenium is provided.

III

Summary. In supplementing swine rations with vitamins, the customary practice is simply to include in the supplement the levels of the critical vitamins that will supply the animal's minimum needs of those vitamins. Any of these vitamins that are present in the feed are over and above the animal's needs and simply serve as a margin of safety. While excesses of vitamins serve no useful purpose, they are not harmful in reasonable amounts. Large excesses, however, may be toxic. The requirements of swine for the different vitamins may be found in the appendix tables.

Table 17–1

VITAMINS IN LIVESTOCK NUTRITION

Name(s) and Molecular Formula(s)	Fat or Water Soluble	Synthesized in Rumen?	Produced Commercially?	Basic Functions	Deficiency Symptoms	Good Farm Sources	Additional Information
Vitamin A $C_{20}H_{30}O$	Fat	No	Yes	Essential for health of epithelial cells. Functions in eyesight—also in bone formation.	Slow growth. Nightblindness. Reproductive disorders. Rough coat. Stiff and/or swollen joints. Total blindness. Low level of liver vitamin A.	Whole milk (carotene).	Not found in plants. Formed in animals from carotene. Formed in wall of intestine, liver, and certain other tissues. Rapidly destroyed by oxygen in heat and light. Reserves stored in liver and body fat. Colorless. May be deficient. Extremely high levels may be toxic.
(Carotene) $C_{40}H_{56}$	Fat	No	Yes	As a percursor of vitamin A.	Possible vitamin A deficiency.	Green pasture, good hay, alfalfa meal, silage, yellow corn, whole milk.	Formed in plants. Used by animals to form vitamin A. Rapidly destroyed by oxygen in heat and light. Reserves stored in body fat. Yellow in color. Vitamin A + Carotene = Vitamin A value. May be deficient. Nontoxic.

continued

Table 17–1(Continued)

Name(s) and Molecular Formula(s)	Fat or Water Soluble	Synthesized in Rumen?	Produced Commercially?	Basic Functions	Deficiency Symptoms	Good Farm Sources	Additional Information
Vitamin D (Antirachitic factor) (Vitamin D_2—ergocalciferol) $C_{28}H_{44}O$ (Vitamin D_3—cholecalciferol) $C_{27}H_{44}O$	Fat	No	Yes	Functions in Ca absorption and in Ca and P metabolism.	Poor growth, rickets, osteomalacia, osteoporosis.	Sun-cured hays, whole milk, sunlight.	Formed by irradiation of sterols with ultra-violet light. Ergosterol + UVL $\rightarrow D_2$ in plants. 7-dehydrocholesterol + UVL $\rightarrow D_3$ in animals. Quite stable. Some storage. Sometimes deficient. Large excesses may be toxic.
Vitamin E (Tocopherols) Alpha $C_{29}H_{50}O_2$ Beta $C_{28}H_{48}O_2$ Gamma $C_{28}H_{48}O_2$ Others	Fat	No	Yes	As an antioxidant. As a metabolic regulator of the cell nucleus. Probably others.	Poor growth. Muscular distrophy— "white muscle" and "stiff-lamb" disease. Reproductive failures?	Cereal grains (especially the germ), green forage, good hay, oil seeds.	Alpha tocopherol most active. Utilization dependent on adequate selenium. Some storage in liver and other tissues. Fairly stable but rapidly destroyed in presence of rancid fat. Seldom deficient if selenium is adequate. Excesses are relatively nontoxic.

continued

Name(s) and Molecular Formula(s)	Fat or Water Soluble	Synthesized in Rumen?	Produced Commercially?	Basic Functions	Deficiency Symptoms	Good Farm Sources	Additional Information
Vitamin K K_1-$C_{31}H_{46}O_2$ phylloquinone K_2-$C_{41}H_{56}O_2$ menaquinone Other napthoquinones (esp. menadione)	Fat Fat Depends on form	No Yes No	Yes	Essential for prothrombin formation and blood clotting.	Prolonged clotting time of blood. Multiple hemorrhages.	Green forage, good hay, rumen and intestinal synthesis.	Fairly stable under normal conditions. Considerable intestinal synthesis. Seldom deficient for livestock. Menadione toxic at high levels.
Thiamin(e) (Vitamin B_1) (Antineuritic factor) (Thiamine hydrochloride) $C_{12}H_{18}ON_4SCl_2$	Water	Yes	Yes	As a coenzyme in energy metabolism.	Poor appetite. Slow growth. Weakness. Hyperirritability.	Whole grains, germ meals, brans, green forage, good hay, milk, rumen synthesis.	Fairly stable under normal conditions. Limited intestinal synthesis. Seldom deficient for livestock. Some storage in pig. Levels several times requirement are nontoxic.
Riboflavin (Vitamin B_2) (Vitamin G) $C_{17}H_{20}N_4O_6$	Water	Yes	Yes	In several enzyme systems related to energy and protein metabolism.	Slow growth. Dermatitis. Eye abnormalities. Diarrhea and leg troubles in pigs.	Milk, skimmed milk, green forages, good hay—especially alfalfa, rumen synthesis.	Fairly stable under normal conditions. Most swine rations not containing milk and/or alfalfa meal will be low in this vitamin. Levels several times requirement are nontoxic.

continued

Table 17–1(Continued)

Name(s) and Molecular Formula(s)	Fat or Water Soluble	Synthesized in Rumen?	Produced Commercially?	Basic Functions	Deficiency Symptoms	Good Farm Sources	Additional Information
Niacin (Nicotinamide) (Niacinamide) $C_6H_6N_2O$ (Nicotinic acid) $C_6H_5O_2N$	Water	Yes	Yes	In enzyme systems related to carbohydrate, protein, and fat metabolism, and tissue respiration.	Digestive disorders, anorexia, dermatitis, and retarded growth in pigs.	Some present in most feeds, but that in corn, barley, wheat, and grain sorghum is largely unavailable. Rumen synthesis.	Quite stable under normal conditions. Some intestinal synthesis. Can by synthesized in body tissues from surplus tryptophan. Swine rations sometimes deficient in available niacin. Levels several times requirement are nontoxic.
Vitamin B_6 (Pyridoxine) $C_8H_{11}O_3N$ (Pyridoxal) $C_8H_9NO_2$ (Pyridoxamine) $C_8H_{11}N_2O$	Water	Yes	Yes	As a coenzyme in amino acid metabolism. In EFA metabolism. In antibody production.	Poor growth, anemia, and convulsions in pigs.	Most common feeds are fair to good sources. Rumen and intestinal synthesis.	Rations for livestock normally require no B_6 supplementation. Levels several times requirement are nontoxic.
Pantothenic acid $C_9H_{17}O_5N$	Water	Yes	Yes	As a part of coenzyme A. In other metabolic reactions.	"Goose-stepping" in pigs—also digestive disorders and unhealthy appearance.	Widely distributed but is it low in corn. Rumen and intestinal synthesis.	Quite stable under normal conditions. Can be deficient for swine. Ca pantothenate usually used as supplemental form. Only d isomer is biologically active. Levels several times requirement are nontoxic.

continued

Name(s) and Molecular Formula(s)	Fat or Water Soluble	Synthesized in Rumen?	Produced Commercially?	Basic Functions	Deficiency Symptoms	Good Farm Sources	Additional Information
Biotin $C_{10}H_{16}O_3N_2S$	Water	Yes	Yes	In enzyme systems related to carbon dioxide fixation and decarboxylation. In fat synthesis.	Dermatitis, loss of hair, and retarded growth in all species.	Widely distributed. Rumen synthesis.	Quite stable. Considerable intestinal synthesis. Amount dependent on nature of diet. Not likely to be deficient for livestock. Rendered unavailable by raw egg white. Levels several times requirement are nontoxic.
Choline $C_5H_{15}O_2N$	Water	Probably	Yes	In transmethylation. In fat metabolism in the liver. In cell structure (not a true vitamin function).	Unthriftiness, incoordination, fatty livers, and poor reproduction in swine.	Most commonly used feeds are fair to good sources.	Some synthesis in body tissues. Not likely to be deficient. Surpluses may be used as a methyl donor to spare methionine. Levels several times requirement are nontoxic.

continued

Table 17–1(Continued)

Name(s) and Molecular Formula(s)	Fat or Water Soluble	Synthesized in Rumen?	Produced Commercially?	Basic Functions	Deficiency Symptoms	Good Farm Sources	Additional Information
Folic acid (Folacin) $C_{19}H_{19}N_7O_6$	Water	Yes	Yes	Functions in the transfer of single carbon units in various biochemical reactions.	Poor growth and various blood disorders in some species.	Most commonly used feeds are fair to good sources. Rumen synthesis.	Some intestinal synthesis. Not likely to be deficient for livestock. Levels several times requirement are nontoxic.
Vitamin B_{12} (Cyanocobalamin) $C_{63}H_{90}O_{14}N_{14}$ PCo	Water	Yes	Yes	As a coenzyme in several biochemical reactions. Especially involved in propionic acid metabolism; also in red blood cell maturation.	Slow growth, incoordination, and poor reproduction.	Protein feeds of animal origin. Fermentation products. Rumen synthesis.	Originally APF. Some intestinal synthesis depending on nature of diet. Cobalt essential for B_{12} synthesis. Can be deficient for swine and early weaned calves. Levels several times requirement are nontoxic.

continued

Name(s) and Molecular Formula(s)	Fat or Water Soluble	Synthesized in Rumen?	Produced Commercially?	Basic Functions	Deficiency Symptoms	Good Farm Sources	Additional Information
Inositol $C_6H_{12}O_6$	Water	Yes	Yes	Not to well understood. Possibly replaces certain other vitamins in short supply.	None demonstrated in livestock.	Widely distributed in farm feeds. Rumen and intestinal synthesis.	Not ordinarily deficient for livestock. Relatively nontoxic.
Para-Aminobenzoic acid $C_7H_7O_2N$	Water	Yes	Yes	Essential for growth of certain microorganisms. May function in rumen and intestinal synthesis of other vitamins.	None demonstrated in livestock.	Not too well known.	Not ordinarily deficient for livestock. Relatively nontoxic.
Vitamin C (Ascorbic acid) (Antiscorbutic factor) $C_6H_8O_6$	Water	No	Yes	Functions in reactions related to the formation and maintenance of collagenous intercellular material.	None demonstrated in livestock.	Not required.	Livestock synthesize vitamin C in body tissues. Never deficient with livestock under normal conditions. Relatively nontoxic.

18 Physiological Phases of Livestock Production

The science of livestock production divides itself into several different physiological phases. Any living animal will always be involved with at least one of these phases (maintenance) but will seldom be involved at any one time with more than about two or three. Each phase has its own set of nutritive requirements and these requirements are additive. While theoretically an animal could be involved in as many as four or five different physiological phases of production at one time, it is next to impossible for an animal to consume, digest, and metabolize sufficient nutrients to carry out the functions associated with so many physiological activities. Every animal should be fed so that the nutritive requirements of each physiological phase of production with which it is involved are met. It is intended for section 18 to recognize these different physiological phases and to review briefly the nutritive requirements of each. These are summarized in table form (Table 18-1.)

Maintenance. Maintenance means *maintaining an animal in a state of well being or good health from day to day. A maintenance ration is the feed required to adequately support an animal doing no nonvital work, making no growth, developing no fetus, storing no fat, or yielding no product.* The nutritive requirements for maintenance are the first to be met. The nutritive needs of an animal for other purposes are for the most part over and above those for maintenance. As much as 100% of an animal's ration may go for maintenance. On the other hand, with full-fed animals as little as 1/3 or even less of an animal's nutrient intake may be required for maintenance purposes. On the average, about 1/2 of all feed fed to livestock goes for maintenance. While it is sometimes profitable to hold animals on a maintenance ration from a period of low prices to a period of high prices, generally speaking, only those nutrients fed over and above the maintenance requirements are available for economic production. The requirements for maintenance are as follows:

A. Energy for the vital functions.

1. This includes energy for the work of the heart, the work of breathing, and the work of other vital functions.

2. This energy must be in the form of net energy.

3. Energy used for the vital functions is ultimately liberated as heat.

4. The heat resulting from the work of the vital organs may go toward maintaining the body temperature.

5. The work of the vital functions is referred to as *basal metabolism.*

6. Basal metabolism is the heat production of an animal while at rest and digesting no food.

7. Basal metabolism may be determined directly or calculated from the O_2 and CO_2 exchange.

8. Basal metabolism is in proportion to the body surface of an animal, not its weight.

B. Heat to maintain body temperature.

1. All farm livestock are warm-blooded animals and normally maintain more or less constant body temperatures.

Horse	100.2°F	Hog	102.6°F
Cow	101.5°F	Sheep	103.5°F

2. Heat for the maintenance of body temperature comes from a variety of sources.

a. Heat from work of vital organs.

b. Heat of nutrient utilization.

c. Heat from work of normal activity.

d. Heat as a by-product of economic work.

e. Heat from the work of shivering.

3. The temperature at which body oxidations must be increased to maintain the body temperature is referred to as the *critical temperature.* It is that temperature at which shivering starts. It is quite variable depending on:

Species of animal   Activity of animal
Hair or wool coat of animal   Air movement
Fatness of animal   Humidity
Level of feed of animal

4. The critical temperature is seldom reached with animals on a liberal feed allowance, except in extremely cold weather.

C. **Protein for the repair of body tissues.**

1. There is a constant breakdown of body tissue protein.

2. The by-products of body protein breakdown are excreted largely in the urine.

3. Because of this constant breakdown and loss of protein from the body, protein must be replenished, and this is equivalent to the maintenance requirement.

4. The nitrogen excreted in the urine during starvation approximates the maintenance requirement of an animal for protein.

5. The protein requirement for maintenance is in proportion to body surface area.

6. Protein for maintenance must be of good quality—that is, it must contain the proper proportion of essential amino acids.

D. **Minerals to replace mineral losses.**

1. There is a constant loss of all of the essential minerals from the body—even calcium and phosphorus from the bones.

2. Maintenance rations must contain sufficient minerals to replace these losses.

3. Most farm feeds contain adequate minerals for maintenance, except for salt.

E. **All the vitamins are essential for maintenance.**

1. All of the vitamins are necessary for life, even if only maintenance is involved.

2. There is a constant destruction and/or loss from the body of all of the different vitamins.

3. Feeding programs must take into account the fact that animals must have a constant source of vitamins even just for maintenance.

F. **Water.**

1. An animal will die more quickly from a lack of water than from a lack of any other nutritive factor.

2. Water is required for essentially all body functions.

3. There is a constant loss of water from the body through urine excretion, feces excretion, perspiration, and respiration.

G. **Certain fatty acids.** While an animal can meet its energy requirements from any of several different sources, certain fatty acids seem to be necessary for maintaining normal health. Two fatty acids in particular, linoleic and linolenic, are referred to as the essential fatty acids (EFA). While arachidonic was originally included in this group, recent work has indicated that it can be synthesized in animal tissue from lenoleic and so can no longer be regarded as an essential fatty acid.

Ⅱ Growth.

A. **Growth** is largely an increase in muscle, bone, organs, and connective tissue. Since meat is basically muscle, then growth is basic for meat production. Also, it is only through growth that an animal is able to attain a mature status. It should be recognized in this connection that the nutritive requirements for growth as outlined below are in addition to those listed above for maintenance. The primary nutritive requirements for growth are as follows:

1. **Protein.**

   a. The dry matter of muscle and connective tissue, and to a considerable degree also that of bone, is primarily protein. Hence, protein is one of the major nutritive requirements of growth.

   b. Protein for growth must be of good quality—that is, it must contain the proper proportions and amounts of essential amino acids at the tissue level.

2. **Energy.**

   a. Animal tissue produced as the result of normal growth ordinarily contains a limited amount of ether extract. Energy in the form of net energy must be provided to meet this need in addition to that in the protein of tissue. Also a certain amount of additional energy is used by the body for growth.

3. **Minerals.**

   a. Since bone formation is a primary activity of growth and since bone is high in calcium and phosphorus content, these two minerals are especially essential for growth.

   b. Other minerals are involved in the digestion and utilization of other nutrients needed for growth.

4. **Vitamins.**

   a. Vitamin D is essential for bone formation.

   b. Certain other vitamins function in various metabolic processes related to nutrient utilization for growth.

5. **Water.**

   a. Fat-free muscle tissue is about 75–80% water. Hence, water is a major requirement for growth. Supplying adequate water for livestock growth, however, is not a major consideration since the amount needed for growth as compared to maintenance is rather inconsequential. However, it behooves the livestock producer to take advantage insofar as possible of the fact that growth, especially muscle growth, is largely water, and water is cheap.

B. Rate of growth.

1. Varies in amount per head daily among the different species—more or less in accordance with the mature size of the species.

2. Varies in amount per 100 lb liveweight daily among the different species, with the larger species in general having the lower rate.

3. Varies among different breeds largely in accordance with the size of mature animals of the respective breeds.

4. Daily growth rate per animal increases until puberty and then decreases until maturity.

5. Daily growth rate per 100 lb liveweight decreases from birth to maturity.

III Fattening.

A. What is fattening? Fattening in an animal is simply the deposition of unused energy in the form of fat within the body tissues. Fattening is of two general types (Figure 18-1).

1. **Abdominal, intermuscular, and subcutaneous deposition.** This is for the most part undesirable but unavoidable if marbling is to be realized.

2. **Intramuscular deposition.** Commonly referred to as *marbling*. This is what is wanted. Difficult to obtain without excessive abdominal, intermuscular, and subcutaneous deposition.

B. Object of fattening. The object of fattening is to make the meat tender, juicy, and of good flavor. While most people do not like fat meat, they like the lean of meat only if it contains a certain amount of fat as marbling. Fat represents the most costly form of gain in livestock. Consequently, livestock are ordinarily fattened only to a point that they will be sufficiently marbled to make their meat acceptable to the consumer.

C. Requirements for fattening.

1. The primary requirement for fattening is energy. This energy must be in the form of net energy. Net energy for fattening may come from any of several forms of feed energy such as:

Starch	Protein
Sugars	Fat
Cellulose	

2. Fattening increases an animal's need for protein over and above that required for maintenance and growth only to the extent that additional protein may be

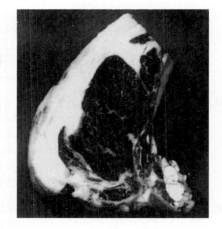

FIGURE 18-1
A prime rib roast showing an abundance of marbling, which
unfortunately is usually accompanied by an excess of outside fat. (Courtesy of the Greenwood Packing Co.,
Greenwood, S.C.)

necessary to promote good digestion. Fattening increases an animal's metabolic requirements for protein little, if at all.

3. Fattening may increase the need for certain of those vitamins related to energy
metabolism.

4. Fattening animals are usually full fed since only that net energy over and above
that required for maintenance and growth is available for fattening. There may
be some loss in feed digestibility with full feeding, but this is more than offset
by an overall increase in the efficiency of feed utilization for growth and fattening.

D. Fattening vs. growth in producing weight gains.

1. Weight gains in animals are derived from the following:

   a. Growth.

   b. Fattening.

   c. Fill or increase in content of feed and water.

2. Growth is a much cheaper form of gain than is fattening.

   a. Gain from growth is primarily in the form of protein tissue and bone. Gain
   from fattening is largely in the form of fat.

   b. Protein tissue is about 25% protein and 75% water. While protein is one
   of the more costly nutrients, water is essentially free. Hence, protein tissue is a less expensive form of gain.

   c. Bone formation in growth requires considerable calcium and phosphorus. Hence, growth utilizes the calcium and phosphorus of feed more completely for weight gain than does fattening. Even if the requirements for growth make it necessary to add calcium and/or phosphorus supplements, these are usually an economical source of nutrients for gain.

   d. Gain from fat is generally relatively expensive. Little, if any, water and minerals are laid down in fattening. In fact, during fattening, fat may actually replace water in the tissues, which makes for a very uneconomical exchange. Also, about 2.25 times as much net energy is required to form a kg of body fat as is required to form a kg of body protein. While protein is usually a costly source of net energy, it is seldom 2.25 times as expensive as other sources.

   e. Since gain from growth is usually more efficient and cheaper than gain from fattening and since animals do most of their growing while young, young animals make more efficient and less expensive gains than do older animals.

   f. Older animals are more easily fattened than are younger animals because a larger percentage of their energy consumption is available for fattening.

IV
Milk production.

   A. We usually are inclined to associate milk production with the dairy cow. However, all animals produce milk upon giving birth to young (Figure 18-2). Milk production simply happens to be the dairy cow's specialty. She produces more milk over longer periods of time than do other animals. However, she certainly has no monopoly on it. In fact, certain other animals may be as efficient as the dairy cow at milk production per kg of feed consumed or per 100 kg of body weight.

   B. **Composition.** On a dry basis, the milk of all animals is fairly similar in composition, being high in protein, fat, NFE, and minerals. The milk of the mare is somewhat of an exception in that it is lower in protein and fat and much higher in NFE than the milk of other farm animals, as shown below:

	% Protein	% Fat	% NFE	% Ash
Mare's milk	21	11	63	4
Milk of other farm animals	27–36	29–36	25–38	5–6

On a fresh basis, the milk of different farm animals varies considerably in percentage of water and dry matter content as shown below:

	% Water	% DM
Cow's milk	87.2%	12.8%
Ewe's milk	80.8	19.2
Goat's milk	86.8	13.2
Mare's milk	90.6	9.4
Sow's milk	79.9	20.1

FIGURE 18-2
While milk production is the dairy cow's specialty, all farm
animals produce milk upon giving birth to young. (Courtesy
of Ralston Purina Co., St. Louis, Mo.)

C. **Milk secretion.**   Milk is produced and secreted by the mammary glands. Nutrients for milk production are carried by the blood to the mammary glands. The nutrients are removed from the blood by the mammary glands, converted into milk, and secreted into the udder. Milk is secreted into the udder more or less throughout the day. Nutrients for milk production must come from the feed, either directly or indirectly via body reserves of nutrients, which came originally from the animal's feed.

D. **Nutritive requirements for milk production.**

1. Are in proportion to the amount of milk produced.

2. Are over and above those for other physiological phases of production such as maintenance, growth, fattening, fetal development, etc.

3. The major nutritive requirements follow:

    a. **Protein.**

    • Must be of good quality at the glandular level.

    • Animals will not produce milk low in protein.

    • If ration is deficient in protein, tissue reserves of protein may be used for milk production.

    • Prolonged shortage of protein will limit milk production.

b. **Energy.** Energy over and above that for milk protein is required for the formation of milk fat and milk sugar.

- Must be in the form of net energy.

- May come from ration carbohydrates, ration fat, or even excess ration protein.

- An animal will not produce milk extremely low in energy.

- If ration is low in energy, body reserves of energy may be used for milk production.

- While ration fat is not essential for milk fat, a small amount of ration fat helps milk production.

- Prolonged shortage of energy will limit milk production.

c. **Calcium and phosphorus.**

- May come from the feed or from supplemental sources.

- Animals will not produce milk that is low in these minerals.

- If ration is deficient in calcium and/or phosphorus, the animal will draw upon its bones for calcium and phosphorus with which to produce milk. This will weaken bones and may result in broken bones.

- Prolonged shortage of calcium and/or phosphorus will limit milk production.

d. **Vitamin A and/or carotene.**

- May not be essential specifically for milk production.

- Animal may produce milk low in these factors.

- Milk rich in these factors is to be preferred.

- Carotene and/or vitamin A for milk production may come from the ration or body reserves.

- Vitamin A and carotene content of milk can be increased by increasing ration vitamin A and carotene.

e. **Vitamin D.**

- Some evidence that vitamin D is essential for assimilation of calcium and phosphorus for milk production.

- Vitamin D content of milk can be stepped up through vitamin D feeding.

f. **Sodium and chlorine (salt).**

- Essential for digestion of nutrients for milk production.

- Milk normally contains some salt.

g. **Other minerals and vitamins.**

- Not too well defined.

- Several involved in nutrient utilization.
- Most essential minerals and vitamins normally in milk; must come from feed, from water, or from rumen or body synthesis.

V

Fetal development. Two major considerations in feeding during gestation are:

A. To provide nutrients for development of fetus and fetal membranes.

   1. Fetal development is basically just prenatal growth. Nutritive requirements for fetal development are qualitatively similar to those for growth—that is, protein, calcium, phosphorus, and vitamin D in particular and others indirectly.

   2. Quantitatively the nutrient requirement for fetal development is not great. A newborn calf weighs on the average about 37 kg (83 lbs) and is about 25% DM, which makes its dry matter content just under 10 kg. This is the nutritive equivalent of about 75 kg of milk or about 3–4 days production for a good cow.

   3. If a ration is deficient in calcium, phosphorus, protein, and/or energy, an animal may draw upon its own body for nutrients to develop the fetus.

   4. If an unbred animal is severely deficient in nutrients, she probably will not come into heat.

   5. If an animal is severely deficient in nutrients during early gestation, she may cease fetal development through resorption or abortion.

B. A second consideration of feeding during gestation is to build up the nutrient reserve of the animal's body to meet the requirements for milk production following parturition.

   1. Few animals can consume, digest, and metabolize nutrients sufficient to meet their needs during heavy lactation. Most must call on body reserves of fat, protein, calcium, phosphorus, vitamin A, and possibly other nutrients for this purpose. Hence, an animal should be fed during late gestation to encourage some buildup of reserves of these nutrients.

   2. Termination of lactation during late gestation will promote this nutrient buildup in the body.

   3. Too liberal feeding, however, during late gestation may result in an oversized fetus and difficult parturition.

VI

Wool production.

A. Wool consists of two fractions.

   1. Wool fiber.

      a. Is practically pure protein.

      b. Is sulfur-containing protein.

    c.   Makes up from 20% to 75% of a fleece (unwashed wool of a sheep).

    d.   Elemental composition:

C—50%	N—18%	S—3%
H—7%	O—22%	

2. **Yolk or grease** consists of the following:

    a.   **Suint.**   Compounds of potassium with organic acid that is water soluble. Makes up from 15% to 50% of unwashed wool.

    b.   **Wool fat.**   Commonly known as *lanolin*. Actually a wax that is not water soluble, but is removed by scouring process. Makes up 8–30% of unwashed wool. Contains carbon, hydrogen, and oxygen.

B. **Nutritive requirements.**

1. Nutritive requirements for wool production are over and above those for other physiological phases of production.

2. With some sheep producing 10 kg (22 lb) or more per year (average of all sheep is around 8 lb), the nutritive requirements for wool production can be considerable.

3. The primary nutritive requirements for wool production are the following:

    a.   **Protein.**   Must be sulfur-containing as fed or as synthesized in the rumen.

    b.   **Energy.**   Sheep must be provided energy that is over and above that in the required protein to produce the yolk in wool. This must be in the form of net energy and can come from any feed energy source.

    c.   **Potassium.**   This mineral is an essential component of the suint in wool. It is more than adequate in most ordinary rations.

    d.   **Other minerals and vitamins.**   While certain other minerals and vitamins are essential either directly or indirectly for wool production, their roles are not well defined in this connection.

VII
Work.

A. **What is work?**   Work is the movement of matter through space.

B. **Types of work.**

1. **Involuntary.**   Such as that of the heart and other vital organs. Essential for life. An essential part of maintenance.

2. **Voluntary.**   A certain amount of voluntary work is an essential part of practical maintenance. That part of voluntary work over and above that involved in maintenance is the present consideration. It is voluntary work for recreation or economic production.

C. The primary nutritive requirement for work is energy.

1. It must be in the form of net energy.

2. Energy for work must be over and above energy for other needs.

3. Energy for work may come from carbohydrates, fats, and/or excess protein.

4. If ration energy is not adequate to meet the needs for work, fat stores of the body may be used for this purpose.

D. The protein, mineral, and vitamin requirements of an animal are increased little, if any, by work.

1. As work may involve increased energy consumption and utilization, the need for those nutrients involved in energy digestion and metabolism may be increased.

2. As work may involve profuse perspiration, there may be a significant loss of certain water soluble nutrients and therefore an increased requirement for such nutrients due to work.

VIII

Egg production.

A. Chickens.

1. Major increases in nutrient requirements of laying hens include energy, protein, and calcium.

2. Diets must be formulated to provide specific amounts of nutrients/day, rather than a percentage of dietary dry matter, since energy density regulates how much a hen will consume.

3. Laying hens fed *ad libitum* may become overly fat.

4. Metabolizable energy need is related to body weight and egg production. A 2.5 kg (6 lb) laying hen producing 90 eggs/100 days should receive more than 40% more kcal ME/day than the same weight hen producing no eggs (371 vs. 259 kcal).

5. Mineral requirements of laying hens are similar to those of nonlayers, except for a greatly elevated calcium need. Calcium requirement will range from 3% to 4% of the diet, depending upon the caloric content; in other words, hens consume less of a higher energy diet than a lower energy diet.

6. Vitamin and trace mineral content of eggs can be increased by feeding higher levels to the hen, which may be desirable for breeder hens.

7. Molting can be "forced" in order to halt egg laying, by feeding an excess of some nutrients, *i.e.,* iodine, zinc, or sodium chloride.

B. Turkey hens. Starting both sexes on feed having the lowest concentration of nutrients for which a balance can be formulated, and continuing this regimen through the breeding period on an *ad libitum* consumption basis, will lessen the likelihood of obesity without affecting performance.

Table 18-1

SUMMARY OF THE PHYSIOLOGICAL PHASES OF LIVESTOCK PRODUCTION AND THEIR NUTRITIVE REQUIREMENTS

	Energy	Protein	Calcium	Phosphorus	Vitamin A	Vitamin D	Other Minerals	Other Vitamins
Maintenance	Yes	Yes	Yes	Yes	Yes	Yes	Yes	Yes
Growth	Mainly as protein	Yes	Yes	Yes	Yes	Yes	Yes	Yes
Fattening	Yes	Only to facilitate digestion	No	No	Possibly	No	Probably	Probably
Milk production	Yes	Yes	Yes	Yes	Not required, but desired	Probably— at least desired	Probably— at least desired	Probably— at least desired
Fetal development	Mainly as protein	Yes	Yes	Yes	Yes	Yes	Probably	Probably
Wool production	Yes	Yes	No	No	No	No	Yes— especially K and P	Probably
Work	Yes	Little, if any	No	No	Possibly	No	Yes	Yes
Egg production	Yes	Yes	Yes	No	No	Yes	No	No

19 Study Questions and Problems

I. **Select the correct answers by number from the terms at the bottom for each of the following. Use terms more than once as necessary.**

1. Iron utilization in the body.
2. Mineral element found in hemoglobin.
3. Most abundant mineral element in wool.
4. Means the opposite of macro.
5. One of the essential amino acids.
6. Prevents white muscle disease in cattle.
7. A deficiency causes goose-stepping in pigs.
8. Thiamine hydrochloride.
9. Same as cobalamin.
10. Crude protein.
11. Second most abundant mineral element in the body.
12. Prevents parakeratosis in swine.
13. A good source of calcium for livestock.
14. A condition sometimes caused by a deficiency of iodine.
15. Converted into vitamin D by sunlight.
16. Hydrolyzes to amino acids.

17. A good source of zinc for livestock.
18. The vitamin that functions in bone formation.
19. Produced in the body by sunlight.
20. Vitamin E.
21. A good source of both calcium and phosphorus for livestock.
22. Same as APF.
23. Mineral compound with which all livestock should be supplemented.
24. Contains 46.67% nitrogen.
25. Functions in vitamin E absorption.
26. A condition in older animals caused by a deficiency of calcium.
27. Necessary to prevent losses in salt-hungry cattle given loose salt.
28. Same as riboflavin.
29. A good source of fluorine for livestock.
30. An excess of this will cause defective teeth formation.
31. Functions in blood clotting.
32. The amino group.
33. Most abundant mineral element in the body.
34. Mineral element found in thyroxin.
35. Mineral element associated with grass tetany in cattle.
36. A condition in young pigs caused by a deficiency of iron.
37. Prevents nightblindness in calves.
38. The percentage of a protein usable as protein in the body.

1.	Pure urea.	18.	Copper.
2.	Defluorinated phosphate.	19.	Fluorine.
		20.	Zinc.
3.	Tocopherols.	21.	Vitamin A.
4.	Selenium.	22.	$N \times 6.25$.
5.	Protein.	23.	NH_2.
6.	Trace.	24.	Biological value.
7.	Phosphorus.	25.	Pantothenic acid.
8.	Trace-mineralized salt.	26.	Osteoporosis.
9.	Vitamin D.	27.	Iron.
10.	Sulfur.	28.	Vitamin K.
11.	Oystershell flour.	29.	Anemia.
12.	Vitamin B_{12}.	30.	Salt.
13.	Arginine.	31.	Goiter.
14.	Iodine.	32.	Vitamin B_1.
15.	Magnesium.	33.	Ergosterol.
16.	Vitamin B_2.	34.	Calcium.
17.	Water.		

II. Select the correct answer for each of the following:

1. Ratio of calcium to phosphorus in bone. (*1:1, 1:2, 2:1, 2:3*)
2. Biological value of a protein void of 2 essential amino acids. (*0, 20, 50, 90*)

3. Percentage of phosphorus in defluorinated phosphate. (*18.0, 33.0, 38.5, 46.7*)

4. Number of vitamins included in the B complex. (*7, 9, 11, 13*)

5. Maximum percentage of protein normally found in soybean meal. (*36, 41, 44, 50*)

6. A protein with a biological value of 60% and a protein with a biological value of 100% are mixed in equal amounts. What is the lowest probable biological value of the mixed protein? (*0, 60, 80, 100*)

7. What is the highest possible biological value of the mixed protein in question 6? (*0, 60, 80, 100*)

8. Percentage of calcium in defluorinated rock phosphate. (*14.5, 32.0, 38.5, 46.7*)

9. Number of IU of vitamin A in one microgram. (*0.3, 1.0, 3.3, 6.7*)

10. Percentage of calcium in alfalfa hay, as fed. (*1.27, 2.29, 4.52, 6.73*)

11. Daily salt consumption of a cow in ounces. (*1, 10, 28, 44*)

12. Highest possible biological value of a protein. (*50, 80, 100, 291.6*)

13. Percentage of salt usually included in a complete swine ration. (*0.5, 1.5, 2.5, 3.5*)

14. Number of fat soluble vitamins that function in animal nutrition. (*4, 11, 16, 24*)

15. Number of water soluble vitamins. (*4, 12, 16, 24*)

16. Total number of vitamins that function in animal nutrition. (*4, 11, 16, 24*)

17. Percentage of nitrogen in crude protein. (*6.25, 16.0, 46.6, 291.6*)

18. Percentage of phosphorus in ground limestone. (*0.02, 14.5, 30.0, 35.9*)

19. Number of mineral elements that function in animal nutrition. (*16, 20, 24, 33*)

20. Number of chemical elements found in vitamin D. (*3, 4, 5, 6*)

21. Number of essential amino acids. (*5, 10, 20, 24*)

22. Maximum percentage of protein normally found in cottonseed meal, as fed. (*36, 41, 44, 50*)

23. Percentage of calcium in pure calcium carbonate. (*14.5, 30.0, 38.5, 40.0*)

24. Total number of naturally occurring amino acids. (*5, 8–10, 20–22, 24+*)

25. Percentage of salt usually included in a mineral mixture for beef cattle on pasture. (*3–5, 5–10, 25–33, 67–75*)

26. Number of trace-mineral elements usually included in trace-mineralized salt. (*6, 9, 16, 20*)

27. Minimum number of chemical elements found in a typical B vitamin. (*3, 4, 5, 6*)

28. Maximum percentage of ground limestone to be included in a mineral mixture for cattle on pasture. (*0, 25, 33, 50, 75*)

29. Percentage of salt usually included in trace-mineralized salt. (*20–30, 33–50, 70–80, 95+*)

30. Number of vitamins containing cobalt. (*0, 1, 3, 5*)

31. Number of macro elements that function in animal nutrition. (*6, 11, 16, 20*)

32. Number of vitamins with which ordinary swine rations are usually supplemented. (*1, 8, 12, 16*)
33. Percentage of crude protein in a steer's ration. (*7, 12, 16, 20*)
34. Number of amino acids found in animal tissue. (*8–10, 20–22, 24–26, 30–33*)
35. The crude protein equivalent of pure urea. (*16.0, 42.5, 46.7, 291.6*)

III. List the ten essential (indispensable) amino acids.

IV. Indicate by number the relative crude protein content of each of the following (as-fed basis):

1. More than 22%.
2. 10.5–22%.
3. 6–10.5%.
4. Less than 6%.

1.	Dried whole milk.	23.	Fresh cow's milk.
2.	Steamed bonemeal.	24.	Alfalfa hay.
3.	Feather meal.	25.	Crimson clover pasture.
4.	Peanut kernels.	26.	Bromegrass hay.
5.	Dried skimmed milk.	27.	White potatoes.
6.	Tankage with bone.	28.	Coastal bermudagrass hay.
7.	Meat and bonemeal.	29.	Alfalfa pasture.
8.	Peanut oil meal.	30.	Cowpea hay.
9.	Brewers dried grains.	31.	Coastal bermudagrass pasture.
10.	Cottonseed meal.	32.	Ground corn cob.
11.	Oat hulls.	33.	Sorgo silage.
12.	Alfalfa leaf meal.	34.	Johnsongrass hay.
13.	Alfalfa meal.	35.	Wheat middlings.
14.	Cottonseed.	36.	Bromegrass pasture.
15.	Wheat straw.	37.	Wheat bran.
16.	Peanut hulls.	38.	Corn silage.
17.	Fish meal.	39.	Lespedeza hay.
18.	Soybean meal.	40.	Sweet potatoes.
19.	Dried beet pulp.	41.	Peanut hay.
20.	Linseed oil meal.	42.	Soybean seed.
21.	Alfalfa stem meal.	43.	Citrus molasses.
22.	Cottonseed hulls.	44.	Ground oats.

45. Cane molasses.
46. Oat straw.
47. Shelled corn.
48. Salt.
49. Grain sorghum.
50. Cottonseed oil.
51. Wheat grain.
52. Corn husks.
53. Cane sugar.
54. Ground limestone.
55. Millet hay.
56. Starch.
57. Cellulose.
58. Lignin.
59. Petroleum oil.
60. Timothy hay.
61. Ground coal.
62. Ground barley grain.
63. Glucose.
64. Oat hay.
65. Alsike clover hay.
66. Stearin.
67. Sericea hay.
68. Peanut oil.
69. Ground snapped corn.
70. Lactose.
71. Riboflavin.
72. Tallow.

V. Indicate by number which of the following are:

1. Water soluble.
2. Synthesized in the rumen.
3. Included in swine supplements.

1. Vitamin A.
2. Vitamin B_1.
3. Folic acid.
4. Vitamin B_6.
5. Vitamin D.
6. Choline.
7. Pantothenic acid.
8. Vitamin C.
9. Vitamin B_2.
10. Vitamin E.
11. Carotene.
12. Vitamin B_{12}.
13. Niacin.
14. Vitamin K.
15. Biotin.
16. Inositol.
17. Para-aminobenzoic acid.

VI. Within each of the following groups of feeds, indicate which is highest in crude protein and which is lowest, as-fed basis.

1. (1) Shelled corn, (2) alfalfa meal, (3) fresh cow's milk, (4) cottonseed meal, (5) soybean meal.
2. (1) Ground snapped corn, (2) cane molasses, (3) dried skimmed milk, (4) wheat bran, (5) bermudagrass hay.
3. (1) Ground oats, (2) corn silage, (3) grain sorghum grain, (4) peanut oil meal, (5) brewers dried grains.
4. (1) Lespedeza hay, (2) Johnsongrass hay, (3) cottonseed hulls, (4) shelled corn, (5) redtop hay.
5. (1) Soybean seed, (2) wheat middlings, (3) soybean meal, (4) cottonseed, (5) dried beet pulp.
6. (1) Fish meal, (2) sweet potatoes, (3) barley grain, (4) cowpea hay, (5) alfalfa stem meal.
7. (1) Tankage with bone, (2) meat scrap, (3) oat hay, (4) wheat grain, (5) rice hulls.

8. (1) Fresh cow's milk, (2) feather meal, (3) linseed oil meal, (4) grain sorghum grain, (5) sericea hay.
9. (1) Citrus molasses, (2) wheat grain, (3) wheat bran, (4) shelled corn, (5) millet hay.
10. (1) Dried citrus pulp, (2) sorgo silage, (3) alfalfa leaf meal, (4) barley grain, (5) peanut hay.
11. (1) Bahiagrass pasture, (2) fescue hay, (3) alfalfa stem meal, (4) ground snapped corn, (5) peanut hulls.
12. (1) Shelled corn, (2) wheat bran, (3) ground oats, (4) barley grain, (5) grain sorghum grain.
13. (1) Cottonseed meal, (2) soybean meal, (3) linseed oil meal, (4) fish meal, (5) peanut oil meal.
14. (1) Soybean seed, (2) dried skimmed milk, (3) soybean meal, (4) dried whole milk, (5) cottonseed meal.
15. (1) Bermudagrass hay, (2) crimson clover pasture, (3) lespedeza hay, (4) oat hay, (5) timothy hay.
16. (1) Fresh cow's milk, (2) tankage, (3) fresh skimmed milk, (4) alfalfa hay, (5) alfalfa pasture.
17. (1) Alfalfa meal, (2) alfalfa leaf meal, (3) alfalfa stem meal, (4) alfalfa hay, (5) alfalfa pasture.
18. (1) Shelled corn, (2) ground snapped corn, (3) corn cobs, (4) corn silage, (5) corn gluten meal.
19. (1) Cottonseed, (2) cottonseed hulls, (3) cottonseed oil, (4) cottonseed meal, (5) dehulled cottonseed.

VII. Indicate by number which each of the following is:

1. A micro mineral.
2. A macro mineral.
3. An essential amino acid.
4. A nonessential amino acid.
5. Other.

1. Riboflavin.	15. Magnesium.	28. Fluorine.
2. Thyroxin.	16. Folacin.	29. Trypsin.
3. Iodine.	17. Creatinine.	30. Biotin.
4. Leucine.	18. Arginine.	31. Choline.
5. Insulin.	19. Selenium.	32. Manganese.
6. Methionine.	20. Para-amino-ben-	33. Proline.
7. Niacin.	zoic acid.	34. Threonine.
8. Ascorbic acid.	21. Inositol.	35. Olein.
9. Cobalt.	22. Tocopherol.	36. Pepsin.
10. Cyanocobalamin.	23. Pyridoxine.	37. Molybdenum.
11. Lysine.	24. Cystine.	38. Potassium.
12. Inulin.	25. Stearin.	39. Lipase.
13. Histidine.	26. Glutamic acid.	40. Thiamine
14. Urea.	27. Phenylalanine.	hydrochloride.

VIII. Using numbers, match the following:

1. Same as tocopherol.	1. Vitamin A.
2. Thiamine hydrochloride.	2. Vitamin B_1.
3. Used to spare methionine.	3. Folic acid.
4. Prevents nightblindness.	4. Vitamin B_6.
5. Same as folacin.	5. Vitamin D.
6. Functions in blood clotting.	6. Choline.
7. Precursor of vitamin A.	7. Pantothenic acid.
8. Resembles a carbohydrate in composition.	8. Vitamin C.
9. Prevents goose-stepping in pigs.	9. Vitamin B_2.
10. Same as pyridoxine.	10. Vitamin E.
11. Originally known as APF.	11. Carotene.
12. Rendered unavailable by raw egg white.	12. Vitamin B_{12}.
13. $C_7H_7O_2N$.	13. Niacin.
14. Same as riboflavin.	14. Vitamin K.
15. Same as nicotinic acid.	15. Biotin.
16. Same as ascorbic acid.	16. Inositol.
17. Anti-rachitic factor.	17. Para-aminobenzoic acid.

IX. Indicate by number which of the following materials, on an as-fed basis, would be among the following:

1. 16 highest in % protein.
2. 16 highest in % phosphorus.
3. 12 highest in % calcium.
4. 12 highest in carotene content/kg.

1. Ground limestone.	19. Cottonseed.
2. Cottonseed hulls.	20. Cane sugar.
3. Orchardgrass pasture.	21. Fresh cow's milk.
4. Tankage with bone.	22. Grain sorghum grain.
5. Ground corn cob.	23. Crimson clover pasture.
6. Peanut kernels.	24. Lespedeza hay.
7. Cottonseed oil.	25. Defluorinated phosphate.
8. White potatoes.	26. Feather meal.
9. Wheat bran.	27. Sorgo silage.
10. Oat straw.	28. Steamed bonemeal.
11. Corn starch.	29. Dried skimmed milk.
12. Wheat grain.	30. Corn silage.
13. Soybean seed.	31. Fish meal.
14. Dried citrus pulp.	32. Barley grain.
15. Cane molasses.	33. Red clover hay.
16. Bermudagrass hay.	34. Alfalfa meal.
17. Alfalfa hay.	35. Sweet potatoes.
18. Soybean meal.	36. Yellow shelled corn.

X. From the figures below, select the proper answer for each of the following:

1. Average percentage of nitrogen in protein.
2. What times nitrogen equals crude protein?
3. Percentage of nitrogen in pure urea.
4. Crude protein equivalent of pure urea.
5. Crude protein equivalent of urea containing 45% nitrogen.
6. Approximate number of essential amino acids.
7. Total number of naturally occurring amino acids.
8. Percentage of protein usually included in livestock rations.
9. Kg of corn required per kg of urea for protein synthesis in the rumen.
10. Maximum biological value of a mixture of equal amounts of two proteins—one with a B.V. of 60% and the other with a B.V. of 100%.

ANSWER POSSIBILITIES: 44 10 16 80 60 6.25 46.67
6.5 291.6 25+ 281+ 8–18 262 8 30–40

XI. Correlate by number the terms at the bottom with their respective definitions. (Based on the glossary of terms at the back of the book. These are only a few of the possibilities.)

1. Near the kidney.
2. Low count of red blood corpuscles.
3. Another name for vitamin C.
4. To waste or cause to waste away.
5. Pertaining to the heart and blood vessels.
6. Areas of tooth decay.
7. Inflammation of the true skin.
8. Difficult parturition.
9. To disperse small drops of one liquid into another liquid.
10. Originating within or inside the cells or tissue.
11. Causes of a disease or a disorder.
12. The undesirable effects produced by taking an excess of a vitamin.
13. A prefix denoting less than the normal amount.
14. Easily destroyed.
15. In nutrition, any chemical substance found in feed.
16. In a dying state, near death.
17. The membrane that lines the passages and cavities of the body.
18. Refers to nerves.
19. The process of forming bone.
20. A compound that can be used by the body to form another compound.
21. Feverish condition.
22. The medical treatment of disease.
23. Movement of matter through space.

24. A yellow organic compound that is a precursor of vitamin A.
25. A technique for separating complex mixtures of chemical substances.
26. Any unhealthy change in the structure of a part of the body.
27. Excessive overweight due to the presence of a surplus of body fat.
28. The colorless fluid portion of the blood, in which the cells are suspended.
29. The energy produced by an individual during physical, digestive, and emotional rest.
30. The efficiency with which a protein furnishes the proper proportions and amounts of the amino acids needed.
31. The coming together of chemical building units to form new materials in the living plant or animal.
32. A chemical compound containing nitrogen, carbon, hydrogen, and oxygen that is present in the urine and results from the metabolism of proteins.
33. Swelling of a part or the entire body due to the presence of an excess of water.
34. A substance belonging to the class of sterols that on exposure to ultraviolet light is converted to vitamin D_2.
35. A protein in the blood that contains iron and carries oxygen from the lungs to the tissues.

1. Hypervitaminosis.	18. Protein.	35. Pyrexia.
2. Work.	19. Putrefaction.	36. Etiology.
3. Ergosterol.	20. Chromatography.	37. Therapy.
4. Adrenal.	21. Atrophy.	38. Obese.
5. Dermatitis.	22. Carcinogen.	39. Antibiotic.
6. Plasma.	23. Caries.	40. Asphyxia.
7. Basal metabolism.	24. Hyper.	41. Buffer.
8. Biological value.	25. Cystitis.	42. Cardiovascular.
9. Biosynthesis.	26. Factor.	43. Chlorophyll.
10. Creatinine.	27. Lesion.	44. Cholesterol.
11. Edema.	28. Aerobic.	45. Labile.
12. Hemoglobin.	29. Mucosa.	46. Fortify.
13. Hydrolysis.	30. Abscess.	47. Hormone.
14. Dystocia.	31. Ossification.	48. Moribund.
15. Anemia.	32. Emulsify.	49. Hypo.
16. Ascorbic acid.	33. Endogenous.	50. Neuritic.
17. Carotene.	34. Precursor.	

XII. Indicate whether each statement is true or false.

1. Young animals make more efficient gains than do older animals.
2. Salt is a mineral that is deficient in most farm-produced feeds.
3. Salt should never be mixed with feed for livestock.
4. A deficiency of iron will produce a condition in young pigs known as anemia.

5. A calcium-phosphorus ratio 1:2 is about right.
6. Salt should never be provided for livestock in the loose form.
7. Copper, cobalt, manganese, and zinc are all required by livestock in micro amounts.
8. A mineral mixture consisting of 2 parts limestone flour and 1 part trace-mineralized salt would be satisfactory for cattle on pasture.
9. Vitamin A is required by all classes of farm livestock.
10. A mineral mixture of 2 parts defluorinated phosphate, 2 parts limestone, and 1 part TM salt would be satisfactory for steers on a fattening ration.
11. Carotene, vitamin A, vitamin D, vitamin K, and vitamin E are all fat soluble.
12. The ultraviolet rays of sunlight convert carotene into vitamin D.
13. Livestock on good pasture should never suffer from a vitamin D deficiency.
14. Skimmed milk is a very poor source of vitamin A.
15. Alfalfa hay is richer in vitamin D than alfalfa pasture.
16. A mature mule that is just being maintained requires no protein in the ration.
17. The feed cost of a pound of growth is greater than the feed cost of a pound of fattening.
18. Alfalfa hay is higher in percentage of calcium than shelled corn.
19. Bermudagrass hay is higher in percentage of carotene than dried skimmed milk.
20. Bonemeal is higher in percentage of calcium than of limestone.
21. All of the water soluble vitamins are referred to as the vitamin B complex.
22. Calcium and phosphorus are not required for maintenance.
23. Sunlight will convert the carotene of plants into vitamin A during the hay making operation.
24. Livestock on pasture frequently suffer from a deficiency in vitamin C.
25. Potassium is a mineral element that is frequently deficient in livestock rations.
26. A ration of cottonseed meal and cottonseed hulls would be deficient in vitamin A.
27. Green pasture is rich in carotene and vitamin D.
28. Defluorinated phosphate is high in both calcium and phosphorus.
29. Nonlegume hays in general are rich in phosphorus.
30. A cow on a ration low in carotene may produce milk that is low in this substance.
31. A cow on a ration low in calcium will produce milk that is low in this mineral.
32. Livestock can produce protein tissue in their body from fat in the ration.
33. Livestock can produce body fat from starch in their ration.
34. Livestock can produce body fat from protein in their ration.
35. The requirements for fattening are similar to those for work.
36. Energy is the principal requirement for fattening.

37. The nutrients required for growth are qualitatively quite similar to those required for fetal development.
38. Working animals require much more protein than do idle animals for repair of their muscle tissues.
39. Heat from the heat increment is valuable for keeping the body warm during very cold weather.
40. Cottonseed meal usually contains more than 36% crude protein.
41. Trace-mineralized salt is a good source of cobalt and copper.
42. Livestock can store considerable quantities of vitamin A in their livers.
43. Most of the B vitamins are synthesized in the rumen of ruminants.
44. Livestock are usually fattened before being sent to market because this is usually a profitable practice.
45. The nutrient requirements of work are similar to those of fattening.
46. The principal nutrient requirement of work is plenty of net energy.
47. Approximately 1/2 of all feed fed to livestock on the average goes for maintenance.
48. The grains in general are good sources of vitamin D.
49. Green pasture is high in vitamin A value.
50. Yellow butter always contains more vitamin A than white butter.
51. Deficiency of vitamin A in cattle may affect rate of gain.
52. A fluorine deficiency in cattle may be avoided by feeding ordinary trace-mineralized salt.
53. The hays usually contain more calcium than phosphorus.
54. A deficiency of calcium in the ration may result in the disease known as osteomalacia.
55. Along with calcium and phosphorus, vitamin D is an important constituent of bone.
56. Skimmed milk is a very good source of vitamin D.
57. Copper, iodine, fluorine, and zinc are all required by livestock in very small amounts.
58. A calcium-phosphorus ratio of 1:2 may result in a calcium deficiency.
59. A cow on a ration low in quality of protein will produce milk that is low in protein quality.
60. The nutrients required for growth are qualitatively similar to those required for maintenance.
61. The percentage of crude protein in a feed may exceed 100% under certain conditions.
62. Mixing two proteins together will always improve their overall biological value.
63. Livestock have no metabolic requirement for nonessential amino acids.
64. A protein is said to be of good quality if it is high in digestibility.
65. Urea should not be fed to swine because it is very toxic to this class of livestock.
66. Cottonseed meal will kill cattle if it is fed to them in large amounts.

67. What used to be referred to as the "animal protein factor" (APF) is now known to be vitamin B_{12}.
68. Phosphorus makes up approximately 1/4 of the mineral content of an animal's body.
69. Almost all farm-produced feeds should be supplemented with salt for livestock.
70. Caution should be used in feeding "hot meal" to pregnant cows since it will sometimes cause abortion.
71. Steamed bonemeal is a good source of protein for livestock.
72. Ground limestone and oystershell flour are about equal in feeding value.
73. Steamed bonemeal, defluorinated phosphate, and oystershell flour are all good sources of phosphorus for livestock.
74. A mixture of equal parts of trace-mineralized salt, steamed bonemeal, and ground limestone would make a good mineral mix for cows on pasture.
75. Legume roughages are, in general, good sources of calcium.
76. Grains are usually higher in phosphorus than in calcium.
77. Hays are usually higher in calcium than phosphorus.

XIII. One hundred pounds of urea is the crude protein equivalent of how many pounds of 44% soybean meal?

SOLUTION:

Crude protein content of 100 lb of 44% SBM $\quad = 44.6$ lb (from the appendix table)

Crude protein equivalent of 100 lb of urea (45%N) $= 45 \times 6.25 = 281.25$ lb

$$\frac{281.25 \times 100}{44.6} = \quad 630.61 \text{ lb}$$

XIV. A feed company has 30 tons of urea (45% N) on hand. This would be the crude protein equivalent of how many tons of 41% cottonseed meal?

Crude protein content of one ton of 41% CSM $\quad = \dfrac{41.0 \times 2000}{100}$

$$= 820.0 \text{ lb}$$

Crude protein equivalent of 30 tons urea (45% N) $\quad = \dfrac{281.25 \times 2000 \times 30}{100}$

$$= 168{,}750 \text{ lb}$$

$$\frac{168{,}750 \text{ lb}}{820 \text{ lb}} = \quad 205.8 \text{ tons}$$

XV. **If urea is $180.00 per ton and shelled corn is $2.24 per bushel, at what price per ton for 44% soybean meal would it become economical to use urea in the ration for finishing steers?**

SOLUTION:

1 lb urea + 6.5 lb corn $\leq$ 7.5 lb SBM

@ $180.00 per ton, urea is 9¢ per lb

@ $2.24 per bu, corn is 4¢ per lb

$$\frac{(1 \times 9) + (6.5 \times 4)}{7.5} = 4.667¢ + \text{ per lb or } \$93.33 + \text{ per ton } (answer)$$

XVI. **If 44% soybean meal is $119.00 per ton, urea (45% N) is $200.00 per ton, and shelled corn is $3.00 per bushel, would it be economical to use urea in a steer-feeding program under these circumstances?**

ANSWER:

Since 7.5 lb of SBM can be bought more cheaply than 1 lb of urea plus 6.5 lb of corn, it would not be economical to use urea in the ration.

XVII. **Calculate the percentage of phosphorus in tri-calcium phosphate.**

SOLUTION:

Chemical formula of tri-calcium phosphate
$$Ca_3(PO_4)_2$$

Atomic weights	
Calcium	40
Phosphorus	31
Oxygen	16

Molecular weight of $Ca_3(PO_4)_2$ = $(3 \times 40) + (2 \times 31) + (2 \times 4 \times 16)$
= 120 + 62 + 128 = 310

% phosphorus in $Ca_3(PO_4)_2$ = $\frac{62}{310} \times 100 = 20.0\%$ (answer)

XVIII. **What is the ratio of calcium to phosphorus in tri-calcium phosphate?**

ANSWER:

From the solution to Problem XVII, it is apparent that the calcium-phosphorus ratio in tri-calcium phosphate is

120:62 or 2:1.03

XIX. Calculate the percentage of calcium in pure calcium carbonate ($CaCO_3$).

ANSWER: 40%

XX. The molecular formula for the amino acid lysine is $C_6H_{14}N_2O_2$. What is the percentage of nitrogen in this amino acid?

SOLUTION:

Atomic weights	
C	12
H	1
N	14
O	16

$$
\begin{aligned}
\text{Molecular weight of lysine} \ &= \ (6 \times 12) + (1 \times 14) + (2 \times 14) + (2 \times 16) \\
&= \quad 72 \quad + \quad 14 \quad + \quad 28 \quad + \quad 32 \\
&= \quad 146
\end{aligned}
$$

$$
\text{\% N in lysine} \ = \ \frac{28}{146} \times 100 = 19.18\% \ (answer)
$$

XXI. Calculate the percentage of nitrogen in the amino acid cystine, which has the following molecular formula: $C_6H_{12}N_2O_4S_2$.

ANSWER: 11.67%

20 Feeds and Feed Groups— General

Numerous different products have been used from time to time over the years for feeding purposes. However, a relatively limited number of these products make up the bulk of the nation's feed supply. This fact is borne out by Table 20-1, which presents annual usage figures for the country's more important feeds.

I

It will be noted that the feed grains make up about 80% of the feed concentrates. Of these, corn is by far the most important with 14 times as much corn being fed as grain sorghum, which is in second place. While oats and barley are important feed grains, both make a rather minor contribution to the overall feed supply. Very little wheat grain is being used as feed under present-day circumstances, but a considerable amount of wheat mill by-products is used for feeding purposes. The molasses also make an important contribution to our overall supply of energy feeds.

II

Soybean meal is by far the most important high-protein feed with about 16 times as much of it being available as cottonseed meal, which is second among the high-protein feeds in tonnage available for feeding purposes. However, tankage and meat

Table 20-1
THE MORE IMPORTANT FEEDS IN THE UNITED STATES[1]
(with figures for the amount of each fed in 1975 contrasted with 1994)

	1975	1994
	Million metric tons	
Feed grains		
Corn	91.0	140.3
Grain sorghum	12.6	10.2
Oats	4.1	4.8
Barley	7.8	3.3
Wheat	0.4	7.4
	Thousand metric tons	
Oil seed meals		
Soybean meal	14,163	24,114
Cotton seed meal	1,148	1,466
Linseed meal	85	104
Peanut meal	284	164
Sunflower meal	108	434
Canola meal	NA	950
Animal protein feeds		
Tankage and meat meal	1,815	2,315
Fish meal and solubles	461	236
Milk products	212	408
Grain protein feeds		
Gluten feed and meal	1,340	1,098[2]
Others		
Wheat millfeeds	4,508	6,591
Rice millfeeds	422	658
Alfalfa meal	1,227	190
Fats and oils	582	1,019
Miscellaneous byproducts	494	1,404

[1]Source, *Feed Situation and Outlook Yearbook,* FDS-1997, Economic Research Service, USDA, March, 1997, Tables 29 and 30.
[2]1993 figure

meals, along with the gluten feeds and meals, make important contributions to the high protein-feed supply. The fish meals and the dried milk products, along with linseed oil meal, peanut oil meal, and copra meal, also make significant contributions.

III

Alfalfa and alfalfa mixtures make up approximately 60% of the nation's hay production. Of the many different types of hay that go to make up the other 40% of the country's overall hay supply, mixtures of clover and timothy, coastal bermudagrass, wild hay, and grain hay are probably the most important from a tonnage standpoint.

IV

On an air-dry tonnage basis, about three times as much hay is produced as silage. Most of the silage is either corn or sorghum, with about 16 times as much corn as sorghum silage being made. While the figures in Table 20-1 for the hays and silages are for production rather than use, essentially all hay and silage produced is ultimately fed to livestock, and very little is ever exported. Consequently then, on the average, annual production of hay and silage is essentially equal to annual use of these products for feeding purposes.

On the following pages are presented some of the more important facts about most of the more important feed materials. In the presentation of these facts the different feeds have been considered by groups, based on certain similarities in composition and feeding value. Within each group the various feeds are taken up in no particular order except that in general the more important feeds within each group have been given first consideration.

21 Air-Dry Energy Feeds

From both a quantitative as well as an economic standpoint the energy nutrients are by far the most important as a group in the practical feeding of livestock. Consequently those feeds that serve as good sources of digestible energy make up the most important feed group and are referred to as the energy feeds. Since the bulk of such feeds are handled through commercial channels, they are normally free of excess moisture or, in other words, are ordinarily in the air-dry form. It is the purpose of section 21 to recognize most of the feeds making up this category and to briefly review some of the more important characteristics of each.

I ____ Shelled corn.

A. Corn is the most extensively produced feed grain (Figure 21-1).

B. It is the most widely grown feed grain crop.

C. It is the most widely used energy feed.

D. It is unexcelled as an energy feed.

E. Of the common feed grains, it excels in pounds of TDN produced per acre.

F. Corn is extremely low in calcium but fair in phosphorus content.

FIGURE 21-1
Most corn grown for grain is harvested with combines today. Note harvested shelled corn in the grain tank; also the
cobs and shucks, as well as the stalks and leaves, remain in
the field. (Courtesy of Ford Tractor Operations, Ford Motor
Co., Troy, Mich.)

G. It is quite deficient in vitamin B_{12} and low in riboflavin and pantothenic acid.

H. It must be supplemented with protein for most classes of livestock.

I. It is especially low in the essential amino acids methionine, lysine, and tryptophan.
See Figure 21-2.

II

Sorghum grain, including hegari grain, kafir grain, milo grain, and most other types and
varieties of grain sorghum.

A. About 10–12% as much grain sorghum as corn is currently produced in the United
States.

B. Most grain sorghum production is located in the semi-arid western regions where
corn does not do well.

C. Most grain sorghum is very similar to shelled corn in chemical composition, except
that most grain sorghum is slightly higher in protein and contains little, if any, carotene.

FIGURE 21-2
The four major feed grains. (Courtesy of the University of
Georgia College of Agriculture Experiment Stations)

D. Grain sorghum may be used to replace 50% or more of the corn in the ration for most livestock without affecting animal performance.

E. If grain sorghum is used to replace all the corn in the ration, feed efficiency and gains may be reduced by as much as 10% or possibly more. However, this can be largely overcome by steam rolling or feeding as high-moisture grain.

F. Grain sorghum should be rolled or ground for most classes of livestock.

G. Grain sorghum is quite drought resistant.

H. It usually presents a harvesting problem. If allowed to dry in the head, much is lost to birds. If harvested early to prevent bird losses, it must be artificially dried or acid treated for safe storage, or stored as high-moisture grain in a silo.

I. Certain bird-resistant varieties of grain sorghum have been developed, but these have proved to be of inferior feeding value because of their low palatability and low digestibility.

III   Oats grain.

A. About 5–6% as much oats as corn is currently fed to livestock in the United States.

B. Oats are widely grown, but a large portion of oats grain is produced in the midwestern and north central states.

C. Oats are higher than corn in crude fiber (10.8% vs. 2.6%, as fed) and accordingly are lower in TDN (68% vs. 77%, for ruminants, as fed).

D. Oats are also somewhat higher than corn in crude protein (10.8% vs. 9.6%, as fed) and a little higher in Ca and P.

E. Oats contain little, if any, carotene.

F. Oats are used extensively in rations for horses, young growing stock, show stock, and breeding animals.

G. Oats are not a good fattening feed and ordinarily are used only to a limited extent, if at all, in fattening rations.

H. Oats are usually rolled, crimped, or ground for feeding.

IV   Barley grain.

A. About 4–5% as much barley as corn is currently fed to livestock in the United States.

B. Most of the barley is produced in the north central and far western states.

C. Barley is similar to oats in protein content and intermediate to oats and corn in content of fiber, TDN, calcium, and phosphorus.

D. Barley is used extensively along with or in the place of oats in rations for horses, young growing stock, show stock, and breeding animals.

E. Barley may be used to replace up to one-half of the corn in rations for fattening animals without materially affecting their performance.

F. When barley is used to replace all the corn in fattening rations, gains may be reduced by as much as 10%.

G. Barley is usually steam rolled (flaked), crimped, or coarsely ground for feeding.

H. Barley is sometimes cooked for beef show animals—supposedly to improve its acceptability.

V Wheat grain.

A. Wheat, like corn, is widely grown in the United States.

B. Wheat currently ranks close to corn in acres produced in the United States but per acre yield is only about 35–40% of corn.

C. Over the years, wheat has been too expensive in most instances to be fed to livestock—too much in demand as a human food.

D. At times, however, for various reasons wheat becomes available at prices that make it competitive with corn and other feed grains as an economical source of energy for livestock.

E. Wheat, except for being considerably higher in protein (14.2% vs. 9.6%, as fed) and having little, if any, carotene, is quite similar to corn in composition and feeding value.

F. For best results, wheat is best mixed at relatively low levels with other grains, especially for feeding cattle and horses.

G. It is usually coarsely ground or cracked for feeding.

H. Because of the high solubility of its starch, feeding wheat to cattle is conducive to acidosis. See Figure 21-3.

VI Ground ear corn (corn and cob meal).

A. Ground ear corn consists of whole ears of corn (grain and cob) ground to varying degrees of fineness.

B. Ear corn has traditionally been considered to be about 80% grain and 20% cob, although modern-day corn sometimes runs somewhat higher than this in percentage of grain to cob.

C. Composition of ground ear corn approximates a weighted average of the composition of shelled corn and the composition of corn cobs.

D. Ground ear corn is somewhat higher in fiber and lower in TDN than ground shelled corn.

E. It is an excellent energy feed for ruminants and for horses and mules, provided other roughages are reduced sufficiently to allow for its content of cob.

F. It should not be fed to growing-fattening pigs or poultry since they can make little, if any, use of the cob, and the cob may irritate the digestive tract and open the way for intestinal infections.

G. It may be used in rations for mature swine, to limit energy intake.

H. Most corn grain is harvested with combines today, leaving the cob in the field; hence, not much ear corn is available for feeding purposes at the present time.

FIGURE 21-3
While most wheat is used for the manufacture of flour for human consumption, some of its by-products—such as wheat bran, as well as wheat shorts and wheat middlings—are widely used for feeding purposes. (Courtesy of the University of Georgia College of Agriculture Experiment Stations)

VII Ground snapped corn.

A. Ground snapped corn consists of whole ears of snapped corn (grain, cob, and shucks).

B. Snapped corn over the years has been found mostly in the southern states but very little is found even there today.

C. Handling snapped corn involves too much labor—most corn for grain is combined today, leaving cob and shuck in the field.

D. Snapped corn has traditionally been considered to be about 70% grain, 17.5% cob, and 12.5% shuck. However, modern-day corn will sometimes vary significantly from these proportions.

E. Composition of ground snapped corn approximates a weighted average of the composition of shelled corn, corn cobs, and corn shucks, depending on the proportion of each that is present.

F. Ground snapped corn is considerably higher in fiber and lower in TDN and protein than ground shelled corn; it is comparable to oats as an energy feed.

G. If available, it is a satisfactory energy feed for ruminants, horses, and mules, provided other roughages are reduced sufficiently to allow for its content of cob and shuck.

H. It should not be fed to growing-fattening pigs or poultry since they can make little, if any, use of the cob or shuck. Both may also irritate the digestive tract and open the way for intestinal infections.

I. It may be fed to mature swine, but it is not too satisfactory for this class of animals.

VIII Wheat bran.

A. Wheat bran consists primarily of the seed coat of wheat that is removed in the manufacture of wheat flour.

B. It is used in livestock feeding:

1. Primarily—

   a. As a source of bulk.

   b. As a mild laxative.

   c. As a source of phosphorus.

2. Secondarily—

   a. As a source of energy.

   b. As a source of protein.

C. It is used primarily in rations for horses, dairy cows, brood sows, and beef show animals.

D. It is seldom used in rations for feedlot steers, growing-fattening pigs, or poultry.

E. Its use in livestock feeding is usually limited to about 10% of the ration in view of its bulkiness, its laxative nature, and its usually relatively high price.

F. It is an excellent source of vitamin E and is excellent for pregnant ewes.

IX Wheat middlings and shorts.

The terms *shorts* and *middlings* have been used over the years alone and in combination with certain other descriptive terms to refer to the various wheat flour by-product feeds. However, the Association of American Feed Control Officials have specified that *wheat middlings* must consist of fine particles of wheat bran, wheat shorts, wheat germ, and wheat flour, and some of the offal from the "tail of the mill" obtained in the usual process of flour milling. It must not contain more than 9.5% crude fiber.

Wheat shorts must consist of fine particles of wheat bran, wheat germ, and wheat flour, and the offal from the "tail of the mill" obtained in the usual process of flour milling. It must not contain more than 7% crude fiber.

A. There are several types of middlings, shorts, and similar products resulting from the manufacture of wheat flour from various types of wheat by different processes and/or processors.

B. All consist of mixtures of fine particles of bran and germ, the aleurone layer, and coarse flour in varying proportions.

C. Such products are ordinarily lower in fiber and higher in TDN than bran and are used primarily as a source of energy in the ration.

D. The fiber content of such feeds will vary from about 2% to 9% with their TDN values varying accordingly.

E. Most such products are used in rations for swine since their content of flour causes them to become gummy upon being eaten, making them unrelished by other classes of livestock.

F. As a component of swine rations, such feeds are amply high in total protein content but require some supplementation of essential amino acids.

X   Dried citrus pulp.

A. Citrus pulp is a by-product of the citrus processing industry, usually consisting of the remains of the fruit after the juice has been removed, and sometimes cull fruit.

B. While sometimes fed in the wet form in the vicinity of the processing plant, it is usually dried in order to facilitate its use at distant points and during off-season periods.

C. Dried citrus pulp has, over the years, been fed mainly to dairy cattle but can also be fed to beef cattle. It is not usually fed to other classes of livestock.

D. While it is relatively high in fiber, it is ordinarily regarded primarily as an energy feed.

E. Usually it is used to make up not more than about 20–25% of the ration.

F. Fed as outlined above, it has a feeding value comparable to that of dried beet pulp, corn and cob meal, rolled barley, and similar feeds (Figure 21-4).

XI   Dried beet pulp.

A. Most of the beet pulp produced in this country is dried to facilitate its handling, shipment, and use.

B. Most of the beet pulp available in this country is fed to dairy cattle but is sometimes also fed to horses and used to fatten cattle and sheep.

C. It is used:

DRIED
CITRUS
PULP

DRIED
BEET
PULP

FIGURE 21-4
Two important by-products widely used for feeding pur-
poses—especially in dairy cow rations. (Courtesy of the
University of Georgia College of Agriculture Experiment
Stations)

   1. Primarily—

     a. As a bulk factor.

     b. As an appetizer.

     c. As a mild laxative.

   2. Secondarily—

     a. As a source of energy.

D. Usually it is used to replace not more than about 20% of the grain in the ration.

E. It is frequently reconstituted with water before being fed to livestock.

XII

Dried sweet potatoes or sweet potato meal.

A. This has been largely an experimental product to date—it is not generally available.

B. It has attracted considerable attention over the years from an experimental standpoint because sweet potatoes give high yields of digestive carbohydrates.

C. Sweet potatoes will outproduce most other crops (even corn) in yield of digestible
carbohydrates per acre—however, they are relatively costly to produce.

D. Dried sweet potatoes are high in content of digestible starch and carotene.

E.   It is low in almost all other nutritive fractions.

F.   When used to replace not more than 50% of the corn in rations for cattle and sheep, dried sweet potatoes have been found to have a feeding value approaching that of corn.

G.   At higher levels, dried sweet potatoes have a feeding value somewhat below that of corn.

H.   Swine do not seem to relish and do not perform particularly well on this product.

XIII   Dried bakery product.

A.   Dried bakery product consists primarily of stale bakery products and certain other bakery wastes that have been blended together, dried, and ground into a meal.

B.   There are companies that make a business of assembling such materials from extensive areas and converting them into the feed that is officially recognized as *dried bakery product*.

C.   Some companies' production of this type of feed amounts to thousands of tons annually.

D.   Where careful control is exercised over the blending operation, a product of considerable uniformity can be produced from such materials.

E.   Dried bakery product is similar to corn in composition, except that it is usually much higher in fat and may contain a considerable amount of salt.

F.   Dried bakery product is an effective substitute for corn in rations for cattle and swine, but in view of its relatively high salt content, its use should be limited to not more than about 20% of the total ration.

XIV   Hominy feed.

A.   Hominy feed is a by-product of the manufacture of hominy, hominy grits, and corn meal for human consumption.

B.   It consists of a mixture of the corn bran, the corn germ (with or without some of the fat removed), and varying amounts of the finer siftings of the starchy portion of the corn grain.

C.   It is similar to corn in composition, although it usually contains slightly more protein and somewhat more fat, and it is higher in fiber.

D.   Hominy feed is about equal to corn in feeding value for various classes of livestock and can be substituted for corn on about a pound-for-pound basis in most livestock rations.

    E. Hominy feed that contains the usual amount of fat will tend to produce soft pork when used as a major component of swine rations.

XV Oat groats.

    A. Oat groats are oats grain from which the hull has been removed—in other words, the oat kernel.

    B. Since most of the feeding value of oats grain is found in the kernel, oat groats are very high in feeding value.

    C. They are usually too expensive for general livestock feeding.

    D. Their use is usually restricted to special diets such as early weaning rations for pigs.

XVI Potato meal.

    A. Cull and/or surplus potatoes are sometimes dried and ground to produce a feed known as potato meal.

    B. Potato meal is satisfactory as a substitute for a part of the grain in beef and dairy rations.

    C. Potato meal produced from potatoes that were thoroughly cooked prior to or during the drying process may be used as a partial substitute for the grain in swine rations.

XVII Rye grain.

    A. Less than 0.5% as much rye as corn is produced in the United States.

    B. A large portion of the rye produced is used for bread making—hence, not much is available for feeding purposes.

    C. Rye grain is similar to corn, wheat, and grain sorghum in composition but is generally less valuable as a livestock feed.

    D. Rye grain is less palatable than other grains and is sometimes contaminated with the fungus ergot, which tends to accentuate this characteristic.

    E. Rye contaminated with ergot is sometimes toxic to livestock—especially swine— if fed at high levels.

    F. High levels of rye in dairy rations tend to produce a hard, unsatisfactory butter.

    G. All things considered, rye has its greatest feeding value when it does not make up more than about one-third of the ration.

H.   Up to this level, it will usually approach the other low fiber grains in feeding value.

I.   Rye grain should be coarsely ground or rolled when fed to livestock.

XVIII   Animal fat, feed grade.

A.   Meat slaughtering, poultry dressing, and rendering plants are frequently burdened with surpluses of animal fat that can be bought at prices that will permit its economical use in livestock feeds.

B.   Since such products are essentially pure fat, they have nutritional value primarily as a source of energy.

C.   Animal fat is used extensively in the manufacture of commercially mixed feeds (Figure 21-5) at the rate of 1% to about 7%, depending on the type of feed, as a source of energy and also for the following reasons:

1.   To reduce dustiness.

2.   To improve color.

3.   To improve texture.

4.   To improve palatability.

FIGURE 21-5
Feed grade animal fat is frequently used at low levels in mixed rations. When used, it is normally added in the melted form. (Courtesy of the University of Georgia College of Agricultural Experiment Stations)

5. To reduce machinery wear.

6. To increase pelleting rate.

D. Beef cattle can make effective use of animal fat up to about 5% of the ration.

E. Swine can make effective use of animal fat up to about 10% of the ration.

F. Animal fat to be used for feeding purposes should be treated with an antioxidant to prevent it from becoming rancid.

XIX. Dried whey.

A. Whey is the portion of milk that remains after most of the casein and fat have been removed for the manufacture of cheese.

B. Dried whey is very high in milk sugar and so must be regarded primarily as an energy feed.

C. It contains a fair amount of protein, which is of excellent quality.

D. It is also relatively high in calcium and phosphorus, as well as several of the B vitamins.

E. Dried whey is used primarily in poultry rations and in early weaning diets for other classes of livestock.

XX. Corn starch and corn oil.

A. These are relatively pure forms of starch and oil, respectively, obtained from the corn grain.

B. They are not generally used as livestock feeds.

C. They are included in the feed composition table (see appendix) to illustrate certain points for instructional purposes only.

D. Corn starch and corn oil are sometimes used as ingredients in purified diets for experimental animals.

XXI. Rice bran.

A. Rice bran consists primarily of the seed coat and germ that are removed from rice grain in the manufacture of polished rice for human consumption.

B. Rice bran, while not as high in protein, is otherwise comparable to wheat bran in feeding value.

C. Rice bran is fed primarily to dairy cows but may be fed to other classes of livestock with satisfactory results.

D. In view of its relatively high content of fat and fiber and its frequent lack of palatability, the use of rice bran is usually limited to not more than approximately one-third of the ration concentrates.

XXII Triticale

A. Triticale is a relatively new small grain crop being grown for winter grazing, silage, and feed grain production on an increasing scale in certain areas from the Lower Coastal Plain of the United States to as far north as Canada.

B. It is a man-made cereal grain produced by crossbreeding wheat (*Triticum*) and rye (*Secale*).

C. The name triticale is derived from combining a part of the generic name of each of the parent grains.

D. While produced by crossing varieties of two separate species, the triticale seed is usually viable and usable for cultural purposes.

E. Like wheat and rye, winter or spring varieties of triticale may be developed.

F. The triticale plant is similar to bearded wheat in appearance but has a somewhat longer, more drooping head.

G. The triticale grain is intermediate in general appearance to wheat and rye, but in some instances, it is larger than either and is less dense and slightly shrunken.

H. A major shortcoming of those triticale varieties presently available is that most are subject to winterkill to a greater or lesser degree depending on the variety and the severity of the winter.

I. Currently available varieties with winter hardiness are subject to severe lodging before the grain matures.

J. Reports to date on the use of triticale for winter grazing, silage, and feed grain purposes are, for the most part, based on limited research; the results are mixed and varied. Some producers, however, have experienced good yields of nutritious forage from triticale, as compared to wheat, for silage production. Also, triticale grain is reported to be higher than wheat in protein and lysine. Final judgment on triticale as a feed crop, however, will have to await the development of well adapted varieties and the evaluation of these varieties through further research for feeding purposes. See Figure 21-6.

Triticale   Wheat

FIGURE 21-6
Triticale grain (Beagle 82) pictured in contrast with wheat
grain. (Courtesy of Otha M. Hale of the University of Georgia
Coastal Plain Experiment Station)

22 High-Protein Feeds

Most of the energy feeds need to be supplemented with a source of protein in order for a ration to be adequate in this nutrient. As a result, limited quantities of one or more feeds high in protein normally must be included in the ration. Since feeds high in protein are usually in limited supply, protein is the most expensive macro-nutrient per pound except for fat, which, because of its high energy content, is frequently higher than protein in price. Consequently, feeds that are good sources of protein are normally thought of as a group under the heading of high-protein feeds. It is the purpose of section 22 to recognize some of the important feeds in this category and to briefly review the more important characteristics of each.

I. Those of animal origin.

A. Dried skimmed milk.

1. Dried skimmed milk is defatted, dehydrated cow's milk.

2. It usually contains around 34% protein.

3. The protein of skimmed milk is of good quality (high biological value).

4. Dried skimmed milk is usually too high priced for general livestock feeding.

5. It is sometimes used in early weaning diets for calves and pigs.

6. It is the basis of veal calf formulations.

B. **Digester tankage.** Also called *wet rendered tankage* or just *tankage.*

1. Tankage is primarily a by-product of the meat packing industry—also available from dead-animal rendering plants.

2. It consists of otherwise unusable animal tissue, including bones, which has been cooked under steam pressure, partially defatted, dried, and ground.

3. It usually contains from 55% to 60% protein.

4. The protein quality varies according to the processing, but it is usually good.

5. Because of its bone content, it is also a good source of calcium and phosphorus.

6. Digester tankage is used primarily in feeds for swine, but because of a low content of certain essential amino acids its use is usually limited to not more than about 5% of the ration.

C. **Tankage with bone.**

1. This product is similar to digester tankage except that it contains a greater proportion of bone, and consequently is higher in calcium and phosphorus and lower in protein.

2. It is used primarily in rations for swine, but because of a low content of certain essential amino acids and a high Ca and P content, its use is usually limited to not more than about 4% of the ration.

D. **Meat scrap or meat meal.**

1. This product is similar to digester tankage in origin except that it is cooked in steam jacketed kettles in its own fat—not under steam pressure.

2. It is similar to digester tankage in overall composition and general feeding value.

3. No dried blood is normally added to meat scrap, making it a more acceptable product than tankage.

4. The trend is toward the production of more scrap and less tankage.

5. Meat scrap is used primarily in rations for swine and poultry, and having somewhat better protein quality than tankage, it can be used a little more liberally in the ration.

E. **Meat and bone scrap or meal.**

1. This product is similar to meat scrap except it contains more bone, and consequently is higher in calcium and phosphorus and somewhat lower in protein. (Figure 22–1.)

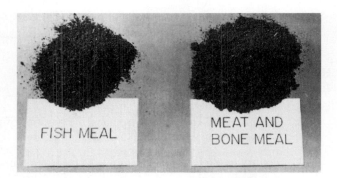

FIGURE 22–1
Two of the more commonly used high-protein feeds of animal origin. (Courtesy of the University of Georgia College of Agriculture Experiment Stations)

2. It is used primarily in rations for swine and poultry, but because of a low content of certain essential amino acids and a high level of Ca and P, its use is usually limited to not more than about 5% of the ration.

F. **Fish meal.**

1. Fish meal consists of fish or fish by-products that have been dried and ground into a meal.

2. There are several types, depending on the type of fish used.

3. The protein content of fish meal is usually around 60%.

4. Fish meal protein is usually of good quality.

5. It is also normally high in calcium and phosphorus.

6. It is used primarily in rations for swine and poultry.

G. **Feather meal.**

1. Feather meal consists of poultry feathers that have been cooked under steam pressure, dried, and ground into a meal.

2. This product is extremely high in protein, usually containing well over 80%.

3. Feather meal protein is more than 75% digestible if it has been properly processed.

4. The amino acid balance is not too desirable for monogastrics, except in minimal amounts.

5. The senior author of this text noted that hydrolyzed poultry feather meal is an excellent source of UIP, and therefore it has become a common ingredient in diets of beef and dairy cattle.

6. This product is used primarily in swine and poultry feeds, but because of a low content of certain essential amino acids, its use is usually limited to about 2.5% of the ration.

H. Poultry by-product meal.

1. This is a by-product of the poultry dressing industry.

2. It is produced from otherwise unusable portions of poultry carcasses that are cooked, dried, and ground into a meal.

3. Poultry by-product meal is similar to meat scrap in appearance, composition, and feeding value.

4. It is used primarily in rations for swine and poultry.

I. Dried whole milk.

1. Dried whole milk is fresh whole milk that has been dried to a powder.

2. It is a good source of excellent-quality protein (25% +) and at the same time is very high in digestible energy.

3. Dried whole milk is so much in demand as a human food that it is usually too expensive to be fed to livestock.

4. It is excellent as a component of milk replacers and early weaning diets for calves and pigs if and when its price will permit its use for these purposes.

J. Blood meal.

1. Blood meal is coagulated packing house blood that has been dried and ground into a meal.

2. It is extremely high in protein (80% +), but the protein is lower in digestibility and quality than most other animal protein feeds.

3. Blood meal is not very palatable to most livestock.

4. For the above reasons blood meal is not very popular as a protein supplement, and its use as a ration component is usually restricted to a relatively low level.

5. Blood meal is especially desirable in ruminant formulations because of its high UIP (undigestible intake protein) content. Early lactation diets for dairy cattle need more than 7% UIP (NRC, 1989).

6. It is one of the richest natural sources of the amino acid lysine (7% or more). Therefore, it is often used in pig and poultry diets.

II Protein feeds of plant origin.

A. Soybean meal (SBM) (Originally referred to as soybean oil meal).

FIGURE 22–2
Two of the more commonly used high-protein feeds of plant
origin. (Courtesy of the University of Georgia College of
Agriculture Experiment Stations)

1. Soybean meal consists of fat-extracted soybeans that have been ground to a meal and sometimes pelleted (Figure 22–2).

2. Most soybean meal is solvent-extracted today.

3. It is of two grades: 44% and 49% protein.

4. 49% SBM is produced from dehulled soybeans; the hull is added back to produce the 44% product.

5. 49% SBM is produced primarily for the broiler industry, secondarily for swine. It is satisfactory for cattle if the price is right.

6. SBM is an excellent source of protein for all livestock classes, except possibly very young animals. Its protein is of high biological value.

7. It is the most widely used high-protein feed, making up approximately two-thirds of the country's high-protein feed supply.

B. Cottonseed meal (CSM).

1. Cottonseed meal consists of dehulled, fat-extracted cottonseed ground to a meal with a certain amount of ground cottonseed hulls added.

2. It is usually of either 41% or 36% crude protein grade, depending on the amount of hull added.

3. 41% protein is a practical maximum—36% protein is the legal minimum in most states.

4. Cottonseed meal is an excellent high-protein feed for ruminants.

5. It may kill growing swine if included in the ration at levels of more than 9% because of a toxic factor known as *gossypol*.

C. **Peanut oil meal (POM) or peanut meal (PNM).**

    1. Peanut oil meal consists of fat-extracted peanut kernels ground to a meal with a certain amount of ground peanut hulls added.

    2. Its composition and feeding value vary considerably depending on the quality of the nuts, the method of fat extraction used, and the amount of hull included.

    3. Good quality peanut oil meal is comparable to soybean meal in feeding value.

    4. It will usually run about 45% or more protein.

    5. It is satisfactory as a source of protein for all livestock classes.

    6. Not very much is available for livestock feed.

D. **Linseed (oil) meal (LSM, LSOM, or LOM).**

    1. Linseed meal consists of fat-extracted flax seed.

    2. It is produced mostly in the Dakotas, Minnesota, and Texas.

    3. It is a satisfactory source of protein for almost all livestock classes, when available, and contains about 35% protein.

    4. Linseed meal is excellent for putting bloom on animals being fitted for show.

    5. Very little linseed meal is used outside of the flax-producing sections except in fitting beef cattle for show

E. **Corn gluten meal.**

    1. Corn gluten meal is a by-product of corn starch and corn oil manufacture.

    2. It is usually about 40% + in protein content.

    3. It is satisfactory as a source of supplemental protein for ruminants (Figure 22–3).

    4. It has a low essential amino acid supplemental value with most grains—hence, it is very poor as a protein supplement for nonruminants.

    5. Corn gluten is one of the richest natural sources of xanthophyll and is an excellent source of yellow coloring in the shanks and body fat of broilers. The yellow coloring is also indicative of beta carotene, the precursor of vitamin A. Although poultry convert beta carotene to vitamin A very efficiently, it is not safe to count on that source to meet the vitamin A needs of poultry, because the beta carotene of yellow gluten meal is very susceptible to destruction by oxidation. Research at Purdue University has shown that very little vitamin A activity remains in corn that is stored accessible to oxygen for more than one year.

F. **Brewers dried grains.**

    1. Brewers grains is the residue that remains after most of the starches and sugars have been removed from the barley and possibly other grains in the brewing process.

FIGURE 22–3
Three high-protein feeds frequently used—especially in rations for dairy cows. (Courtesy of the University of Georgia College of Agriculture Experiment Stations)

2. Some may be fed in the wet form near the brewery; however, most brewers grains is dried to facilitate handling, shipping, and storage.

3. Brewers dried grains is fed mainly to dairy cows (Figure 22–3), primarily as a source of protein and secondarily as a source of energy, up to about one-third of the concentrate mix.

4. It sometimes is used to replace a part of the grain in rations for horses, but it is too high in protein to be used very extensively for this class of livestock.

5. In view of its high fiber content, brewers dried grains is seldom fed to swine and poultry.

G. Distillers dried grains.

1. Distillers grains consist of the residue that remains after the alcohol has been distilled off and other liquid materials have been removed from grain processed for alcohol production.

2. It is sometimes fed in wet form near the distillery; however, it is usually dried to facilitate handling, shipping, and storage.

3. Distillers dried grains is fed mainly to dairy cattle (Figure 22–3), primarily as a source of protein and secondarily as a source of energy.

4. It is also satisfactory as a source of protein for beef cattle and sheep.

5. In view of its fiber content, distillers dried grains are seldom fed to swine and poultry.

H. Copra meal.

1. Copra meal is what remains after the dried meats of coconuts have been subjected to fat extraction, and ground.

2. While it is regarded as a protein supplement, it is rather low in protein (21% +) compared with other high-protein feeds.

3. The quality of its protein is not as good as that of soybean meal.

4. The relatively low quality of its protein and its relatively high content of fiber restrict its use in rations for swine and poultry.

5. Copra meal makes an excellent dairy feed and is largely fed to this class of livestock.

I. Safflower meal without hulls.

1. This is what remains after most of the hull and the oil have been removed from safflower seed.

2. It is fairly high in protein (40% +), but the quality of its protein is not as good as that of soybean meal.

3. The relatively low quality of its protein and its relatively high fiber content restrict the use of this product for swine and poultry.

4. It is a satisfactory source of supplemental protein for other classes of livestock.

J. Sesame oil meal.

1. Sesame oil meal is produced from what remains following the production of oil from sesame seed.

2. It is comparable to soybean meal in protein content, but the protein is not as good as soybean meal for supplementing farm grains for nonruminants.

3. When available, it makes a very satisfactory source of supplemental protein for ruminants.

K. Sunflower meal or sunflower oil meal.

1. Sunflower meal is produced from what remains following the production of oil from dehulled sunflower seed.

2. It is comparable to soybean meal in protein content, but it is not as good as soybean meal for supplementing farm grains for nonruminants.

3. When available, it makes a very satisfactory source of supplemental protein for ruminants.

L. Rapeseed Meal (RSM)*.

1. Rapeseed meal is what remains following the extraction of the oil from rapeseed.

2. Rapeseed is produced extensively in Canada where the meal is widely used as a protein feed for livestock and poultry.

3. While three different processes (expeller, prepress solvent, and direct solvent) are currently employed for extracting the oil from rapeseed, the prepress solvent is presently the most widely used.

4. Rapeseed meal contains from 36% to 39% protein, which compares favorably with soybean meal protein in amino acid balance. Rapeseed meal protein is somewhat lower in lysine and higher in methionine than the protein of soybean meal.

5. Rapeseed meal has been limited in its use as a high-protein feed over the years.

   a. Meals produced from the earlier grown varieties of rapeseed were high in glucosinolates, the hydrolytic products of which (i.e., isothiocynanates, oxyzolidinethiones and nitrites) are more or less toxic to most livestock, causing lowered performance and possibly goiter when used as a major source of protein in the ration.

   b. Since most or all of the hulls of rapeseed remain in the meal with the extraction methods followed to date, rapeseed meal is relatively high (11.0–13.0%) in crude fiber and low in digestible energy.

   c. Problems of acceptability by certain classes of livestock have been experienced with certain batches of rapeseed meal.

6. Considerable progress has been made during the past few years toward overcoming the shortcomings of rapeseed meal as a livestock feed.

   a. New low-glucosinolate varieties of rapeseed have been developed that have made it possible to nearly eliminate the thyrotoxicity problem of rapeseed meal. Tower, Candle, and Regent are all such varieties.

---

* The information in this section was obtained for the most part from Publication No. 51, Rapeseed Meal Association of Canada.

b. New, improved procedures that provide better control of moisture, temperature, time, and pH during oil extraction tend to keep the toxic forms of glucosinolates in rapeseed meal at nonharmful levels.

c. New varieties of rapeseed are being developed that have a thin yellow hull as contrasted to the thicker darker hull of the older varieties, thus making it possible to produce an attractive, lower-fiber, higher-energy meal.

d. Also, methods are being developed for removing a portion of the hull from the rapeseed before subjecting it to the oil extraction process, thereby producing a meal that is lower in fiber and higher in digestible energy. The crude protein level is also improved.

7. While the extent to which rapeseed meal may be used in livestock rations in the future will depend on its glucosinolate and fiber content. With the use of new improved varieties and oil extraction procedures it appears that this meal may be used as follows:

a. Rapeseed meal may serve as the sole source of supplementary protein for beef cattle and sheep.

b. It may make up as much as 25% of the grain mixture for lactating cows and a similar percentage of the starter ration for young calves.

c. Good quality rapeseed meal may be used in swine rations as follows:

For early weaning   —up to 20% of the ration
For growing & finishing   —up to 15% of the ration
For breeding stock   —up to 10% of the ration

d. No reports on the use of rapeseed meal as a horse feed are available to date.

8. With meals that are known to be high in glucosinolates or for which glucosinolate levels are not known, the usage of such meals should be reduced to about 50% of the above levels to be on the safe side.

M. Soybeans, cottonseed, and peanuts (Figure 22–5).

1. Soybeans, cottonseed, and peanuts are usually subjected to oil extraction before being used for feeding purposes. (Figure 22–4.)

2. Sometimes under circumstances of low vegetable oil prices and high protein meal prices, it becomes more economical to use the unextracted beans, seed, or nuts as a source of supplemental protein than it is to use the oil meal. (Figure 22–5.)

3. While not as high as the respective oil meals in crude protein, the oil seeds are still fairly potent sources of protein, and with a higher fat content they contribute more energy to the ration on a pound-for-pound basis than do the meals.

FIGURE 22–4
Mature soybeans being combined at harvest time. (Courtesy
of Sperry-New Holland, New Holland, Pa.)

SOYBEANS   COTTONSEED   PEANUTS

FIGURE 22–5
The three major oil seeds. (Courtesy of the University of
Georgia College of Agriculture Experiment Stations)

4. Whole soybeans (39.2% crude protein) have been found to be a very satisfactory source of supplemental protein for beef cattle, horses, and sheep and also for dairy cattle except that large amounts of soybeans will tend to produce a soft butter.

5. Whole soybeans are not a satisfactory source of supplemental protein for swine unless the beans are thoroughly cooked, which significantly improves the digestibility and quality of the protein. Raw soybeans contain a product that prevents the normal functioning of the enzyme trypsin in the hog's digestive tract. Cooking destroys this trypsin inhibitor.

6. Whole soybeans, even though they are cooked, will tend to produce soft carcasses in growing-fattening pigs if they make up more than about 10% of the ration.

7. Whole cottonseed (22.0% crude protein) has been used with satisfactory results as a source of supplemental protein for beef cattle and presumably could be used in a similar manner for dairy cattle, horses, and sheep.

8. Whole cottonseed is not a satisfactory feed for swine because of its high fiber content (19.1%) and because of its tendency to produce gossypol poisoning.

9. Harvested peanuts (28.4% crude protein) are usually too valuable as a human food to be fed to livestock, but there is no apparent reason why they could not be used as a source of supplemental protein for all classes of livestock, should the price be attractive.

10. If used at levels above about 5% of the ration, peanuts are likely to cause dairy cows to produce soft butter and growing-fattening pigs to produce soft carcasses.

11. Peanuts are sometimes used as a swine feed through a hogging-off program, in which case they serve very effectively as the animal's primary source of both protein and energy except that soft carcasses will almost invariably result.

III   Commercial protein supplements.

A. These are specially formulated protein supplements prepared by commercial companies.

B. Each is usually for a particular class of livestock.

C. They are usually blends of animal and vegetable high-protein feeds, and probably include urea, if for ruminants.

D. They may also include minerals, vitamins, and/or antibiotics.

E. Those for swine will usually contain 30–40% crude protein while those for ruminants may run 2–3 times this high in crude protein as the result of a relatively high content of urea.

IV   Hot meals.

A. These are usually mixtures of a high-protein feed and plain salt, with the latter added at levels ranging from 10% to 50%, but usually from 20% to 30%.

B. They are used primarily for beef brood cows on low-protein pasture.

C. The salt serves as a consumption regulator, thus permitting free choice feeding.

D. The level of salt is adjusted to bring about the desired level of consumption.

E. With such a feeding program, a cow will consume up to 5 times, and possibly more, its normal salt consumption.

F. The extra salt will cause no harmful effects as long as plenty of water is available.

G. The cost of the extra salt is justified by the feeding convenience provided. Only a short trough is required with this method of feeding; trough space per animal is much less. All animals share in the feed on an equal basis.

H. Some users of hot meal report an increase in appetite for low-grade roughage from its use.

I. This method of supplying supplement does not facilitate checking the herd at feeding time.

V Liquid supplements.

A. This is simply an alternative form in which to provide supplemental nutrients.

B. Nutrients are supplied either in water solution or in suspension.

C. Minerals and vitamins as well as protein may be included.

D. Blackstrap molasses and urea are usually the primary components of liquid cattle supplements, which usually carry a crude protein equivalent content of about 30% to 35%.

E. Such supplements are used most extensively for beef brood cows on low-protein forage.

F. They are usually provided to such animals in a manner whereby they are consumed by licking—either from a wheel turning in a tank containing the supplement or from a rough-bottomed trough covered with a thin layer of the supplement.

G. Their main advantage for brood cows is convenience of feeding while their main disadvantage is overconsumption, frequently making the cost prohibitive; also, since there is no definite feeding time, they do not facilitate the checking of the herd at the time of feeding as is possible with something like range cubes.

H. Liquid supplements, for reasons of economy and convenience, are also frequently used for adding supplemental nutrients to rations for steers in the feedlot.

VI Range cubes.

A. Available on the market today are numerous products designed to be used as supplements for beef brood cows (Figure 22–6) and stocker animals running on low-quality grazing such as dead growth of summer pasture or on low-grade hay.

FIGURE 22–6
Range cubes being fed as a supplement to beef brood
heifers being wintered on low-quality pasture. (Courtesy of
Ralston Purina Co., St. Louis, Mo.)

B. Such products are frequently provided in the form of so-called range cubes, which are simply large pellets, about 3/4″–1″ in diameter and 1 1/2″–2 1/2″ in length.

C. They are usually fed on the ground in clean areas, thus making troughs unnecessary and facilitating sufficient scattering to permit all animals to get their proportionate share.

D. Since some brands of range cubes are designed to provide supplemental energy as well as protein, they sometimes run as low as 20% crude protein.

E. They are usually fed at a level of 1–3 lb per head daily, depending on the quality and availability of the forage material.

F. Range cubes have some definite advantages over liquid supplements and hot meals.

    1. No facilities are required for feeding range cubes.

    2. The level of consumption of range cubes can be precisely controlled.

    3. The use of range cubes facilitates bringing the herd together at regular intervals for easy checking.

    4. Range cubes may be used to facilitate moving the herd from one pasture to another.

VII "Liquid" Blocks.

A. A recent addition to the products (forms) available for providing supplemental nutrients is the *liquid* block. (Figure 22–7.)

B. *Liquid* blocks are basically a fortified molasses product that has been treated in a manner that causes it to take on a semi-solid form.

C. In addition to 20–25% protein (usually less than 25% equivalent protein from non-protein nitrogen) and energy (mainly from cane molasses) the blocks are ordinarily fortified with the critical minerals and vitamins.

D. They are usually marketed in 50-, 250-, and 500-lb units.

E. Consumption varies from 0.5–5.0 lb per cow per day depending on formulation used, availability of forage, and quality of forage. Normal consumption would probably be from 1–3 lb per head daily.

F. Their main advantage for brood cows is convenience of feeding; however, since there is no definite feeding time, blocks do not facilitate checking of the herd at feeding time as is possible with something like range cubes.

G. They are usually fed on the ground, thus making troughs unnecessary. Compared to liquid supplements, no "lick" tank is required. Blocks are resistant to damage by sun, rain, or snow.

H. Consumption of blocks should be monitored to ensure proper intake.

VIII The decision as to which high-protein feed to purchase will depend on the following:

A. Cost per unit (lb) of protein.

B. Energy content of the high-protein feed.

C. Convenience of feeding, including labor.

D. Equipment required.

E. The other feedstuffs being utilized with the high-protein feed.

F. Whether the high-protein feed is to be mixed in a total ration or fed as an individual feed.

G. Class of animal being fed.

(a)

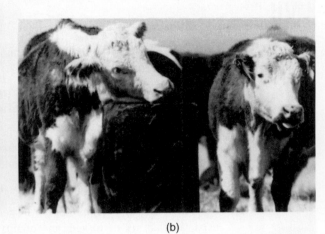

(b)

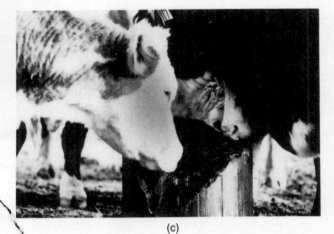

(c)

FIGURE 22–7
Shown in the above pictures are (a) a 50-lb block of the fortified solidified molasses product, (b) a 500-lb block of the
same product, (c) a partially consumed 500-lb block.
(Courtesy of Purina Mills, St. Louis, MO.)

23 Air-Dry Roughages

Roughages are usually grouped according to whether they are legume or nonlegume. Legume roughages of course come from leguminous crops and nonlegume roughages from nonleguminous crops. A legume differs from a nonlegume in that the former is able to provide itself with nitrogen from the air through the action of nitrifying bacteria harbored in nodules on its roots. Legumes as a group are considerably higher than nonlegumes in both protein and calcium and because of this are usually regarded as superior to nonlegumes in feeding value. Certain by-product materials such as cobs, husks, stover, hulls, and straws, which are frequently used as feed roughage, are usually not designated as legume or nonlegume but because of their fibrous nature and inferior feeding value are usually referred to simply as low-quality roughages. It is the purpose of section 23 to recognize the more important air-dry roughages in the different categories and to review some of the special characteristics of each. Characteristics of most of these roughages are presented in Table 27–1 at the end of this section.

I   **Legume hays and meals.**

    A. Alfalfa hay.

       1. Alfalfa is the most widely produced hay crop.

2. It is a perennial that, under favorable conditions, will maintain a good stand for several years.

3. It will provide 3 or more cuttings per year (Figure 23–1).

4. It is one of the highest yielding hay crops.

5. It is the most nutritious of all hays.

6. It is relished by all hay-eating animals.

7. It is very drought resistant.

8. Alfalfa is grown extensively throughout the midwestern and western states.

9. Over the years, not much alfalfa has been grown in the southern coastal states.

   a. It requires fertile, well drained soil with a low water table, so it is not adapted to much of the Coastal Plain.

   b. Insects and diseases have deterred its production in other sections of the South.

   c. Coastal bermudagrass has proven to be a very satisfactory hay crop for the region.

FIGURE 23–1
Alfalfa being mowed, conditioned, and wind-rowed in one operation, preparatory to being baled for hay. (Courtesy of Sperry-New Holland, New Holland, Pa.)

10. Recent developments, however, have brought about an increased interest in the use of alfalfa as a hay crop in the Southeast.

   a. The increased cost of nitrogen fertilizers has brought about an increased interest on the part of farmers in leguminous crops for hay production.

   b. The development of new strains and varieties of alfalfa with resistance to troublesome diseases and parasites has opened new opportunities for using this crop.

   c. Improved cultural practices have been developed over the past few years for expanded alfalfa production.

   d. There has been a growing demand on the part of dairymen in the area for a high-quality hay, such as alfalfa, to use in their feeding programs.

   e. Exceptionally high-quality alfalfa hay can be conducive to bloat in cattle.

B. **Red clover hay.**

   1. Red clover is one of the more important hay crops throughout much of the Corn Belt and the northeastern United States.

   2. It does not require soil as fertile as alfalfa does, it is less drouth resistant than alfalfa, and accordingly it is not ordinarily as high yielding as alfalfa.

   3. Red clover is usually grown in combination with timothy and/or other non-legume crops.

   4. Red clover-timothy mixtures will vary from almost all red clover the first year to almost all timothy after the first couple of years.

   5. Hay that is predominantly red clover will approach alfalfa in feeding value if harvested under favorable conditions.

C. **Alsike clover hay.**

   1. Alsike clover, in many respects, is similar to red clover as a hay crop.

   2. It somewhat resembles red clover in appearance but is less rank growing and has a lighter colored blossom.

   3. It is grown in much the same area of the country as red clover but on a much less-extensive scale.

   4. It too is usually seeded with timothy and other grasses for hay production.

   5. It will thrive on soils that are too wet and/or too acid for red clover.

   6. While not normally as high yielding as red clover, alsike clover hay compares favorably with red clover hay in feeding value.

D. Lespedeza hay.

1. Good quality lespedeza hay approaches alfalfa hay in feeding value.

2. At one time, lespedeza was widely grown as a hay crop throughout the southern states but not much of this crop is grown today.

   a. Lespedeza is usually good for only one cutting per season.

   b. Lespedeza is relatively low yielding.

   c. Lespedeza hay often is not of good quality as the result of shattering and a high content of weeds.

   d. Other more desirable hay crops have been developed—especially coastal bermudagrass.

E. Sericea hay.

1. Sericea is a perennial lespedeza grown in the southern states.

2. Top quality sericea hay approaches alfalfa hay in feeding value.

3. To make top quality sericea hay, it must be mowed when very immature and harvested in a manner to avoid shattering the leaves.

4. While there is presently a considerable acreage of sericea to be found throughout the southern states, mainly on land formerly in the government soil bank, very little of it is used for hay-making purposes.

5. Sericea hay that is overmature and badly shattered is very low in feeding value and not readily eaten by livestock because of its stemmy nature and high content of tannin.

F. Peanut hay.

1. Peanut hay usually consists of what remains of the peanut plant after the nuts, which grow underground, have been harvested.

2. Its quality varies greatly, depending on the harvesting procedure and conditions.

3. Top quality peanut hay will approach alfalfa hay in feeding value.

4. Most peanut hay is medium to low in quality and much inferior to alfalfa hay.

5. While it was once an important hay crop in the peanut sections, not much peanut hay is made today.

   a. New harvesting procedures do not lend themselves to making peanut hay.

   b. Most people rely on other crops, such as coastal bermudagrass and bahiagrass, for hay in those sections where peanut hay was formerly used.

G. Sweet clover hay.

1. Sweet clover is not ordinarily seeded as a hay crop but is sometimes harvested for hay.

2. Sweet clover generally is too coarse and stemmy, and shatters its leaves too readily to make a quality hay.

3. Also, sweet clover tends to harbor a mold that causes a condition known as *sweet clover poisoning* in which an animal will bleed to death as the result of consuming excessive levels of dicoumerol present in some sweet clover hay.

H. Cow pea and soybean hays (discussed together since they are similar in several respects).

1. Both crops are annuals and must be freshly seeded for each cutting.

2. Neither crop is used very extensively as a hay crop.

3. Both are frequently used as emergency hay crops.

4. The feeding value of both crops varies greatly, depending on harvesting conditions.

5. Both are inclined to lose their leaves badly during harvest.

6. Both crops normally contain considerable amounts of coarse stems, which livestock tend to refuse.

7. The overall consumption of both types of hay is greatly improved by fine chopping or coarse grinding.

I. Dehydrated alfalfa meal.

1. Dehydrated alfalfa meal (Figure 23–2) is alfalfa that has been harvested at an early stage of maturity, artificially dried, and ground into a meal.

2. It is produced mostly in the western states.

3. It is used for the most part in diets for poultry and swine, in early weaning diets, and in horse rations.

4. It is used especially as a source of various minerals and vitamins and certain unidentified nutritive factors—also as a source of yellow pigments (carotene and xanthophyll) in layer (poultry) diets.

5. It is usually used at levels not to exceed about 5% of the overall diet.

6. The most common grade of dehydrated alfalfa meal is that containing not less than 17% crude protein.

J. Alfalfa leaf meal and alfalfa stem meal.

1. Sometimes dehydrated alfalfa is processed whereby the leaves are separated from the stems to produce alfalfa leaf meal.

FIGURE 23–2
Dehydrated alfalfa meal. (Courtesy of the University of
Georgia College of Agriculture Experiment Stations)

2. The remaining stems are then ground to produce alfalfa stem meal.

3. Alfalfa leaf meal is much lower than alfalfa meal in fiber content and is used primarily in certain special diets in which the nutritive content of alfalfa is desired but a low fiber level is required.

4. Alfalfa stem meal is produced strictly as a by-product of alfalfa leaf meal production.

5. While alfalfa stem meal consists almost entirely of alfalfa stems, this product has fairly good nutritive value as dry roughages go.

II. Nonlegume hays.

A. Coastal bermudagrass hay.

1. Coastal bermudagrass is the most extensively produced hay crop in much of the Deep South (Figure 23–3).

2. It is a medium-height perennial that, when properly fertilized, maintains its stand indefinitely.

3. Under favorable conditions, it will provide 3 or more cuttings per year and is one of the highest yielding hay crops.

4. When adequately fertilized and cut at the proper stage, coastal bermudagrass will make a top quality nonlegume hay.

5. Coastal bermudagrass is a hybrid and does not produce viable seed—hence, it must be reproduced vegetatively from root stolons.

6. It will sometimes suffer from winterkill in the colder sections.

FIGURE 23–3
Large round bales of coastal bermudagrass hay remain in the field amid regrowth of grass without significant damage from the weather. (Courtesy of E. R. Beaty of the University of Georgia College of Agriculture Experiment Stations)

7. When allowed to become overmature before being cut, it loses much of its feeding value.

8. It is recommended that for quality hay production, coastal bermudagrass be cut about every 4–6 weeks.

9. Four new giant bermudagrass hybrids (coastcross 1, Midland, Tifton 78, and Tifton 44), which are reported to be more digestible and/or more winter hardy than Coastal, are now available as possible alternatives to Coastal as a hay crop.

B. Timothy hay.

1. Timothy is a medium-tall perennial grass.

2. It is one of the more widely grown grasses through the Corn Belt and the northeastern states.

3. It is usually seeded in combination with red clover but normally predominates the meadow after the second year.

4. Timothy is especially popular as a hay for horses, but a combination of timothy with red clover or alfalfa is generally preferred for other classes of livestock.

5. Like most hay crops, it should be cut not later than the early bloom stage for maximum yield of digestible nutrients.

C. Redtop hay.

1. Redtop is a fairly short-growing perennial grass that is found over much of the southern Corn Belt and most of the northeastern states.

2. It is used extensively and primarily in pasture mixtures for these sections.

3. It will tolerate a wide range of soil and moisture conditions and is sometimes grown for hay where conditions are not favorable for more desirable hay crops.

4. While redtop cut at the proper stage and harvested under favorable conditions makes a fairly satisfactory nonlegume hay, the hay is not very palatable and is usually produced only as a last resort.

5. Redtop gets its name from the fact that the mature heads lend a reddish tinge to a meadow made up of this crop.

D. Bromegrass hay.

1. Bromegrass is a medium-tall perennial that is grown extensively throughout most of the north central states but usually in combination with alfalfa and/or other legumes.

2. It is especially valuable for preventing bloat in cattle and sheep that are grazing on such mixtures.

3. When such mixtures are harvested for hay, bromegrass makes up an important part of the nutritive value of the overall forage combination.

4. While bromegrass is not grown extensively in pure stand for hay production, when grown and cut at the proper stage it makes a very satisfactory nonlegume hay crop.

E. Orchardgrass.

1. Orchardgrass is a medium-tall perennial that is widely grown in many sections.

2. It is grown mostly in that area lying between the northern and southern border states.

3. Orchardgrass is used primarily as a component of pasture mixtures but makes a very satisfactory hay if harvested no later than the early bloom stage.

4. As with most forage crops, overmature orchardgrass makes a low-quality hay.

F. Reed canarygrass hay.

1. Reed canarygrass is a tall, rank-growing coarse-stemmed perennial.

2. It is grown primarily in the northern states.

3. It thrives on land too wet for most other hay crops.

4. If harvested just prior to heading, it makes a fairly satisfactory hay.

5. It is not one of the more extensively grown hay crops.

G. Bermudagrass (common) hay.

1. Bermudagrass is a low-growing perennial found extensively throughout the South.

2. While it is widely used for pasture, it is used only to a limited extent for hay.

3. When it has been well fertilized, bermudagrass hay approaches coastal bermudagrass hay in feeding value.

4. Bermudagrass is not as drought resistant or as high yielding as coastal bermudagrass.

5. Much so-called bermudagrass hay is surplus pasture growth that has been cut for hay; such hay is usually overmature and low in digestibility and feeding value.

H. Bahiagrass hay.

1. Bahiagrass is a medium-tall perennial that is widely grown over much of the southern Coastal Plain.

2. It is grown primarily for pasture, secondarily for hay.

3. Good-quality bahiagrass hay approaches good-quality coastal bermudagrass hay in feeding value.

4. Since most bahiagrass hay is made from surplus pasture growth, it is usually overmature and low in quality.

5. Bahiagrass hay is used primarily as a wintering feed for the beef breeding herd.

6. Bahiagrass has the advantage over coastal bermudagrass of producing viable seed and so being reproducible from seed.

7. It will not live through the winter above the southern Coastal Plain and so is found mostly in the Coastal Plain area.

I. Fescue hay.

1. Extensive areas of alta and Kentucky 31 fescue are found in various sections of the United States—especially in the southern states above the Coastal Plain.

2. These two strains of fescue are very similar and are used almost interchangeably with each other.

3. They are both medium-tall perennials that remain green throughout the winter in many sections.

4. They are used primarily for pasture—especially for winter grazing.

5. Other varieties of tall fescue are found in some sections, and while similar in composition to alta and Kentucky 31, they are as a rule less palatable.

6. Excess growth of fescue pasture is sometimes harvested for hay in the late spring.

7. The quality of fescue hay varies greatly, depending on the stage of maturity at harvest and harvesting conditions.

8. Most fescue hay is of low quality, being high in fiber and low in protein and TDN.

9. The fescue hay that is made is usually used as a wintering feed for the beef breeding herd.

10. Much of the fescue grown throughout the United States is infected with endophyte fungus, which interrupts normal performance of cattle. This shows up in both pasture and hay. Symptoms of so-called "fescue toxicity" include loss of appetite, elevation of body temperature, poor growth rate, decreased milk production and impaired reproductive performance. In advanced cases, "fescue foot" may appear in which hooves actually may slough off. Some relief from this condition can be achieved by mixing legumes into the pasture mixture, explained possibly as only a dilution factor.

J. Dallisgrass hay.

1. Dallisgrass is a perennial bunch grass grown primarily as a pasture crop throughout much of the South.

2. Surplus growth of Dallisgrass pasture is sometimes harvested for hay.

3. Such growth is usually rather mature when cut for hay—hence, Dallisgrass hay is usually of only fair quality.

K. Johnsongrass hay.

1. Johnsongrass, a tall, rank-growing southern perennial, over the years has been regarded as a weed and extensive acreages throughout the South have been infested with this plant.

2. While seldom, if ever, planted as a hay crop, it has been used for such frequently when available and needed.

3. Johnsongrass has coarse, sappy stems that make it difficult to cure into hay.

4. Livestock tend to refuse a large portion of the stems of Johnsongrass hay.

5. Johnsongrass is about as low in protein as any nonlegume crop.

6. Johnsongrass is higher in calcium than most nonlegumes.

7. Johnsongrass is usually found only on fertile soil.

8. Johnsongrass can be eliminated from a field by continuous close grazing over a period of 2–3 years.

9. Very little Johnsongrass hay is made today.

L. Sudangrass hay.

1. Sudangrass is a tall-growing annual that is frequently used as an emergency hay and/or pasture crop.

2. As to hay crop, sudangrass in many respects is similar to Johnsongrass.

    a. Both are members of the sorghum family.

    b. They resemble each other in appearance.

    c. Both have coarse, sappy stems that present curing problems.

    d. Livestock tend to refuse the coarse stems of both.

    e. The two hays are similar in feeding value.

3. Sudangrass hay is normally available in very limited quantities and is usually fed on the farm where it is produced.

M. **Millet hay.**

1. The millets are rapid-growing summer annuals of several types and varieties.

2. Millet of one type or another is grown in most areas as a temporary pasture and/or hay crop.

3. Millet cut at an early stage of maturity makes a satisfactory dry roughage for cattle but is not too satisfactory for lambs.

4. Millet hay is not recommended for horses since it sometimes causes swollen joints and lameness in this class of livestock.

5. Millet is strictly an annual but may produce more than one cutting per season.

6. Not very much millet is grown for hay.

N. **Oat hay.**

1. Oat hay is oats that have been cut while still green, usually in the dough stage, and made into hay.

2. The grain remains as a part of the hay.

3. If cut at a fairly early stage of maturity, oats make a very satisfactory hay crop.

4. More often than not, it is harvested as an emergency hay crop when it is apparent that other hay sources are in short supply.

5. Oat hay should not be confused with the oat straw, which remains after the grain has been harvested from the mature oat plant and which is much lower than oat hay in feeding value.

6. Only a limited amount of oat hay is normally made in most areas.

O. **Prairie and meadow hays.**

1. Extensive acreages of native grasses are harvested for hay in many sections of the western states.

2. Such forages are referred to as prairie hay or meadow hay, depending on the section.

3. When cut at the proper stage and harvested under favorable conditions, such forages can be used to make hay of excellent quality, suitable for feeding all classes of hay-consuming livestock.

4. Some such forages, however, are allowed to become overripe before being cut and, as a result, produce hay that is low in feeding value.

P. **Kentucky bluegrass hay.**

1. While Kentucky bluegrass is grown extensively throughout the north central and the northeastern states as a pasture and lawn grass, it is quite low yielding as a hay crop and is used very little for this purpose.

2. Most Kentucky bluegrass that is harvested for hay is surplus pasture growth that is overmature and consequently low in feeding value.

III **Miscellaneous air-dry roughages.**

A. Corn cobs, husks, and stover.

1. Recent data by Lowrey of the Georgia Station indicate that the dry matter of the present-day mature corn plant divides itself approximately as follows:

Grain	38%
Cobs	7%
Husks	12%
Leaves	13%
Stalks	30%

2. With an annual production of corn grain of about 125 million tons, theoretically there could be available in this country annually for feeding purposes as much as the following:

Cobs	23 million tons
Husks	39 million tons
Leaves	43 million tons
Stalks	99 million tons
	204 million tons

3. While corn cobs, husks, and stover are at best all low in feeding value, they do contain considerable amounts of digestible fiber and NFE and, if properly supplemented with protein, minerals, and vitamins, can serve as a useful roughage in rations for at least certain classes of ruminant animals.

4. The extent to which cobs and shucks have been used in the past for feeding purposes has been for the most part as components of ground snapped corn and ground ear corn or corn and cob meal.

5. Very little corn is harvested as snapped corn or ear corn today; most corn is presently combined, leaving the cob and husks in the field where they are, for the most part, unused.

6. The use of corn stover has been limited, in general, to grazing stalk fields with stocker animals.

7. For the most effective use of corn cobs, husks, and stalks, all should be finely chopped or coarsely ground to facilitate their consumption.

8. The use of such products may be facilitated also by ensiling.

B. Soybean hulls and soybean mill feed.

1. In the manufacture of 49% soybean meal and in the manufacture of soybean flour or grits, the outer coat of the soybean seed is removed and made available as soybean hulls for feeding purposes.

2. In some instances, soybean hulls are steam rolled into a flake and marketed as "soybean flakes."

3. In other instances, varying amounts of other by-products of soybean manufacture are mixed with the soybean hulls and the mixture marketed as "soybean mill feed."

4. Soybean hulls have an average as-fed content of:

Crude protein	11.0%
Crude fiber	36.4%
TDN	70.0%

5. As may be noted from the above, soybean hulls are fairly digestible considering their high fiber content.

6. The composition of soybean mill feed will vary, depending on the amount of soybean by-products other than hulls present, but it is normally somewhat lower in fiber and higher in protein and TDN than soybean hulls.

7. Since soybean mill feed usually consists chiefly of soybean hulls, the two products are ordinarily not too different in composition and feeding value.

8. The principal use of both products is in rations for dairy cows where they serve primarily as a source of bulk and secondarily as sources of protein and energy.

C. Oat, barley, and wheat straws.

1. These are what remains after the grain has been harvested from the mature oat, barley, and wheat plants, respectively.

2. While there are tremendous tonnages of all three straws available in the United States each year, they are used more extensively for bedding than for feeding purposes.

3. Oat, barley, and wheat straws are sometimes used in maintenance rations for cattle and horses, but all three are very low in feeding value and must be supplemented with protein and minerals, and possibly vitamin A, for satisfactory results.

4. Oat, barley, or wheat straw may be used as a satisfactory source of long roughage for fattening steers otherwise receiving a low roughage ration.

D. Cottonseed hulls, peanut hulls, and oat hulls.

1. There is a considerable tonnage of all three of these products available in the United States each year.

   a. Most cottonseed is subjected to oil extraction with cottonseed hulls as a by-product.

   b. Peanut hulls are a by-product of the manufacture of peanut butter, peanut oil, and shelled peanuts.

    c. Oat hulls result as a by-product of the manufacture of oat groats and rolled oats from oats grain.

  2. All are usually mixed with more palatable feeds to induce their consumption.

  3. All three are very low in nutritive value and are used, when fed, primarily as a bulk factor in the ration (Figure 23–4).

  4. They may provide limited nutritive value to the animal if properly supplemented.

  5. Of these three feeds, cottonseed hulls is the most palatable and has the greatest feeding value.

E. Rice hulls.

  1. Rice hulls is one of the by-products from the manufacture of polished rice from rice grain.

  2. This material is one of the lower-quality feeds fed to livestock.

  3. Rice hulls are usually ground and mixed with more palatable feeds to induce their consumption.

  4. About the only feeding value of rice hulls is as a bulk factor or filler in the ration.

FIGURE 23–4
Two commonly used low-quality roughages. (Courtesy of the University of Georgia College of Agriculture Experiment Stations)

Table 23–1

CHARACTERISTICS OF SEVERAL DRY ROUGHAGES
(90% DRY MATTER EQUIVALENT, OR AS-FED BASIS)[1]

Material	Ash %	Crude Fiber %	Ether Extract %	Nitrogen-Free Extract %	Crude Protein %	Total Digestible Nutrients %
Leguminous products						
Alfalfa hay	8.6	28.2	1.7	35.9	16.0	51
Alfalfa meal, dehydrated	9.7	24.0	2.8	37.4	17.4	55
Alsike clover hay	7.8	26.2	2.4	39.1	12.4	51
Cowpea hay	10.5	24.4	2.6	35.1	17.7	54
Lespedeza hay	4.7	28.4	2.5	39.4	13.8	44
Peanut hay	8.2	39.3	3.3	39.1	9.9	48
Red clover hay	6.7	27.1	2.5	38.2	13.0	52
Soybean hay	7.2	30.6	2.3	35.0	14.1	49
Sweet clover hay	7.8	28.9	1.9	38.4	9.5	49
Nonleguminous products						
Bahiagrass hay	5.4	30.5	1.5	49.2	4.3	48
Bermudagrass hay, common	8.0	24.8	1.8	43.7	9.2	43
Bromegrass hay	8.1	28.4	3.2	44.5	9.8	51
Coastal bermudagrass hay	6.3	27.0	2.0	43.9	11.7	49
Dallisgrass hay	7.9	30.8	1.9	41.1	9.2	49
Fescue hay	5.9	32.6	2.0	41.4	7.2	48
Kentucky bluegrass hay	5.9	26.9	3.0	44.3	9.1	54
Millet hay	8.5	26.3	1.7	41.8	8.8	48
Oat hay	7.2	29.1	2.2	43.6	8.6	52
Orchardgrass hay	6.5	31.0	2.8	39.7	9.4	51
Prairie and meadow hays	7.2	30.7	2.1	45.2	5.8	46
Redtop hay	6.0	28.4	2.8	47.4	7.4	50
Reed canarygrass hay	7.3	30.2	2.7	40.0	9.1	36
Sundangrass hay	10.7	26.2	1.8	41.7	10.9	51
Timothy hay	4.8	30.3	2.4	47.3	8.8	53
Miscellaneous roughages						
Corn cobs	1.0	32.2	0.8	52.7	2.8	44
Corn husks	3.2	30.0	0.8	51.2	3.3	55
Corn stover	8.1	29.3	1.1	43.2	5.4	51
Cottonseed hulls	2.8	43.2	1.5	39.3	3.8	42
Oat straw	7.2	37.2	2.8	41.8	4.1	46
Rice hulls	19.0	38.9	1.0	30.3	3.0	11
Soybean hulls	4.0	36.2	2.0	37.0	10.8	69

[1]To convert to 100% dry matter equivalent, multiply any of the values in the table by 0.90.

24 The Molasses

There are several types of molasses. All are concentrated water solutions of sugars, hemicelluloses, and minerals obtained usually as by-products of various manufacturing operations involving the processing of large amounts of the juices of extracts or plant materials.

I

The various types of molasses are:

A. **Cane or blackstrap molasses.** This product is obtained as a by-product of the manufacture of cane sugar from sugar cane. It is what remains after as much sugar has been removed from the sugar cane juice as is possible following standard procedures. Considerable amounts of cane molasses are produced in the southeastern United States; it is also imported from Hawaii, the Philippines, and Africa.

B. **Beet molasses.** Beet molasses is obtained as a by-product of the manufacture of beet sugar from sugar beets. It is produced primarily in the sugar beet sections of the western states, and its use is largely restricted to that area of the country.

C. **Citrus molasses.** Citrus molasses is produced from the juice of citrus wastes. At one time it was available in significant quantities in the citrus-producing sections.

Citrus molasses is not as available as it once was since increasing amounts of it are being used in the production of citrus pulp with molasses, dried.

D. Wood molasses. In the manufacture of paper, fiber board, and pure cellulose from wood, there results an extract that contains the more soluble carbohydrates and minerals of the wood material. In some instances this extract is processed into a molasses suitable for livestock feeding purposes.

II Use of molasses in livestock feeding.

A. The different types of molasses are similar in feeding value per pound of dry matter.

B. Cane molasses is used by far the most extensively. (Figure 24–1.)

C. Molasses is usually used in rations for cattle, sheep, and horses for the following reasons.

   1. To improve ration acceptability.

   2. To improve rumen microbial activity.

   3. To reduce dustiness of ration.

   4. As a binder for pelleting.

   5. As a source of energy.

   6. As a source of unidentified factors.

D. Molasses is usually used in amounts not to exceed about 10–15% of the ration.

   1. It has its greatest feeding value at or below these levels.

   2. More than 15% molasses will cause a ration to become sticky and difficult to handle.

   3. High levels of molasses will tend to disrupt microbial activity.

   4. Used at levels not to exceed 10–15% of the ration, molasses approaches corn in feeding value; at higher levels its feeding value decreases considerably.

E. The different types of molasses are sometimes available in dehydrated forms.

   1. Dehydration is usually accomplished by spray drying, where a mist of molasses is sprayed into a blast of hot air.

   2. The use of dehydrated forms will simplify ration mixing.

   3. However, most forms of dehydrated molasses are quite deliquescent and must be used accordingly.

FIGURE 24–1
A 4,000-gallon storage tank such as one needs to receive
bulk deliveries of molasses, which usually come in 3,000-
gallon tank-truck loads. (Courtesy of the University of
Georgia College of Agriculture Experiment Stations)

    4. The same limitations apply to the use of dehydrated or dried molasses as apply to the use of the liquid product.

F. Molasses is seldom used in rations for swine since it tends to cause scouring in this class of livestock.

III   Molasses Brix.

A. *Brix* is a term commonly used in referring to the composition of molasses. Brix is expressed in degrees and was originally used to indicate the percentage by weight of sugar in sucrose solutions, with each degree Brix being equal to 1% of sucrose.

B. One way to determine the Brix of a sucrose solution is to measure its specific gravity and then refer to a conversion table to arrive at the degrees Brix or the level of sucrose present. The Brix of a molasses is determined in the same way. However, because molasses contains considerable amounts of soluble materials other than sucrose that influence its specific gravity, the Brix value is not a true measure of the sugar content of molasses. It does, however, reflect the relative level of sugar present and so has, over the years, been used as a convenient basis for expressing molasses quality.

Generally accepted Brix values for cane, beet, and citrus molasses, along with corresponding specific gravities and weights per gallon for these products, are shown below:

	Degrees Brix	Specific Gravity (20°C)	Wt per Gallon
Beet molasses	79.5	1.4109	11.75 lb
Cane molasses	79.5	1.4109	11.75 lb
Citrus molasses	71.0	1.3558	11.29 lb

25

Other High-Moisture Feeds

I   **High-moisture grain.**

    A.   Frequently it is necessary or desirable to harvest feed grains, especially corn and grain sorghum, at moisture levels too high to permit safe bin storage without artificial drying. In some instances, the moisture level is so high as to make artificial drying uneconomical. In such cases, the grain is sometimes stored in an airtight silo and preserved in a semi-ensiled state. The moisture level of such grain will usually run from about 22% to 30%, and it is commonly referred to as high-moisture grain—not that it is extremely high in moisture content, but it is simply somewhat higher than that normally stored under bin conditions. High-moisture grain has been found to be equal to and in some instances superior to air-dry grain as a feed for most classes of livestock on a dry matter basis. This is especially true for grain sorghum.

    B.   Since the dry matter content of high-moisture grains varies considerably, when using such feeds for ration formulation it is usually best to work on a dry matter basis. If composition figures for high-moisture grain as such are not available, the dry basis composition figures for the air-dry grain can be used in balancing the ration. The dry weights must then be converted to the moisture level of the product as fed before carrying out the actual feeding operation. In order to do this, it is necessary, of course, to determine the dry matter content of the particular high-moisture grain on hand. However, by working on a dry matter basis, only the dry matter content and not the complete analysis of the high-moisture grain being used needs to be known.

II Haylage.

A. Frequently weather conditions are not favorable for harvesting green forages for hay. On the other hand, such crops often are too high in water content to make good silage. A practice that is sometimes followed in such instances is to make the crop into what has become known as haylage. Haylage is the feed produced by storing in an airtight silo a forage crop that has been dried to a moisture level of about 45% to 55%. Such forages undergo no seepage in storage and usually produce a silage-like product of excellent quality. On a dry matter basis, haylages have been found to be equal, and sometimes superior, to hay made from the same crop.

B. Since the moisture content of haylages varies considerably between batches, when such feeds are used in ration formulation (as with high-moisture grains), it is usually best to proceed on a dry matter basis. If composition figures for the particular haylage under consideration are not available, they can usually be closely approximated on a dry basis from the dry basis composition of hay from the same forage.

C. In any event, the dry weights, obtained from the ration-balancing process, must be converted to an as-fed basis in carrying out the actual feeding operation. In order for this to be done, however, it is necessary that the moisture level of the haylage be determined.

III

Silages. On many livestock farms, forages are harvested in a high-moisture state and stored under anaerobic conditions to produce a feed commonly known as silage. Almost any crop can be made into silage, although some are superior to others for this purpose. Conventional silage is a roughage feed and must be used accordingly in the feeding program. Since silage normally contains only about 25% to 35% dry matter, about 3 lb of silage must be used to replace 1 lb of air-dry hay in the feeding program. While this sometimes requires an animal to consume a considerable poundage of feed per head daily, a large part of this feed is moisture, the intake of which the animal compensates for by drinking less water. For further details about silos, silage making, and silage utilization, refer to section 28, "Silage and Haylage—Production and Use."

IV Fresh forages.

A. Green forages are utilized for feeding in the fresh form primarily as pasture. Harvested green forages are usually made into hay, haylage, or silage for use in the feeding program. However, sometimes crops are harvested in the green state and fed in the fresh form as so-called *greenchop*. This method of forage utilization has been confined to dairy operations for the most part in the past. It has been used very little in beef cattle feeding programs. It could be, however, that problems of pollution, production efficiency, and/or energy conservation might necessitate a wider use of this method of forage utilization in feeding cattle of all kinds in the future.

B. On the other hand, until this method of forage feeding is more generally followed and the use of the different forage crops for this type of feeding program has been studied more thoroughly, it does not seem appropriate to attempt to discuss in detail the use of various types of greenchop for feeding purposes. It suffices for the present simply to say that the dry matter of greenchop should have a feeding value comparable to that of hay, haylage, or silage made from the same crop. It should be kept in mind, however, that individual forage crops will vary in composition from day to day, depending on weather conditions and stage of maturity. Especially variable is the percentage of dry matter in the fresh forage material, which tends to complicate deciding just how much fresh forage to feed on any certain day to provide the desired amount of forage dry matter.

V **Wet by-products.** Under air-dry feeds, reference was made to such products as dried citrus pulp, dried beet pulp, brewers dried grains, and distillers dried grains. These same materials are sometimes available at their point of production in the fresh or wet state and may be used in this form for feeding purposes. Their feeding value in the wet form is not materially different on a dry basis from that of the air-dry products. However, in view of their high water content it is uneconomical to transport such feeds any considerable distance. Also, they cannot be stored effectively, and, consequently, fresh supplies must be obtained daily. Since, as with high-moisture grain and haylage, the moisture content of such products is not consistent, it is usually best to work with such feeds on a dry basis composition when using them in the feeding program.

VI **Roots and tubers.**

A. Extensive use has been made of roots and tubers for livestock feeding in northern Europe over the years. However, roots and tubers have never been used extensively for livestock feeding in the United States, and the amount of such crops used for feeding purposes in this country today is almost insignificant.

B. When used for feeding, roots and/or tubers are generally used as a substitute for silage. Sometimes such crops are used simply as a relish for livestock being fed for top performance or being fitted for show. However, barring major changes in our nation's agricultural situation, it is not likely that roots and tubers will be significant in this country's livestock feeding in the foreseeable future. Consequently, it does not seem worthwhile to discuss these crops individually from the standpoint of the different classes of livestock.

The compositions of a representative group cf such crops have been included in the appendix table. This has been done not because these crops are now or are apt to become important livestock feeds but only so that the information might be available to the teacher and student for instructional purposes.

VII Fresh milk.

A. All farm livestock normally produce significant quantities of milk upon giving birth to young. Only the milk of dairy cows, however, is ever available in quantities that permit its use for feeding purposes other than nursing. Even then, it is rarely fed in any significant amount in the fresh form. Usually any fresh milk that is in surplus of that needed for current human consumption is dehydrated and held in powdered form for later use, primarily as food for humans and secondarily as a component of certain specialty diets for livestock. Consequently very little fresh milk, other than that consumed by nursing young, ever finds its way into the feeding program.

B. However, all of our farm livestock do produce milk upon giving birth to young, and milk of one kind or another is the primary feed of most newly born farm animals for the first several weeks. Also, surplus milk is sometimes used in other ways for feeding. In other words, a large part of the feed fed to livestock goes to support the production of milk—milk that in turn is used to a very significant degree as a feed for farm animals. Consequently, it behooves a student of feeds and feeding to become aware of the important role that milk plays either directly or indirectly in most livestock production programs and to prepare for coping with the various aspects of its production and/or utilization.

C. Since different farm animal species produce milk similar in composition and since quantitatively cow's milk is by far the most important, composition figures only for cow's milk in its different forms and its by-products have been included in the appendix table.

26 Identifying Feeds from Their Composition

While tables of composition providing information on the composition and nutritive value of individual feeds are usually available, students of feeds and feeding will find it helpful to be able to relate individual feeds to a certain general composition or to associate some specific composition with a particular feed group without the aid of composition tables. In Table 26–1, the authors have attempted to summarize some of the more distinguishing composition characteristics of the various feed groups. A knowledge of the information provided in this table will greatly facilitate one's ability to work with feeds and feed composition figures.

In using this information, however, it should be realized that in many instances there is considerable overlapping of different feed groups with respect to their content of the various nutritive fractions, and frequently there is no single point that will provide a clearcut line of separation of the feed groups with respect to certain nutritive components. Even so, information such as is presented in Table 26–1 can be most helpful to a student of feeds and feeding as he or she attempts to become familiar with the composition and feeding value of different feeds, and to relate them to each other.

An interesting and helpful learning exercise in this connection is the systematic use of such information in associating a particular composition with a specific feed group or possibly even with an individual feed.

Table 26–1

**SUMMARY OF SOME DISTINGUISHING COMPOSITION
CHARACTERISTICS OF DIFFERENT FEED GROUPS
(ALL COMPOSITION FIGURES ON AN AS-FED BASIS)**

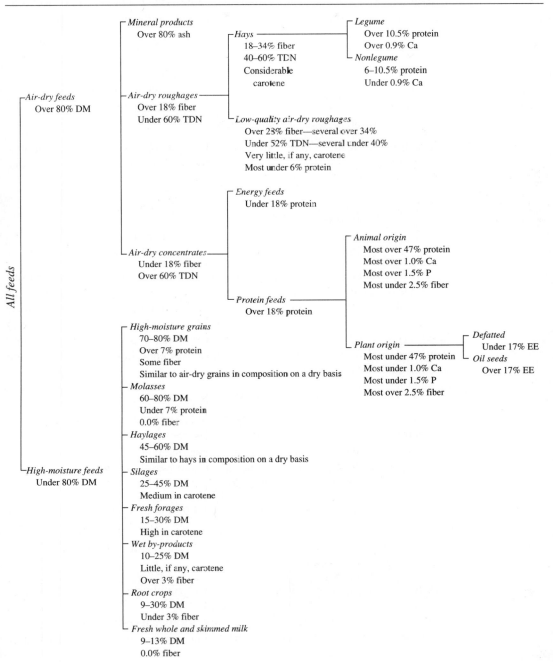

All feeds

Air-dry feeds
Over 80% DM

  Mineral products
  Over 80% ash

  Air-dry roughages
  Over 18% fiber
  Under 60% TDN

    Hays
    18–34% fiber
    40–60% TDN
    Considerable
    carotene

      Legume
      Over 10.5% protein
      Over 0.9% Ca
      Nonlegume
      6–10.5% protein
      Under 0.9% Ca

    Low-quality air-dry roughages
    Over 28% fiber—several over 34%
    Under 52% TDN—several under 40%
    Very little, if any, carotene
    Most under 6% protein

  Air-dry concentrates
  Under 18% fiber
  Over 60% TDN

    Energy feeds
    Under 18% protein

    Protein feeds
    Over 18% protein

      Animal origin
      Most over 47% protein
      Most over 1.0% Ca
      Most over 1.5% P
      Most under 2.5% fiber

      Plant origin
      Most under 47% protein
      Most under 1.0% Ca
      Most under 1.5% P
      Most over 2.5% fiber

        Defatted
        Under 17% EE
        Oil seeds
        Over 17% EE

High-moisture feeds
Under 80% DM

  High-moisture grains
  70–80% DM
  Over 7% protein
  Some fiber
  Similar to air-dry grains in composition on a dry basis

  Molasses
  60–80% DM
  Under 7% protein
  0.0% fiber

  Haylages
  45–60% DM
  Similar to hays in composition on a dry basis

  Silages
  25–45% DM
  Medium in carotene

  Fresh forages
  15–30% DM
  High in carotene

  Wet by-products
  10–25% DM
  Little, if any, carotene
  Over 3% fiber

  Root crops
  9–30% DM
  Under 3% fiber

  Fresh whole and skimmed milk
  9–13% DM
  0.0% fiber

Air-dry vs. high-moisture feeds. The first step in carrying out such an exercise is to look at a feed's dry matter content to determine if it is an air-dry feed or a high-moisture feed.

A. **Air-dry feed.** If a feed contains more than 80% DM, it should be regarded as an air-dry feed and would probably fall into one of the following feed groups:

Mineral products	Air-dry protein feeds of plant
Legume hays	origin, defatted
Nonlegume hays	Air-dry protein feeds of plant
Low-quality air-dry roughages	origin, not defatted
Air-dry energy feeds	(oil seeds)
Air-dry protein feeds of animal origin	

B. **High-moisture feed.** On the other hand, if a feed contains less than about 80% DM, it should be regarded as a high-moisture feed and would probably fall into one of the following feed groups:

High-moisture grains	Fresh forages
Molasses	Wet by-products
Haylages	Root crops
Silages	Fresh whole or skimmed milk

C. **Mineral products.** If a feed, on the basis of its composition, falls into the air-dry feed category, the next step is to decide whether it is a roughage, a concentrate, or a mineral feed. Mineral feeds or products are readily recognized as such from their composition because they are ordinarily quite high in ash (usually more than 80%) and, except for steamed bone meal, contain very little, if any, of the organic nutrients. They may or may not be high in calcium and/or phosphorus, depending on the product under consideration.

D. **Air-dry roughages vs. air-dry concentrates.**

1. An air-dry nonmineral feed may be classed as a roughage or a concentrate by referring to its crude fiber and/or its TDN content. As a general rule air-dry roughages will run over 18% in crude fiber and contain less than 60% TDN. Air-dry concentrates, on the other hand, will usually contain less than 18% crude fiber and more than 60% TDN.

2. Air-dry roughages may be further divided on the basis of their composition into the following groups:

Legume hays	Low-quality air-dry roughages
Nonlegume hays	

E. **Legume vs. nonlegume vs. low-quality air-dry roughages.** The hays are characterized by having 18% to 34% crude fiber, 40% to 60% TDN, and a considerable amount of carotene, whereas the low-quality dry roughages are, on the average, higher in fiber and lower in TDN with little or no carotene and a protein content of less than 6.0%, except for peanut hulls at 7.1%. The legume hays are easily distinguished from the nonlegume hays on the basis of composition because the former ordinarily contain more than 10.5% crude protein and more than 0.9% calcium, while the latter usually contain 6% to 10.5% protein and less than 0.9% calcium.

F. **Energy concentrates vs. protein concentrates.** If a feed on the basis of its composition turns out to be an air-dry concentrate, it must then be decided whether it is a protein concentrate or an energy feed, since both on the basis of TDN and crude fiber contents qualify as air-dry concentrates. Air-dry concentrates with more than about 18% crude protein are usually classed as protein concentrates and those with less than 18% crude protein as energy feeds. The protein concentrates may be further subdivided into those of animal origin and those of plant origin.

G. **Protein supplements of animal vs. plant origin.** Those of animal origin, except for dried whole milk and dried skimmed milk, all run more than 47% crude protein, whereas those of plant origin, except for 49% soybean meal, all run less than 47% protein. Also, most of those of animal origin have more than 1.0% calcium and 1.5% phosphorus and less than 2.5% fiber, while those of plant origin usually contain less than 1.0% calcium and 1.5% phosphorus and more than 2.5% fiber.

The protein concentrates of plant origin may be divided into the defatted meals, on the one hand, and the oil seeds, on the other. These two groups are easily distinguished from each other in that the defatted meals will always run less than 7% and usually much lower in ether extract, whereas the oil seeds all exceed 17% in this nutritive fraction.

II

High-moisture feeds. If a feed contains less than 80% DM, it will normally be classed as one of the high-moisture feeds, the subgroups of which were listed previously. The separation of these different subgroups on the basis of their composition is not nearly as clearcut as is that for the air-dry feeds. There are, however, numerous composition differences among some of these various subgroups that the student will find interesting and helpful.

A. For example, if the composition of a feed shows it to contain between 60% and 80% DM, the feed is probably either a high-moisture grain or a molasses. To which of these two groups the feed belongs may be determined from its protein and/or crude fiber content. The molasses normally contain no crude fiber, whereas high-moisture grains will always show from small to significant amounts of this feed constituent. Also, the molasses normally contain less than 7% crude protein, unless some product such as urea has been added, while the high-moisture grains usually exceed this level of crude protein, even on a wet basis.

B. Any feed containing between 45% and 60% DM will almost always be one of the haylages. The other high-moisture feeds including the silages, the fresh forages, the wet by-products, the root crops, and the fresh milks all normally contain less than 45% DM. However, there is considerable overlapping of the latter groups in DM content. Silages generally range from about 25% to 45% in DM content and, being more mature, usually surpass fresh forage crops in this regard. Also, they usually contain less carotene. Otherwise, there is little basis for distinguishing between these two feed groups on the basis of the composition information usually available.

C. Wet by-product feeds might be confused with fresh forages, on the one hand, and root crops, on the other, based on their content of DM. However, the fresh forages can usually be distinguished from the wet by-products on the basis of carotene content, if this information is available, since the wet by-products usually contain little, if any, carotene, whereas the fresh forages normally are quite high in this regard. On the other hand, root crops can be distinguished from wet by-product feeds from their composition in that the latter usually contain more than 3% crude fiber on a wet basis, while the root crops usually run below this figure in crude fiber content.

D. While some of the root crops might be confused with the fresh milks on the basis of DM content, it is easy to distinguish between these two groups on the basis of composition in that the root crops always contain at least a small amount of fiber, whereas the fresh milks always exhibit a 0.0% fiber content.

III

Identifying individual feeds. With more experience it is frequently possible to distinguish between two or more feeds within the same feed group on the basis of their composition. For example, fresh whole milk is easy to distinguish from fresh skimmed milk on the basis of fat content. Ground snapped corn, ground ear corn, and ground shelled corn can be separated on the basis of their relative contents of crude fiber. Ground snapped corn or ground ear corn can be distinguished from wheat grain or grain sorghum grain on the same basis. Yellow shelled corn can be distinguished from the other feed grains by its content of carotene. Meat scrap is distinguished from meat and bone meal through the higher bone content and accordingly higher calcium and phosphorus contents for the latter product. Solvent-extracted meals are always lower in ether extract than are the mechanically extracted meals in view of the greater efficiency of the solvent process for fat extraction. Alfalfa meal, alfalfa leaf meal, and alfalfa stem meal are easily distinguished from each other since the leaves of the alfalfa plant are higher in protein and lower in crude fiber than are the stems, and the three meals vary in composition accordingly.

The above are only a few of the more obvious illustrations of how feeds can be distinguished from each other on the basis of logical differences in composition. Many others that might not be so clearcut could be cited. In fact, with experience a student of feeds and feeding can develop proficiency along this line to a point where he or she can usually recognize a certain composition as being that of any one of about two or three feed possibilities, and frequently he can even specify the particular feed.

27 Hay and Hay Making

Object in hay making: To reduce the moisture content of green forage crops sufficiently to permit their safe storage without spoilage or serious loss of nutrients.

A. The maximum permissible water content for baling hay is about 18% to 22%, depending on the fineness of the hay, the tightness of the bale, the prevailing humidity, and the amount of air movement.

B. Freshly mown hay that is stored before it is sufficiently dry can be destroyed by fire resulting from spontaneous combustion.

C. Hay that is stored containing excessive moisture may tend to mold, making it unsuitable for feeding purposes.

D. Hay that is improperly harvested may suffer losses due to:

1. **Shattering.** This is the loss of leaves, which represent the most nutritious part of the hay plant. Legumes tend to shatter badly.

2. **Leaching.** Rain on hay during the curing periods tends to leach out and cause the loss of the more water-soluble nutrients.

3. **Bleaching.** While a certain amount of exposure to sunlight is essential and desirable for making good hay, excessive exposure to sunlight will cause heavy losses of certain nutrients—especially carotene.

II Proper hay-making procedures.

A. A forage crop to be harvested for hay should be mowed just as soon after reaching an early bloom stage of maturity as circumstances will permit. Undue delay in harvesting hay will result in low-quality hay. However, it is sometimes necessary to delay the harvesting of a hay crop because of extended periods of rain or forecasts of rain and because of other extenuating farm circumstances.

B. Every effort should be made to select periods of rain-free weather for harvesting hay. A minimum of one and normally about two days of good drying weather are required for curing hay. Weather forecasts can frequently be of great assistance in this regard.

C. "Conditioning" will reduce the curing time required for certain crops, especially crops with coarse sappy stems, by as much as 50%. To "condition" a hay crop, it is passed through a set of rollers to crack open the stems and thereby facilitate the drying process. (Figure 27–1.)

D. A forage crop being harvested for hay should be raked before it is completely dry to avoid excessive shattering and overexposure to the sun.

FIGURE 27–1
Coastal bermudagrass being mowed, conditioned, and windrowed in one operation preparatory to being harvested for hay. (Courtesy of E. R. Beaty of the University of Georgia College of Agriculture Experiment Stations)

E. Turning of a windrow, when necessary to facilitate drying, should be done when the dew is on—especially with hays subject to serious shattering.

F. Baling should be carried out just as soon as the hay is sufficiently dry. Square bales should be stored as soon as possible and in any event before they are rained on. Round bales, however, will shed rain and may be left in the field for extended periods without serious damage to the hay. (Figure 27–2.)

G. The high cost of baling, storing, and feeding conventional hay bales during the past few years has prompted many producers to start putting hay into large round bales or into mechanically formed stacks.

    1. Such bales and stacks vary greatly in size, depending on the circumstances, with the bales ranging from about 500 lb up to about 2,000 lb in weight and the stacks ranging to as much as 3 tons or more.

    2. Both bales and stacks are ordinarily stored in the open and normally do not suffer excessive weather damage from such exposure.

    3. Grass hays usually suffer less damage than do legume hays from outside storage.

FIGURE 27–2
A modern hay-making operation in progress. Note that as the bales are formed, they are automatically ejected into the wagon at the rear. (Courtesy of Ford Tractor Operations, Ford Motor Co., Troy, Mich.)

4. Stacks completely eliminate the use of twine, and the large round bale reduces its use to a minimum.

5. The harvesting and feeding of both large bales and stacks are usually completely mechanized. (Figure 27–3).

6. Where cattle are simply turned in to such bales and stacks, there is usually considerable loss of hay from trampling.

7. Large round bales (Figures 27–4 through 27–7) lend themselves to being unrolled for rack or pasture feeding, and there are mechanical devices available to facilitate this practice.

8. The equipment for making and handling large bales and stacks (Figures 27–8 and 27–9, 27–10) is relatively expensive and can be justified only where a considerable volume of hay is involved.

III Special characteristics of hay as a feed.

A. Hay is classed as a dry roughage and is used primarily as a source of bulk in the ration, and secondarily as a source of nutrients.

B. Hay of average quality will usually run from 25% to 32% crude fiber and 45% to 55% TDN.

FIGURE 27–3
Feed handling is largely mechanized today. Chain-drag type elevators such as the one shown are widely used for handling almost all types of feeds, ranging from small grains to baled hay. (Courtesy of Kewanee Machinery & Conveyor Co., Kewanee, Ill.)

FIGURE 27–4
A large-round-bale baler in operation. (Courtesy of Vermeer
Manufacturing Company, Pella, Ia.)

FIGURE 27–5
A large-round-bale unroller in operation. Such a procedure is
sometimes followed for feeding large round bales on the
ground (or snow). (Courtesy of Vermeer Manufacturing Co.,
Pella, Ia.)

FIGURE 27–6
Hereford brood cows eating hay from large round bales that
have been unrolled on the snow. Wastage from trampling is
sometimes excessive with such a practice—especially if the
hay is fed too liberally. (Courtesy of Vermeer Manufacturing
Co., Pella, Ia.)

FIGURE 27–7
Offering large round bales of hay without some sort of protective ring to limit access will lead to excessive wastage of
the hay.

FIGURE 27–8
Large round baling machines can be equipped with wrapping
devices to wrap the bales in a protective cover. This will cut
down greatly on loss of nutrients in hay stored outdoors.

FIGURE 27–9
A hay stacker being unloaded. (Courtesy of Farmhand, Inc.,
Hopkins, Minn.)

(a)   (b)

FIGURE 27–10
A stack mover in operation. Such equipment may be used
for moving either (a) stacks or (b) large round bales.
(Courtesy of Farmhand, Inc., Hopkins, Minn.)

C. Hay is primarily a cattle, horse, and sheep feed. Very little hay of any kind is ever fed to swine. Dehydrated alfalfa meal is sometimes included in swine rations up to 5% to 10% of the ration, but its use for this purpose is on the decline.

IV   Use of hay preservatives.

A. Effective hay making over the years has been dependent on drying the forage material either by natural or artificial methods to about 18% to 22% moisture depending on the prevailing circumstances as mentioned under paragraph IA.

B. During recent years research efforts by college experiment stations and industry have been directed toward the use of various products as preservatives that, if added to forage material at baling time, might permit the forage to be safely baled with as much as 25% or more moisture.

C. Some of the products that offer promise of being used for this purpose are sodium diacetate, proprionic acid, acetic acid, and anhydrous ammonia.

D. If, through the use of such products, the safe storage level for moisture in forage could be increased from 18–20% up to 25% or more, improved hay making could be realized through the following:

1. More prompt baling, thereby reducing losses from rain.

2. Greater percentage of leaves saved, thereby increasing the total yield of more nutritious hay.

E. It appears that the expanded use of such a practice at the farm level will depend on the development of procedures for the effective application of the preservative at a cost the farmer can afford.

28

Silage and Haylage— Production and Use

I

What is silage? What is haylage? These two terms refer to crops that are stored at moisture levels higher than approximately 20%. The principle of such stored feeds is to exclude oxygen—or to store feeds anaerobically. Under anaerobic conditions, suitable microorganisms ferment available carbohydrates to produce organic acids, primarily acetic and lactic, with lesser amounts of butyric acid. Such fermentation continues until a pH of from 3.6 to 5.0 is reached, after which the fermentation action ceases. As long as oxygen is kept from such material, no further action will take place. *Silage* is usually made from crops such as corn or sorghum, but can be made from almost any crop that is palatable to livestock. Usually, such crops contain from 55% to 65% moisture. *Haylage* is a special favorite of dairymen because such forages as alfalfa make excellent haylage. Such crops are allowed to wilt in the swath—or windrow—until the moisture content has dropped to 35% to 45%. Such feed is lighter in weight per unit of measure and is well relished by livestock. Wilting of such crops removes the necessity of added carbohydrates for enhancement of the fermentation process. Wilted forage must be chopped fine enough so that it will pack to prevent trapping oxygen.

II

What is a silo? A silo is an airtight to semi-airtight structure designed for the storage and preservation of high-moisture feeds as silage.

Silos are of several types.

A. Horizontal.

    1. Trench.

        a. Usually consists of a trench dug into the side of a hill—sometimes all the way through the top of a hill.

        b. There are many different sizes but usually range from 10 ft to 40 ft in depth, 15 ft to 50 ft in width, and 100 ft to 300 ft in length.

        c. They are usually narrower at the bottom than at the top to facilitate effective packing of silage material.

        d. The floor and walls may vary from plain dirt to all concrete. Usually at least the floor is concreted in order to avoid the problem of mud as the silage is removed.

        e. The floor is usually sloped toward the downhill end to permit drainage. (Figure 28–1.)

        f. Such silos are usually filled from the upper end to the lower end using dump wagons and/or trucks.

        g. A tractor is usually used to pack the silage as the silo is filled.

        h. When the silo is filled, the silage is usually covered over with a heavy-weight polyethylene film which is weighted down and held in place with

FIGURE 28–1
(a) A partially filled trench-type silo with dirt floor and exterior walls, divided down the center with a concrete partition. (b) A nearly empty trench-type silo with concrete floor and sprayed-cement walls. A tractor with front-end loader is being used for filling silage wagon. (c) The lower end of a trench type silo with concrete floor and sprayed-cement walls. (Courtesy of the University of Georgia College of Agriculture Experiment Stations)

soil, boards, posts, old tires, or some other material. As an alternative to plastic film, a thin layer of cane molasses has been used to provide an effective seal.

i. Trench silos are usually emptied from the lower end to the upper end using a tractor with a front-end loader. (Figure 28–2.)

j. Trench silos are sometimes provided with movable stanchions and used on a self-feeding basis for livestock.

k. A trench silo that extends all the way through the top of a hill would be treated much as two trench silos end to end with each other.

2. **Bunker.**

a. This type of silo is sometimes used on very flat, rocky, and/or pervious soil not well suited to a trench silo.

b. The side walls are usually made with posts and boards lined inside with building paper or plastic film. The floor is usually concreted, and the ends are left open.

c. A bunker silo is used much in the same manner as a trench silo, the main difference being that the former is above ground and the latter below ground.

3. **"Weenie bags"** (Figure 28–3).

a. A temporary silo, good for one use.

FIGURE 28–2
A large concrete bunker-type silo being filled with corn silage. (Courtesy of Dr. W. E. Wheeler, USDA, Roman L. Hruska Meat Animal Research Center, Clay Center, Nebr.)

FIGURE 28–3
So-called "weenie bags" can provide a one-time storage facility for silage or haylage. The machine shown will pack a 100-foot long (35 m) plastic tube. Such a tube reportedly will hold 100 tons of silage, with no spoilage as long as the plastic is not punctured.

    b.   Fairly expensive alternative.

    c.   Keep it "air-tight" and it will preserve its contents.

B.   **Vertical or upright**—sometimes called *tower silos.*

    1.   **Conventional upright** (semi-airtight).

        a.   Most present-day conventional upright silos are constructed of reinforced poured concrete or of concrete staves.

        b.   All upright silos are cylindrical in shape. (Figure 28–4.)

        c.   Conventional upright silos may or may not have a roof; they usually do, primarily to protect the unloading equipment.

        d.   A conventional upright silo is ordinarily equipped with a series of doors about 2 ft square located approximately every 4 ft up one side of the silo. These are closed as the silo is filled and opened as the silo is emptied.

        e.   Conventional upright silos vary greatly in size but are usually from about 18 ft to 50 ft in diameter and from 40 ft to 80 ft in height.

        f.   This type of silo is normally unloaded from the top—usually with a mechanized unloader.

FIGURE 28–4
Shown in the picture are two upright, concrete stave, top un-
loading silos along with a completely mechanized feeding
setup. (Courtesy of Badger Northland Inc., Kaukauna, Wis.)

    g. Conventional upright silos are usually emptied before being refilled, but a partially empty silo of this type may be refilled.

    h. For effective preservation of ensilage in a conventional upright silo, the ensilage should contain from 25% to 40% dry matter.

2. "Airtight" or sealed silos (Limited oxygen).

    a. To date these have all been of an upright type.

    b. They are constructed of protected metal with rubber-cemented joints, or of monolithic concrete. (Figure 28–5.)

    c. When properly constructed, they are completely airtight.

    d. They vary in size from about 18 ft to 40 ft in diameter and from about 40 ft to 100 ft in height.

    e. Forages varying in dry matter content from about 25% to 75% may be effectively preserved and stored in an airtight silo.

    f. There are several makes of limited-oxygen silo.

       • The A. O. Smith *Harvestore* (Figure 28–5) was the first and is probably the most widely used limited-oxygen silo. In many respects, this

FIGURE 28–5
Limited-oxygen type silos equipped with a completely mechanized feeding system. (Courtesy of A. O. Smith Harvestore Co., Arlington Heights, Ill.)

silo approaches the ideal in silage preservation and storage. It holds ensiling losses to a minimum, facilitates continuous harvesting and feeding, has a breather bag in the top to allow for interior gas expansion and contraction, and unloads from the bottom by means of a mechanical unloader. It has two major drawbacks: (1) its initial cost is relatively high, and (2) the feeder is completely dependent on the mechanical unloader for silage. This is a steel-glass fused structure.

• The Piedmont Silo Company builds an airtight silo (the *Profit Center*) that is of monolithic concrete construction (Figure 28–6). It is filled through a door on the side of the silo at the top. Also at the top of the silo are an access door, a vent door, and an automatic gas pressure valve. The silo comes equipped with a flail-type bottom unloader.

IV — Kinds of crops used for silage.

A. Practically any crop may be made into silage, provided it contains an appropriate level of moisture, adequate amounts of readily fermentable carbohydrates, and adequate levels of other nutrients, and provided it can be sufficiently packed.

B. The most commonly used silage crops are as follows:

1. Corn.

FIGURE 28–6
Limited-oxygen silos of monolithic concrete construction.
(Courtesy of Piedmont Silo/Covington, Georgia.)

    a. Most extensively used silage crop—about 16 times as much corn as sorghum is grown for silage (Figure 28–7).

    b. Corn silage is unexcelled in quality.

    c. Corn is not as high yielding as some of the forage sorghums.

2. Forage sorghum (sorgo).

    a. Included here are the traditional sweet sorghums as well as some new hybrid sorghum varieties.

    b. Most are very high yielding but do not produce the best silage; such silage is inclined to contain excess water and acid, and hence is not very palatable or nutritious.

    c. Probably not more than about 5% of the silage produced in the United States is forage sorghum.

3. Grain sorghum.

    a. Probably not more than about 5% of the silage produced in the United States is grain sorghum.

FIGURE 28–7
A forage harvester being used for chopping corn in the field
for making silage. Corn is by far the most widely used silage
crop. (Courtesy of Allis-Chalmers, Milwaukee, Wis.)

b.   Grain sorghum silage is between corn silage and forage sorghum silage in palatability but because of a higher dry matter content is superior to sorgo in nutritive value.

c.   The grain sorghums are generally low yielding.

4.   **Small grains and hay crops.**

a.   None of these are used extensively as a silage crop outside of the southeast.

b.   The quality of silage resulting from these materials is quite variable, depending on such factors as stage of maturity, moisture content, and additives used, if any.

c.   Yields are variable but are usually below corn and the sorghums.

V

Preparation of forage for making silage and haylage.

A.   Most crops to be used for silage are permitted to mature or field dry to a moisture level of 65% to 75% (25% to 35% DM). For corn this is about the early dent stage of maturity and for grain sorghum the late dough stage at the earliest. This is when the moisture level is about right for good silage formation. Silage materials containing less than 25% dry matter (more than 75% moisture) form a very sour silage and usually lose large amounts of silage juices during storage, involving a considerable loss of nutrients. Silage materials with more than 35% dry matter do not pack well and frequently develop spots of mold during storage as the result of excess entrapped oxygen. This is especially true for nonairtight silos.

B. Silage crops are usually chopped into fairly small pieces for making silage. The pieces will usually vary from a fraction of an inch to more than an inch in length. This permits good packing and facilitates the mechanization of silage handling (Figure 28–8.) Ground shelled corn, cane molasses, lime-stone, urea, and/or various commercial preparations are sometimes added to reduce fermentation losses, improve silage quality, and/or increase nutritive value.

C. Crops intended for haylage making must be dried in either the swath or windrow until the moisture has declined to 30% to 40%, before it is chopped and collected for ensiling.

VI

How silage is formed.

A. Basic for silage formation is the early establishment and maintenance of a relatively oxygen-free (anaerobic) environment.

1. All silos are constructed to be as oxygen-limiting as practicable with the type of silo being used.

2. When practicable, silage material is packed as it goes into the silo to reduce air spaces to a minimum and eliminate as much oxygen as possible.

3. Most silage is made from green to semi-green forage material that uses up the oxygen present during the fermentation process.

B. Following storage, the forage material normally undergoes an anaerobic fermentation to change it into silage.

1. The fermentation normally begins within hours after storage.

2. The heat produced by the living cells of the forage material during the fermentation process tends to create a favorable environment for the anaerobic bacteria responsible for the fermentation process.

3. The fermentation process accelerates during the first 2 to 3 days, then levels off, and normally will, for the most part, terminate itself within the first 2 to 3 weeks, when a pH of 3.6 to 5.0 is reached.

4. Products of the fermentation are the following:

   a. Organic acids—primarily lactic, acetic, and butyric.

   b. Ethanol in varying amounts.

   c. Fermentation gases—primarily CO_2, CH_4, CO, NO, and NO_2.

   d. Water.

   e. Heat.

5. The fermentation is normally terminated by the accumulation of certain by-products of bacterial metabolism, which tend to preserve the forage material indefinitely unless air is permitted to enter.

FIGURE 28–8
Field chopped corn fodder being moved from a field wagon
through a blower into a silo in the process of making silage.
(Courtesy of Sperry-New Holland, New Holland, Pa.)

6. If air is able to permeate the silage, a prolonged aerobic fermentation, and possibly molding and spoilage, may result.

C. **Silage additives**—Over the years, different materials or products have been added to forage prior to ensiling. The goal of these additives is to produce an ensiled feedstuff with a greater nutritive value, greater acceptance by the animal, and/or greater reduction of losses associated with ensiling. Additives can be classified into two major groups.

1. **Fermentation inhibitors.**

   a. Acids may be employed for direct acidification. Examples are formic acid, used extensively in Europe, and propionic acid, used in the United States. These materials are caustic and corrosive, and tend to be expensive but have potential for improving certain types of forage crops.

   b. Preservatives such as sodium chloride, antibiotics, sodium metabisulfate, formalin, formaldehyde, etc. have been investigated. Formaldehyde has shown excellent promise but is not cleared for use in feeds by the FDA.

2. **Fermentation stimulants.**

   a. Nutrient sources such as grain, molasses, whey, citrus pulp, beet pulp, urea, ammonia, and limestone have been added to forage at ensiling. Adding molasses or ground feed grains to forage crops with low fermentable carbohydrates (legume and grass crops) can improve the fermentation and silage quality, but the additional cost may outweigh the advantage gained in silage quality. Using large amounts of dry grains, citrus pulp, straw, hay, etc. is good to increase the dry matter content of a forage that is too high in moisture but is generally uneconomical. The addition of 10 lb of urea per ton of corn silage at ensiling time has become a common practice. This will increase the protein of the dry matter by about 50%. Also anhydrous ammonia can be used to increase the protein content of corn silage (8–9 lb per ton of silage). Limestone added to corn silage at the rate of 10 lb per ton of forage may increase the feeding value for beef cattle but not for lactating dairy cows, because the increased lactic acid in the silage results in lower butterfat content of the milk.

b. In recent years, a large number of products have come on the market that, when added to the forage, supplies microbes or enzymes. The use of microbial inoculants has given variable results depending on the type of forage crop and the moisture content of the crop at ensiling. Under certain conditions the inoculants have reduced dry matter loss and resulted in slight increases in gain and feed efficiency of cattle. The decision to use one of these products should be determined by the potential benefit in feeding value compared to the cost of the material. A silage additive should never take the place of good ensiling practices, including the following:

- Harvesting the crop at the proper stage of maturity.

- Harvesting and filling the silo rapidly.

- Chopping finely enough to insure adequate silo packing (exclude air).

- Covering and protecting from the elements.

- Using rapidly and regularly once opened.

VII

Use of silage and haylage as a livestock feed.

A. Silage and haylage are used primarily in beef and dairy feeds, as part—or all—of the roughage. These feeds, especially haylage, may be fed to sheep, but neither fit well into horse-feeding programs. Research has demonstrated that feeding pregnant sows in drylot only corn silage and a balancing supplement will aid in controlling weight gains.

B. About 2.5 lb to 3.0 lb (1.1 kg to 1.4 kg) silage, or 2.0 lb to 3.0 lb haylage will be required to replace 1 lb hay.

C. From 2 in to 5 in of ensiled material should be fed off the exposed surface in order to keep the ensiled material "fresh." The warmer the weather, the more it will be necessary to feed. With a limited-oxygen storage structure, this is not as critical—but such structures must be kept closed except when ensiled material is being withdrawn.

D. Once feeding has begun from a conventional silo, it is best to continue feeding from that silo on an uninterrupted basis until it is empty; otherwise, there will be considerable spoilage of the exposed surface.

E. Silage is usually used when pastures are short.

F. In addition to serving as a roughage, silage is a good appetizer and tends to keep cattle on feed during hot weather.

VIII Advantages of silage and haylage.

A. More TDN per acre. When only the grain of a crop is harvested, up to 50% of the TDN is left in the field.

B. Maintains feed in a succulent form. Dry feed must become saturated with water before it can be digested. Silage never loses its wetness.

C. No losses from shattering, leaching, or bleaching.

D. No waste in feeding. Even the cobs, shucks, and coarsest stems are eaten.

E. Even weedy crops make good silage.

F. Silage can be made in almost any kind of weather.

G. Permits early re-use of land. Facilitates double cropping.

H. Less fire hazard with silage than with hay and other dry crops.

I. Less internal parasite trouble with silage than with pasture.

J. A silage program facilitates complete mechanization of forage harvesting and feeding.

IX Disadvantages of silage and haylage.

A. Extra labor is required at silo-filling time.

B. Considerable costly equipment is required for the harvesting, storing, and feeding of silage.

C. Must handle considerable water as a part of the silage material.

D. Not well suited for intermittent use.

E. Silage in less than oxygen-limiting silos may undergo considerable deterioration if held for extended periods.

29 Problems of Feed Storage

Having feed available for livestock throughout the year ordinarily involves the storage of large amounts of feed. Most feed crops are seasonal and consequently must be held in some form of storage from harvest to feeding. This responsibility is ordinarily shared by the feed producers, the elevator operators, and the livestock feeders. In some instances, the producer and the feeder may be one and the same. In any event, it is essential that the feed material be properly stored from harvest time to feeding time in order to avoid damage to the feed with loss of nutrients and possible harm to consuming livestock.

I. One of the major factors influencing the effectiveness of a feed storage operation is that of moisture content.

    A. This is especially true for feeds stored in a supposedly air-dry form. While it is assumed that air-dry feeds will sooner or later attain a moisture level of about 10%, they will frequently vary on either side of this figure. Feeds with considerably less than this amount of moisture are usually feeds that have been subjected to artificial drying. Feeds are never too dry for bin or shed storage. However, feeds that appear to be sufficiently dry for bin or shed storage sometimes contain more moisture than will permit their safe storage.

B. An excess amount of moisture will cause such feeds to heat and, if not given prompt attention, they will mold and become unsuitable for feeding purposes. The amount of moisture that a feed can safely tolerate will vary depending on the atmospheric temperature, the humidity, the amount of air circulation through the feed, and possibly other factors.

C. Ground feeds will tolerate up to about 11% moisture under average conditions without molding during storage. Moisture levels in ground feeds higher than 11% will probably result in heating, caking, and/or molding during storage over an extended period. On the other hand, unground feed grains such as corn, whole oats, barley, and unprocessed grain sorghum will tolerate somewhat higher levels of moisture in storage without spoilage. In view of this, whole grains containing more than about 11% moisture should not be used in the production of ground mixtures to be held in storage for more than just a few days, especially during warm weather. Small grains such as barley, oats, and grain sorghum will tolerate up to about 13% moisture in storage, while shelled corn will tolerate up to about 15.5%.

D. Anytime a grain exceeds its permissible moisture level for safe storage, it should be dried down to this level before being placed in storage. For each percentage point of moisture that a grain contains over the permissible level, about a 2% reduction, or a little more, in the price of the grain is justified; 1.2% covers the lower level of dry matter it contains and the rest covers the cost of drying the grain down to a permissible level.

E. As the cost of energy increases, the latter charge will naturally increase in magnitude; and, as the energy crunch becomes greater over the years ahead, the need for avoiding the use of fossil fuels for artificially drying feed grains will become ever more pressing.

F. Fortunately, alternative procedures for expediting the use of grains for feeding purposes are available or are being developed.

1. One alternative procedure that might be used is to allow the crop to remain in the field longer before being harvested thereby allowing it to dry enough through field drying for storage. Such a practice, however, subjects a crop to the hazards of weather damage and predators during the drying period. It also extends the time a crop occupies a field, thus making it more difficult to carry out a double cropping program, regarded by many as an essential part of a profitable farming operation.

2. Progress is being made in the use of solar energy to replace fossil fuels in carrying out a feed drying operation. To what extent such a development will find its way into the feed drying business will be determined by future circumstances.

G. Alternatives to the drying of feeds to facilitate their safe storage and use are available.

1. One such alternative is treating them with a preservative. Propionic acid sprayed at the rate of 0.4% to 0.5% on corn grain carrying up to 24% moisture has been found to provide mold-free storage in open but weather protected bins at 23°C for several months. Acetic, formic, and isobutyric acids used in appropriate strengths and combinations have also been found to be effective for this purpose. The acid treatments tend to pickle the grain, thereby preventing fungi formation. The higher the moisture level, the greater the amount of acid required. Other grains may be similarly preserved. Grains so treated seem to have an equal or slightly superior DM feeding value than those that have been brought to air dryness before feeding.

2.   Another alternative for the storage and use of high-moisture grains is to ensile them. This is most effectively carried out with high-moisture grains in a so-called airtight or oxygen-limiting solo. Grains having from about 22% to 30% moisture are usually used. With less than about 22% moisture, ensiling takes place so slowly and to such a slight extent that a limited amount of molding will occur, supported by the oxygen entrapped in the grain at storage time. Again, grains so handled have usually been found to be equal to or slightly superior to air-dry grain for livestock feeding purposes in per lb of DM.

H.   Moisture also must be considered in the harvesting, storage, and use of forage crops.

1.   For a forage crop to be stored and used as an air-dry forage, it must be brought down to about 18% to 22% moisture before being baled or stacked for storage. The precise level of moisture that can be tolerated will depend on the coarseness of the material, the size and tightness of the bale or stack, and the temperature, humidity, and movement of the surrounding air. Should the moisture level in the forage material be too great in relation to other factors, the forage will build up an interior heat of considerable magnitude associated with molding or browning. Under certain conditions temperatures may go so high as to produce spontaneous combustion causing fire with loss of feed and property.

2.   If a drying problem in the harvesting of a forage crop is anticipated, it may be handled by storing the crop as haylage or as silage. If the crop carries from 60% to 75% moisture, it can usually be stored effectively as silage. Almost any conventional silo can be used for this purpose. If the forage contains more than about 75% moisture, it would be best to allow it to dry in the swath to a lower moisture level or mix it with a low-moisture material before putting it in the silo. Forages with more than about 75% moisture tend to lose nutrients through seepage and to produce an overly acid silage.

3.   Should it be feasible to dry a forage crop in the swath down to 40% to 60% moisture, but anticipated inclement weather prevents the crop from being made into hay, one might choose to make it into "haylage." Haylage is a feed that is intermediate between hay and silage in moisture content and is preserved as a low-moisture silage in a silo. For most effective preservation it is best to store it in a limited-oxygen silo. Haylage also may be made in a conventional upright silo provided the moisture level is not too low and the material is well packed. It is not recommended that one try to make haylage in a

trench or bunker silo since such silos are not ordinarily sufficiently airtight or the haylage sufficiently well packed to prevent a certain amount of mold formation.

4. Hay, haylage, and silage made from the same forage crop will not differ greatly in feeding value per lb of dry matter. However, harvesting losses may at times be excessive because of drying problems in the making of hay, especially during inclement weather. Also, certain crops, mainly the legumes, will shatter their leaves badly if permitted to become too dry during the haying operation, resulting in a serious loss of valuable nutrients.

II

While not a factor in the storage of forage crops (such as hay, haylage, and/or silage), insects pose a serious problem in the storage of most grains and certain other feed materials.

A. Several different species of moths, beetles, and weevils are potentially involved in this problem. Both the adult and larval stages can cause damage of one type or another to feed materials.

B. Damage may be in the form of actual destruction of the grain or feed or through contamination with insect eggs and feces or other insect excrement.

C. Most feed-inhabiting insects range from about 1/16 in to 5/16 in in length although one or two species fall outside this range.

D. Infestations are most extensive in the southern states but occur more or less in all parts of this country and in most other countries throughout the world.

E. The type and degree of insect damage in any individual lot of feed material will depend on several factors.

1. A major factor is the degree of infestation of the material with insect eggs and/or with developing and/or mature individuals at the time it is stored. To a large degree, this will be determined by the level of infestation in the area where the feed is produced.

2. The accessibility of the feed to insect invasion prior to and during harvest is also a major consideration.

3. Undue delay in harvesting and the use of inadequate holding facilities during harvest are major factors in this connection. The proper treatment of grain with a chemical protectant at the time of harvest will greatly help prevent insect invasion. Also, insect infestations are sometimes destroyed completely or in part by high drying temperatures.

F. At any particular level of insect infestation at which grains are placed in extended storage, the extent of potential insect damage will be determined largely by the length of time in storage and the environmental temperature during storage.

1. Since most grain insect species require only 1 to 2 months during favorable circumstances to complete their life cycles, and since most insect populations increase more or less in geometric progressions, the shorter the period of storage, the less the damage from insects will usually be.

2. In all instances where grain is to be held in storage for several months every effort should be made to start out with a minimum of initial insect infestation and to prevent recurring infestations from taking place during storage.

G. In any event, grain held in storage for extended periods during the summer, and especially grain held for more than a year, should be inspected at intervals for the presence of insects or evidence of insect damage.

1. Should any insects or insect damage be found, immediate steps should be taken to fumigate such grain.

2. Someone who is regarded as an authority on fumigating procedures should be consulted in this matter.

30 Processing Feeds

Mechanical processing of grains.

    A.   Most grains are ground, rolled, or crimped before being fed to livestock. (Figure 30–1.)

            1.   Grinding is usually accomplished with a *hammer mill* (Figures 30-2 and 30–3). A hammer mill grinds by beating grain until it is fine enough to pass through a screen. The size of screen will determine the degree of fineness obtained.

            2.   Rolling is accomplished by passing grain between a closely fitted set of rollers. This leaves the grain in the form of a flake, and the process is sometimes referred to as *flaking*. Cattle seem to prefer flaked grain to ground grain and usually do somewhat better on the flaked compared to the ground product, especially steam-flaked; steam-flaking accomplishes some gelatinization of the starch.

            3.   Crimping is similar to rolling except that instead of rollers with smooth surfaces being used, rollers with corrugated surfaces are used. The end result is much the same.

    B.   Mechanical processing of grains is especially recommended for the following:

            1.   Very young animals before their teeth are fully developed. Grain is usually ground for such animals.

FIGURE 30-1
A typical feed-processing plant such as accompany most big feeding operations. (Courtesy of Dr. W. E. Wheeler, USDA, Roman L. Hruska Meat Animal Research Center, Clay Center, Nebr.)

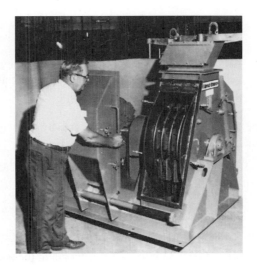

FIGURE 30-2
A commercial hammer mill with interior exposed. A perforated metal screen that encloses the grinding chamber has been removed to reveal the rotor, hinge pins, and hammers. The inlet at the top includes a magnet for the removal of tramp metal. (Courtesy of Sprout-Waldron, Muncy, Pa.)

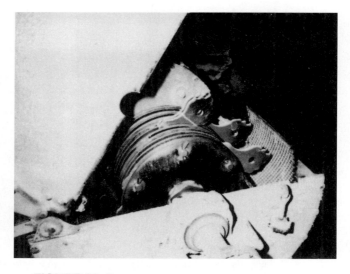

FIGURE 30–3
An inside view of a farm-type hammer mill with some of the hammers and a section of the screen exposed. (Courtesy of the University of Georgia College of Agriculture Experiment Stations)

2. Very old animals with badly worn teeth. For such animals grain is also usually ground.

3. All cattle over six months of age unless the roughage content of the ration is very low. Grinding, rolling, and flaking are all used extensively in processing grain for cattle, horses, and sheep.

C. The following classes of livestock usually chew grain fairly thoroughly and therefore do not need mechanically processed grain.

Calves 2–6 months of age	Horses and mules
Sheep and lambs	Fattening pigs

However, even with these classes, the improvement of feeding value from processing will usually offset the processing costs.

D. Grinding to a medium fineness (equivalent to about a coarse meal) is usually best. Finely ground feeds (flour fine) are dusty and unpalatable to livestock. Fine grinding might be desirable for grains containing small weed seeds, the viability of which should be destroyed. Coarse grinding or cracking may be more trouble than it is worth.

E. Naturally grinding is conducive to the production of more homogeneous mixtures.

II   **Grinding hays.**

   A.   It is not necessary to grind good-quality hay to realize its effective use.

   B.   Grinding coarse, stemmy hays will encourage their total consumption by livestock but will not improve their digestibility.

   C.   Hay must be ground for incorporation into complete ration mixtures for livestock.

   D.   The coarser the hay is ground, the more it retains its bulk value.

   E.   Grinding their hay will cause a drop in the butterfat level of dairy cows.

III   **Pelleting feeds.**

   A. Pelleting grains and other concentrates.

   1.   Feeds to be pelleted are usually ground first—hence, they enjoy the benefits of grinding.

   2.   Pelleting returns the feed to a free-flowing form, thus facilitating the mechanization of handling it and also its use in a self-feeder.

   3.   Pelleted feeds (Figure 30–4) are usually less dusty and more palatable, and livestock consume more.

   4.   Pelleting reduces storage space requirement of a feed.

   5.   Pelleting of ground concentrate mixtures will add to the cost of the feed.

   B. Pelleting hay and other roughages.

   1.   Hays must be ground prior to pelleting—hence, pelleting has most of the advantages and disadvantages of grinding.

   2.   Pelleting of hays and other roughages converts them into a free-flowing form that can be handled mechanically and fed in a self-feeder.

   3.   Pelleting of hay and other roughages reduces the space requirement for storage by as much as 75%.

   4.   Pelleting of hay and other roughages increases consumption and performance in beef cattle.

   5.   Pelleting eliminates the air spaces in hays and other roughages, and increases their density, thus causing them to lose much of their roughage value.

   6.   Pelleting reduces dustiness of hay and other roughages.

   7.   Pelleting roughages is about twice as costly as pelleting concentrates.

FIGURE 30–4
Pelleted feeds have some definite advantages as a form in
which to provide feed for livestock. (Courtesy of Ralston
Purina Co., St. Louis, Mo.)

8. Pelleting greatly reduces the roughage value of hays and other fibrous feeds.

C. **Pelleting complete rations.**

1. Same advantages as with individual feeds.

2. Pelleting of complete rations for cattle destroys the roughage value of any hay or roughage in the ration.

3. Unpelleted roughage must be fed along with the pellets for good results with cattle and probably sheep.

4. Pelleting of complete rations for cattle and sheep will usually not be economical. It may be practical for horses and swine.

D. **Cubing supplements for brood cows on low-quality pasture.**

1. "Cubes" are nothing more than large pellets.

2. Cubes may be fed on the ground in clean pastures; no troughs are needed.

3. The cost of the cubing is usually more than offset by the added convenience.

IV **Cooking feeds.**

    A. Cooking may take on different forms.

        1. Conventional cooking with or without steam pressure.

        2. Short period steaming—usually in connection with rolling or flaking.

        3. Roasting and "popping."

    B. While cooking will frequently improve the palatability and nutritive value of feeds, it has not proved to be a profitable practice for feeds in general, except with a few large feedlots.

    C. Cooking seems to have its greatest benefit with

        Irish potatoes   Field beans
        Soybeans

    D. Barley is frequently cooked for show cattle to improve ration acceptability.

    E. Garbage from homes and feeding establishments has been widely used over the years as a feed for swine. This practice has been greatly curtailed during recent years, however, as the result of a multi-state outbreak of VE disease (Vesicular Exanthema) in swine in the 1950s. This is a serious disease of swine, which is spread primarily through the consumption of uncooked or inadequately cooked flesh from hogs that had the disease when slaughtered. In order to prevent the undue spread of this and other diseases among swine through the feeding of garbage, the USDA ruled that any garbage to be fed to swine must be cooked for at least 30 minutes at a temperature of 212°F. Some fifteen or more states have passed laws completely prohibiting the feeding of garbage to swine, cooked or uncooked. It may still be fed in other states provided USDA cooking requirements are met. Operators feeding their own domestic table scraps are not affected by garbage feeding regulations.

V **Wetting feeds.** Not usually considered to be practical except possibly with

    Very dusty feeds   Rations for show cattle
    Oats for horses and mules

VI **Soaking feeds.** Not usually considered to be practical except possibly with

    Very hard grains that are not   Dried beet pulp, soybran flakes,
        mechanically processed   and similar feeds normally
    Tender- or sore-mouthed horses   fed in the wet form.
        and mules.

VII   **Fermenting feeds (other than silage).** No benefit has ever been demonstrated experimentally from the fermenting of feeds. Efforts to improve ration digestibility and protein quality and to detoxify certain feeds by fermentation have been to no avail.

VIII   **Germinating grains.** Not generally considered to be practical. It may serve as a means of providing the grass juice factor for dairy cows when pasture is not available, but this has not been demonstrated experimentally to be of practical value.

IX   **Liquid feeds and slurries.**

    A. Supplements are sometimes provided to brood cows and developing heifers on pasture through feeders in the liquid form. This is a convenient way of feeding supplements to such animals (Figure 30–5).

    B. Supplements in the liquid form are sometimes added to complete ration mixes for livestock. This is a satisfactory way of adding supplemental nutrients to the ration, provided it is more convenient and/or cheaper than the dry form and does not add excessive moisture to the mix.

FIGURE 30–5
Brood cows on pasture utilizing pasture supplement from a lick tank. (Courtesy QLF Feeds, Dodgeville, WI.)

C. Some swine producers follow a practice of feeding the complete ration in the form of a slurry or gruel. Some of the reasons given for following such a practice are:

1. It helps mechanize the feeding operation.

2. It eliminates dust.

3. It reduces feed wastage.

4. The animals prefer wet feed to dry feed.

5. Early weaned pigs do better on wet feed than they do on dry feed.

6. Permits limited calorie intake, where desirable.

D. Has proven popular in many European feeding operations.

31 Uniform State Feed Bill

Except for certain federal regulations dealing primarily with the use of new drugs in animal feeds, the manufacture and distribution of commercial feeds are, for the most part, regulated by the respective State Departments of Agriculture in accordance with their state laws. Fortunately, the feed laws, rules, and regulations for most of the states are quite similar. To a very large degree this has come about as the result of the preparation and publication of a so-called "Uniform State Feed Bill" by the Association of the American Feed Control Officials (AAFCO). This uniform bill is an important part of every feed manufacturers library and allows them to observe the regulations contained therein. The student of this text would not be expected to memorize all the regulations regarding a "safe" product. However, a feed manufacturer would undoubtedly maintain a good understanding of the regulations of AAFCO, because there are penalties for those who do not comply.

I

The purpose of this section is to give the student a general understanding of the Uniform State Feed Bill so that as either a purchaser or manufacturer of feeds, he or she will appreciate the legal aspects of feed manufacturing. The following represents the author's abbreviated interpretation of the bill and should not be used as a guideline by a feed manufacturer.

A. **An Act.** This bill of regulations has been passed into law in most states. The first part contains definitions of some words and phrases used in the bill.

1. *Distribute* means to offer for sale; a *distributor* is any person who offers for sale.

2. *Commercial feed* means all materials that are used as feed or for mixing in feed, except whole seeds unmixed, or physically altered entire unmixed seeds (when not adulterated). Some materials may be excluded from this category, such as hay, straw, stover, silage, cobs, husks, hulls, and individual chemical compounds or substances, when such commodities are not intermixed with other materials.

3. *Feed ingredient* means each of the constituent materials making up a commercial feed.

4. *Mineral feed* is a commercial feed intended to supply primarily mineral elements.

5. *Drug* is any article intended for use in the diagnosis, cure, mitigation, treatment, or prevention of disease in animals other than man, and articles other than feed intended to affect the structure and function of the animal body.

6. *Manufacture* means to grind, mix, blend, or further process a commercial feed for distribution.

7. *Brand name* identifies one feed manufacturer's products from that of another.

8. *Label* is a display of written, printed, or graphic material upon or affixed to the container in which a commercial feed is distributed, or on an invoice or delivery slip.

9. *Ton* is 2000 lb, avoirdupois; *percent* is percentage by weight.

10. *Official sample* is one taken by a designated feed-control official.

11. *Pet food* is for domesticated animals maintained near the owners residence; *specialty pet food* refers to pets normally maintained in a cage or tank, such as (but not limited to) gerbils, hamsters, fish, goldfish, snakes, etc.

B. **Registration.** No person is to manufacture a commercial feed unless the proper forms have been filed with the state regulatory agency. Each location of business within that state must be registered.

C. **Labeling.** A commercial feed must be labeled with very specific details, including the following:

1. Net weight.

2. Product name and brand name.

3. Guaranteed analysis for all ingredients for which a claim is made. (These analyses should be conducted by methods set forth by the Association of Official

Analytical Chemists [AOAC] because that is how any State Feed Control Official will verify such claims.)

4. The common or usual name of each ingredient used in the manufacture of the commercial feed.

5. Name and principal mailing address of the manufacturer or the person responsible for the distribution of the feed.

6. Adequate directions for use of all commercial feeds containing drugs, such as level of feeding per day, potential hazards for other classes of animals that might consume it, and whether there is a withdrawal time before such animals may be marketed.

D. **Misbranding.** A commercial feed will be deemed *misbranded* if the labeling is false, if it is distributed under the name of another commercial feed, if it is not labeled as required, if any word that should be in the label is missing or altered.

E. **Adulteration.** This aspect is pretty straightforward. A commercial feed shall be deemed *adulterated* if it bears or contains poisonous or deleterious substances that may render it injurious to health. Furthermore, under this category, if it contains "raw" ingredients that contain poisonous substances such as herbicides or insecticides that should have been removed, it is adulterated. If a less-valuable substance has been substituted for a more-valuable ingredient, it falls under the category of adulteration. If it contains viable weed seeds in amounts exceeding the limits that have been stated, it is adulterated.

F. **Prohibited acts.** Naturally, it is illegal to manufacture or distribute misbranded or adulterated feeds. It is also illegal to manufacture and distribute feeds without registering with the State Feed Control Office, and paying registration and inspection fees.

G. **Inspection fees and reports.** An inspection fee at an established rate per ton is assessed on all commercial feeds. In the case of specialty pet food that is distributed only in packages of one pound or less, an annual fee is assessed. The feed manufacturing official responsible for reporting files a periodic report (usually 4 times per year) setting forth the number of pounds (tons) distributed in the state during the specified time period. At that time, the reporting person must pay the assessed fees.

H. **Inspection, sampling, and analysis.** Upon presentation of proper credentials, designated officers of the State Control Office may enter, during normal business hours, any factory, warehouse, or establishment within the state, in which commercial feeds are manufactured, processed, packed, or held for distribution, or enter any vehicle being used to transport or hold such feeds. If the officer making the inspection decides to sample any commercial feed, a receipt describing the samples taken is given to the owner. After analysis, the owner of the commercial feed is given a copy of such analysis. If inspection of feed indicates mislabeling or adulteration, the owner must be given a portion of the original sample.

I. **Detained commercial feed.** When the representative has reason to believe (by analyses, etc.) that the commercial feed being distributed is in violation of the State Feed Control Law, he may issue and enforce a written or printed "withdrawal from distribution" order.

J. **Penalties.** According to the bill, "Any person convicted of violating any of the provisions of this Act or who shall impede, hinder, or otherwise prevent, or attempt to prevent said Officer in performance of his duty in connection with the provisions of this Act shall be adjudged guilty of a misdemeanor and shall be fined."

K. **Publications.** Normally the State Feed Control Office will publish an annual report indicating the results of analyses of commercial feeds sold within the state.

II Official Rules and Regulations.

This section deals with such specifics as definitions and terms, and label format.

A. **Label format.**

1. Guaranteed analysis of feeds.

   a. Minimum percent protein; maximum or minimum percent protein from non–protein nitrogen.

   b. Minimum percent of crude fat.

   c. Maximum percent of crude fiber.

   d. Minerals to include in the following order:

      - Minimum and maximum percentage of calcium [Ca].

      - Minimum percentage of phosphorus [P].

      - Minimum and maximum percentage of salt [NaCl].

      - Other minerals.

      Guarantees are not required for minerals when there are no specific label claims and when the commercial feed contains less than 6 1/2% of calcium, phosphorus, sodium and chloride.

   e. Total sugars as invert on dried molasses products being sold primarily for their sugar content.

   f. Guarantees for vitamins are not required when the feed is neither formulated for, nor represented as, a vitamin supplement.

2. All the information concerning the feed manufacturer, the type of feed, and name and address.

3. The word "protein" is not permitted in any feed that contains urea.

4. Vitamin content must be in "milligrams per pound" except

    a. Vitamin A (other than precursors thereof) must be listed as International or USP units/lb.

    b. Vitamin D in products for poultry feeding are listed in ICU (International Chick Units [vitamin D_3]) per lb.

    c. Vitamin E shall be listed as IU or USP units per pound.

5. Feeds containing any added non–protein-nitrogen NPN must indicate the percent of total claimed protein coming from NPN; plus it must indicate the species of animal for which the feed is intended. If more than one-third of the total protein is from NPN, the label must state, "CAUTION: USE AS DIRECTED."

B. Adulterants. The following are all adulterants: Fluorine-bearing ingredients, such as raw rock phosphate; soybean meal from which the oil has been extracted with any chlorinated solvents; presence of sulfur dioxide or sulfurous acid in products purported to be a source of vitamin B_1; and viable weed seeds that have not been ground fine enough to destroy viability (note here that "ground grain screenings can contain a certain amount of foreign material").

C. Warnings. The following statements are typical of warnings that must be on tags where applicable.

1. If the product contains drug residues, then the label must state, "WARNING: THIS PRODUCT CONTAINS DRUG RESIDUES. DO NOT USE WITHIN 15 DAYS OF SLAUGHTER AND DO NOT USE 15 DAYS PRIOR TO OR DURING THE FOOD PRODUCTION OF DAIRY ANIMALS OR LAYING HENS."

2. If the product contains high levels (25 ppm or greater) of copper, a maximum guarantee of the copper and the following statement is required, "WARNING: CONTAINS HIGH LEVELS OF COPPER: DO NOT FEED TO SHEEP."

3. If the product derives one-third or more of the guaranteed total crude protein from NPN, the label must provide adequate directions for safe use of the product and the precautionary statement (as indicated in a previous paragraph): "CAUTION: USE ONLY AS DIRECTED."

Note to the student: Once again, the authors wish to point out this section is presented only to give the student an appreciation of the strict control over feed manufacturing, uniformly, throughout the United States. This section is not meant as an authoritative set of rules for feed manufacturers or distributors.

32 Study Questions and Problems

I. **Indicate by number whether each of the following feeds would be classified as:**

1. An air-dry energy feed.
2. An air-dry protein feed, animal origin.
3. An air-dry protein feed, plant origin.
4. An air-dry legume roughage.
5. An air-dry nonlegume hay.
6. An air-dry low-quality roughage.
7. An air-dry oil seed.
8. An air-dry mineral feed.
9. A high-moisture feed.

1. Peanut kernels.
2. Cottonseed meal.
3. Ground oystershells.
4. Meat and bonemeal.
5. Dried skimmed milk.
6. Cowpea hay.
7. Grain sorghum grain.
8. Corn silage.
9. Defluorinated phosphate.
10. Tankage with bone.

11. Fresh turnips.
12. Bermudagrass hay.
13. Red clover hay.
14. Shelled corn.
15. Peanut hulls.
16. Milo grain.
17. Ground corn cob.
18. Orchardgrass pasture.
19. Steamed bonemeal.
20. Soybean seed.
21. Fresh cow's milk.
22. Dried bakery product.
23. Sericea hay.
24. Coastal bermuda greenchop.
25. Poultry by-product meal.
26. Hominy feed.
27. Bahiagrass hay.
28. Oat straw.
29. Linseed meal.
30. Fescue hay.
31. Ground limestone.
32. Meat scrap.
33. Alfalfa meal.
34. Beet molasses.
35. Alfalfa hay.

36. Wheat grain.
37. Lespedeza hay.
38. Oat hulls.
39. Cottonseed.
40. Millet hay.
41. Soybean hay.
42. Tankage.
43. Brewers dried grains.
44. Oats grain.
45. Peanut oil meal.
46. Dried beet pulp.
47. Cottonseed hulls.
48. Coastal bermudagrass haylage.
49. Fish meal.
50. Corn stalks.
51. Dried citrus pulp.
52. Corn gluten meal.
53. Ground snapped corn.
54. Oat hay.
55. Rice hulls.
56. Johnsongrass hay.
57. Barley grain.
58. Wheat straw.
59. Feather meal.
60. Triticale grain.

II. **Indicate by number whether each of the compositions below is the composition of:**

1. An air-dry energy feed.
2. An air-dry protein feed, animal origin.
3. An air-dry protein feed, plant origin.
4. An air-dry legume roughage.
5. An air-dry nonlegume hay.
6. An air-dry low-quality roughage.
7. An air-dry oil seed.
8. An air-dry mineral product.
9. A molasses.
10. A high-moisture grain.
11. A haylage.
12. A silage.
13. A fresh forage.
14. A wet by-product.
15. A root crop.
16. Fresh whole or skimmed milk.

	DM %	Ash %	Fiber %	EE %	Prot %	TDN %	CA %	P %	Caro mg/kg
1.	94.2	24.9	2.5	9.4	54.9	62.0	8.49	4.18	*
2.	95.7	80.4	1.9	1.9	7.1	*	30.92	14.01	*
3.	89.0	3.0	5.3	1.7	11.6	71.9	0.07	0.40	*
4.	86.9	6.0	36.2	1.7	3.6	38.2	0.31	0.09	*
5.	11.3	0.5	3.4	0.2	1.3	7.2	0.10	0.01	*
6.	91.0	4.5	29.6	1.5	9.0	49.1	0.31	0.14	179.3
7.	12.9	1.2	1.2	0.2	1.3	10.6	0.04	0.04	114.9
8.	9.3	0.7	0.0	0.1	3.3	8.5	0.13	0.10	*
9.	12.6	0.7	0.0	3.6	3.5	16.1	0.12	0.09	0.9
10.	17.6	1.7	4.9	0.6	3.0	11.8	0.24	0.05	50.0
11.	90.0	7.1	29.1	2.3	13.2	52.7	1.30	0.20	16.1
12.	75.0	1.0	1.7	3.3	7.7	68.0	0.03	0.24	3.6
13.	50.0	4.9	15.3	0.9	8.5	26.8	0.71	0.10	35.7
14.	92.7	3.5	16.9	22.9	23.1	86.6	0.14	0.68	*
15.	92.7	6.1	10.9	5.6	41.4	73.6	0.19	1.09	*
16.	85.4	1.5	7.1	3.4	8.0	72.6	0.04	0.23	3.5
17.	91.4	19.0	0.6	9.8	60.4	67.8	5.14	2.91	*
18.	23.8	1.1	3.8	1.5	5.5	15.9	0.07	0.12	*
19.	99.9	96.8	0.0	0.0	0.0	0.0	35.85	0.02	0.0
20.	91.5	4.4	59.8	1.2	6.6	16.7	0.25	0.06	*
21.	91.5	4.5	13.1	1.2	47.4	70.3	0.20	0.65	*
22.	94.8	2.4	2.8	47.7	28.4	131.1	0.06	0.43	*
23.	23.1	1.1	0.6	0.1	2.2	18.5	0.01	0.05	*
24.	30.7	2.3	7.3	0.9	3.0	17.6	0.08	0.06	5.0
25.	88.5	2.1	2.3	3.1	8.9	71.3	0.03	0.29	*
26.	28.2	2.2	6.6	0.7	1.9	16.1	0.06	0.09	7.3
27.	67.7	5.4	0.0	0.2	5.7	52.5	1.20	0.12	*
28.	88.6	4.5	30.2	2.3	6.3	48.8	0.36	0.15	47.3
29.	87.8	6.3	38.3	1.4	3.2	42.8	0.14	0.07	*
30.	17.7	2.1	2.8	0.6	5.0	11.8	0.25	0.09	26.3

*None for all practical purposes.

III. Match by number the materials listed below with the following composition (as-fed basis).

	DM %	Ash %	Fiber %	EE %	Prot %	TDN %	CA %	P %	Caro mg/kg
1.	89.0	1.3	2.6	3.8	9.6	77.0	0.03	0.26	2.0
2.	88.0	2.3	5.0	1.9	11.9	74.0	0.04	0.34	2.0
3.	89.0	3.1	10.8	4.8	11.8	68.0	0.07	0.33	*
4.	75.0	9.8	0.0	0.1	4.4	54.0	0.75	0.08	*
5.	75.0	1.0	1.7	3.3	7.7	68.0	0.03	0.24	3.6
6.	99.0	0.0	0.0	0.0	281.2	*	0.0	0.0	0.0
7.	93.0	3.5	1.4	2.9	84.9	65.0	0.26	0.67	*
8.	92.0	21.5	2.0	8.9	59.4	67.0	5.86	3.07	*
9.	93.0	28.2	2.2	12.8	46.6	63.0	11.16	5.41	*
10.	90.0	5.8	3.4	0.9	49.7	78.0	0.26	0.63	*
11.	93.0	6.1	11.9	4.6	41.0	72.0	0.19	1.08	*
12.	92.0	5.1	5.3	17.2	39.2	83.0	0.25	0.60	1.0
13.	95.7	80.4	1.9	1.9	7.1	*	30.92	14.01	*
14.	100.0	100.0	0.0	0.0	0.0	0.0	32.0	18.00	0.0
15.	99.9	96.8	0.0	0.0	0.0	0.0	34.0	0.02	0.0
16.	0.0	0.0	0.0	0.0	0.0	0.0	0.0	0.0	0.0
17.	99.0	99.0	0.0	0.0	0.0	0.0	0.0	0.0	0.0
18.	99.0	*	0.0	0.0	0.0	93.0	0.0	0.0	0.0
19.	99.9	0.0	0.0	99.9	0.0	0.0	0.0	0.0	0.0
20.	100.0	0.0	0.0	100.0	0.0	172.8	0.0	0.0	0.0
21.	92.0	9.7	24.0	2.7	17.3	55.0	1.40	0.23	120.0
22.	90.0	8.2	23.4	2.3	15.3	52.0	1.27	0.22	59.0
23.	91.0	8.4	27.8	1.8	8.9	42.0	0.43	0.16	53.0
24.	91.0	2.6	43.3	1.5	3.7	43.3	0.13	0.09	*
25.	30.0	1.7	7.5	1.0	2.5	21.0	0.09	0.08	13.0
26.	21.0	2.1	4.9	0.6	4.3	13.0	0.41	0.06	38.8
27.	23.0	1.1	0.6	0.1	2.2	19.0	0.01	0.06	*
28.	12.0	0.8	0.0	3.6	3.3	16.0	0.12	0.09	0.9
29.	10.0	0.7	0.0	0.1	3.0	9.0	0.13	0.10	*
30.	50.0	4.9	15.3	0.9	8.5	26.8	0.71	0.10	35.7

*None for all practical purposes.

1.	Feather meal.	9.	Yellow shelled corn.
2.	Oats grain.	10.	Dehydrated alfalfa meal.
3.	Defluorinated phosphate.	11.	Cottonseed hulls.
4.	Soybean seed.	12.	Fresh skimmed milk.
5.	Bermudagrass hay.	13.	Pure glucose.
6.	Fresh cow's milk.	14.	White potatoes.
7.	Urea.	15.	Water.
8.	Alfalfa pasture.	16.	Tankage.

17. Mineral oil.
18. Ground limestone.
19. Steamed bonemeal.
20. Cane molasses.
21. Tankage with bone.
22. Barley grain.
23. 49% soybean meal.

24. Alfalfa hay.
25. 41% cottonseed meal.
26. High-moisture shelled corn.
27. Alfalfa haylage.
28. Corn silage.
29. Corn oil.
30. Sand.

IV. Supply in the space provided, the correct answer for each of the following questions:

1. The most extensively produced feed grain crop in the United States. _____

2. The second most extensively produced feed grain crop in the United States. _____

3. The third most extensively produced feed grain crop in the United States. _____

4. The fourth most extensively produced feed grain crop in the United States. _____

5. The most extensively used energy feed in the United States. _____

6. The most extensively used high-protein feed in the United States. _____

7. The most extensively used high-protein feed of plant origin in the United States. _____

8. The second most extensively used high-protein feed of plant origin in the United States. _____

9. The most extensively used source of non-protein nitrogen in the United States. _____

10. The most extensively used silage crop in the United States. _____

11. The most extensively used silage crop in the Southeast. _____

12. The most extensively used hay crop in the United States. _____

13. The most extensively used hay crop in the Southeast. _____

14. The most commonly used type of molasses in the Southeast. _____

15. The most extensively grown oil seed crop in the United States. _____

16. The feed grain most like shelled corn in composition and feeding value. _____

17. The feed grain with the greatest percentage of fiber. _____

18. Corn harvested with the cob, shuck, and grain intact.

_____

19. The seed coat from wheat in the manufacture of flour.

_____

20. The most ideal energy feed for finishing livestock.

_____

21. Peanut hulls are sometimes used in cattle finishing rations primarily as a source of this.

_____

22. The two primary components of most liquid cattle supplements.

_____

23. Hot meals always contain a relatively high level of this.

_____

24. Forage crops stored in a silo containing 45–55% moisture.

_____

25. The feed grain that produces the greatest yield of TDN per acre.

_____

26. The feed grain that presents special problems at harvest time.

_____

27. The feed grain crop that is the most drought resistant.

_____

28. The feed grain considered best for horses, young growing stock, and breeding animals.

_____

29. A high-protein feed that is usually toxic to growing-fattening pigs.

_____

30. An air-dry energy feed that would require no salt supplementation.

_____

31. Oats grain from which the hull has been removed.

_____

32. Dried sweet potatoes are especially high in what nutrient other than carotene?

_____

33. The most widely used source of calcium for livestock feeding.

_____

34. The most widely used source of phosphorus for livestock feeding.

_____

35. The predominant nutrient in feather meal.

_____

36. High-protein feed that excels for putting bloom on show animals.

_____

37. Name a crop that has the ability to harbor nitrifying bacteria in nodules on its roots.

_____

33

Balancing Rations— General

I   **What is a balanced ration?**

    A.   A balanced ration is one that will supply the different nutrients in such proportions as will properly nourish a given animal when fed in proper amounts.

    B.   A balanced daily ration is one that will supply the different nutrients in such proportions and amounts as will properly nourish a given animal for a 24-hour period.

II   **What is needed for balancing a ration?**

    A.   The first requirement for balancing a ration is a feeding standard. This is a table that states the amounts of nutrients that should be provided in rations for farm animals of various classes in order to secure the desired results. These requirements may be expressed as the following:

        1.   Amounts per animal per day, or

        2.   Percentage of overall feed mixture or amount per kg of ration.

    B.   In order to be of use, feeding standards must be accompanied by and used with feed composition tables that provide information concerning the nutritive composition of the feeds to be used in balancing the ration.

III Brief history of balanced rations.

A. Thaer of Germany was among the first to move toward a scientific approach to livestock feeding. In 1810 he published a table of "Hay Equivalents." These were the amounts of different feeds equivalent to 100 lb of meadow hay. Such an approach naturally prompted much difference of opinion among authorities.

B. Grouven in 1859 proposed the first feeding standard. However, through necessity, it was based on total nutrients, rather than digestible nutrients or some other more precise nutrient evaluation, and consequently was not very accurate.

C. Wolff, another German, in 1864 presented the first feeding standard based on digestible nutrients. It was naturally quite limited in scope.

D. The Wolff standards were revised in 1896 by another German named Lehman and published as the Wolff-Lehman standards. While they were widely used at the time, they contained many inaccuracies and are now out of date.

E. Dr. W. A. Henry of the University of Wisconsin came out with the first edition of his *Feeds and Feeding* in 1898. It presented the Henry feeding standards. Various other standards have been introduced over the years, but none has persisted as have those he originated. Beginning with the 10th edition of *Feeds and Feeding,* which came out in 1910, Dr. Henry was assisted by Dr. F. B. Morrison, a former student. Following the death of Dr. Henry in the early 1930s, Dr. Morrison continued to keep the book revised through the 22nd edition, published in 1956. However, Dr. Morrison died in 1958, and there have been no revisions of Morrison's *Feeds and Feeding* published since the 22nd edition.

F. Back in the 1940s, the National Research Council of the National Academy of Sciences initiated a series of publications on the nutrient requirements of various species of farm animals. In 1959, this same organization published a rather comprehensive set of tables on feed composition. From time to time all of these publications have been revised, and today they are regarded as the final authority on nutritive requirements and feed composition by workers in the field.

IV Factors to be considered in balancing rations.

A. Nutritional.

   1. Dry matter.

      a. A certain minimum amount of dry matter is essential to satisfy an animal's appetite and to promote the proper functioning of its digestive tract.

    b. Animals have certain physical and physiological limitations of dry matter consumption beyond which they cannot go.

2. Protein.

    a. All animals require protein. The amount will depend on the physiological processes of the animal. Many feeds are deficient in protein. No other nutrient can take the place of protein. The essential amino acid requirement must be met in this connection.

    b. Adequacy of protein may be based on

Total protein	Digestible protein plus
Digestible protein	certain critical essential
	amino acids

3. Energy.

    a. All animals require energy. The amount will depend on the physiological processes of the animal. Except for keeping the body warm, an animal needs its energy in the form of net energy. This energy may come from sugar, starch, cellulose, fat, and/or excess protein.

    b. Adequacy of energy may be based on:

Digestible energy	Metabolizable energy
Total digestible nutrients	Net energy
(TDN)	

4. Calcium.

    a. All animals require calcium.

    b. The amount of calcium required will depend on the physiological processes of the animal.

    c. No other nutrients can substitute for calcium.

    d. Calcium needs and allowances are ordinarily based on total calcium.

    e. Proper Ca:P ratio is important.

    f. Excess calcium may be harmful by interfering with the availability of other nutrients.

5. Phosphorus.

    a. All animals require phosphorus.

    b. The amount of phosphorus required will depend on the physiological processes of the animal.

    c. No other nutrient can substitute for phosphorus.

    d. Phosphorus needs and allowances are ordinarily based on total phosphorus.

     e.   Proper Ca:P ratio is important.

     f.   Excess phosphorus may be harmful by rendering other nutrients unavailable.

6. **Vitamin A.**

     a.   All animals require vitamin A as preformed vitamin A, or carotene.

     b.   Amount will vary, depending on the circumstances.

     c.   Vitamin A may be provided as preformed vitamin A or as carotene.

     d.   Vitamin A needs and allowances are based on total "vitamin A value" (vitamin A and carotene combined).

     e.   Carotene is sometimes very inefficiently converted to vitamin A, resulting in a vitamin A deficiency in the presence of ample carotene.

7. **Other minerals.**

     a.   Calcium and phosphorus are usually the only two macro minerals that are likely to be in critical supply. Ordinarily, no attempt is routinely made to balance allowances precisely against requirements of other macro minerals.

     b.   Minimum needs for critical trace minerals are usually met through the use of trace-mineralized salt. Ordinarily, no attempt is routinely made to balance allowances precisely against requirements.

8. **Other vitamins.**   The need for other vitamins that may be in short supply is ordinarily met by adding to the overall ration the animal's minimum daily requirement for these vitamins, letting the natural content of these vitamins in the feed serve as a margin of safety. Ordinarily no attempt is routinely made to balance allowances precisely against requirements.

B. **Economic.**

1. Feeds are not always priced in accordance with their nutritive value. Some feeds may be a cheaper source of nutrients than other feeds with any given set of prices.

2. To compare feeds as economical sources of nutrients, it is not sufficient to compare them in terms of price per bushel or even per Cwt. Different feeds have different weights per bushel as well as different contents of nutrients per Cwt.

3. High-energy feeds are usually compared pricewise on the basis of the cost per lb of TDN or per unit of energy.

4. High-protein feeds are usually compared pricewise on the basis of the cost per lb of protein, either total or digestible.

5. For example, shelled corn, sorghum grain, and barley grain could be compared as sources of TDN as follows:

	Prevailing Price	Prevailing Price per Cwt	TDN per Cwt	Cost per Lb TDN
Shelled corn	$3.00 per bushel	$5.36	77.0	$.070
Sorghum grain	$6.00 per Cwt	$6.00	78.0	$.077
Barley grain	$2.60 per bushel	$5.42	74.0	$.073

Since in the above calculation corn is lowest in cost per lb of TDN, it would be the cheapest source of energy at the prevailing set of prices.

6. High-protein feeds may be compared in a similar manner as sources of protein simply by inserting total protein or digestible protein for TDN in the calculation.

7. In establishing the prevailing price, care should be exercised to make sure the price used is FOB the buyer's feedlot and that the feeds are in comparable physical form.

8. For feeds that are intermediate in protein content between that of high-protein feeds, on the one hand, and high-energy feeds, on the other, the above procedure will not suffice. For such feeds, the Petersen method of evaluating feeds as outlined in section 34 is recommended.

34

The Petersen Method of Evaluating Feeds

An alternative method to that outlined previously for comparing different feeds with respect to their nutritive worth was devised several years ago by Petersen of the University of Minnesota. The Petersen method is superior to the one outlined earlier in that it is more precise and is applicable to all feeds and not just to high-energy feeds, on the one hand, or high-protein feeds, on the other. By using the Petersen method, appropriate weight is given to both the protein and the energy contents of a feed in establishing its feeding value, whereas the relative value of a feed as established by the preceding method is based entirely on its content of protein, or energy, but not both.

I. Establishing constants.

A. The Petersen method involves the establishment of a set of constants for each feed to be considered as a potential ration component. These constants are calculated for each feed to reflect the relative value of each feed's content of both digestible protein (DP) and usable energy expressed as TDN.

B. In most instances, protein is much more costly per unit of TDN than are carbohydrates and fats. Consequently, high-protein feeds are ordinarily more expensive than are low-protein feeds of comparable energy content. It is obvious then that the actual nutritive worth of most feeds will usually depend largely on their content of DP and TDN and the prevailing costs of these two nutritive fractions.

C. The most widely used high-energy feed and the most widely used high-protein feed are normally used as base feeds for establishing protein and energy worth. Ground shelled corn and 44% soybean meal would logically qualify as base feeds under prevailing conditions. Other feeds might be used, however, as base feeds, should they become more widely used as sources of protein or energy than the above.

D. Constants of the nature referred to above have been calculated for the more important feeds and are included in Table 34-1. In calculating these constants, ground shelled corn and 44% soybean meal were used as the base high-energy feed and the base high-protein feed, respectively. Digestible protein was used in calculating protein worth, and TDN was used in calculating energy worth. Constants were calculated for both cattle and swine.

While the procedure followed in the use of DP and TDN figures for calculating the various constants is unique and most interesting to the scientifically minded student, knowing the procedure is not essential for the use of the constants, and so is not included here. Anyone who desires to become familiar with this procedure is referred to the following publication: Petersen, Wm. E. "A Formula for Evaluating Feeds on the Basis of Digestible Nutrients." *Journal of Dairy Science* (1932) 15:293.

II Use of constants for feed evaluation.

A. In using the constants of Table 34-1 for calculating the actual worth of a feed based on its content of DP and TDN, one first determines the prevailing price per ton of the 2 base feeds—ground shelled corn and 44% soybean meal. These prices are then multiplied by the respective constants for a particular feed to be evaluated, and the 2 products are added together to obtain the actual worth of that feed for feeding purposes in dollars per ton. By comparing this calculated worth per ton with the prevailing price per ton for any particular feed, one can readily determine whether that feed is a good buy at prevailing prices.

B. To illustrate the evaluation of a feed by the Petersen method, let us assume that ground shelled corn costs $135.00 per ton and 44% soybean meal costs $250.00 per ton. One wishes to determine the actual worth per ton of wheat middlings as a concentrate feed for a milking herd.

The calculation should proceed as follows, using constants for cattle from Table 34-1 for the feed in question.

$135.00 (price of corn per ton) × .887 (corn constant for wheat middlings) = $119.75

$250.00 (price of SBM per ton) × .127 (SBM constant for wheat middlings) = $ 31.75

TOTAL $151.50

From the above, it is apparent that at the prevailing prices for corn and SBM, wheat middlings has a worth of $151.50 per ton as a feed for a milking herd. If this feed can be bought at a price below its worth, then it would be considered to be a good buy compared with corn and SBM as sources of protein and energy. This would not, however, rule out the possibility of some other feeds being an even better buy as determined by a similar evaluation.

III

Negative constants. In some instances, the constants are negative values.

A. Negative corn constants are always associated with feeds extremely high in the ratio of DP to nonprotein TDN content. This is because an increase in the price of corn with the price of soybean meal remaining constant actually tends to lower the value of DP per lb and to increase the value per lb of nonprotein TDN. As a result, those feeds extremely high in protein and low in nonprotein TDN are actually lowered more in overall nutritive worth by the reduced value of DP than they are raised by the simultaneous increase in the value per lb of nonprotein TDN.

B. A similar situation prevails with those feeds that have negative SBM constants except that negative SBM constants are always associated with feeds having an extremely low ratio of DP to nonprotein TDN. In any case, when negative constants are involved, they are supposed to have a negative effect on the overall worth of a feed and should be so handled in carrying out the feed evaluation calculation.

IV

Limitations of the Petersen method. In the use of feed evaluations calculated by the Petersen method, appropriate consideration should be given to the fact that TDN from different sources does not necessarily have the same energy value for the different body functions. For example, the TDN of roughages approaches the TDN of concentrates in energy value for maintenance but has only a fraction of the energy value of concentrates for body gain. Other similar relationships also prevail. Consequently, in comparing feed worth figures as calculated following the Petersen procedure, such comparisons should be restricted to between roughages, on the one hand, and between concentrates, on the other. In other words, one should not attempt to substitute roughage for concentrates in the ration of animals in heavy production simply on the basis of favorable feed worth values as calculated by the Petersen method.

Table 34–1
CONSTANTS FOR EVALUATING FEEDS

	Based on Content of			
	DP and TDN for Cattle		DP and TDN for Swine	
Name of Feed	44% SBM	Shelled Corn	44% SBM	Shelled Corn
Dehydrated alfalfa meal	.281	.467	.155	.275
Alfalfa hay	.185	.459	.127	.287
Dehydrated alfalfa leaf meal	.310	.478	.254	.413
Dehydrated alfalfa stem meal	.121	.507	—	—
Fresh alfalfa forage	.091	.131	—	—
Animal fat, feed grade	−.439	2.652	—	—
Blood meal	—	—	1.641	−.848
Meat scrap	1.359	−.459	1.197	−.437
Digester tankage	1.468	−.515	—	—
Meat and bonemeal	1.211	−.341	1.049	.037
Bahiagrass hay	−.101	.745	—	—
Dried bakery product	−.018	1.077	.056	1.036
Barley straw	−.084	.568	—	—
Barley grain	.076	.852	.069	.830
Beet molasses	−.039	.819	—	—
Dried beet pulp	.046	.879	.069	.926
Wet beet pulp	0.000	.092	—	—
Dried beet pulp/molasses	.011	.858	−.111	.987
Bermudagrass hay	−.010	.582	—	—
Dehydrated coastal bermudagrass meal	.192	.522	—	—
Coastal bermudagrass hay	.024	.607	—	—
Fresh coastal bermudagrass	.045	.207	—	—
Kentucky bluegrass hay	.013	.706	—	—
Fresh Kentucky bluegrass forage	.044	.198	—	—
Bromegrass hay	.068	.576	—	—
Fresh bromegrass forage	.085	.196	—	—
Cabbage heads	.024	.084	—	—
Reed canarygrass hay	.066	.540	—	—
Carrots	−.006	.141	−.001	.135
Dried whey	.107	.899	.198	.761
Dried whole milk	.440	1.005	.356	1.352
Fresh cow's milk	.059	.152	.058	.143
Fresh skimmed milk	.072	.042	.068	.052
Dried skimmed milk	.732	.317	.735	.378
Fresh citrus pulp	−.026	.219	—	—
Citrus pulp/molasses, dried	−.131	1.108	—	—
Dried citrus pulp	−.124	1.096	−.046	.643
Citrus molasses	−.057	.726	—	—
Alsike clover hay	.104	.576	—	—
Fresh crimson clover forage	.034	.120	—	—

| | Based on Content of | | | |
| | DP and TDN for Cattle | | DP and TDN for Swine | |
Name of Feed	44% SBM	Shelled Corn	44% SBM	Shelled Corn
Fresh ladino clover forage	.072	.107	—	—
Red clover hay	.106	.578	—	—
Fresh red clover forage	.045	.146	—	—
Fresh white clover forage	.076	.081	—	—
Copra meal	.371	.602	.288	.656
Corn stover	−.060	.714	—	—
Ground corn cob	−.107	.673	—	—
Corn fodder silage	−.007	.259	—	—
Corn stover silage	−.018	.275	—	—
Corn ear silage	−.015	.411	—	—
Corn bran	−.057	.926	—	—
Ground ear corn	−.065	.991	−.012	.885
Ground snapped corn	−.059	.910	—	—
Hominy feed	.001	1.088	—	—
Corn oil	−.433	2.615	−.434	2.599
Corn starch	−.214	1.297	−.216	1.303
Corn gluten meal	.825	.205	—	—
Yellow shelled corn	0.000	1.000	0.000	1.000
Cottonseed hulls	−.136	.721	—	—
Ground cottonseed	.224	.904	—	—
36% cottonseed meal	.691	.222	.701	.241
41% cottonseed meal	.829	.178	.850	−.021
41% solvent-extracted cottonseed meal	.854	.090	—	—
Cowpea hay	.202	.497	—	—
Dallisgrass hay	−.046	.720	—	—
Fresh dallisgrass forage	.022	.174	—	—
Fescue hay	.006	.579	—	—
Fresh fescue forage	.041	.215	—	—
Fish meal	1.323	−.352	1.421	−.467
Linseed meal	.772	.231	.745	.175
Restaurant garbage	.031	.258	.049	.252
Brewers dried grains	.427	.379	.477	.059
Wet brewers grains	.081	.129	.068	.210
Distillers dried grains	.462	.553	.392	.751
Kudzu hay	.192	.476	—	—
Lespedeza hay	.115	.574	—	—
Sericea hay	.146	.384	—	—
Millet hay	.019	.611	—	—
Meadow hay	−.023	.741	—	—
Prairie hay	−.073	.684	—	—
Oat hay	−.025	.721	—	—
Oat hulls	−.045	.454	—	—
Oat straw	−.081	.671	—	—
Oat grain	.096	.761	.115	.675
Oat groats	.120	1.050	.182	.887

| | Based on Content of | | | |
| | DP and TDN for Cattle | | DP and TDN for Swine | |
Name of Feed	44% SBM	Shelled Corn	44% SBM	Shelled Corn
Orchardgrass hay	.042	.639	—	—
Fresh orchardgrass forage	.040	.188	—	—
Peanut hay	.047	.704	—	—
Peanut hulls	.049	.169	—	—
Peanut oil meal	1.116	−.129	1.109	−.237
Peanut kernels	.384	1.326	—	—
Potato meal	−.020	.938	.044	.924
White potatoes	−.010	.247	−.029	.273
Feather meal	—	—	1.594	−.699
Poultry by-product meal	—	—	1.144	−.086
Redtop hay	−.035	.763	—	—
Rice hulls	−.034	.160	—	—
Rice bran	.106	.657	.113	.727
Fresh rye forage	.066	.129	—	—
Rye grain	.026	.892	.063	.927
Fresh ryegrass forage	.047	.146	—	—
Safflower meal without hulls	.855	.102	—	—
Sesame oil meal	.889	.110	.980	.090
Grain sorghum silage	−.005	.230	—	—
Grain sorghum grain	−.026	.938	−.019	1.013
Hegari grain	−.015	.937	−.013	1.048
Johnsongrass hay	−.039	.726	—	—
Kafir grain	.042	.795	.020	.993
Milo grain	.008	.907	.025	.952
Sorgo silage	−.026	.230	—	—
Sudangrass hay	−.019	.698	—	—
Fresh sudangrass forage	.028	.158	—	—
Soybean hay	.130	.467	—	—
Soybean hulls	.064	.695	—	—
Soybean seed	.823	.305	.656	.534
44% soybean meal	1.000	0.000	1.000	0.000
49% soybean meal	1.219	−.147	1.164	−.203
Cane molasses	−.108	1.047	—	—
Sunflower meal	1.111	−.248	1.028	−.112
Sweetclover hay	.174	.525	—	—
Sweet potato meal	−.156	1.076	—	—
Sweet potatoes	−.055	.376	—	—
Timothy hay	−.044	.667	—	—
Turnips	−.003	.109	−.007	.099
Wheat straw	−.096	.637	—	—
Wheat bran	.214	.606	.198	.532
Wheat shorts	.150	.794	.232	.619
Wheat middlings	.127	.887	.210	.722
Wheat grain	.084	.926	.103	.887
Wood molasses	−.189	.859	—	—

35

Calculating a Balanced Daily Ration—General

I

The first step in calculating a balanced daily ration for any given animal is to determine the amount of each critical nutrient required by the animal by referring to the appropriate table of nutrient requirements in the appendix of the text. The nutrients that are usually regarded as critical and that are always considered, either directly or indirectly, in any ration balancing process are the following:

Dry matter	Calcium
Protein	Phosphorus
Energy	Carotene (vitamin A)

With swine, consideration must also be given to the matter of essential amino acid balance as well as to the adequacy of several of the other vitamins. In addition to vitamin A, the following vitamins are frequently deficient in unfortified swine feeds under present-day conditions: niacin, riboflavin, pantothenic acid, and vitamins B_{12}, D, and E.

II

The second step in calculating a balanced daily ration for any given animal is to formulate a suitable combination of feeds that will provide the critical nutrients in the amounts needed to satisfy the requirements.

A. Various approaches and techniques may be used in determining which feeds to use and the quantity of each for balancing the ration.

1. The cheapest source of each critical nutrient should be determined as discussed previously if cost of production is to be given consideration.

2. If a calculator is available and preciseness is desired in the ration-balancing process, the use of certain mathematical formulas and equations as discussed subsequently is recommended.

3. If a calculator is not available and/or if considerable preciseness is not required, the use of feeding guides in conjunction with a trial-and-error approach may serve the purpose.

B. The requirements should always be met within acceptable limits by the ration formulated.

1. Any nutritive allowance that is not more than about 3% below the minimum requirement is usually considered to be within acceptable limits, although it is always best to at least meet the requirement whenever practicable.

2. The energy allowance should not be permitted to exceed the requirement by more than about 5% since an animal is definitely limited in its energy utilization capacity.

3. A protein allowance in excess of the requirement by as much as 5–10% is sometimes good insurance against feeds of below normal protein content, especially if protein feeds are not too expensive.

4. A protein allowance greatly in excess of the requirement will normally not harm the animal but will usually cause the cost of the ration to be higher than necessary.

5. Excesses of calcium and phosphorus are sometimes difficult to avoid and are permissible, provided there has been no undue use of mineral feeds and the Ca:P ratio is held between 1:1 and 2:1.

6. Large excesses of carotene are often impractical to avoid, are not normally detrimental to the health of an animal, and are usually acceptable as a part of a balanced ration.

7. In the routine balancing of rations for livestock, the rations are not ordinarily evaluated for vitamin content, other than for vitamin A. If other vitamins are to be added to the ration, they are usually added at the minimum daily requirement level or at some lower level such as experience has shown to give satisfactory performance.

8. DM requirements, when listed, should be interpreted in accordance with their intended purpose.

a. With full-fed animals they usually represent a maximum DM consumption capacity on the part of the animal, not to be exceeded by more than about 3%.

    b. With other animals, they usually indicate the approximate amount of feed to be fed and can be varied in either direction to a considerable degree should circumstances seem to warrant.

9. Normally rations today are calculated by any one of many available computer programs, which will calculate the least-cost combination of ingredients that will meet the nutrient specifications given it.

36

Balancing a Daily Ration for a Growing and Finishing Steer

I Introduction.

A. Definition. A balanced ration is one that furnishes available nutrients in such proportions and amounts that it will nourish a given animal for 24 hours. In addition, the required nutrients must be contained in the amount of dry matter such an animal is able to consume in the 24-hour time period. Therefore, it is critical to set forth the nutrient requirements and then attempt to incorporate such nutrients within the dry matter intake of the animal.

B. No one nutrient is "more" essential than another. In almost every balanced feed formulation, energy is present in the greatest quantity. Even though this large energy requirement is critical to the animal, it is not considered to be any more important than the 15 to 70 gm (0.5 to 3 oz) of calcium nor the 4 to 30 gm (0.14 to 1 oz) of phosphorus. However, because of the much larger proportion of energy needed, usually that is where one starts in the formulation of a ration.

C. Computer formulation. Even though the student will be exposed to the fundamentals of ration formulation, there are computer programs available in which one enters the animal conditions (weight, desired gain, etc.) and the feedstuffs available, and the computer calculates the precise formulations to meet the needs specified—all in a matter of a few seconds.

D. **Net energy_{maintenance} plus Net energy_{gain, or lactation}; TDN.** Energy calculation utilizing the National Research Council (NRC) Requirements are being converted from Total Digestible Nutrients (TDN) to the Net Energy (NE) as newer standards are being published. This is not a great handicap, however, since computer diskettes are available that will go through the steps required—and quickly. There are drawbacks and advantages to both systems for deriving the ultimate balanced ration. The NE system is more precise, since it takes into account that animals use feedstuffs with different efficiencies for maintaining body functions, than for production, for example, weight gain or milk production. However, even though the NE system has worked out maintenance values and production values, the calculation of such is dependent upon having the entire ration intact before its adequacy can be checked. In other words, with the NE system, some advanced knowledge of ration formulation is needed in order to set up a combination of feedstuffs and then follow through to find out what the respective NE for maintenance and NE for production would be, and then predict what the performance would be. In contrast, the TDN does not differentiate between maintenance and production requirements in evaluating feedstuffs utilization. At first attempt, the TDN method might appear to be simpler.

E. **For a 300 kg (660 lb) finishing steer to gain 1.36 kg (3.03 lb) per day.** As a starting point in formulation, let us calculate a daily ration for a 300 kg finishing steer that we want to gain 1.36 kg per day. (Permit the author to assure the student not to panic at the use of the metric system. The NRC has put increasing emphasis on this system, but it is easy to follow one system, and then convert to the other at any time. For example, one merely multiplies the metric figure by 2.2 to get the English equivalent, i.e., the 300 kg steer $\times$ 2.2 = 660 lb. The author hopes that the student can become adept at going from one system to the other.) We will employ the "100% dry weight basis" in the calculation and then convert back to the "as-is" or "air-dry" basis when the problem is solved so that the solution may be utilized under typical feeding conditions. The reason for this is that the moisture of a feedstuff in the arid Phoenix, Arizona, area might be greatly different from that in the more humid area of Indianapolis, Indiana, in the summertime.

II

Setting up the requirements. First of all, one must know the animal—a 300 kg yearling finishing steer to gain 1.36 kg per day. The first step is to find the requirements for the expected performance in the beef performance tables in the appendix (in this case Table Beef–1, and Table Beef–2). The beginning student should start with Table Beef–2. When the weight of the animal (300 kg) and the rate of expected gain (1.36 kg/d) are located in the table, most of the components of a ration that will achieve that goal are shown. (The current NRC Beef Bulletin utilizes both the TDN and NE approaches to ration formulation.):

Bodyweight	Type of Ration	Dry Matter Intake/d, kg	Expected Daily Gain, kg
300	A	7.9	0.32
	B	8.4	0.89
	C	8.2	1.36
	D	7.7	1.69
	E	7.1	1.90

From the above section of Table Beef–2, one was able to identify that a 300 kg steer to gain 1.36 kg/d should consume 8.2 kg of ration type C described at the top of that table. The characteristics of a type C ration are as follows:

a. *Neutral detergent fiber* (NDF), 30% of the dry matter (DM).
b. *TDN*, 70% of the DM.
c. $NE_{maintenance}$, 1.67 Mcal/kg DM.
d. NE_{gain}, 1.06 Mcal/kg of DM.
e. *Crude protein*, 12.6% of the DM.
f. *Calcium*, 0.48% of the DM.
g. *Phosphorus*, 0.24% of the DM.

III

Making the formulation.   First, one must determine which feedstuffs are available. The approach to be used here does not solve for least-cost calculation. That must be determined by linear programming, which necessitates a program adapted to a calculator. Typically, such available feedstuffs will need to include roughage (silage, hay); energy feeds (corn, milo, barley); protein concentrates (soybean meal, cottonseed meal, urea); vitamin A; and perhaps growth stimulants.

As a starting point, we will use corn silage and orchard grass as roughages, shelled corn as energy feed, soybean meal as supplemental protein source, dicalcium phosphate as calcium and phosphorus source, limestone as a calcium source, trace mineralized salt, and 0.5% of the mixture for additives such as growth stimulants. In order to assemble the feedstuffs needed in the right proportions, one must refer to the Feed Analysis Table in the appendix. (Note that this table lists the analyses in both the "as-is" and the "100% dry matter" basis. As indicated previously, we will employ the 100% dry matter basis in calculating the formulation, then convert back to the as-is basis, and convert the kg to lb, for direct feedlot application.)

COMPOSITION OF SELECTED FEEDSTUFFS (APPENDIX TABLE)

	Dry Matter Basis							
Feedstuff	*Dry Matter %*	*TDN %*	*Crude Protein %*	*Ca %*	*P %*	*NDF %*	*NE_m Mcal/kg*	*NE_{gain} Mcal/kg*
Orchard grass hay	91	54	8.4	.39	.35	65	1.16	0.43
Corn silage (well-eared)	30	69	8.3	.29	.25	46	1.57	0.94
No. 2 shelled corn	89	87	10.9	.03	.29	9.0	2.09	1.42
Dehulled soybean meal	90	87	54	.26	.29	7.9	2.09	1.45
Ground limestone	90			36				

This animal requires 0.48% calcium in its diet. Since none of the feedstuffs in the table contain that amount, special consideration must be made to provide supplemental calcium. One must start by including about 1.1% ground limestone (36% Ca—see table) to be sure there is adequate calcium. Cattle diets should also contain a minimum of 0.3% salt, and finally, there should be room for a premix (0.4%) to add vitamin A, plus any additives needed. So, 1.1% limestone + 0.3% salt + 0.4% premix = 1.8% of the ration that is needed that contributes neither energy nor protein. Subtracting 1.8% from 100% leaves 98.2% of the diet to provide essentially energy and protein, plus some calcium and maybe all the phosphorus.

A logical starting point in solving this problem is to approximate the energy (TDN) needs. Once we have made such calculations, then the NDF and NE can be checked to validate the formulation (certainly all of the nutrient requirements will be checked). This first calculation can formulate a ration built around only three sources of energy quite common in the Corn Belt—namely, orchard grass and hay corn silage as the roughages, and shelled corn as the concentrate. Much more complicated diets can be formulated subsequently, just as long as the formulation is palatable and well balanced.

A. **The algebraic method of determining proper proportions to achieve an average value.** If the roughage (say equal parts—dry matter basis—of orchard grass hay and corn silage) contains 61.5% TDN (average of 54% for orchard grass hay and 69% for corn silage) and the corn contains 87% TDN, the proportion of the corn is X and the proportion of the roughage is equal to $1 - X$. An equation can be set up and solved for the mixture (roughage and corn) contributing 70% TDN (or any other predetermined level) of TDN. However, one must take into account that some of the formulation has been assigned to limestone (1.1%), salt (0.3%), and a possible premix (0.4%), so that only 98.2% of the formulation (8.42 kg of dry matter × 98.2% = 8.26 kg of corn plus orchard grass-corn silage mixture) must provide the 70% of TDN specified by the NRC. Therefore, this becomes 70% TDN divided by 98.2% of the ration = 71.28% TDN in the corn-roughage mixture.

$$87X + 61.25 (1 - X) = 71.28$$
$$87X + 61.25 - 61.25X = 71.28$$
$$25.75X = 10.03$$
$$X = .3895, \text{ or } 38.95\% \text{ of the mix as shelled corn.}$$

100% – 38.95% = 61.05% of the mix as equal parts hay and silage
8.26 kg diet dry matter × 38.95% shelled corn = 3.22 kg shelled corn
8.26 kg diet dry matter × 61.05% = 5.04 kg roughage (2.52 kg each, hay and silage.)

B. **Test for accuracy of calculation.** Next, we need to test our calculation to see how nearly we met the TDN requirements, which was the objective for this part, and then to see how nearly our calculation meets other designated needs for the formulation.

TESTING THE DIET FOR ADEQUACY

Ingredient	Dietary level kg/d	Protein kg	NDF %	Ca %	P %	TDN %	NE_m Mcal	NE_g Mcal
Orchard grass hay	2.52	0.21	19	.12	.10	16	.35	.13
Corn silage	2.52	0.21	14	.09	.07	21	.47	.28
Shelled corn	3.22	0.35	3	.01	.11	33	.80	.54
Ground limestone	0.09			.38				
Salt	0.03							
Premix	0.04							
Total	8.42	0.77	36	.60	.28	70	1.62	.95
NRC requirement	8.42	1.061	30	.48	.23	70	1.67	1.06
Deficiency		−0.291					−0.05	−0.11

% crude protein = .77 kg/8.42 kg diet = 9.14% (NRC asks for 12.6%)
NE_m, Mcal/kg = 1.62 Mcal/kg diet (NRC asks for 1.67)
NE_g, Mcal/kg = 0.95 Mcal/kg diet (NRC asks for 1.06)

Now that we have made a trial run and analyzed our results, note that we are deficient in protein, NE_m, and NE_g. We see from the "Table of Selected Feedstuffs," presented earlier that dehulled soybean meal and shelled corn have quite similar analyses for energy, either TDN or NE. So it is possible to bring up the protein by substituting soybean meal for corn without having to consider any energy difference. But first, substitute some corn for some of the roughage.

Note that the NE_g is the most deficient of the two NE values, so we will work on that one. The table shows respective NE_g values as follows: silage, 0.94; orchard grass hay, 0.43 (the average for equal parts silage and orchard grass hay is .685) and corn, 1.42. Our NRC goal is 1.06 Mcal/kg diet). Let X = level of shelled corn, while $1 - X$ is the proportion of roughage.

$$1.42X + 0.685(1 - X) = 1.06$$
$$1.42X + 0.685 - .685X = 1.06$$
$$.735X = .375$$
$$X = .375/.735 \quad = .51 \text{ or } 51\% \times 8.26 = 4.21 \text{ kg shelled corn}$$
$$100\%-51\% = 49\% \times 8.26 = 4.04 \text{ kg roughage (2.02 kg each silage and hay)}$$

Now, let us check the NE$_g$ of the new mix (4.21 kg shelled corn; 6.11 kg roughage)

$$4.21 \text{ kg corn} \times 1.42 \text{ Mcal NE}_g/\text{kg} = 5.98 \text{ Mcal}$$
$$2.02 \text{ kg corn silage} \times 0.94 \text{ Mcal NE}_g/\text{kg} = 1.90 \text{ Mcal}$$
$$2.02 \text{ kg orchard grass hay} \times 0.43 \text{ Mcal NE}_g/\text{kg} = 0.87 \text{ Mcal}$$
$$5.98 + 1.90 + 0.87 = 8.75 \text{ Mcal divided by 8.42 kg dry matter} = 1.04 \text{ Mcal NE}_g/\text{kg}$$
dry matter.

Note how the NE method can improve upon the TDN approach to get a more exacting figure. In other words, the TDN solution did not contain quite enough energy to get the 1.36 kg daily gain anticipated.

We have one additional calculation to make—we must bring the total protein into line. Note that our first formula contained insufficient protein for optimal growth—the diet needed an additional .29 kg of protein. First, we will see how much protein our energy-revised formula does contain.

$$4.21 \text{ kg shelled corn} \times 10.9\% \text{ protein} = .46 \text{ kg protein}$$
$$4.04 \text{ kg hay-silage mix} \times \text{ave } 8.35\% \text{ protein} = .34 \text{ kg protein}$$
$$.46 + .34 = 0.80 \text{ divided by 8.42 kg diet} = 9.5\% \text{ protein (NRC calls for}$$
$$12.6\% \text{ protein)}$$
$$12.6\% \text{ protein in 8.42 kg diet} = 1.06 \text{ kg protein.}$$

Now we can use algebra to solve the protein deficiency. Soybean meal contains 54% protein and shelled corn contains 10.9% protein. Let X equal the soybean meal and $1 - X$ equal the corn:

$$54X + 10.9(1 - X) = 17$$
$$54X + 10.9 - 10.9 X = 17$$
$$43.1X = 6.1$$
$$X = .1415, \text{ or } 14.15\% \text{ soybean meal, plus } 85.85\% \text{ shelled corn}$$
$$14.15\% \times 4.21 \text{ kg grain mix} = 0.6 \text{ kg soybean meal} \times 54\% \text{ protein} = .32 \text{ kg protein}$$
$$85.85\% \times 4.21 \text{ kg grain mix} = 3.6 \text{ kg shelled corn} \times 10.9\% \text{ protein} = 39 \text{ kg protein}$$
There is a total of .71 kg protein in the new grain mix + .46 kg protein
from roughage = 1.07 kg.

Now we have balanced for the deficient NE$_g$ and the deficient protein, and the finalized formulation looks like:

	Dry Matter	Quantity					
		100% Dry Matter Basis			Air Dry (as-is) Basis		
Ingredient	*%*	*kg*	*lb*	*%*	*kg*	*lb*	*%*
Orchard grass hay	91	2.02	4.44	24	2.22	4.88	16
Corn silage	30	2.02	4.44	24	6.73	14.81	48
Shelled corn	89	3.60	7.92	42.8	4.04	8.90	29
Soybean meal (dehull)	90	0.60	1.32	7.1	0.67	1.46	4.8
Limestone	90	0.09	0.19	1.1	0.10	0.22	.7
Salt (TM)	90	0.03	0.07	0.3	0.03	0.07	.2
Premix	90	0.04	0.09	0.5	0.04	0.09	.2

37 Balancing a Steer Ration with Minimum Use of the Metric System

For those who prefer to work with lb and % rather than kg, kcal, Mcal/kg, g/kg, mg/kg, etc., the following procedure is recommended for ration balancing purposes. Assuming a ration is to be balanced for a 661 lb finishing yearling steer gaining 2.9 lb daily, typical of cattle in Fig. 37–1, one must first obtain the requirements of such an animal from the appendix table. This is done utilizing TDN rather than ME for evaluating energy adequacy. Also, the requirement values for DM, TP, TDN, Ca, and P are listed in lb rather than kg, while carotene is left in mg, as shown below:

	Roughage DM	TP	TDN	Ca	P	Caro
	lb	lb	lb	lb	lb	mg
Requirements for a 661 lb finishing yearling steer gaining 2.9 lb daily	2.4	1.83	13.2	0.064	0.051	40

(If the requirements in the tables available are listed in kg but not in lb, kg may be converted to lb by multiplying by 2.205. If some of the requirements are listed in grams

FIGURE 37–1
Most cattle today are finished in large feedlots such as the one shown above. (Courtesy of Ralston Purina Co., St. Louis, Mo.)

but not in kg and/or lb, grams may be converted to kg by dividing grams by 1000—in other words by moving the decimal point three places to the left. Kilograms can then be converted to lb by multiplying by 2.205.)

II

The animal's roughage requirement is the first to be met. This is done by including in the ration at the outset 2.4 lb of roughage dry matter, which we have decided in this illustration will be coastal bermudagrass hay. The as-fed composition of coastal bermudagrass hay is then obtained from the appendix table.

	DM	TP	TDN	Ca	P	Caro
	%	%	%	%	%	mg/lb
Coastal bermudagrass hay	90.0	5.4	49.0	0.39	0.18	43.1

It will be noted that all of the above composition figures, except carotene, are given in %. This is in keeping with the goal of this illustration—that is, to eliminate the use of metric figures as much as possible from the ration balancing process.

III

The amounts of the various critical nutrients in 2.4 lb of coastal bermudagrass hay DM or 2.67 lb ($2.4/90.0 \times 100 = 2.67$) lb of as-fed coastal bermudagrass hay are then determined by multiplying each of the above figures by 2.67, with each percentage figure first being divided by 100. The results of these calculations are as follows:

	TP	TDN	Ca	P	Caro
	lb	lb	lb	lb	mg
2.67 lb of CBG hay	0.14	1.31	0.010	0.005	115.1
Nutrients still needed	1.69	11.89	0.054	0.046	0.0
(Requirements minus the nutrients in the hay)					

IV

If it is again assumed that the energy (TDN) still needed is to be provided primarily through ground shelled corn, then the amount of corn required can be calculated by dividing the TDN still needed (11.89 lb) by the amount of TDN per lb of ground shelled corn (0.77). Based on this calculation, 15.44 lb of corn would be required.

A tentative ration of 2.67 lb of coastal bermudagrass hay and 15.44 lb of ground shelled corn is then checked for nutritional adequacy, as follows:

	TP	TDN	Ca	P	Caro
	lb	lb	lb	lb	mg
2.67 lb CBG hay	0.14	1.31	0.010	0.005	115.1
15.44 lb ground shelled corn	1.48	11.89	0.005	0.040	13.9
TOTALS	1.62	13.20	0.015	0.045	129.0

V

Upon comparing the above totals with the requirements, it is apparent that the tentative ration, while adequate in roughage, TDN, and carotene, is still deficient in TP as well as in Ca and P. Our next concern is to correct the TP shortage. This can be done by substituting a small quantity of some high-protein feed such as soybean meal for an equal amount of corn. In this way the protein shortage is corrected without materially altering the TDN content of the ration, since the two feeds are similar in TDN content and only a small amount of each feed is involved.

The amount of 44% soybean meal needed as a substitute for corn can be calculated by determining the total amount of protein that must come from a mixture of corn and soybean meal. The total protein requirement is 1.83 lb, and the hay contributes 0.14 lb; therefore 15.44 lb of corn-soy mix must contribute 1.69 lb (1.83 lb needed − 0.14 lb in hay = 1.69 lb). This means the corn-soy mix must contain 10.94% protein (1.69 lb protein divided by 15.44 lb corn-soy mix). Using the simple algebraic approach, let X equal amount of soybean meal and $1 − X$ equal corn:

$$44X - 8.5(1 - X) = 10.94$$
$$44X - 8.5 - 8.5X = 10.94$$
$$35.5X = 2.44$$
$$X = .068, \text{ or } 6.8\% \times 15.44 = 1.05 \text{ lb soybean meal}$$
$$1 - X = 93.2\% \text{ corn} \times 15.44 = 14.39 \text{ lb corn}$$

A tentative ration containing 14.39 lb shelled corn and 1.05 lb soybean meal as the grain mix, along with 2.67 lb coastal bermudagrass hay is then ready to check for nutritional adequacy, as follows:

	Protein	TDN	Ca	P	Carotene
	lb	lb	lb	lb	mg
2.67 lb CBG hay	0.14	1.31	0.010	0.005	115.1
14.39 lb ground shelled corn	1.22	11.08	0.004	0.039	13.4
1.05 lb SBM	0.46	0.81	0.003	0.007	—
Totals	1.82	13.20	0.017	0.051	128.5
Deficiency	0	0	0.047	0	0

VI

Upon comparing the latter totals with the steer's requirements, it is apparent that the ration is adequate in all critical nutrients except Ca, in which it is still quite deficient. The next step then is to correct the Ca shortage. This can be done by adding a small amount of some appropriate Ca source, such as ground limestone. The amount of ground limestone to be added can be calculated by dividing the amount of Ca still needed (0.047 lb) by the amount of Ca in 1 lb of ground limestone (0.34 lb). On the basis of this calculation (0.047/0.34 = 0.126), a need for 0.126 lb of ground limestone is indicated. This amount of ground limestone would add the following amounts of critical nutrients to the ration:

	TP	TDN	Ca	P	Caro
	lb	lb	lb	lb	mg
0.126 ground limestone	—	—	0.043	—	—
New TOTALS	1.82	13.20	0.064	0.051	128.5
Percentage of requirements	100.0%	100.0%	100.0%	100.0%	311.3%

VII

Upon comparing the latter totals with the steer's requirements for the various critical nutrients as listed earlier, it is apparent that the ration is now adequate in all nutritive factors.

The final ration then is as follows:

> 2.67 lb of coastal bermudagrass hay
> 14.39 lb of ground shelled corn
> 1.05 lb 44% soybean meal
> 0.126 lb ground limestone

Since the above ration is already in lb of feed as fed, it requires no conversion from kg to lb, or from an oven dry to an as-fed basis.

38

Balancing a Ration for a Steer Using a High Level of Roughage

In "stocker" type cattle production, the manager wishes to cause weanling calves (7 to 8 months of age) to gain approximately .7 to .9 kg/d (1.5 to 2.0 lb/d). A rather high roughage level diet can accomplish this goal, if it is balanced properly with minerals and possibly some supplemental protein. This might be the logical program to follow when producing replacement heifers; or when the manager does not have much corn to finish out cattle and wants to get some economical growth on steers and then sell them as feeder cattle to a finishing cattle feeder. The purpose of this section is to balance a diet for such cattle and to see that it contains the nutrients necessary to do just that.

I

The first thing we need to do in formulating any feeding program is to identify the nutrient needs for a calf in such a program. From Table Beef–1 in the appendix select a 250 kg calf (550 lb) to gain 0.9 kg/d (2.0 lb/d). (Note from Table Beef–1, there is no figure that fits gaining 0.9 kg per day, but this will give experience in interpolation between two requirement figures, e.g., gains of 0.5 and 1.0 kg/d.)

NUTRIENT REQUIREMENTS FOR A 250 KG CALF TO GAIN 0.9 KG PER DAY

	NE_m Mcal/d	NE_g Mcal/d	MP g/d	Ca g/d	P g/d
Maintenance	4.84	—	239	8	6
Gain (0.9 kg/d)	—	2.80[1]	27	22	9
Total	4.84	2.80	510	30	15

[1]Since the daily gain (0.9 kg/d) is 4/5 of the way between the posted 0.5 and 1.0 kg/d in Table Beef–1, all the figures in this row are calculated to reflect that.

In Table Beef–2, figures are posted for the steer to gain 0.9 kg (0.89 kg)/d. Since there is no @50 kg category listed in Table Beef–2, we will have to use the requirements for a 300 kg animal, which is a B ration. Under that category, the steer should consume 8.4 kg of a diet containing 43% of the DM as neutral detergent fiber (NDF), 60% TDN, 1.35 Mcal NE_m/kg, 0.77 Mcal NE_g/kg, and 10% crude protein containing 79% digestible intake protein (DIP).

As a starting point, test equal parts oat hay (4.2 kg) and timothy hay (4.2 kg) as a total diet:

	kg day	NE_m Mcal	NE_g Mcal	Protein kg	TDN kg	NDF kg	Ca g	P g
Oat hay	4.2	5.58	2.90	.39	2.56	2.64	10	9
Timothy hay	4.2	5.29	3.82	.34	2.44	2.58	18	8
Total	8.4	10.87	6.72	.73	5.00	5.22	28	17
Requirement	8.4	11.34	6.46	.84	5.04	5.06	30	13

Now evaluate the formula. It is deficient in NE_m, but since it is over in NE_g, the animal's body will juggle those two amounts of NE sufficiently to take care of that. It is also 0.11 kg (0.24 lb) deficient in protein, and that is critical for growth. One-half lb of soybean meal would take care of that, but it is difficult to divide small quantities among several animals. One might want to consider using a lick tank and letting the cattle self-feed themselves. Finally, even though the minerals Ca and P, are just about correct, the matter of salt has not been taken into account. Especially on growing diets, it is a good idea to provide free-choice mineral (protected from the weather, as in Figure 38–1) consisting of two parts dicalcium phosphate to one part of ruminant trace mineralized salt. Figure 38–1 shows a shelter for providing supplemental minerals or protein, and a lick tank among the heifers.

II

Solving a hypothetical problem, algebraically. Our approach with equal parts oat hay and timothy hay worked out pretty well, but the student needs to work out a problem where balancing is essential. So let us set up a problem that we must solve algebraically. Suppose we have a roughage to offer our stocker calves that is only 50% TDN, and the requirements from Table Beef–2 says the diet should contain 60% TDN. In figuring the TDN, let us use shelled corn (87% TDN) and a 50% TDN roughage. (We went through this in section 36, but a review will not hurt.) Let X equal the proportion of corn and $1 - X$ equal the proportion of roughage. The equation becomes

$$87X + 50(1 - X) = 60$$
$$87X + 50 - 50X = 60$$
$$37X = 10$$
$$X = .27, \text{ or } 27\% \text{ of the mix is shelled corn}$$

In checking the calculation, $27\% \times 87\%$ TDN = 23.49
$73\% \times 50\%$ TDN = 36.50
Total TDN = 59.99% TDN

III

What about bluegrass pasture?   Still consuming 8.4 kg dry matter as bluegrass pasture, the following nutrient intake would be expected: protein, 14.9%; TDN, 64%; NE_m,

FIGURE 38–1
Stocker heifers on pasture need equipment (feeder trough) for presenting supplemental feeds when the forage will not support the performance desired. Note that beside the watering trough, there is a lick tank for providing protein, minerals, and trace minerals.

1.41 Mcal/kg; NE_g, 0.78 Mcal/kg; NDF, 55%; Ca, .33%; P, .34%. In the previous problem, the diet requirements for the steer to gain at the expected rate were as follows: protein, 10%; TDN, 60%; NE_m, 1.35 Mcal/kg; NE_g, .77 Mcal/kg; NDF, 43%. We will not figure the adequacy of Ca and P, because we should supply a free-choice mineral consisting of 2 parts dicalcium phosphate to 1 part ruminant trace mineralized salt.

There is one potential hazard to following the above set of data. They represent the most ideal bluegrass pasture conditions and when such herbage matures or when there is not adequate rainfall to keep the herbage lush and tender, the analyses presented will fall apart rapidly and so will the gain of the cattle unless supplementation of deficient nutrients is made.

39

The California System for the Net Energy Evaluation of Rations for Growing-Finishing Cattle

I

The metabolizable energy (ME) of feeds is used by livestock much more efficiently for maintenance than it is for gain. Also, feeds vary greatly in the proportion of their ME content that is usable by livestock for gain. Concentrates, for example, are in general superior to roughages in the gain-producing capacity of their ME energy, and good-quality roughages are superior to low-quality roughages. Consequently, the ME content of a ration is sometimes not an accurate measure of its capacity to produce gain. This is especially true when a relatively high proportion of the ME comes from roughages, and particularly so when these roughages are of low quality. Roughages in general, and low-quality roughages in particular, yield a high proportion of their ME in the form of heat, which is useful to the extent needed for keeping the animal warm, but of no value for producing gain. In fact, excess heat may actually interfere with an animal's gain-producing capacity.

II

In an effort to provide a more accurate basis for evaluating feeds from a gain-producing standpoint, various sets of so-called net energy (NE) values have been proposed over the years. However, some of the earlier NE systems have been justifiably criticized for not separating body functions in the establishment of net energy values.

Recognizing the need for such a separation, workers at the California Station have proposed a system that is designed to accomplish this goal. This system separates an animal's energy requirements into NE for maintenance (NE_{main} or NE_m) and NE for gain (NE_{gain} or NE_g); the requirements of various classes of steers and heifers for each are given in Tables for beef in Appendix. Also, individual feeds have been assigned dual net energy values, one for maintenance (NE_m) and another for gain (NE_{gain}), and these values have been included in the appendix table.

In evaluating any particular ration, then, from the standpoint of its energy adequacy using the latter method, separate NE_m and NE_{gain} values are calculated for the overall ration. The NE_m value is used with as much of the overall ration as is needed to meet the animal's NE_m requirements. The remainder of the ration is then available for gain based on the calculated NE_{gain} value.

III

To illustrate the use of the California system, let us proceed to calculate the gain-producing potential of the following ration.

Col. 1 Final Ration (as-fed basis)		Col. 2 (From Appendix) NE_m	Col. 3 (Col. 1 × Col. 2) Total NE_m	Col. 4 (From Appendix) NE_{gain}	Col. 5 (Col. 1 × Col. 4) Total NE_{gain}
		Mcal/kg	Mcal	Mcal/kg	Mcal
4.00	kg coastal bermudagrass hay	1.05	4.20	0.39	1.56
4.41	kg ground shelled corn	1.85	8.16	1.26	5.57
0.42	kg 44% SBM	1.79	0.75	1.20	0.50
0.011	kg defluorinated phosphate	0.00	0.00	0.00	0.00
0.021	ground limestone	0.00	0.00	0.00	0.00
8.862	kg total ration	(1.48)	13.11	(0.86)	7.63

Calculated NE_m value for overall ration (13.11/8.86 = 1.48 Mcal/kg)
Calculated NE_{gain} value for overall ration (7.63/8.86 = 0.86 Mcal/kg)
Requirement of 300 kg finishing steer for NE_m (from requirement table) = 5.55 Mcal
Amount of overall ration needed for maintenance (5.55/1.48 = 3.75 kg)
Amount of overall ration available for gain (8.86 – 3.75 = 5.11 kg)
Amount of NE_{gain} in 5.11 kg of ration available for gain (5.11 × 0.86 = 4.39 Mcal)
Amount of gain to be expected from 4.39 Mcal of NE_{gain} for 300 kg steer (from requirement table) = approximately 1.02 kg

On the basis of the California system, about 1.02 kg of daily gain could be expected from a 300 kg steer receiving the above ration. This is a somewhat lower gain than normally would be expected for such an animal receiving a supposedly balanced ration. However, there was a relatively high level of hay used in balancing the above ration, and with this amount of hay the above predicted level of gain is probably about right.

IV

For the sake of comparison, it is interesting to use the California system to evaluate from the standpoint of its gain-producing capacity a ration that has been balanced on the basis of ME but with more of the ration's ME coming from grain than was true with the previous ration see following ration. This ration and its evaluation using the California system would be as follows:

Ration (As-fed basis)		NE_m	Total NE_m	NE_{gain}	Total NE_{gain}
		Mcal/kg	Mcal	Mcal/kg	Mcal
1.22	kg coastal bermudagrass hay	1.05	1.28	0.39	0.48
6.01	kg ground shelled corn	1.85	11.12	1.26	7.57
0.42	kg 44% SBM	1.79	0.75	1.20	0.50
0.011	kg defluorinated phosphate	0.00	0.00	0.00	0.00
0.050	kg ground limestone	0.00	0.00	0.00	0.00
7.711	kg total ration	(1.71)	13.15	(1.11)	8.55

Calculated NE_m value for overall ration = 1.71 Mcal/kg
Calculated NE_{gain} value for overall ration = 1.11 Mcal/kg
Requirement of 300 kg finishing steer for NE_m = 5.55 Mcal
Amount of overall ration needed for maintenance = 3.25 kg
Amount of overall ration available for gain = 4.46 kg
Amount of NE_{gain} in 4.46 kg of ration available for gain = 4.95 Mcal
Amount of gain to be expected from 4.95 Mcal of NE_{gain} for 300 kg steer = approximately 1.14 kg

It will be noted from the above that the latter ration, even though it contains 1.15 kg less total feed, actually supplies 0.56 Mcal more NE_{gain} and should support about 0.12 kg greater average daily gain. In other words, even though the two rations were balanced to provide essentially the same amounts of ME, the latter ration, because of its greater proportion of concentrates to hay, supplies more NE_{gain} and in turn would support a greater daily gain.

40

Balancing a Daily Ration for a Dairy Cow

I
The balancing of a ration for a dairy cow differs in certain respects from the balancing of a ration for a beef animal.

A. In the first place, the requirements cannot be taken directly from the table but must be calculated by adding to the requirements for maintenance (including growth and fetal development, if involved) an additional set of requirements for milk production based on the amount of milk produced and its fat content.

B. Secondly, the roughage requirement is met simply by following the general practice of feeding not less than 1.5% and not more than 2% of the cow's liveweight as air-dry roughage daily or the equivalent of silage.

C. Also, since the energy requirements of dairy cows have been estimated and published in terms of net energy for lactating cows ($NE_{lactating\ cows}$), and since $NE_{lact\ cows}$ (also sometimes abbreviated as NE_l) values have been made available for most of the more common dairy feeds, dairy cow rations are frequently balanced for energy on the basis of $NE_{lact\ cows}$ rather than on the basis of metabolizable energy (ME) or total digestible nutrients (TDN).

302

Otherwise, the balancing of a dairy cow ration is similar to the balancing of a steer ration, and most of the same techniques can be applied. After the requirements of a dairy cow have been established, the ration-balancing process is very similar to that followed in balancing a ration for a steer, using a high level of roughage.

For the sake of illustration, let us balance a ration for a 500 kg mature dairy cow in early gestation producing 20 kg of 4% (fat) milk daily. The daily requirements for such a cow would be calculated as follows:

	Total Protein kg	NE_{lact} Mcal	Ca g	P g
Requirements*				
Maintenance, 500 kg cow	.364	8.46	20	14
Production, 20 kg 4% milk	1.920	14.80	64	40
Total	2.284	23.26	84	54

*Taken from Table Dairy – 1, which is an adaptation of NRC (Dairy), sixth ed., 1989.

On the basis of the general guide above, a 500 kg cow should be fed somewhere between 7.5 kg and 10 kg of air-dry roughage daily or the equivalent of silage. In this instance, let us assume that the lower level of roughage (7.5 kg) is fed in the form of 2.5 kg alfalfa hay and 15 kg of corn silage (15 kg of silage to replace 5 kg of air-dry roughage).

A. The first step is to determine the amount of each nutrient provided by the above amounts of alfalfa hay and corn silage, and then calculate by difference the amount of each nutrient that must come from the concentrates in the ration.

	Total Protein kg	NE_{lact} Mcal	Ca g	P g
2.5 kg alfalfa hay	0.382	2.92	32	6
15 kg corn silage	0.375	7.05	14	12
Total	0.757	9.97	46	18
Requirement	2.284	23.26	84	54
Needed from grain mix	1.527	13.29	38	36

B. The next step is to determine the approximate amount of concentrates that will need to be fed with the above roughages to meet the $NE_{lact\ cows}$ requirement. This is done by dividing the 13.29 Mcal $NE_{lact\ cows}$ needed from the concentrates by the average $NE_{lact\ cows}$ content of the concentrates to be fed. If it is assumed that 1.75 Mcal

$NE_{lact\ cows}$/kg would be about right for the approximate average $NE_{lact\ cows}$ content of concentrates fed to dairy cows, then 13.29/1.75 would equal the approximate amount of total concentrates to be fed (7.59 kg). A ration such as the one that follows is then set up and tested for adequacy. The 3 kg of corn and 4.59 kg of barley used are purely arbitrary. Other combinations of these and/or other concentrates could be used to meet the energy needs.

	Total Protein kg	NE_{lact} Mcal	Ca g	P g
2.5 kg alfalfa hay	0.382	2.92	32	6
15 kg corn silage	0.375	7.05	14	12
7.59 kg grain mix				
3 kg ground shelled corn	0.288	5.34	1	8
4.59 kg ground barley	0.546	7.85	2	16
Total	1.591	23.16	49	42
Requirement	2.284	23.26	84	54
Still deficient	0.693	adequate	35	12

C. It is apparent from the above that the latter ration is low in TP, as well as calcium and phosphorus. The next step then is to correct the shortage of TP by substituting some high-protein concentrate such as soybean meal for a part of the barley. The amount of barley to replace with soybean meal is arrived at by dividing the amount of additional TP needed (2.172 kg – 1.591 kg = 0.581 kg) by the increase in TP brought about by each kg of barley replaced by 44% soybean meal (0.446 kg – 0.119 kg = 0.327 kg). Upon dividing 0.581 kg by 0.327 kg, it is determined that 1.78 kg of barley will need to be replaced with soybean meal. (While simultaneous equations may be used for making the above calculations, it is usually simpler and more expeditious to use the substitution procedure where more than one major energy source is involved.)

The ration is then revised accordingly and rechecked for adequacy as follows:

	TP kg	$NE_{lact\ cows}$ Mcal	Ca kg	P kg	Caro mg
2.5 kg alfalfa hay	0.382	2.92	0.032	0.006	147.5
15.0 kg corn silage	0.375	7.05	0.014	0.012	195.0
3.0 kg ground shelled corn	0.288	5.34	0.001	0.008	6.0
2.81 kg ground barley	0.334	4.81	0.001	0.010	9.2
1.78 kg 44% SBM	0.794	3.08	0.005	0.011	0.0
TOTALS	2.173	23.20	0.053	0.047	357.7
Percentage of requirement	100.0%	99.7%	73.6%	92.2%	376.5%

D. In view of the fact that the above ration is still low in both calcium and phosphorus, it is desirable that additional amounts of these nutrients be added. Since most of the commonly used phosphorus supplements are also high in calcium, it is best to correct the phosphorus shortage next and then add some calcium, if needed.

By subtracting the amount of phosphorus in the latter ration above from the phosphorus requirement (.0510 – .047), it is determined that an additional .004 kg of phosphorus needs to be added to the ration. If .004 kg is divided by the amount of phosphorus in 1 kg of defluorinated phosphate (0.18 kg), the amount of defluorinated phosphate that needs to be added to correct the phosphorus shortage is determined. This figures out to be .022 kg. This amount of defluorinated phosphate is then added to the above ration, which is rechecked for adequacy as follows:

	TP kg	$NE_{lact\ cows}$ Mcal	Ca kg	P kg	Caro mg
2.5 kg alfalfa hay	0.382	2.92	0.032	0.006	147.5
15.0 kg corn silage	0.375	7.05	0.014	0.012	195.0
3.0 kg ground shelled corn	0.288	5.34	0.001	0.008	6.0
2.81 kg ground barley	0.334	4.81	0.001	0.010	9.2
1.78 kg 44% SBM	0.794	3.08	0.005	0.011	0.0
.022 kg defluorinated phosphate			0.007	0.004	
TOTALS	2.173	23.20	0.060	0.051	357.7
Percentage of requirement	100.0%	99.7%	83.3%	100.0%	376.5%

It will be noted that the defluorinated phosphate that was added not only corrected the phosphorus shortage but also, to a large degree, corrected the shortage of calcium. To make this ration adequate in calcium, .035 kg of ground limestone [(.072 – .060)/.34] should be added to it.

E. Before being usable under most present-day farm conditions, however, the above ration will need to be converted to lb by multiplying each of the individual quantities of feed by 2.205. The final ration then would be:

> 5.51 lb alfalfa hay
> 33.08 lb corn silage
> 6.62 lb ground shelled corn
> 6.20 lb ground barley
> 3.92 lb 44% soybean meal
> .049 lb defluorinated phosphate
> .077 lb ground limestone

In balancing the above ration, as-fed weights and composition figures were used for the various feeds. Essentially the same results could have been realized using dry feed weights and composition figures. Also, ME or TDN could have been used rather than $NE_{lact\ cows}$ for evaluating energy without significantly affecting the overall results.

As with the balancing of a steer ration earlier, those who prefer to avoid the use of the metric system may, for the most part, do so by using TDN as the measure for evaluating energy adequacy. However, as with the steer, the requirements for all the critical nutrients except carotene must be converted from kg to lb by multiplying by 2.205 unless this has already been done in the table.

IV

The above formulation was balanced to meet many of the established requirements. However, newer knowledge has put emphasis on additional measurements, i.e., neutral detergent fiber (NDF, primarily cell walls), and undigestible intake protein (UIP). It is worthwhile to check out the 1989 Dairy NRC Bulletin (sixth rev. ed.), as to requirements. First, it is recommended that a 500 kg (1100 lb) dairy cow yielding 20 kg (44 lb) 4% butterfat milk/d should receive a diet containing 28% NDF and about 5.2% UIP. How does the above diet measure up to those standards?

Ingredient	Dry Matter %	NDF %	UIP %
Alfalfa hay (2.5 kg)	90	38	22
Corn silage (15 kg)	35	46	30
Shelled corn (3 kg)	90	11	55
Ground barley (2.81 kg)	88	18.1	27
44% soybean meal (1.78 kg)	91	10.3	34

It would appear that the NDF level will exceed the recommended 28% level; however, it would appear that the UIP level may be borderline to deficient. (Bear in mind that this cow's diet should contain about 15.5% protein in the total diet and that the 5.2% UIP listed above will represent just about 1/3 of the 15.5% total protein.) Thus, when one starts calculating the total protein requirement, some of the above ingredients do not have 1/3 of their protein content as UIP. In the higher-producing dairy cow, a deficiency of UIP probably could limit her production. Such feedstuffs as hydrolyzed feather meal, blood meal, fish meal and meat meal are fairly good sources of UIP and may need to be considered.

V

A 365-day overview of the higher milk producer.   The dairy cow production year can be broken into about 4 or 5 logical programs reflecting changes in her nutritional needs:

A.   Calving to peak milk production.   Milk production in dairy cattle normally follows a curve in which it increases for 80 or 90 days after the birth of her calf. This tendency must be aided by the feeding of higher-energy feedstuffs. It is at this time that the term *lead-feeding* comes into use. From a scientific point of view, one probably should calculate a high-producing dairy cow's needs once per week— admittedly, this may not be practical. Nevertheless, as the cow is climbing the ladder of production during her first 3 months of production, it would be beneficial to calculate her production needs and add a little extra nutrients. If the nutrition needs were calculated on last week's production, they would be inadequate for next week's needs during this period of increasing production. Feeding a bit more, then, is called lead feeding. Naturally, once her peak of production has been reached, she should be fed in accordance with her production—and later pregnancy—needs.

In order to accomplish this, one should feed some of the best grade of hay possible; grain level should be increased; and attention should be given to utilization of UIP (undegradable intake protein) to meet as much as 1/3 of the total protein. If the diet gets too concentrated and there is danger of acidosis, sodium bicarbonate can be added as a rumen buffer.

B. **Flat-top in production.** This is the period when the cow has reached her peak of production and the manager hopes to be able to hold her as close to this peak as possible. Naturally there will be a gradual decline from her peak of production at about 80 days to day 210. Nutrient intake should be tied closely to production because lead-feeding during this period would not pay very good dividends. This is a period when the cow can start to build back some of the weight loss she has suffered during the earlier phase. By now, she is pregnant and so nutrients are beginning to be divided among four needs, namely, (1) maintenance, (2) milk production, (3) forming her unborn calf, and (4) regaining parturition and early lactation weight losses.

C. **Easing-off period.** For the next 100 days, the cow is preparing seriously for her next calf and lactation. In doing this, she does quite well on much less concentrated rations and, thus, more forage and less concentrates are needed. Although one would not practice lead-feeding during this period of decline in milk production, it is a good idea to keep close tabs on production and feed only enough concentrate to meet her needs for production, body condition, and age.

D. **Rest period.** During this nearly two-month period the cow is not producing milk. She utilizes loose, long hay, and little or no concentrates are needed unless she needs to gain just a little more body condition.

E. **Getting ready for next lactation.** During this last two weeks of pregnancy, she needs to be reintroduced gradually to her lactation diet. In other words, start getting her back on concentrate feeding by feeding up to 0.5% of her body weight. Dry, long hay should be kept in the diet at this time to avoid potential digestive disturbances.

41

Balancing Daily Rations for Horses

_____I_____ The balancing of daily rations for horses is similar in most respects to balancing rations for cattle, and most of the same general methods and techniques may be used.

    A. As in balancing rations for other livestock classes, the first step is to determine the requirements for the particular animal to be fed as given in the appropriate NRC table. The critical nutrients to be considered are essentially the same for horses as for other animals. As with other classes of livestock, no specific DM requirement is indicated. While a "Daily Feed" intake is given in the NRC tables of requirements for horses, this is based on the assumption that the ration used contains a normal proportion of roughage to concentrates for the particular class of horses in question and does not apply for rations and mixtures of other energy concentrations such as might be used for feeding this class of animals.

    B. Consequently, in order to be certain that a horse is provided with the proper DM allowance—that is, with a ration which contains sufficient bulk, on the one hand, but does not exceed the feed-consuming capacity of the animal, on the other—a minimum of 1% and a maximum of 2% of the animal's liveweight as air-dry roughage is ordinarily included in the ration. Grain and supplement are then added as needed to balance the ration.

FIGURE 41–1
Quarter horse mares and foals have a nearly complete diet on lush grass pasture. (Photo courtesy of Dr. Gary Potter, Texas A. & M. University)

C. The rate of roughage feeding should be adjusted so that the animal's total daily consumption of air-dry feed does not exceed about 2.25% to 2.50% of the animal's liveweight. This is roughly the upper limit of a horse's capacity to consume feed.

II

Following the general guidelines listed previously, a formulation designed to fit a specific horse feeding situation is in order. A typical feeding situation would be that for a 500 kg mature mare in the first 3 months of lactation. First, the nutrient requirements should be obtained from the Horse Nutrient Requirement Tables presented in the appendix and which, for this situation, are as follows:

	Digestible Energy Mcal	Crude Protein grams	Calcium grams	Phosphorus grams	Magnesium grams
Required, daily	28.3	1,427	56	36	10.9

There are requirements for lysine, but the feed table lists no lysine analysis for timothy hay, which would be a constituent. A requirement for potassium is listed but the amount provided exceeds the requirement so greatly that it is hardly necessary to go through the mechanics of calculating such.

Feedstuffs that can be expected to find their way into a horse formulation would be as follows:

	Digestible Energy Mcal/kg	Crude Protein %	Calcium %	Phosphorus %	Magnesium %
Timothy hay	2.56	9.1	0.48	0.20	0.16
Oats, grain	3.40	13.3	0.07	0.38	0.14
Soybean meal, 44	3.70	49.4	0.30	0.68	0.30
Dicalcium phosphate	—	—	22.5	18.5	—
Limestone	—	—	38.0	—	—

The data for the above feed analyses were taken from the Ruminant Feed Table. This table lists all feed ingredients on a 100% dry matter basis. In order to convert to an air-dry basis, one may divide "100% dry matter basis" by 89% or 90% to obtain air-dry values for most feeds except silages and fresh pasture crops.

A lactating mare probably is going to require some concentrates to supply added energy and protein for the lactation process, and, naturally, she will require roughage. As in calculating most formulations, a logical starting point is the energy requirements. Let us assume that the horse will need 1% of her body weight, minimum, as hay. Five kg of timothy hay provides the following: 2.56 Mcal of DE/kg × 5 kg = 12.8 Mcal of DE. The diet needs 28.3 Mcal of DE, therefore. 28.3 Mcal − 12.8 Mcal provided by timothy hay = 15.5 Mcal, yet to be provided. Oats contains 3.40 Mcal of DE/kg; 15.5 Mcal needed divided by 3.4 Mcal of DE/kg in oats = 4.6 kg oats, which fulfills the energy requirement. Next, let us set the timothy hay and oats into a table to see what else is needed:

	Quantity kg	Digestible Energy Mcal	Crude Protein g	Ca g	P g	Mg g
Timothy hay	5.0	12.8	455	24	10	8
Oats grain	4.6	15.5	611	3	17	6
Totals	9.6	28.3	1066	27	27	14

Since energy is adequate but protein, calcium, and phosphorus are deficient, we can work on the protein next by substituting some soybean meal for some of the 4.6 kg of oats. A 4.6 kg mixture of oats (13.3% protein) and soybean meal (49.4% protein) must contain 972 grams of protein (1,427 needed − 455 supplied by the timothy hay), or 21.13% protein (927g/4,600 g). In the following equation, X = soy and 1 − X = oats grain.

$$49.9X - 13.3(1 - X) = 21.13$$
$$49.4 \, X - 13.3 - 13.3X = 21.13$$
$$36.1 \, X = 7.83$$
$$\text{(Soy) } X = 21.69\% \times 4.6 \text{ kg} = 1 \text{ kg soy} \times 49.4\% \quad = 494 \text{ g protein}$$
$$\text{(Oats, } 1 - X) \; 4.6 \text{ kg} - 1 \text{ kg} = 3.6 \text{ kg oats} \times 13.3\% = 479 \text{ g protein}$$
$$\text{Total protein} = 973 \text{ g}$$

Thus, 455 g protein from hay, plus 973 g from soy – oats = 1,428 total protein.

Finally, minerals must be adjusted. Ca needed = 56 g, and P needed is 36 g, The addition of soy brings total P to 31 g. Calcium now needed is 24 g, which, when divided by 38% Ca in limestone = 63 g limestone needed. So, to the 5 kg timothy hay + 3.6 kg oats grain + 1 kg soybean meal, add 63 g limestone + 7 g trace mineralized salt, and the formulation meets the 500 kg lactating mare's nutrient needs for one day.

___

III

___

There are several alternatives that might have been used in balancing the above ration, but regardless of the alternative pursued, the procedure and the results would have been essentially the same. Some of these alternatives include the following:

A. Total digestible nutrients (TDN) rather than digestible energy (DE) might have been used for balancing the ration with respect to energy.

B. Pounds rather than kilograms might have been used as the unit of feed weight. However, this would have necessitated multiplying all of the requirements, except carotene, by 2.205 to change them from kg to lb. Also, this would make it necessary for energy to be expressed as lb of TDN rather than Mcal of DE.

C. Hays other than timothy, grains other than oats, and a protein supplement other than soybean meal could have been used with equally effective results.

42

Balancing Daily Rations for Sheep

I

Sheep are probably the best grazing domesticated animal. If there is any browse at all, sheep seem to have a propensity to survive. This fact was responsible for some range wars in the early days of the settlement of the western United States. Because sheep have thinner lips than cattle, they can graze herbage so close to the ground that cattle cannot find any material to graze. Therefore, men who grazed cattle did not like the sheepmen. The survival alluded to above is not enough in agricultural production today, because there is so much capital invested in livestock that each animal must be pushed to the maximum of their genetic potential most of the time. A key to maximum production is providing balanced rations along with optimal management, of course.

II

Establishing the nutrient requirements of sheep. Obviously, one must first set forth the nutrient requirements for a specific situation. In this case, an "idle" 70 kg ewe; then a 60 kg ewe with a single nursing lamb in the first 8 weeks of lactation. (Refer to tables in the appendix for establishing nutrient requirements for sheep.)

	70 kg (154 lb) Ewe Maintenance		60 kg (132 lb) Ewe Lactating, One Lamb, First 8 Weeks	
		%		%
Dry matter/d, kg	1.2		2.6	
Roughage/d, kg	1.2		1.74	
Energy				
TDN/d kg	0.66	55	1.69	65
ME, Mcal	2.38		6.10	
Total protein/d, kg	.107	8.9	.299	11.5
Dig. prot/d, kg	.058	4.8	.187	7.2
Calcium/d, gm	3.2	.27	13	.50
Phosphorus/d, gm	3.0	.25	9.4	.36

Next, let us look at some feedstuffs and their analyses to determine which might be used in the above ewe formulations.

	TDN %	Protein %	Calcium %	Phosphorus %
Alfalfa hay, mid bloom	63	15.3	1.27	0.22
Bermudagrass hay, sun cured	49	5.4	0.39	0.18
Kentucky bluegrass, fresh	64	14.9	0.33	0.34
Fescue hay, sun cured	50	8.3	0.51	0.36
Lespedeza hay, sun cured	42	12.8	1.14	0.21
Oat hay	61	8.5	0.24	0.22
Orchardgrass hay, sun cured	54	8.4	0.39	0.35
Timothy hay, sun cured	58	8.1	0.43	0.20
Soybean meal, 44%	87	49.7	0.29	0.70
Shelled corn	87	10.9	0.03	0.29
Dicalcium phosphate	—	—	26.5	20.5
Ground limestone	—	—	38	—

From the preceding table, it is quite obvious that the Kentucky bluegrass pasture would meet all the needs of maintenance; alfalfa hay would meet all the requirements, but with too much protein, and possibly not enough phosphorus. Sheep, like beef cattle, should also be provided a free choice mineral mix, but such mixture is different for sheep than for cattle. A mineral mixture consisting of equal parts dicalcium phosphate, ground limestone, and ruminant-type trace mineralized salt would make an excellent offering for sheep. This mixture contains less phosphorus, which decreases the incidence of urinary calculi. If the mineral requirements are handled in this manner, we can now concentrate on meeting only energy and protein requirements.

III

Needs of the lactating ewe. The nutritional needs of the lactating ewe are much more sophisticated than those for the maintenance ewe. First, suppose we have orchardgrass hay (54% TDN; 8.4% protein), soybean meal (87% TDN; 49.7% protein), and shelled corn (87% TDN; 10.9% protein) available.

Since the protein shortage of the orchardgrass hay is obvious, we will balance the 2.6 kg of dry matter to contain 11.5% protein (algebraically). Let X = soybean meal and $1 - X$ = orchardgrass hay.

$$49.7X - 8.4(1 - X) = 11.5$$
$$49.7X - 8.4 - 8.4X + 11.5$$
$$41.3\,X = 3.1$$
$$X = 7.5\%\ \text{soy} \times 2.6\ \text{kg} = 0.2\ \text{kg soybean meal}$$
$$92.5\% \times 2.6 = 2.4\ \text{kg orchardgrass hay}$$

Let us check the adequacy of the diet formulated:

	TDN, kg	Protein, kg
2.4 kg orchardgrass	1.30	.2016
0.2 kg soybean meal	.17	.0994
Total	1.47	.3010
Divided by 2.6 kg =	56.5%	11.6%

We met the protein requirement by substituting 0.2 kg soybean meal for 0.2 kg orchardgrass hay; however, also note that the TDN requirement is for 65% and our new formula contains only 56.5%. So, let's make an energy adjustment. Note that both corn and soybean meal contain 87% TDN. Therefore, we can balance 87% TDN with orchardgrass hay (54% TDN) and from the solution, subtract the 0.2 kg of soybean meal from the corn total without changing the TDN balance. Let X = corn and $1 - X$ = orchardgrass in the formula.

$$87X - 54(X - 1) = 65$$
$$87X - 54X - 54 = 65$$
$$33X = 11$$
$$X = .33\ (33\%) \times 2.6\ \text{kg} = .86\ \text{kg corn}\ (-0.2\ \text{kg soy}) = .66\ \text{kg corn}$$
$$67\% \times 2.6\ \text{kg} = 1.74\ \text{kg orchardgrass hay}$$

With the corn added, the totals are as follows:

	TDN, kg	Protein, kg
.66 kg corn	.57	.0719
1.74 kg orchardgrass hay	.94	.1461
.2 kg soybean meal	.17	.0994
Total	1.68	.3174
Divided by 2.6 kg diet	65%	12.2%

Note that the energy is correct but by substituting corn (10.9% protein) for hay (8.4% protein) in the energy adjustment, we added extra protein. However, by replacing 0.05 kg soybean meal with 0.05 kg shelled corn, we now have 11.5% protein.

.71 kg shelled corn × 10.5% protein = .0773
1.74 kg hay × 8.4% protein = .1461
.15 kg soybean meal × 49.7% = .0745
Total   .2979
Divided by 2.6 kg = 11.458% protein

Now, to convert kg to lb:

.71 kg shelled corn × 2.2 = 1.56 lb corn divided by 89% dm = 1.76 lb air dry
1.74 kg orchard gr. hay × 2.2 = 3.83 lb. divided by 91% dm   = 4.2 lb air dry
.15 kg soy meal × 2.2 = .33 lb divided 89% dm   = .37 lb air dry

IV

Application to other sheep-feeding situations. Any other aspects of sheep formulations can be calculated in like manner. Remember to do the following:

1. Describe the feeding situation, i.e., lactating ewe, ewe last 6 weeks gestation, etc.
2. Define the requirements using the tables in the appendix.
3. List and characterize the feeds you have available.
4. Then solve the problem and check to see whether your formulation meets the requirements.

43

Balancing Daily Rations for Swine

Corn and soybean meal. Balancing rations for swine is quite similar to balancing rations for any other specie discussed in this text, except that swine may be a bit simpler, because there are not as many choices of feedstuffs. Corn and soybean meal dominate most of the swine formulas. Corn is the most abundantly available energy grain and it fits well into monogastric nutrition. Wheat is the grain of preference for swine—they prefer wheat to corn, if given a choice, and wheat is worth approximately 5% more as a feed for swine than is corn. However neither quantities nor economics favor wheat. Soybean meal is the most readily available source of supplemental protein, which fits well with corn. In fact, when supplemented with the proper minerals and vitamins, a combination of corn and soybean meal that contains the designated crude protein level will normally meet both the amino acid level requirement and the energy requirement for almost any swine-feeding program. Furthermore, when submitted to least-cost formulation, the result probably will be a "fortified corn-soy" diet, implying that it is the most economical type of diet under almost any circumstances. Naturally, there may be exceptions to this and the swine manager must note fluctuations in prices. At one time, for example, it was noted that there was an apparent excess of wheat middlings on the market causing the price of this product to go down. Putting this product to the least-cost test resulted in wheat middlings replacing corn in the formulation. Swine-feeding tests showed this substituted formula worked as well as one in which the energy source was corn. However, it is rare for wheat middlings to be at a price that is competitive with corn.

II Types of swine formulations.

A. Economics usually tend to favor mixing the entire formula—except, perhaps, for the vitamin and trace mineral premix packages. Antibiotic additions usually come in one of the premix packages listed. Therefore, the mixer needs ground corn, soybean meal (either 44 or 49% protein), a source of phosphorus (dicalcium phosphate, defluorinated rock phosphate, steamed bonemeal, etc.), a source of calcium (ground limestone, ground oyster shells, etc.), salt, and the two premixes referred to above. Naturally, other sources of energy may be used as long as they fit into the formulation and are competitive, pricewise, such as wheat, milo, wheat middlings, barley, and oats. Other sources of supplemental protein may be used, such as animal proteins and limited amounts of cottonseed meal (being careful with the gossypol content of cottonseed meal) as long as the amino acids balance out and as long as such substitutes are competitive in price with soybean meal. Once the protein, calcium, phosphorus, and salt balance out (and the premix vitamins, antibiotic, and trace minerals have been added), the first limiting amino acids, namely lysine and possibly tryptophan and methionine/cystine must be checked. Finally, one needs to make sure the energy content analysis is adequate.

   Some swine feeders like to add limited amounts of dehydrated alfalfa meal, especially in drylot sow. However, for growing/finishing swine, alfalfa will not add much to the nutritional value of the diet and probably will add expense. If alfalfa is added, it should be limited to no more than 10% or 15%, and it is essential that the metabolizable energy content of the diet is checked.

B. Protein supplements are available for mixing with corn at a prescribed level to provide a complete mixed formulation. Such commercial supplements might contain 40% protein, 3.5% to 4.0% calcium, 1.2% phosphorus, 2.5% to 3.0% salt, and adequate vitamins and trace minerals so that when it is mixed with corn to provide a 13 to 16% protein mixture (depending on the pig's size) it contains the proper balance of amino acids and adequate energy. This approach has the advantage of being convenient and saving time. Some feeders prefer to place ground shelled corn in one compartment of the self-feeder and a 40% protein supplement in another compartment of the self-feeder, letting the pigs eat cafeteria-style. This method presumes that the pigs will balance their daily needs without overeating on the much more expensive protein supplement. If the pigs tend to overeat on the free-choice supplement, they may be seeking additional minerals. This tendency may be alleviated by adding a free-choice mineral mixture in another compartment, consisting of equal parts limestone, dicalcium phosphate and salt. Usually, this will not be necessary.

C. Mineral and vitamin supplements including the major minerals, calcium, phosphorus, and salt; trace minerals and vitamins; and antibiotics are available. With this type supplement, all the swine feeder has to do is to mix ground shelled corn, soybean meal, and the mineral-vitamin-supplement in the proper prescribed proportions to produce a balanced ration. This differs from A, only in that this premix package contains the major minerals needed, whereas in A the mixer had to add the major minerals when mixing the total formulation.

III Balancing formulations.

A. Mixing the entire ration from "scratch." Naturally, the first thing a feed mixer needs to know is what the requirements are for the animal for which the formula is being prepared. Such requirements are set up in table form in the appendix. Nevertheless, for the purpose of solving this problem, a sample of such recommendations might be listed as follows: for an 88 lb (40 kg) pig, 18% protein, 0.60% calcium, 0.50% phosphorus (remember that the phytate phosphorus availability of grains ranges only from 15% to 25%), 0.25% salt, 0.95% total lysine, and 1,480 kcal metabolizable energy/lb (3,265 kcal/kg). The nutrient content of the feedstuffs being used is as follows:

NUTRIENT CONTENT OF SOME TYPICAL SWINE FEEDS (AS-FED BASIS)

Feed	Metabolizable Energy kcal/lb	Protein %	Lysine %	Calcium %	Phosphorus %	(Available P, %)
Corn	1459	8.3	0.26	0.03	0.28[a]	0.04
Soybean meal	1675	47.5	3.02	0.34	0.69[b]	0.16
Dicalcium phosphate	—	—	—	22.0	18.5	
Ground limestone	—	—	—	38.0	—	

[a]Because of phytate bonding, corn P is about 14% available.
[b]Approximately 23% available. (It should be noted that with the addition of the enzyme *phytase,* organic sources become increasingly available.)

In swine diets formulated with corn and soybean meal, those two ingredients contribute about 97.5% of the total diet. The remaining 2.5% consists of mineral supplements and carrier mixes containing vitamins, trace minerals, and additives. Since corn and soybean meal are so similar in metabolizable energy content, any combination of these two will result in a relatively high-energy diet.

The first step in formulation is presented in the following equation where C is the percentage of corn and S is the percentage of dehulled solvent-extracted soybean meal in the diet. $C + S = 97.5$ or $S = 97.5 - C$. Lysine is expected to be the first limiting amino acid in corn-soybean meal diets, but one can manipulate the proportions of corn and dehulled soybean meal to meet the required concentration of the amino acid and be reasonably certain that the concentration of the other amino acids in the diet will be adequate.

To balance the corn-soy diet on the basis of lysine requirement, the above formula can be altered to the following: $(A \times C) + [(B \times (97.5 - C))] = (L \times 100)$, where A is the percentage of lysine in corn in the diet, B is the percentage of lysine in soybean meal, $97.5 - C$ is the percentage of soybean meal in the diet, and L is the lysine requirement of the 40-kg pig, as expressed as a percentage of the diet. Values for A, B, and L are then substituted into the equation, leaving only one unknown (C). The percentages of corn and soybean meal in the diet can be solved as follows: $0.26C + 3.02 (97.5 - C) = (0.90 \times 100)$, where C is 74.1% corn in the diet. Because S is $97.5 - C$, then S is 23.4% $(97.5 - 74.1)$% soybean meal in the diet.

The next step is to add an ingredient to supply inorganic phosphorus to complete the requirement (0.50%) for total phosphorus. If dicalcium phosphate, which contains 18.5% phosphorus is selected, the next equation will show how much dicalcium phosphate (DP) to include in the diet:

$(18.5 \times DP) = (0.50 \times 100) - (74.1 \times \%\text{ P in corn}) - (23.4 \times \%\text{ P in soybean meal})$. $(18.5 \times DP) = (0.50 \times 100) - (74.1 \times 0.28) - (23.4 \times 0.69)$. $DP = 0.71\%$ dicalcium phosphate in the diet.

The next step is to add an ingredient to supply calcium (0.60%). Ground limestone is the most economical source and it contains 39.8% calcium. The following equations show how much limestone (GL) should be added:

$(38 \times Gl) = (0.60 \times 100) - (74.1 \times \%\text{ Ca in corn}) - (23.4 \times \%\text{ Ca in soybean meal})$ $- (0.71\%\text{ Ca in dicalcium phosphate})$.

$(38 \times GL) = (0.60 \times 100) - (74.1 \times 0.03) - (23.4 \times 0.34) - (0.71 \times 22.0)$

$GL = 0.90\%$ ground limestone in the diet.

One can fortify the diet completely by adding 0.25% salt; a vitamin premix that supplies the recommended levels of vitamins A, D, E, K, B_{12}, riboflavin, niacin, and pantothenic acid; a trace mineral mix supplying iron, zinc, copper, manganese, iodine, and selenium; and, if desired, a premix that contains one or more microbial agents. The fortified diet is shown in the following table. It is made up to 100% by increasing the corn level to 74.4%.

FORTIFIED SWINE DIET

Nutrient	% of Formula
Yellow corn	74.44
Soybean meal, dehulled	23.40
Dicalcium phosphate	0.71
Ground limestone	0.91
Salt	0.25
Vitamin premix	0.10
Trace mineral premix	0.10
Antimicrobial premix	0.10
Total	100.00

Formulation on a true or apparent digestible lysine basis is essentially the same procedure as described above except that the true or apparent digestible lysine values for corn or soybean meal are substituted in the calculations. The figures for true or apparent digestible lysine in any feedstuff are obtained by multiplying the total lysine value by the digestion coefficient figures published in the 1998 National Research Council, Nutrient Requirements for Swine.

Similar procedures can be used to calculate the diets on an available phosphorus basis. However, it appears there may be an adjustment in available phosphorus for swine and poultry. This is due to the addition of the commercially available phytase enzyme that releases the phytic acid-phosphorus bond, making the phosphorus of such feedstuffs more available to monogastric animals. Due to the potential for environmental pollution from feeding excessive phosphorus to compensate for less-available phytin-bound phosphorus, conceivably regulations might impose a requirement for utilizing the phytase enzyme in such formulations.

Since the above formula is given in percentages, naturally one can utilize either kilograms or pounds in preparing to mix a formulation.

B.   Use of a "complete supplement".   Commercially mixed supplements may come with fortifying minerals, vitamins, and even antimicrobials. However, the steps of the mixing procedure will be the same as the "starting-from-scratch" procedure listed above. Of course, the final test will be on how the pigs consuming the formulation perform. Therefore, one should double check the formulation to be reasonably certain every nutrient is present in the proper proportions.

44 Formulating Balanced Rations Mixtures—General

Most swine and poultry are fed on complete ration mixtures rather than specific daily amounts of certain feeds as arrived at by calculating a balanced daily ration. When complete ration mixtures are used (see Figures 44–1 and 44–2 for types of feed mixers), it is essential that such mixtures be formulated to meet the animal's minimum requirements for the various critical nutrients based on the amount of feed consumed. In order to calculate the concentration of any given nutrient needed in a ration mixture to meet an animal's daily requirement, it is necessary to know the animal's daily feed consumption. When animals are hand-fed, this can be predetermined on the basis of calculated needs. When animals are self-fed, their anticipated feed consumption must be estimated. While self-fed animals eat all they want, an animal has certain physical and physiological limits to its capacity for feed consumption. This capacity is reasonably well established for most animals and can be obtained by referring to the dry matter column of the daily nutrient requirement table for the respective class of livestock. Once the anticipated daily feed consumption has been found, the concentration of any given nutrient needed in this amount of feed to meet the animal's daily requirement can be calculated. These concentrations have been calculated for the various classes of livestock and are reported in the respective NRC tables on nutritive requirements.

Variations of several general procedures may be followed in formulating ration mixtures for use in livestock feeding.

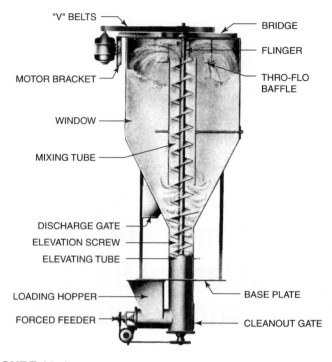

"V" BELTS — BRIDGE

FLINGER

MOTOR BRACKET — THRO-FLO BAFFLE

WINDOW —

MIXING TUBE —

DISCHARGE GATE —
ELEVATION SCREW —
ELEVATING TUBE —

LOADING HOPPER — BASE PLATE
FORCED FEEDER — CLEANOUT GATE

FIGURE 44–1
Feed mixers are of two general types—horizontal and vertical. Shown is a drawing of a vertical type mixer with the different parts and interior action indicated. (Courtesy of Sprout-Waldron, Muncy, Pa.)

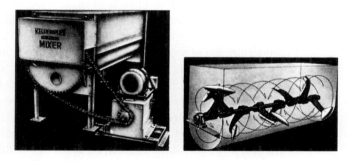

FIGURE 44–2
Feed mixers are of two general types—horizontal and vertical. Shown is a typical horizontal type mixer with a schematic drawing of its interior action. (Courtesy of The Duplex Mill and Mfg. Co., Springfield, Ohio)

Formulating a Feed Mixture Based on a Balanced Daily Ration or a Portion Thereof

45

I. Formulating a complete ration mixture based on a balanced daily ration.

A. In formulating a complete ration mixture following this procedure, one simply calculates a balanced daily ration for a single animal. Once this has been done, the figures for this daily ration are then converted to a percentage (per hundred weight) basis as follows:

$$\frac{\text{Wt of each feed in daily ration}}{\text{Total wt of daily ration}} \times 100 = \% \text{ of that feed in ration mix}$$

B. After the percentage or amount per hundred weight (cwt) of each feed in the daily ration has been determined, the amount of each feed per mix batch is found by multiplying the amount of each feed per cwt by the number of cwt per mix batch. This method, as applied to a balanced daily ration, would be as follows:

	Amount in Daily Ration		Amount Per 100 lb of Ration Mix		Amount Per Ton (2000 lb) of Ration Mix
	kg	lb	lb		lb
Ground CBG hay	1.22	2.69	15.74	× 20 =	314.8
Ground shelled corn	6.01	13.25	77.53	× 20 =	1550.6
44% SBM	0.42	0.93	5.44	× 20 =	108.8
Defluorinated phosphate	0.011	0.024	0.140	× 20 =	2.8
Ground limestone	0.050	0.110	0.644	× 20 =	12.9
TM salt	0.039	0.086	0.503	× 20 =	10.1
TOTALS	7.750	17.090	99.997		2000.0

II

While the above procedure works out satisfactorily when one desires to formulate a complete ration mix, it is sometimes necessary or desirable to develop a feed mixture that includes only a portion of the ration ingredients. The other ration component(s), such as the roughage portion, is fed separately from the rest of the ration.

A. For example, in the above illustration it might not be feasible to grind the hay to be fed. As a result, the hay might be fed in the long form in the amount specified with the remainder of the ration being fed as a separate mix. To formulate the mix excluding the hay, one would simply list the ingredients other than hay in the daily ration with the amount of each used. These amounts are converted to a % or per cwt basis, as in the previous example and then converted to the size of batch desired as follows:

	Amount in Daily Ration, Excluding Hay	Amount Per 100 lb of Mix	Amount Per Ton (2000 lb) of Mix
	lb	lb	lb
Ground shell corn	13.25	92.01	1840.2
44% SBM	0.93	6.46	129.2
Defluorinated phosphate	0.024	0.167	3.34
Ground limestone	0.110	0.764	15.28
TM salt	0.086	0.597	11.94
TOTALS	14.40	99.998	1999.96

B. The latter procedure is probably more useful in feeding dairy cows than other classes of livestock, since the roughage and concentrate portions of the dairy cow's ration are probably more often fed as separate entities. For example, a balanced daily ration for a dairy cow has been calculated as follows:

	lb
Alfalfa hay	5.51
Corn silage	33.08
Ground shelled corn	6.62
Ground barley	6.20
44% SBM	3.92
Defluorinated phosphate	0.051
Ground limestone	0.077

Since the roughage portion of the dairy cow ration is usually fed separately from the concentrate portion, it is a common practice to make a single mix out of the concentrates and minerals and feed it separately from the roughages with appropriate allowance being given to variations in milk production. The concentrate mix would be developed following the procedure used in II A above, as follows:

	Amount in Daily Ration, Excluding Hay	Amount Per 100 lb of Mix		Amount Per Ton (2000 lb) of Mix
	lb	lb		lb
Ground shell corn	6.62	38.925	× 20 =	778.50
Ground barley	6.20	36.456	× 20 =	729.12
44% SBM	3.92	23.049	× 20 =	460.98
Defluorinated phosphate	0.051	0.300	× 20 =	6.00
Ground limestone	0.077	0.453	× 20 =	9.06
TM salt	0.139	0.817	× 20 =	16.34
TOTALS	17.007	100.00		2000.00

46

Formulating Feed Mixtures by the Use of the Square Method

Frequently it is necessary to blend two or more feeds together into a mixture containing a certain definite percentage of some major nutritive factor. For this purpose a procedure generally referred to as the Square Method may be used.

I

When only two feeds are involved. Sometimes it is necessary to determine what combination of two feeds will give a mixture with a certain content of some particular nutrient. For example, a swine producer may need to know what combination of ground shelled corn and 40% complete pig supplement will provide a mix suitable for self-feeding to a group of pigs averaging about 27.5 kg (61 lb) in weight. From the appendix table, it is determined that pigs of this weight should receive a 16% crude protein ration. What combination of corn and 40% supplement will provide a 16% crude protein mix may be quickly, easily, and precisely calculated by using the so-called Square Method. The steps in the use of the Square Method for the above purpose are as follows:

A. Draw a square at the left side of the page.

B. Insert the % crude protein desired in the final mixture (16%) in the middle of the square.

APPLICATION OF THE SQUARE METHOD
USING ONLY TWO FEEDS

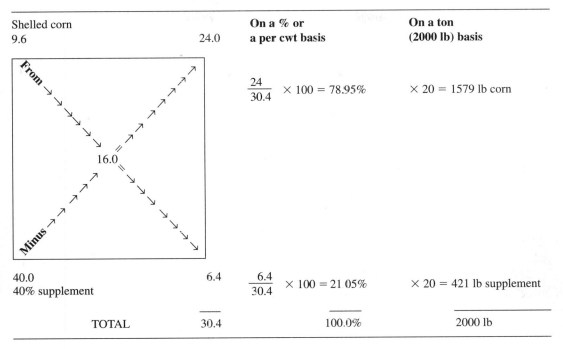

Shelled corn
9.6   24.0

**On a % or
a per cwt basis**

**On a ton
(2000 lb) basis**

$$\frac{24}{30.4} \times 100 = 78.95\%$$

$$\times 20 = 1579 \text{ lb corn}$$

16.0

40.0   6.4
40% supplement

$$\frac{6.4}{30.4} \times 100 = 21\ 05\%$$

$$\times 20 = 421 \text{ lb supplement}$$

TOTAL   30.4   100.0%   2000 lb

C. Place "corn" with its % crude protein (9.6) on the upper left corner and "40% supplement" with its % crude protein (40.0) on the lower left corner. (For this method to work, one feed must be above the desired level of protein and the other below.)

D. Subtract the % crude protein in corn (9.6) from the % crude protein desired in the mix (16), and place the difference (6.4) on the corner of the square diagonally opposite from the corn. This amount is supplement.

E. Subtract the % crude protein desired in the mix (16) from the % crude protein in the supplement (40), and place the difference (24) on the corner of the square diagonally opposite from the supplement. This amount is corn.

F. The above remainders represent the proportions of the two feeds that will provide a mix containing the desired % crude protein. The amounts are then converted to a percentage or a per hundred weight basis and then to other weight bases as desired for mixing purposes.

II

When three or more feeds are involved. Frequently, it is desirable to use more than two feeds in formulating a feed mixture. For example, a swine producer may desire to use a mixture of corn, oats, and a 40% complete supplement in formulating a 12% crude

APPLICATION OF THE SQUARE METHOD
USING THREE OR MORE FEEDS

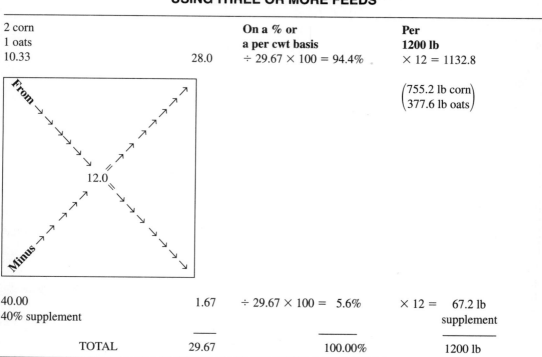

2 corn		**On a % or**	**Per**
1 oats		**a per cwt basis**	**1200 lb**
10.33	28.0	÷ 29.67 × 100 = 94.4%	× 12 = 1132.8

$$\begin{pmatrix} 755.2 \text{ lb corn} \\ 377.6 \text{ lb oats} \end{pmatrix}$$

12.0

40.00	1.67	÷ 29.67 × 100 = 5.6%	× 12 = 67.2 lb
40% supplement			supplement
TOTAL	29.67	100.00%	1200 lb

protein mix for his pregnant sows. The following steps would be involved in the use of the Square Method in this connection.

A. Draw a square as in the previous example.

B. Place the % crude protein desired (12) in the middle of the square.

C. Separate the feeds into two groups, specify the proportion of each feed in each group, and calculate the weighted average % protein in each group. For this example, let us assume that corn and oats were grouped together in the proportion of 2:1 with the supplement being used alone. The average % protein in the corn and oats must then be calculated as follows:

$$2 \times 9.6 = 19.2$$
$$1 \times 11.8 = \underline{11.8}$$
$$31.0/3 = 10.33\%$$

D. Place "2 corn + 1 oats" with its calculated % of crude protein (10.33) on the upper left corner of the square and "40% supplement" (40.0) on the lower left corner. (For the method to work, the figure on one left-hand corner of the square must be above and the one on the other below the level of protein desired.)

E. Subtract diagonally and proceed with the calculations as in the previous example.

F. Divide the final figure for "corn + oats" into 2/3 corn and 1/3 oats. (The proportion of each feed in each group must always be indicated initially and complied with in the final mixture.)

III

With a fixed percentage of one or more ration components. Sometimes a feeder desires to formulate a mixture containing a certain percentage of some nutrient, such as protein, but with a fixed percentage of one or more ration components.

A. For example, a swine producer might wish to formulate a 12% crude protein mixture for pregnant sows using corn, oats, soybean meal, and a mineral and vitamin supplement. He wants to include in the mixture exactly 20% oats and 3% mineral and vitamin supplement. He then needs to know what combination of corn and soybean meal can be used to make up the other 77% of the mixture and give an overall mixture that contains exactly 12% crude protein. An adaptation of the Square Method may be used for this purpose.

B. It is known that a crude protein level of 12% is desired for the overall mixture. This means that there is to be 12 lb of protein per 100 lb of mixture. Since 20 lb of each 100 lb of the mixture is oats (20%) then the oats in each 100 lb of the mixture would supply 2.36 lb of crude protein (11.8% of 20 lb). The mineral and vitamin supplement is essentially protein-free and so would contribute no protein of consequence to the mix. Hence, the oats (20 lb) and the mineral and vitamin supplement (3 lb) per 100 lb of mix would provide 2.36 lb crude protein. The remainder of the 12 lb of crude protein needed per 100 lb of mixture (12.00 lb − 2.36 lb = 9.64 lb) must then come from the 77 lb of corn and soybean meal per 100 lb of the overall mixture. In order to determine what combination of 77 lb of corn and soybean meal will provide the 9.64 lb of needed protein, an adaptation of the Square Method may be used.

C. To do this, it is first necessary to calculate what % protein will be needed in the corn and soybean meal combination to provide 9.64 lb of protein per 77 lbs, as follows:

$$9.64 \div 77 \times 100 = 12.52\%$$

This figure is then used in conjunction with the Square Method as follows:

Per 77.0 parts
(calculated by ratio)

Corn

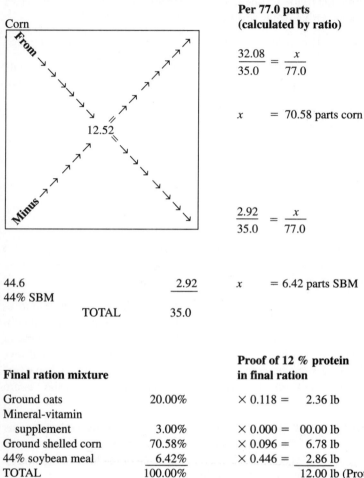

$$\frac{32.08}{35.0} = \frac{x}{77.0}$$

$x \quad = 70.58 \text{ parts corn}$

$$\frac{2.92}{35.0} = \frac{x}{77.0}$$

44.6	2.92	$x \quad = 6.42 \text{ parts SBM}$
44% SBM		
TOTAL	35.0	

Final ration mixture

Proof of 12 % protein
in final ration

Ground oats	20.00%	× 0.118 =	2.36 lb
Mineral-vitamin			
supplement	3.00%	× 0.000 =	00.00 lb
Ground shelled corn	70.58%	× 0.096 =	6.78 lb
44% soybean meal	6.42%	× 0.446 =	2.86 lb
TOTAL	100.00%		12.00 lb (Protein per
			100 lb mix)

47

Use of Algebraic Equations in the Formulation of Feed Mixtures

Those who enjoy a mathematical approach to the solution of problems may find the use of algebraic equations preferable to the Square Method for formulating feed mixtures. In fact, appropriate algebraic methods may be used in the place of the Square Method in each of the preceding illustrations, as outlined below.

When only two feeds are involved. Algebraic equations may be used in place of the Square Method for determining what combination of two different feeds will give a mixture containing a certain definite percentage of some particular nutrient. To illustrate, let us return to the swine feeder who wanted to know what combination of ground shelled corn and 40% complete pig supplement would provide a mixture that contained exactly 16% crude protein. Through the use of the Square Method, it was determined that a mixture of 78.95% corn and 21.05% supplement would give the desired level (16%) of protein. Algebraic equations can be used to make these same determinations as follows:

$$x = \text{lb corn per 100 lb mix}$$

$$y = \text{lb supplement per 100 lb mix}$$

$$x + y = 100 \text{ lb mix}$$

$$0.096x + 0.400y = 16.0 \text{ (lb protein/100 lb mix)}$$

$$\text{(subtract) } 0.096x + 0.096y = 9.6$$

$$0 + 0.304y = 6.4$$

$$y = 6.4/.304 = 21.05\% \text{ supplement in mix}$$

$$x = 100 - 21.05 = 78.95\% \text{ corn in mix}$$

II

When three or more feeds are involved.   To use algebraic equations to formulate a feed mixture containing a definite percentage of some nutrient using three or more feeds, it is first necessary, as with the Square Method, to separate the feeds into two logical groups, to specify the proportion of each feed in each group, and to calculate the weighted average % of the nutrient under consideration in each group. One is then ready to set up the appropriate equations and solve for the unknowns.

Using the same problem as was used in illustrating the Square Method in this connection (formulating a 12% crude protein mix using corn and oats in the ratio of 2:1 along with a 40% supplement), one would then proceed as follows:

$$x = \text{lb corn} + \text{lb oats per 100 lb mix}$$

$$y = \text{lb supplement per 100 lb mix}$$

$$x + y = 100 \text{ lb of mix}$$

$$0.1033x + 0.4000y = 12.00 \text{ (lb protein/100 lb mix)}$$

$$\text{(subtract) } 0.1033x + 0.1033y = 10.33$$

$$0.2967y = 1.67$$

$$y = 1.67/0.2967 = 5.6\% \text{ supplement in mix}$$

$$x = 100 - 5.6 = 94.4\% \text{ corn and oats (2:1) in mix}$$

III

With a fixed percentage of one or more ration components.   Let us assume that algebraic equations are to be used for solving the same problem as was used to illustrate the use of the Square Method in this connection (formulating a 12% crude protein mix using corn and 44% soybean meal along with a fixed 20% oats and 3% mineral and vitamin supplement). As with the Square Method, the amount of protein provided by the 20 lb of oats and 3 lb of mineral and vitamin supplement per 100 lb of mix must first be calculated (2.36 lb) and subtracted from the 12 lb desired in the overall mix (12.00 – 2.36 = 9.64 lb). This is the amount of protein that must come from the 77 lb (100 – 23) of corn and soybean meal per 100 lb of overall mix. The lb of corn and soybean meal, respectively, are then calculated using appropriate equations as follows:

$$x = \text{lb of corn per 77 lb of corn and SBM and per 100 lb of overall mix}$$

$$y = \text{lb of SBM per 77 lb of corn and SBM and per 100 lb of overall mix}$$

$$x + y = 77.0 \text{ lb of corn and SBM}$$

$$0.096x + 0.446y = 9.64$$

$$(\text{subtract}) \; 0.096x + 0.096y = 7.39 \; (77.0 \times .096)$$

$$0.35y = 2.25$$

$$y = 2.25/.35 = 6.43 \text{ lb SBM/77 lb corn and SBM}$$

$$x = 77 - 6.43 = 70.57 \text{ lb corn/77 lb corn and SBM}$$

Final ration mixture

Ground shelled corn	70.57%
44% soybean meal	6.43%
Ground oats	20.00%
Mineral and vitamin supplement	3.00%
TOTAL	100.00%

IV

When definite amounts of two feed nutrients from two feed sources are desired.

A. Frequently it is desirable to calculate what combination of two feeds or two feed groups will provide a required amount of each of two feed nutrients. Usually the two feed nutrients involved will be protein and energy. The respective amounts of each nutrient may represent the total amount of each nutrient required in the overall ration, as in most swine finishing rations, or they may represent the amounts of protein and energy still required after the respective amounts of these nutrients in the roughage portion of the ration have been deducted from the total dietary needs, as with the rations previously balanced for finishing steers.

B. While only a single major energy source and a single major protein source may be involved in making such a calculation, more than one energy feed and/or more than one protein feed may go in the formulation as the major source of each nutrient. In such instances the average weighted protein and energy content of each group of feeds (the high-protein and the high-energy feeds, respectively) would be used in making the calculations. After the amount of each feed group needed to provide the required amounts of protein and energy has been determined, the amounts of the different feeds in the protein and energy groups, respectively, are then calculated.

C. This procedure might be used in determining the amounts of corn and soybean meal that should be used to meet the daily requirements of a pig for protein and energy. For example, the daily nutrient requirements of a certain growing-finishing pig are as follows:

TP	ME	Ca	P	Caro
kg	kcal	kg	kg	mg
0.280	6320	0.011	0.009	10.4

Simultaneous equations can be used to calculate the amounts of corn and soybean meal needed to provide the above amounts of TP and ME as follows:

$$x = \text{kg of corn}$$

$$y = \text{kg of soybean meal}$$

$$(TP)\ 0.096x + 0.446y = 0.28$$

$$(ME)\ 3300x + 2817y = 6320$$

$$3300x + 15331y = 9625$$

$$-12514y = -3305$$

$$+12514y = +3305$$

$$y = 0.2641\ (\text{kg 44\% SBM, as fed})$$

$$x = \frac{0.28 - (0.446 \times 0.2641)}{0.096}$$

$$x = 1.690\ (\text{kg corn, as fed})$$

Proof of correct answers:

$$1.690 \times 0.096 = 0.162$$

$$0.264 \times 0.446 = \underline{0.118}$$

$$\text{TOTAL} \quad 0.280\ (\text{kg TP required})$$

$$1.690 \times 3300 = 5577$$

$$0.264 \times 2817 = \underline{744}$$

$$\text{TOTAL} \quad 6321\ (6320\ \text{kcal ME required})$$

D. To determine a balanced daily ration for the pig in question, one would need to add to the above quantities of corn and SBM an amount of a mineral and vitamin supplement to meet the requirements. This will usually amount to about 3% of the total ration. In other words, a daily ration for the above pig might consist of the following:

Ground shelled corn	1.690 kg
Soybean meal	0.264
Mineral & vitamin supplement	0.060
TOTAL	2.014 kg

E. The preceding daily ration might then be converted into a balanced ration mixture by converting each ingredient to a percentage or a per cwt basis and then multiplying each per cwt figure by the number of cwt desired in each mix batch, as follows:

Daily Ration per Head		% or per cwt	Mix Batch of 1 Ton
Ground shelled corn	1.690	83.91 × 20 =	1678.2
44% SBM	0.264	13.11 × 20 =	262.2
Minerals & vitamins	0.060	3.00 × 20 =	60.0
TOTALS	2.014	100.02	2000.4

48

Use of the Square Method for Formulating a Balanced Ration Mixture for a Finishing Steer

---

I. Let us assume that we are going to use the Square Method to develop a feed formula containing 11% crude protein, as-fed basis, for a 300 kg yearling steer gaining 1.3 kg daily, and that we are using ground shelled corn, 44% soybean meal, ground peanut hulls, and minerals.

A. First let us decide which variation of the Square Method it will be best to use in making this calculation. Since it is desirable to use a fairly definite percentage of roughage (15%) and of minerals (tentatively 0.5% each of defluorinated phosphate, ground limestone, and trace-mineralized salt) in this mixture, then it would seem that the method under section 47, III involving a "fixed percentage of one or more ration components" would be the one to use.

B. The next step then is to calculate how much of the total percentage of protein per 100 lb of the mix (11%) will come from the fixed ingredients per 100 lb of mix as follows:

$$
\begin{array}{llll}
15.0 \text{ lb ground peanut hulls} & \times & 0.071 = 1.065 \\
0.5 \text{ lb defluorinated phosphate} & \times & 0.000 = 0.00 \\
0.5 \text{ lb ground limestone} & \times & 0.000 = 0.00 \\
\underline{0.5 \text{ lb trace-mineralized salt}} & \times & 0.000 = \underline{0.00} \\
16.5 \text{ lb} & & 1.065 \text{ lb}
\end{array}
$$

From the above, it is apparent that 1.065 lb of protein would come from the 16.5 lb of peanut hulls and supplemental minerals per each 100 lb of overall mix. The other 9.935 lb (11.0 – 1.065) of protein needed per 100 lb of overall mix will have to come from the 83.5 lb of corn and soybean meal per 100 lb of mix.

C.   If 9.935 lb of protein must come from the 83.5 lb of corn and SBM, then a mixture of these two feeds containing 11.9% protein (9.935/83.5 × 100 = 11.9%) would need to be used. To determine how much of each to use, one might employ the Square Method in its simplest form as follows:

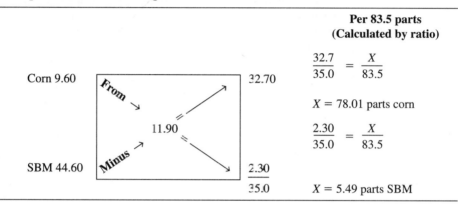

Per 83.5 parts
(Calculated by ratio)

$$\frac{32.7}{35.0} = \frac{X}{83.5}$$

$X = 78.01$ parts corn

$$\frac{2.30}{35.0} = \frac{X}{83.5}$$

$X = 5.49$ parts SBM

D.   The tentative final ration and proof of correct protein content.

Tentative Final Ration Mixture/100 lb		% Protein	lb Protein
Ground peanut hulls	15.0 × 0.071	=	1.06
Defluorinated phosphate	0.5 × 0.000	=	0.00
Ground limestone	0.5 × 0.000	=	0.00
Trace-mineralized salt	0.5 × 0.000	=	0.00
Ground shelled corn	78.01 × 0.096	=	7.49
44% soybean meal	5.49 × 0.446	=	2.45
TOTALS	100.0		11.00 lb

E.   Evaluation of calcium and phosphorus adequacy of tentative final ration.

Tentative Final Ration Mixture/100 lb		Ca %	Ca lb	P %	P lb
Ground peanut hulls	15.0	0.24	0.036	0.06	0.009
Defluorinated phosphate	0.5	32.00	0.160	18.00	0.090
Ground limestone	0.5	34.00	0.170	0.02	0.000
Trace-mineralized salt	0.5	0.00	0.000	0.00	0.000
Ground shelled corn	78.01	0.03	0.023	0.26	0.203
44% soybean meal	5.49	0.30	0.016	0.62	0.034
TOTAL lb/100 lb mix			0.405		0.336
Requirement from the appendix			0.41		0.32
Adequacy of above mixture			98.8%		105.0%

It may be noted that the ration at this point is 1.2% short of the requirement in Ca (0.405 lb vs 0.41 lb) and 5% over the requirement in P (0.336 lb vs 0.32 lb). While a 300 kg yearling steer would probably perform satisfactorily on the levels of Ca and P provided by the ration as it now stands, it would be simple to adjust the level of each to the exact requirement by removing a small amount of defluorinated phosphate to lower the P level, and then adding ground limestone as necessary to meet the Ca requirement. The ration has 0.016 lb of excess P (0.336 − 0.32); to remove this amount of P from the ration it would be necessary to remove 0.089 lb of defluorinated phosphate. This would also remove 0.028 lb of Ca from the ration, which when added to the 0.005 lb the ration already lacked, makes a total of 0.033 lb of Ca that needs to be added. This means that the ground limestone in the mixture should be increased by 0.097 lb. The final ration, as adjusted, would be as follows:

Final Ration	*lb/100 lb or %*
Ground peanut hulls	15.00
Defluorinated phosphate (0.500 lb − 0.089 lb)	0.411
Ground limestone (0.500 lb + 0.097 lb)	0.597
Trace-mineralized salt	0.500
Ground shelled corn	78.01
44% soybean meal	5.49
Total	100.008

The above ration mixture, after adjusting the P and Ca levels, for all practical purposes still comes to a total of 100 lb. Should the quantity of defluorinated phosphate and/or ground limestone that is added or removed amount to more or less than 1% (the space allotted to these products in the initial ration), a total weight of 100 lb for the overall mix may be retained, without altering the make-up or composition of the ration significantly, by increasing or decreasing the amount of corn used as necessary to give a total weight of 100 lb.

II
Fortifying steer ration mixtures with vitamin A.

A. For some reason, in present day cattle feeding operations Vitamin A deficiencies are sometimes experienced with cattle on rations containing what would be regarded as an abundance of carotene for adequate Vitamin A nutrition. Consequently, most cattle feeders currently follow the practice of fortifying their finishing rations with approximately 1000 IU of so-called preformed Vitamin A per lb of ration dry matter. This would be an especially desirable practice with the above ration since the roughage is peanut hulls, which contain practically no carotene or Vitamin A value. The roughage of most rations, if of good quality, is usually a major source of this nutrient.

B. In order to accomplish the fortification of a mixture with Vitamin A, one would need to obtain a supply of a Vitamin A supplement, one preferably manufactured by a reliable firm specializing in such products. One such product on the market carries 4,000,000 IU of Vitamin A per lb. In any case it would be necessary to calculate the amount of the product to use in order to accomplish the level of fortification desired. The product that carries 4,000,000 IU of Vitamin A per lb would contain enough Vitamin A per lb to fortify 4000 lb of ration DM. Since most air-dry rations are approximately 90% DM, then 1 lb of the product would fortify approximately 4444.4 lb of air-dry ration. In other words, 0.45 lb of the Vitamin A supplement would need to be added per ton of air-dry ration mix. Other than to provide supplemental Vitamin A, this small amount of Vitamin A would have no practical effect on the composition of the ration and would simply be added to the ration mix over and above the other ration ingredients.

Subcutaneous injection with 5 million IU of Vitamin A will provide adequate vitamin A for up to six months. Such vitamin A will move to the liver for storage.

49

Formulation of a Balanced Ration Mixture for Lactating Sows

Let us assume that we are going to formulate a ration mix for lactating sows. We have already decided to include 10% ground oats and to reserve 3% of the space for minerals and vitamins. Therefore, the remaining 87% of the mix will contain such proportions of ground shelled corn and dehulled, solvent-extracted soybean meal to provide energy and protein—with modern day emphasis on the content of the amino acid, lysine.

The first step in making this determination is to list the nutrient requirements of lactating sows in percentages of the various critical nutrients of the ration as follows (based on 1998 Nutrient Requirements of Swine as set forth in the swine nutrient requirements listed in the appendix):

NUTRIENT REQUIREMENTS FOR A LACTATING SOW
(175 KG [385 LB] SOW, NURSING
10 PIGS IN A 21-DAY LACTATION)

Crude protein, %	17.0
Total lysine, %	0.89
Metabolizable energy, kcal/100 lb	148,409
Calcium, %	0.75
Phosphorus, %	0.60

II _____ The next step is to refer to the monogastric feed table in the appendix to determine how much of the total 17.05 protein required is furnished by the 10% ground oats. This will prove to be 1.14% protein.

III _____ Next, calculate the amount of protein still needed in the mix by subtracting the amount supplied by 10% oats (1.14%) from the total amount needed (17%.)

  17.00% in total mix – 1.14% supplied by 10% oats = 15.86% protein still needed

It is assumed that the mineral and vitamin mix added to the diet will contribute no protein; therefore, the 15.86% protein still needed must come from the 87% corn and soybean meal. The amounts of corn and soybean meal needed can be calculated by the algebraic approach.

IV _____ We calculated that 87% corn-soy must contain the 15.86% of protein needed in the final mix; $15.67 \div 87 = 18.22\%$ protein in the corn-soy mix. The appendix monogastric feed tables tell us that average corn contains 8.5% protein (on an as-fed basis) and that dehulled, solvent-extracted soybean meal contains 48.5% protein. Let X = amount of soybean meal and $1 - X$ = corn. The equation becomes as follows:

$$48.5X - 8.5 (1 - X) = 18.22$$

$$48.5X - 8.5 - 8.5X = 18.22$$

$$40X = 9.72$$

$$X = 21.14 \text{ (soy)} \times 48.5\% = 10.25\%$$

$$87 - 21.4 = 65.86 \text{ (corn)} \times 8.5 = 5.60$$

$$10\% \text{ oats} = 1.14$$

$$\text{Total protein} = 16.98\%$$

V _____ The preceding solution demonstrated what ratio of corn and soybean meal, along with 10% oats, was needed to supply 17% protein. Now, it is necessary to calculate the other critical nutrients—ME, lysine, Ca, and P—by calculating how much of each is contributed and what the total is.

	Protein %	Lysine %	ME Kcal	Ca %	P %
Ground oats, 10%	1.14	0.04	11,930	.01	.03
Ground corn, 65.86%	5.60	0.17	103,879	.01	.19
Soybean meal, 21.14%	10.25	0.63	23,878	.10	.14
Total	16.98	0.84	139,687	.12	.36
Requirement	17.00	0.89	148,409	.75	.60

Note that the total protein was calculated to be 17% and undoubtedly rounding off has resulted in a slight mathematical deficiency. The lysine is slightly low and, had the problem been calculated on the basis of lysine, that could have been made exact, but total protein would then have been a little high. Metabolizable energy (ME) is low, but the sow can compensate by eating slightly more feed. The glaring deficiencies are calcium and phosphorus.

VI

Attention should be given next to the phosphorus deficiency. We will shore up the phosphorus deficiency first because many of the phosphorus supplements carry calcium, also. The oats, corn, and soybean meal carry 0.36% phosphorus, whereas a total of 0.60% P is required. Therefore, the P supplement must supply 0.24% P (0.60% − 0.36%). Since dicalcium phosphate contains 18.5% P, 1.29% dicalcium phosphate must be added to the formulation (0.24% needed divided by 18.5% P in dical = 1.29% dical). Dicalcium phosphate contains 22% calcium, also, so 1.29% dical adds 0.24% P plus 0.28% Ca. Let us pull down the calcium and phosphorus columns from the above table to see how we stand.

	Calcium, %	Phosphorus, %
From oats, corn, and soybean meal	0.12	0.36
Add 1.29% dicalcium phosphate	0.28	0.24
Total	0.40	0.60
Still needed	0.35	

Since limestone contains 38% calcium, 0.92% limestone will supply the 0.35% Ca needed.

VII

The entire formulation now needs to be presented with total analysis:

Ingredient	%	Protein %	Lysine %	ME kcal	Ca %	P %
		Nutrients Provided				
Ground oats	10.00	1.14	0.04	11,930	.01	.03
Ground corn	65.86	5.60	0.17	103,879	.01	.19
Soybean meal	21.14	10.25	0.63	23,878	.10	.14
Dical. phos.	1.29	—	—	—	.28	.24
Ground limestone	.92	—	—	—	.35	-
Iodized salt	.29					
Premix	.50					
Totals	100.00	16.99	0.84	139,687	.75	.60
Requirement	100.00	17.00	0.89	148,409	.75	.60

VIII

The formulation will be adequate for the lactating sow; the NRC Tables (see appendix) indicate the sow should consume 4– kg (8.8 lb) per day with differing amounts for sows with more or fewer pigs nursing. So, in a matter of management, ME can be added by feeding slightly more feed/day.

Since the formula is calculated on a percentage basis, it can be adapted to multipliers for calculating pounds or kilograms in a large mix, i.e., ton, long ton, etc.

50

Calculating and Altering Levels of Nutrients in Feed Mixtures

As indicated previously, certain classes of livestock are frequently fed completely, or in part, on feed mixtures. The nutritional adequacy of such mixtures is normally based on the levels or concentrations of the various critical nutrients in the mixture, expressed as a percent or as an amount per lb or per kg or per ton. Consequently, in order to properly evaluate such mixtures from the standpoint of their nutritional adequacy, one must have information on the levels of the respective nutrients in the mixture. While such information can be obtained by subjecting the mixture to complete laboratory analyses and metabolism studies, it is usually determined by arithmetic calculations, using average table values for the composition of the various ration components.

To illustrate how such calculations are made, let us assume that a swine feeder is planning on using a ration mix to feed out a group of pigs from an initial weight of about 20 kg to a final weight of 100 kg. He has obtained from a neighbor a feed mix formula that he is thinking about using for his feeding program. Before doing so, however, he wants to check out its nutritional adequacy. To do this, he must first calculate the levels of the respective critical nutrients in the mix and then compare these with the established requirements for pigs of the size being fed (see appendix).

Since the necessary trace minerals and vitamins are assumed to be supplied by the trace-mineralized salt and vitamin supplement in the mix, the critical nutrients in this illustration are energy, protein, calcium, and phosphorus. In order to determine the levels of these nutrients in the mix, the amounts of each nutrient supplied by the respective mix components are first calculated and the total amount of each nutrient per batch of mix obtained. These totals are then compared to the requirements per ton for the respective nutrients to determine the percentage of adequacy of each, as shown in the table on page 346.

III

Upon comparing the levels of the different nutrients in the mix with the requirements of 20–35 kg pigs for these nutrients, it may be noted that the mix meets the requirements within acceptable limits for all the critical nutrients except protein. The mix as fed provides only 273.2 lb of crude protein per ton (13.7%), whereas the required level is 320.0 lb per ton or 16%. This means that the original mix should be adjusted to provide a 16% level of protein while changing the level of other nutrients as little as possible. This might be done by replacing some of the corn in the original mix with an equal amount of soybean meal since the latter is much higher than corn in crude protein and the two feeds are similar in metabolizable energy.

The question then arises as to how much corn should be replaced with soybean meal to realize the needed increase in the percentage of crude protein. This amount can be found by first calculating the increase (in lb) of protein needed in the 2000 lb of mix. One then calculates how much net increase in protein is realized with each lb of corn replaced with soybean meal and then divides the total increase needed by the net increase realized with each lb of corn replaced.

In each lb of soybean meal there is 0.446 lb of crude protein. However, for each lb of soybean meal added, a lb of corn is removed and a lb of corn contains 0.096 lb of crude protein. Consequently in replacing a lb of corn with a lb of soybean meal, a net increase of only 0.35 lb (0.446–0.096) of crude protein is realized.

A 2000-lb batch of the mix containing 13.7% of crude protein contains 273.2 lb of protein. For the mix to have 16% crude protein, it would need to contain 320 lb of protein in each 2000 lb batch. This amounts to an increase of 46.8 lb (320.0–273.2) of protein per batch. By dividing 46.8 lb by 0.35 lb, it is determined that the corn in the original formula should be reduced by 133.7 and the soybean meal increased by the same amount. In other words, the following adjusted mix should meet the protein requirement and still be adequate or close to adequate in energy, calcium, and phosphorus. The adjusted mix and a recalculation of the nutrients therein follows on the next page.

IV

It will be noted that the adjusted mix contains 320 (319.9) lb of crude protein as required. While the energy level is slightly below that listed in the requirements, it is well within the acceptable limit (within 1% of the requirement) and would be considered adequate for all practical purposes. The calcium and phosphorus are both a little high following the soybean meal for corn substitution, and the mix might be improved in this regard. The mix, as adjusted, contains 10.4 lb of phosphorus per 2000 lb whereas, based on a

Amount of Each Nutrient/Ton of Mix

Mix Ingredients	Per Ton lb	ME kcal/lb	CP %	Ca %	P %	ME Mcal	CP lb	Ca lb	P lb
Ground shelled corn	1711	1497	9.6	0.03	0.26	2561	164.3	0.51	4.45
Soybean meal	234	1278	44.6	0.30	0.62	299	104.4	0.70	1.45
Defluorinated phosphate	22	0	0.0	32.00	18.00	0	0.0	7.04	3.96
Ground limestone	13	0	0.0	34.00	0.02	0	0.0	4.42	0.0
TM salt	10	0	0.0	0.0	0.0	0	0.0	0.0	0.0
Vitamin supplement*	10	1278	44.6	0.30	0.62	13	4.5	0.03	0.06
TOTAL	2000					2873	273.2	12.70	9.92
Requirement/2000 lb (20–35 kg pigs)						2880	320.	12.0	10.00
Percentage of requirement in mix						99.8	85.4	105.8	99.2

Composition

*Since it was known that soybean meal had been used as the vitamin carrier and since quantitatively the carrier makes up most of a vitamin supplement, the same composition was used for the vitamin supplement as was used for the soybean meal. If the nature of the carrier is not known, it is best to assume that the vitamin supplement contains no critical nutrients other than vitamins.

required concentration of 0.5%, it should contain only 10 lb. In other words, it should contain 0.4 lb less phosphorus per 2000 lb batch. Such a reduction in phosphorus could be realized by reducing the defluorinated phosphate in the mix by 2.2 lb, thus lowering the phosphorus content in the mix by 0.4 lb (2.2 × 0.18) to 0.5%, which is the requirement.

Mix Ingredients	lb	ME Mcal	CP lb	Ca lb	P lb
		Amount of Each Nutrient/Ton of Altered Mix			
Ground shelled corn	1577.3	2361	151.4	0.47	4.10
Soybean meal	367.7	470	164.0	1.10	2.28
Defluorinated phosphate	22.0	0	0.0	7.04	3.96
Ground limestone	13.0	0	0.0	4.42	0.00
TM salt	10.0	0	0.0	0.00	0.00
Vitamin supplement	10.0	13	4.5	0.03	0.06
TOTALS	2000.0	2844	319.9	13.06	10.40
Requirements/2000 lb		2880	320.0	12.00	10.00
Percentage of requirement in mix		98.8	100.0	108.8	104.0

V

In removing 2.2 lb of defluorinated phosphate from the mix to correct the phosphorus level, however, 0.7 lb (2.2 × 0.32) of calcium was also removed, lowering the amount of calcium in 2000 lb of mix to 12.36 lb (13.06 – 0.70) or a level of 0.618%, whereas the requirement is 0.6%. In other words, the calcium level is still 103% of the requirement, so it requires no corrective action.

VI

In correcting the phosphorus level in the mix, 2.2 lb of defluorinated phosphate was removed, leaving the overall mix 2.2 lbs short of 2000 lb or 1 ton. This might be corrected without materially changing the nutrient balance of the mix by adding another 2.2 lb of corn, making the final adjusted mix and its content of critical nutrients as follows:

Makeup of Mix Ingredients	lb	ME Mcal	CP lb	Ca lb	P lb
		Amount of Each Nutrient Provided by			
Ground shelled corn	1579.5	2365	151.6	0.47	4.11
Soybean meal	367.7	470	164.0	1.10	2.28
Defluorinated phosphate	19.8	0	0.0	6.34	3.56
Ground limestone	13.0	0	0.0	4.42	0.00
TM salt	10.0	0	0.0	0.00	0.00
Vitamin supplement	10.0	13	4.5	0.03	0.06
TOTALS	2000.0	2848	320.1	12.36	10.01
Requirement/2000 lb		2880	320.0	12.00	10.00
Percentage of requirement in mix		98.9	100.0	103.0	100.1

VII

The previous mix should be satisfactory for feeding the pigs up to 35 kg of weight. However, at about this weight, as shown below, the pig's requirements change.

| | Requirements Per kg of Air-Dry Diet | | | |
	ME kcal/lb	CP %	Ca %	P %
20–35 kg pigs	1440	16.0	0.60	0.50
35–60 kg pigs	1447	14.0	0.55	0.45

Note that there is very little difference in the energy requirement of the two classes of pigs, but the protein requirement drops from 16% to 14%, the calcium requirement drops from 0.6% to 0.55% and the phosphorus requirement drops from 0.5% to 0.45% as the pigs attain the heavier weight. Once again, for economic as well as nutritional reasons, adjustments should be made in the ration mix. This would be done in much the same manner as illustrated previously. One would proceed to alter the protein level of the mix first, but, since in this instance protein level is to be lowered, corn would be substituted for soybean meal rather than soybean meal for corn. The basic calculations, however, would be the same. After appropriate adjustments have been made to alter the protein level, the amounts of defluorinated phosphate and ground limestone used will be adjusted to bring the calcium and phosphorus levels in line with the requirements. The phosphorus level should be corrected first. This is done by adjusting the amount of defluorinated phosphate used in the mix. The amount of ground limestone used is then adjusted as necessary to bring the calcium level in line with the calcium requirement.

VIII

In adjusting feed mixtures to alter nutrient levels, energy level is usually the first to be given attention. Should the energy level have to be altered, this is usually accomplished by substituting a certain amount of roughage for a like amount of concentrates, or vice versa, depending on whether the energy level is to be raised or lowered. Increasing the concentrates and reducing the roughage will, of course, tend to raise the energy level, whereas increasing the roughage and reducing the concentrates will tend to lower it. Energy levels may be influenced also by varying the proportion of fat or high-fat feeds used in the mix. As would be expected, high fat levels tend to increase the energy density of the mix.

IX

Once the energy level of a mix has been brought in line, the protein level is normally the next consideration. The protein level is usually altered by substituting a high-protein feed for an equal amount of low-protein feed, or vice versa, depending on whether the protein level is to be raised or lowered. Urea is sometimes useful in altering the crude protein level of feed mixtures for ruminants provided the mixture contains ample amounts of readily fermentable carbohydrates. Only a small amount of urea is ordinarily required and it is usually substituted for a like amount of almost any noncritical mix component.

X

After the necessary adjustments have been made in the mix to bring the energy and protein levels in line, attention is then directed to the levels of calcium and phosphorus. Since most economical sources of phosphorus for livestock also contain calcium, the phosphorus level is usually considered first. It is ordinarily brought in line by simply raising or lowering the amount of the phosphorus supplement used.

The calcium level of the mix is then checked and the amount of calcium supplement (usually ground limestone or oystershell flour) in the mix is raised or lowered as necessary to bring the calcium level in line.

In adjusting the amounts of the phosphorus and calcium supplements included in the mix, there may be a small increase or decrease in the combined amount of these two materials used. This may be compensated for by adding or removing a pound or so of one of the least critical ration components as needed to maintain the overall weight of the mix at 2000 lb, or at whatever size of mix batch is desired.

51 Formulating and Using Premixes

I

Frequently in mixing rations for livestock, it is more convenient to mix the components of the ration that are required in relatively small amounts into a premix that can be used as a single ingredient as individual batches of the overall ration are mixed. Such a practice will usually allow more efficient use of time and labor at ration-mixing time by reducing the number of individual weighings involved. Also, after the premix has been prepared, the mixing of separate batches of the overall ration can be handled by less well-trained personnel. Such premixes may include all or just part of the supplemental materials. Certain of the supplemental materials, such as the trace minerals and/or vitamins, may already be in the form of premixes. Some premixes are more suitable for on-the-farm preparation than are others, depending on the level of technical know-how and the equipment required in their formulation and preparation.

II

A type of premix that is suitable for on-the-farm preparation is one that includes several supplemental materials in relatively small amounts compared to the major ingredient or ingredients. An illustration of such a ration mix would be one for growing pigs such as follows:

	%
Ground shelled corn	79.53
49% soybean meal	18.00
Ground limestone	0.75
Defluorinated phosphate	1.00
Trace-mineralized salt	0.50
Vitamin supplement	0.20
Antibiotic	0.02

A. Let us assume that a swine feeder is using the above ration for his growing pigs. He wants to combine all the ingredients other than the ground shelled corn and soybean meal into a single premix for use with the corn and soybean meal at ration-mixing time.

The premix would consist of the following ingredients in the proportions indicated:

Ground limestone	0.75
Defluorinated phosphate	1.00
Trace-mineralized salt	0.50
Vitamin supplement	0.20
Antibiotic	0.02
TOTAL	2.47

B. The levels of these ingredients per hundred weight of premix would then be calculated as follows:

Ground limestone	$0.75/2.47 \times 100 =$	30.36 lb
Defluorinated phosphate	$1.00/2.47 \times 100 =$	40.49 lb
Trace-mineralized salt	$0.50/2.47 \times 100 =$	20.24 lb
Vitamin supplement	$0.20/2.47 \times 100 =$	8.10 lb
Antibiotic	$0.02/2.47 \times 100 =$	0.81 lb
TOTAL		100.00 lb

C. The amounts at the right above would be the amounts of the respective ingredients to use per 100 lb of premix. To prepare 1000 lb of premix, 10 times the amounts of the various premix ingredients should be used. In order to mix a ton of the complete ration, the following would be included:

Ground shelled corn	1590.6 lb
49% soybean meal	360.0 lb
Premix	49.4 lb
TOTAL	2000.0 lb

D. It is apparent that for each 100 lb of ration, 2.47 lb of the premix would be required, or for each ton of ration, 49.4 lb (2.47 × 20) of premix would be needed. In other words, 1000 lb of the premix would suffice for mixing 20 tons of the overall ration with 12 lb of premix left over.

III

Another situation where premixing might be used would be in the feeding of a herd of lactating beef cows. Let us assume that a farmer has a herd of beef brood cows with young calves for which he has no pasture. He must winter them on the following ration per head daily:

> 12 lb of coastal bermudagrass hay
> 33 lb of corn silage
> 0.5 lb of 44% soybean meal
> 0.05 lb of steamed bonemeal
> 0.05 lb of trace-mineralized salt
> 47,000 IU vitamin A

The hay is fed in a rack and the silage in a trough. He would like to spread the other ingredients over the silage at feeding time. Weighing out the appropriate amount of each of these ingredients for each feeding becomes a rather tedious chore and he decides to combine these ingredients into a single premix to reduce the number of weighings. This can be done by calculating the proportionate amounts of each ingredient per 100 lb of premix as follows:

	Amount to be Fed Per Head Daily	*Amount of Each Ingredient Per 100 lb of Premix*
	lb	*lb*
44% soybean meal	0.50/0.612	× 100 = 81.70
Steamed bonemeal	0.05/0.612	× 100 = 8.17
Trace-mineralized salt	0.05/0.612	× 100 = 8.17
Vitamin A supplement	0.012*/0.612	× 100 = 1.96
TOTAL	0.612	

A vitamin A supplement that contains 4,000,000 IU per lb is used. It is calculated that 0.012 lb of this supplement would be needed to provide the 47,000 IU of vitamin A required per cow daily. This was arrived at by dividing the daily requirement (47,000 IU) by the IU of vitamin A per lb of supplement (4,000,000).

Once the various ingredients have been combined in the above proportions, they are ready to be used as a single supplement material. In the instance cited above, the premix would be used at the rate of 0.612 lb per cow daily. If there are 100 cows in the herd, a total of 61.2 lb would be fed daily on top of the silage. Should the silage be fed in two feedings, the daily supplement allowance would be divided accordingly.

IV

Premixes of the types illustrated above are rather simple to formulate and prepare; however, the formulation and preparation of vitamin and/or trace-mineral premixes are considerably more involved and are only infrequently carried out on the farm. Even so, a student of feeds and feeding should become familiar with the techniques of the formulation and use of even the more complicated premix preparations.

V

In Table 51–1 are summarized some general guides for the formulation of premixes of the more involved types. The application of these general guides in the formulation of a vitamin-antibiotic premix for growing swine is illustrated in Table 51–2. In Table 51–3 is illustrated the application of these same principles in the formulation of a micromineral premix for fattening steers. It will be noted that in the formulation of a trace-mineral premix, arriving at the amount of the various minerals per gram of the respective source materials is somewhat more involved than it was with the formulation of the vitamin-antibiotic premix. Otherwise the basic steps are essentially the same.

Table 51–1
GUIDE TO THE FORMULATION OF PREMIXES

Column 1	Column 2		Column 3	Column 4	Column 5
Name of vitamin, mineral, or other nutrient or medication to be added to the ration	Amount of nutrient or medication to be added through premix per unit of ration		Concentration of vitamin, mineral, or other nutrient or medication in source material	Grams of source material required per lb of ration	Grams of source material to be added per ton of ration
	(a) per kg	(b) per lb	Amt. per g		
	List here the requirement of each vitamin, mineral and/or medication to be included as given in the NRC tables, or the desired portion thereof	Calculate by dividing previous column by 2.2	Express amount in same units as used in preceding column. Obtain from manufacturer's container or if a chemical compound, by calculation. If concentration shown on container is per lb or kg, convert to a per g basis	Divide column 2b by column 3	Multiply previous column by 2000

Use amounts shown in column 5 of the respective source materials along with an appropriate carrier to formulate premix. Use sufficient carrier to permit the use of the overall premix at not less than 10 lb per ton of ration. Thoroughly mix source materials with carrier and thoroughly mix premix into the ration. Carrier may be almost any finely ground feed material—preferably one of the basic ration ingredients. Soybean meal is frequently used for this purpose.

Table 51-2
FORMULATION OF A VITAMIN-ANTIBIOTIC PREMIX FOR 100-LB GROWING SWINE

Vitamin or Antibiotic	Amount of Vitamin or Antibiotic to be Added Through Premix Per Unit of Ration		Amount of Vitamin or Antibiotic Per g of Source Material	Grams of Source Material Required Per lb of Ration	Grams of Source Material to be Added Per Ton of Ration
	Per kg	Per lb			
Vitamin A	1300 IU	591 IU	30,000 IU/g	0.019700	39.400*
Vitamin D	125 IU	56.8 IU	6,000 IU/g	0.009467	18.933*
Riboflavin	2.2 mg	1.0 mg	1,000 ug/g	0.001000	2.000*
Niacin	10 mg	4.5 mg	1,000 mg/g	0.004500	9.000*
Pantothenic acid	11 mg	5.0 mg	1,000 mg/g	0.005000	10.000*
Vitamin B_{12}	11 mcg	5.0 mcg	(60 mg/lb) 132 mcg/g	0.037878	75.756*
Vitamin E	11 mg	5.0 mg	(20,000 IU or mg/lb) 44 IU or mg/g	0.113636	227.272*
Antibiotic	22 mg	10.0 mg	(50 g/lb) 110 mg/g	0.090909	181.818*
				TOTAL or	564.179 g 1.24 lb

*These amounts of the respective vitamin and antibiotic sources (totaling 564.179 g or 1.24 lb) should be thoroughly mixed with 8.76 lb of carrier to produce 10 lb of premix to be added per ton of ration. Should one desire to prepare a large quantity of this premix, each item in the above formula, including the carrier, would be multiplied by the appropriate factor.

Table 51–3

FORMULATION OF A MICROMINERAL PREMIX FOR INCLUSION IN A COMPLETE RATION FOR FATTENING STEERS

| Mineral | Amount of Mineral to be Added Through Premix Per Unit of Ration | | Source Material | Calculation of Milligrams of Mineral Per Gram of Source Material | | | | Grams of Source Material Required Per lb of Ration | Grams of Source Material to be Added Per Ton of Ration |
	Per kg	Per lb		Molecular Weight of Source Material	Atomic Wt of Mineral	Percent Mineral in Source Material	Mg of Mineral Per g of Source Material		
Iodine	0.075 mg	0.034 mg	Ca (IO$_3$)$_2$	389.9	127(2)	65.1	651	0.000052	0.104*
Iron	10 mg	4.55 mg	FeCO$_3$	115.8	55.8	48.2	482	0.009440	18.880*
Copper	4 mg	1.8 mg	Cu$_2$O	143	63.5 (2)	88.8	888	0.002027	4.054*
Cobalt	0.075 mg	0.034 mg	CoCO$_3$	118.9	59	49.6	496	0.000069	0.138*
Manganese	5 mg	2.3 mg	MnO	70.9	55	77.6	776	0.002964	5.928*
Zinc	25 mg	11.4 mg	ZnO	81.4	65.4	80.3	803	0.014197	28.394*
								TOTAL	57.498 g
								or	0.13 lb

*These amounts of the respective mineral sources (totaling 57.498 g or 0.13 lb) should be thoroughly mixed with 9.87 lb of carrier to produce 10 lb of micromineral premix to be added per ton of ration. The materials used as sources of the respective minerals will depend on their prices and suitability. Should one desire to prepare a large quantity of this premix, each item in the above formula, including the carrier, would be multiplied by the appropriate factor.

52

Computerized Least-Cost Rations

A great deal of interest has been demonstrated by feed manufacturers and livestock feeders over the past few years in the computer formulation of so-called least-cost rations. Such rations differ from conventionally formulated rations only in that a computer is used in their formulation. As are conventionally formulated rations, computerized least-cost rations are supposedly balanced with respect to nutritional adequacy using the most economical, satisfactory sources available for supplying the various critical nutrients in the amounts needed. The advantage of the computer is that it will do almost instantly what would otherwise be a long, laborious, almost impossible task done by hand calculation.

I

Basic requisites. There are several basic requisites necessary for the computer formulation of least-cost rations. They are:

- Computer facilities
- Personnel trained in the use of computer facilities
- Information on nutritional requirements
- Information on feed suitability
- Information on feed composition
- Information on prevailing prices of available feeds

A. Computer facilities and their use.   Under prevailing conditions, only the major feed companies and the larger livestock operations are in a position to justify having their own computer facilities and the personnel to operate them. On the other hand, anyone with a telephone can have computer services available by making the necessary arrangements with an organization or firm that provides this type of service. Such organizations or firms ordinarily have the personnel required for establishing a working arrangement between the livestock feeder and the computer facility. The Agricultural Extension Services in most states are in a position to provide farmers with computer service.

   While it is not in any way the intent of this book to provide basic instruction in computer operation, the authors are pleased to provide such nutritional information as might be required by a computer operator for use in the computer formulation of least-cost rations for various classes of livestock.

B. Nutritional considerations.   From the standpoint of nutritive requirements, such information takes on the form of so-called restrictions. These restrictions for the various critical nutrients are expressed as minimum and/or maximum values, depending on the nutrient under consideration. Generally speaking, the fewer the restrictions and the wider the restriction ranges consistent with acceptable ration quality, the better.

   For certain critical nutrients such as salt, the trace minerals, and the vitamins, which are usually supplied at minimum allowance levels independent of the amounts of these nutrients provided by the major ration components, no formal restrictions are ordinarily provided. The rations are simply formulated with consideration being given to protein, energy, fat, fiber, calcium, and phosphorus, and the trace minerals, critical vitamins, and salt are added per ton as needed to meet the minimum requirements. In fact, sometimes the calcium and phosphorus restrictions are also omitted, and supplemental sources of these two nutrients are added as required based on the amounts of each supplied by the major ingredients as calculated by the computer.

C. Individual feed restrictions.   In formulating rations, whether it be by conventional or computer procedures, it is frequently necessary or desirable to place certain limitations on the use of one or more potential ration components. These may be either minimum or maximum limitations, depending on the circumstances. For example, sometimes a feed may be used in the ration for a particular class of livestock up to a certain level, but it cannot be used in unlimited amounts without possible complications. In other instances, circumstances sometimes dictate that a certain minimum amount or some fixed amount of a feed be used in a ration. As with the various critical nutrients, such limitations are effected in computerized ration formulation through the use of appropriate restrictions. As indicated earlier, these restrictions may be in the form of minimum and/or maximum values. Should it be desirable to include a fixed level of some nutrient or feed in the ration, this would ordinarily be accomplished by using minimum and maximum restriction values of the same magnitude for the feed or nutrient in question.

D. **Feed composition and prices.** Once appropriate restriction values have been established for the various critical nutrients and feeds, this information is fed into the computer along with information on the composition and price of the various alternative feeds available in the area. In this connection it should be remembered that the list of alternative feeds and ingredients must be of such a combination as will permit the computer to comply with the program restrictions on a reasonable basis in the formulation of the desired ration. In other words, a sufficient number of alternative sources of each critical nutrient should be provided to permit the computer considerable latitude in its selection process. Also, of course, the composition figures used for the various feeds and products under consideration must be accurate for the computer to turn out meaningful results.

For effective least-cost ration formulation, there must also be available reliable, realistic prices on potentially available feed ingredients. Such prices must, of course, be F.O.B. at the user's base of operation. Also, they must include the cost of any processing to which a feed must be subjected to put it in a form comparable to that of other alternative feeds. In addition, only products should be included in the list of alternative feed possibilities that can be handled with the user's existing facilities. Otherwise, the cost of providing any additional facilities to use such an ingredient should be given appropriate consideration.

II

Table of restrictions. In Table 52–1 are listed some proposed restrictions for use in the computer formulation of rations for various classes of livestock. It should be noted in this connection that under certain circumstances it might not be necessary to use all of the restrictions listed for a particular livestock class. In other instances, additional restrictions may need to be included. Also, in some instances, the restrictions are not hard and fast, and might be varied either up or down from that given, should circumstances seem to warrant it. While these restrictions are designed primarily for use in computer ration formulation, the student will frequently find these of assistance in connection with other feeds and feeding endeavors.

III

Limitations to computer formulation. In the use of the computer for ration formulation, one can always be certain that a computerized least-cost ration is the cheapest balanced ration that can be formulated from the feeds available at the prices used. Also, a ration so formulated will always be balanced nutritionally insofar as the computer has been instructed with respect to nutritional considerations. It will not, however, always be the ration that will produce the greatest and/or the most efficient gains or production. There are some attributes of feeding value, such as palatability or acceptability, on which it is difficult to place a numerical value. In some instances, certain feeds tend to complement each other so that the feeding value of both feeds used together is above the average of the feeding value of the two feeds fed separately. Also, the feeding value of some feeds tends to vary with the level fed.

Table 52–1
SOME PROPOSED RESTRICTIONS FOR USE IN THE COMPUTER FORMULATION OF LEAST-COST RATIONS FOR SEVEN DIFFERENT CLASSES OF LIVESTOCK PER 100 LB (AIR-DRY BASIS)

	Finishing Steer Calves		Finishing yrlg Steers		Growing-Fattening Pigs (40–130 lb)		Growing-Fattening Pigs (130–220 lb)		Lactating Sows		Lactating Dairy Cows		Pleasure Horses	
	Min	Max	Min	Max	Min	Max	Min	Max	Min	Max	Min	Max	Min	Max
Protein														
TP, lb	11.0	—	10.0	—	16–14	—	13.0	—	15.0	—	13.5	—	7.7	—
Energy (use any one)														
TDN, lb	72.0	—	72.0	—	75.0	—	75.0	—	75.0	—	60.0	—	56.1	—
DE, Mcal	—	—	—	—	150.0	—	150.0	—	150.0	—	120.0	—	112.2	—
ME, Mcal	118.0	—	118.0	—	141.0	—	141.0	—	141.0	—	103.0	—	—	—
NE$_{\text{tact cows}}$, Mcal											62.0	—		
Fat, lb	2.0	7.0	2.0	7.0	2.5	10.0	2.5	10.0	2.5	10.0	1.8	5.0	2.0	3.5
Crude fiber, lb	7.0	—	7.0	—	—	5.0	—	5.0	—	8.0	15.3	—	11.0	—
Calcium, lb	0.40	0.80	0.35	0.70	0.60	0.80	0.50	0.70	0.75	0.90	0.45	0.86	0.27	0.60
Phosphorus, lb	0.30	0.40	0.25	0.35	0.50	0.65	0.40	0.50	0.50	0.75	0.32	0.43	0.18	0.30
Molasses (any type), lb	—	10.0	—	10.0	—	0.0	—	0.0	—	0.0	—	10.0	—	10.0
Dried citrus pulp, lb	—	20.0	—	20.0	—	0.0	—	0.0	—	0.0	—	20.0	—	0.0
Dried beet pulp, lb	—	15.0	—	15.0	—	0.0	—	0.0	—	0.0	—	15.0	—	—
Wheat, lb	—	40.0	—	40.0	—	—	—	—	—	0.0	—	25.0	—	10.0
Wheat bran, lb	—	10.0	—	10.0	—	0.0	—	0.0	—	—	—	10.0	—	25.0
Cottenseed meal, lb	—	—	—	—	—	9.0	—	9.0	—	10.0	—	—	—	10.0
Corn gluten meal, lb	—	—	—	—	—	—	—	—	—	9.0	—	—	—	—
Urea, lb	—	1.0*	—	1.0*	0.0	0.0	0.0	0.0	0.0	0.0	—	1.0*	—	0.0

*Should be used at this level only after animal has been adapted to urea feeding.

Furthermore, in the use of computerized least-cost rations for ruminants, considerable discretion will need to be exercised in switching from one ration to another. It is generally recognized that from 2 to 3 weeks are required for the rumen microflora to become completely adjusted to an entirely new ration, and something less than top performance is usually experienced during an adjustment period. Consequently, should the computer indicate a need for a major shift in the components of a ruminant ration, every effort should be made to accomplish such a shift in a series of at least 3 or 4 minor ration adjustments extending over a period of about 2 weeks. Otherwise, various digestive disturbances and something less than top performance may result.

While certain large feedlot operators and feed manufacturers may find the computer helpful in making selections between alternate feeds, the computer is not an absolute necessity for carrying out an effective program of livestock feeding. An ordinary hand calculator will meet the needs of the average livestock producer for making most necessary feeding calculations. Furthermore, it is not within the scope of this book, and the course for which it is intended to be used as the text, to cover the high-tech aspects of least-cost ration formulation. Most livestock producers, primarily because of the transportation, processing, and brokerage costs carried by purchased feeds, try to use home and locally produced materials in so far as practical in their feeding operations. Such a policy tends to reduce the number of available feed alternatives to choose from and limit the necessary feeding calculations to those that can be made on an ordinary hand calculator. Where circumstances lend themselves to the use of the computer and seem to justify the trouble and expense of such an approach to a feeding problem, the computer is a tool that, if used with discretion, can do much to minimize feed costs for feed manufacturers and large feedlot operators. There is not room, however, for the science and technology of least-cost ration formulation to be made a part of a course in feeds and feeding. Such subject matter might be the basis of a course in applied computer technology to be taken after a student has had a course in feeds and feeding and in computer technology.

53 Estimating Feed Requirements

In the development of plans for carrying on a livestock operation, it is essential that one be able to make fairly accurate estimates of feed requirements. Various approaches may be used for making such estimates, and the one to use will depend on the circumstances.

One procedure that might be used is to develop a balanced daily ration such as might be used for a particular group of animals. The amounts of the various feeds making up this ration are then multiplied by the number of animals involved and in turn by the number of days the animals are to be fed on this ration. Each major change in the ration and/or in the number of animals to be fed will require a separate feed-use period.

To illustrate this approach to estimating feed requirements, let us assume that a dairyman needs an estimate of his feed use for the coming year. It is his hope to keep an average of 75 cows on the milking line during the year with an average production per cow of 14,000 lb of about 4% milk. The cows will weigh, on the average, 1100 lb. For about 6 months out of the year, he plans to use an average daily ration of approximately the following

5.5 lb alfalfa hay	4.6 lb 44% soybean meal
33.0 lb corn silage	0.13 lb defluorinated phosphate
6.6 lb ground shelled corn	0.15 lb trace-mineralized salt
4.9 lb ground barley	0.03 lb ground limestone

For the other six months, he is planning to use pasture to meet about half of his roughage needs. The daily feed needs per lactating cow during this period then would be the same as above except that the hay and silage requirements would be reduced by 50%.

In order to maintain a milking herd of 75 cows, the producer plans to have on hand throughout the year an average of approximately 15 dry cows to which he plans to feed about the same daily roughage allowance per cow as is fed to the milking cows but only one-third as much concentrates.

For the sake of simplifying this illustration, it will be assumed that with this operation, replacement cows are brought in from an outside source.

The annual feed needs of the operation would then be calculated according to table 53–1.

How closely the above totals will approximate the producer's actual feed usage will, to a large degree, depend on how closely he follows his planned program of operation. Also, there will no doubt be miscellaneous feed ingredients required that are not covered in the above calculation. On the whole, however, such estimates should be adequate for developing crop production and feed-purchasing plans and for developing an operating budget on which to base necessary financial arrangements.

II

A somewhat different approach is sometimes used in estimating feed requirements for a steer finishing operation and other similar enterprises. This approach, rather than being based on animal days involved and average daily feed consumption, is based on pounds of gain to be produced and the pounds of feed required per pound of gain. For example, a farmer may have on hand 100 head of steer calves that he would like to finish for market. The steers are currently averaging approximately 500 lb in weight, and he would like to put them on the market at about 900 lb. This means that about 400 lb of gain is to be put on each steer or a total of 40,000 lb on the 100 head. He plans to self-feed these animals on the following ration mixture:

	%
Ground coastal bermudagrass hay	20.00
Ground shelled corn	70.00
44% soybean meal	9.00
Defluorinated phosphate	0.33
Ground limestone	0.33
Trace-mineralized salt	0.33
TOTAL	100.00

Table 53–1
ESTIMATE OF ANNUAL FEED REQUIREMENTS FOR A 90-COW DAIRY HERD

	Alf Hay	Corn Sil	Sh Corn	Barley	44% SOM	Defl Phos	TM Salt	Grnd LS
				(Pounds)				
Milking herd—6 mth, no past								
Per cow/day	5.5	33.0	6.6	4.9	4.6	0.13	0.15	0.03
For 75 cows, 183 days	75,488	452,925	90,585	67,253	63,135	1785	2059	412
Milking herd—6 mths with past								
Per cow/day	2.75	16.5	6.6	4.9	4.6	0.13	0.15	0.03
For 75 cows, 183 days	37,744	226,462	90,585	67,253	63,135	1785	2059	412
Dry cows—6 mths, no past								
Per cow/day	5.5	33.0	2.2	1.6	1.5	0.04	0.05	0.01
For 15 cows, 183 days	15,098	90,585	6,039	4,392	4,118	110	137	27
Dry cows—6 mths with past								
Per cow/day	2.75	16.5	2.2	1.6	1.5	0.04	0.05	0.01
For 15 cows, 183 days	7,549	45,292	6,039	4,392	4,118	110	137	27
TOTAL	135,879	815,264	193,248	143,290	134,506	3790	4392	878
	67.9 tons	407.6 tons	3451 bu	2985 bu	67.3 tons	38 cwt	44 cwt	9 cwt

Experience has proven that steers of this age and weight will require, during an average finishing period, approximately 7 lb of the above ration per lb of gain. This means that a total of about 280,000 lb of feed with the preceding composition would be required for finishing the 100 steers with amounts of the individual feeds as calculated:

$$280,000 \times 0.20 \quad = 56,000 \text{ lb hay (28 tons)}$$

$$280,000 \times 0.70 \quad = 196,000 \text{ lb corn (3500 bu)}$$

$$280,000 \times 0.09 \quad = 25,200 \text{ lb SBM (12.6 tons)}$$

$$280,000 \times 0.0033 \quad = 933 \text{ lb defluorinated phosphate (0.5 ton)}$$

$$280,000 \times 0.0033 \quad = 933 \text{ lb ground limestone (0.5 ton)}$$

$$280,000 \times 0.0033 \quad = 933 \text{ lb trace-mineralized salt (0.5 ton)}$$

With older steers and with steers carried to somewhat heavier weights, the feed requirement per lb of gain will increase to as much as 8 or 9 lb of air-dry feed per lb of gain, and possibly more. Reducing the roughage to concentrates ratio will tend to lower the feed required per lb of gain, whereas increasing the roughage to concentrates ratio will tend to increase the feed requirement per lb of gain. The above figures are based on the use of available growth-promoting products. If such products are not used, then feed requirements will need to be increased by about 10%.

III

An approach similar to that used for steers might also be used for estimating the feed requirements for a swine-feeding enterprise. Pigs being finished from weaning to market weight will, on the average, require approximately 3.25 lb of air-dry concentrates per lb gain.

IV

In calculating feed requirements for pigs, as well as for steers, it should be pointed out that the feed required per unit of gain does not remain constant throughout the feeding period. In fact, the air-dry feed requirement per lb of gain will increase from the beginning of the finishing period to the end from around 2.5 lb up to about 4 lb for pigs and from about 5.5 lb up to around 8.5 lb for steer calves on a 20% roughage ration.

54 Weights, Measures, Volumes, and Capacities

I Bushel.

A. A bushel is the volumetric equivalent of a cylinder 18.5″ in diameter and 8″ high. It is 2150.42 in^3. For practical purposes it is assumed to be 1.25 ft^3. Standard weights per bushel have been established for most grains, seeds, fruits, roots, etc.

B. Standard weights per bushel (in lb).

Wheat bran	20
Peanuts in hull	22
Cottonseed	32
Oats grain	32 (actual wt will vary from 25–40)
Barley grain	48 (actual wt will vary from 44–50)
Shelled corn	56
Sorghum grain	56
Rye grain	56
Wheat	60
Soybeans	60
Cowpeas	60
Sweet potatoes	55
Irish potatoes	60

Ear corn	70	(is 2.5 ft^3 or amount required to shell out 56 lb)
Snapped corn	80	(is 3.5 ft^3 or amount required to shell out 56 lb)
Water	77.6	(62.5 lb per ft^3)

II

Gallon (liquid).

A. **A liquid gallon** is 231 in^3.

B. **Standard weights per gallon** (in lb).

Vegetable oils	7.7
Water	8.34
Milk	8.6
Molasses	11.75

III

Miscellaneous.

A. **Silage** weighs:

30–40 lb per ft^3 in a horizontal silo
40–50 lb per ft^3 in an upright silo

B. **Baled hay.**

Approximate size of bale	= 14 in $\times$ 18 in $\times$ 36 in
Approximate volume per bale	= 5.26 ft^3
Approximate weight per bale	= 60 lb
Weight per ft^3	= 11.4 lb

IV

Calculating volumes.

A. **Rectangular** bins, etc.

Volume = Length $\times$ Width $\times$ Depth
(All must be in a common unit of measure)

B. **Cylindrical** bins, silos, tanks, drums, etc.

Volume = Area of circular base or end $\times$ Height or length of cylinder

$$A = \pi r^2$$

$$\pi = 3.1416$$

$$r = \text{radius} = 1/2 \text{ diameter}$$

(a)   (b)

FIGURE 54–1
(a) A round metal bin such as is widely used for the storage
of grains and for which it is frequently necessary to calculate
capacity. (b) Round feed bins with cone-shaped bottoms for
which it is frequently necessary to calculate capacity.
(Courtesy of the University of Georgia College of Agriculture
Experiment Stations)

C. **Cone-shaped** bin, etc.

Volume = Area of circular end × Height to point of cone × 1/3

V

Calculating capacities.

A. **Capacity in bushels.**

$$\frac{\text{Volume in ft}^3}{1.25} = \text{Capacity in bushels}$$

B. **Capacity in gallons.**

$$\frac{\text{Volume in in}^3}{231} = \text{Capacity in gallons}$$

C. **Capacity in lb.**

No. of gallons × Wt per gal in lb = Capacity in lb
No. of bushels × Wt per bus in lb = Capacity in lb
No. of ft^3 × Wt per ft^3 in lb = Capacity in lb

VI

Conversion of various measurements into the metric system. See table of equivalents
in appendix.

55 Feeding Beef Cattle in Drylot

I General.

A. Cattle are fed according to class and individual.

B. Breeding and stocker classes are fed primarily on roughages with supplements as needed to meet the nutritive requirements.

C. Finishing classes are fed primarily on concentrates with only as much roughage as is necessary for good performance, usually from 10%–20% of the ration.

D. Cattle of any age may be full fed at any time on any type of roughage without harmful effects to the animal except that certain fresh legume forages will sometimes cause cattle to bloat.

E. Cattle of all ages must be brought onto high grain feeding gradually, starting them out on just 1–2 lb per head daily and increasing this amount about 0.5–1.0 lb per head daily until they are on full feed, otherwise digestive disturbances may occur causing acidosis, founder, and possibly death.

F. Cattle on full feed will consume about 3% of their body weight as air-dry feed daily or the equivalent dry matter as haylage, silage, or high-moisture grain. Cattle on a

Beef Consumption per Capita per Year (Boneless Equivalent) in the United States, 1960–1998

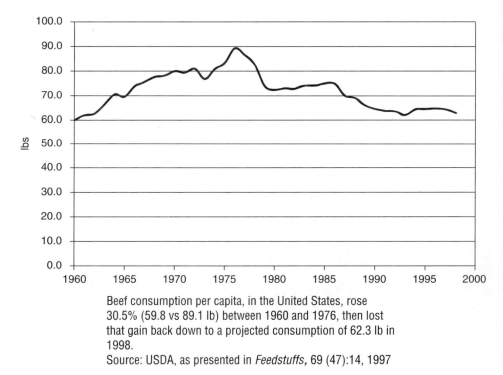

Beef consumption per capita, in the United States, rose
30.5% (59.8 vs 89.1 lb) between 1960 and 1976, then lost
that gain back down to a projected consumption of 62.3 lb in
1998.
Source: USDA, as presented in *Feedstuffs,* 69 (47):14, 1997

maintenance ration will require about 2% of their body weight daily as air-dry roughage or the equivalent in dry matter as haylage or silage.

G. Cattle may be self fed on any type of ration, but for rations containing significant levels of grain, the animals should be gradually adjusted to such a feeding program.

H. If limited grazing is available, the drylot ration allowance should be reduced accordingly.

I. While the following guides will help one to approximate the needs of the different classes of beef cattle, in the end all animals should be fed according to their nutritive requirements as indicated in the respective NRC tables.

II

Wintering dry brood cows in drylot. (See Section 65 for feeding on pasture or range.)

A. Feed such animals as much as possible on roughage feeds: hay, haylage, silage, and/or low-quality roughage.

B. Feed a little less than 2% of the body weight daily as air-dry roughage (or the equivalent dry matter as haylage and/or silage) plus concentrates as needed to balance the ration (usually less than 2 lb/hd daily will be needed).

1. Such animals may be fed almost any type and combination of roughage feeds available.

2. Almost any combination available of high-protein and/or high-energy feeds may be used as needed to balance the ration of such animals for protein and energy.

C. Such animals should be provided on a free-choice basis, a mineral supplement consisting of the following:

> 1 part trace-mineralized salt
> 2 parts defluorinated phosphate
>   or dicalcium phosphate
>   or steamed bone meal

D. Many breeders like to provide dry brood cows with 20,000 to 30,000 IU of vitamin A per head daily, or inject them subcutaneously with 5 million IU in the fall of the year.

E. The daily roughage allowance of such animals is usually provided in either one or two feedings, with any supplement being fed at a single feeding. At least half of the ration is best fed in the evening during cold weather.

F. Should partial grazing sometimes be available for the brood cow herd, the above allowances should be reduced accordingly.

Sample rations for drylot feeding of pregnant beef cows are listed below and are based on an 1100 lb (500 kg) British-type cow, calving in early March, and capable of producing 15 lb (6.8 kg) of milk/day. Although vitamin A needs are not indicated, such cows should have been injected subcutaneously with 5 million IU of vitamin A, probably in November or December. (These guidelines were formulated by workers at Kansas State University, Manhattan, Simms, D. D., et al. Beef Cow Nutrition Guide Nov. 1993.)

III

Feeding beef cows that are nursing calves in drylot. (See Section 65 for feeding on pasture or range.)

A. The breeding and pasture programs should be planned so that the drylot feeding of beef brood cows that are nursing calves more than 30 days old is, for the most part, unnecessary.

1. Beef brood cows should be bred so they will calve not more than about 30 days before good pasture is available.

2. Almost any mature beef cow in good flesh will produce enough milk on almost any adequate dry cow wintering ration for almost any calf under about 30 days of age. Thin cows and first-calf heifers should be cut out and fed more liberally during this period (Figure 55–1).

RATIONS FOR THE LAST TRIMESTER OF PREGNANCY
(DECEMBER THROUGH FEBRUARY)

		Amount Per Head Per Day (lbs)					
Ration	*Feedstuff*	*lbs*		*Feedstuff*	*lbs*	*Mineral Req.*
1[a]	Corn stover	18	+	Alfalfa hay	8	P + TM Salt
2	Corn stover	20	+	Wheat midds	7	TM Salt
3	Milo stover	18	+	Alfalfa hay	7	P + TM Salt
4	Milo stover	19	+	Wheat midds	7	TM Salt
5[b]	Dry winter grass	16	+	Alfalfa hay	7	P + TM Salt
6[b]	Dry winter grass	20	+	Commercial cubes[c]	4	P + TM Salt
7	Dry winter grass	16	+	Wheat midds	9	TM Salt
8	Brome hay	26				TM Salt
9	Prairie hay	16	+	Alfalfa hay	9	TM Salt
10	Prairie hay	20	+	Wheat midds	5	TM Salt
11	Wheat straw	15	+	Alfalfa hay	10	TM Salt
12	Wheat straw	17	+	Wheat midds	8	Ca + TM Salt
13	Forage sorghum hay	25				TM Salt

[a]Under the conditions described, cows consuming this ration would lose approximately ⅓ body condition score. Consequently, if this loss is unacceptable, this ration should be adjusted.
[b]Under the conditions described, cows consuming these rations would lose approximately ½ body condition score from December 1st to March 1st. Thus, additional supplementation may be necessary.
[c]20% CP.

RATIONS FOR LACTATION (MARCH THROUGH APRIL)

		Amount Per Head Per Day (lbs)					
Ration	*Feedstuff*	*lbs*		*Feedstuff*	*lbs*	*Mineral Req.*
1	Brome hay	28		—		TM Salt
2	Corn stover	15	+	Alfalfa hay	14	P + TM Salt
3	Corn stover	22	+	Commercial cubes[a]	8	TM Salt
4	Milo stover	14	+	Alfalfa hay	14	P + TM Salt
5	Milo stover	21	+	Commercial cubes[a]	8	TM Salt
6	Wheat straw	15	+	Alfalfa hay	14	P + TM Salt
7	Prairie hay	16	+	Alfalfa hay	11	P + TM Salt
8	Forage sorghum hay	20	+	Alfalfa hay	8	TM Salt
9	Winter native grass	19	+	Commercial cubes[a]	8	TM Salt
10	Winter native grass	14	+	Alfalfa hay	13	TM Salt

[a]20% CP.

FIGURE 55–1
These brood cows have access to large round bales of hay
to provide extra energy and some extra protein to enhance
their milk production.

3. Every effort should be made to move beef brood cows and their calves to good pasture by the time the calf is about 30 days old.

4. A beef brood cow with a nursing calf should be able to meet most of her nutritive needs on good pasture. The exceptions are salt; possibly certain other minerals, especially magnesium during late winter and spring; calcium and phosphorus on thin soils; and perhaps certain trace minerals, depending on the circumstances.

B. Should it be impossible to move a beef brood cow to good pasture by the time her calf is about a month old, then she should be separated from the dry cows and fed a more liberal ration.

1. The protein supplied to such a cow will need to be approximately doubled and the TDN increased by about 50% over the dry cow ration.

   a. Adequate protein and energy may be provided by using a roughage of good quality and by increasing the allowance by about 50%, or to approximately as much as the cow will consume.

   b. The needed protein and energy might also be provided by leaving the roughage allowance unchanged and adding enough grain and protein meal (4 to 6 lb total) to meet the increased protein and energy needs.

2. The calcium and phosphorus needs will increase but should be taken care of for cows that are nursing calves with a free-choice mineral mix such as follows:

> 1 part trace-mineralized salt
> 2 parts defluorinated phosphate
> or dicalcium phosphate
> or steamed bonemeal

3. If there are any indications of a need for vitamin A supplementation then, for the sake of the health of both the cow and calf, this vitamin should be supplied to cows that are nursing calves at about 30,000–40,000 IU per cow per day.

4. The daily roughage allowance of such cows is usually provided in either one or two feedings, while the concentrate allowance is usually fed in a single feeding. At least half of the ration is best fed in the evening during cold weather.

5. To creep feed nursing calves, use almost any combination available of whole or rolled corn, oats, barley, and/or grain sorghum, and add about 12% soybean meal or other suitable protein meal to the mix as the calves reach about four months of age (Figure 55–2).

FIGURE 55–2
A creep that provides the opportunity for suckling calves to obtain primarily energy feeds while nursing protein and mineral-rich milk.

IV

Feeding stocker steers and heifers in drylot. (See Section 63 for feeding on pasture or range.)

A. Feed such animals largely on roughage feeds: hay, haylage, silage, and/or low-quality roughage.

B. Feed stocker animals a little more than 2% of their body weight daily as air-dry roughage (or equivalent dry matter as haylage and/or silage) plus concentrates as needed to balance the ration (amount to be fed depends on gain desired but usually will not exceed about 4 lb per head daily).

    1. Such animals may be fed almost any type and combination of roughage feed available. If there is a choice, the better roughage should be fed to the younger animals.

    2. Almost any combination available of high-protein and/or high-energy feeds may be used as necessary to meet the needs of such animals for protein and energy, but replacement heifers should not become too fat (Fig. 55–3).

FIGURE 55–3
Heifers fitted for show and sale often carry too much body fat for good reproductive performance. Good managers will be able to help over-conditioned heifers gradually lose some of such finish. (Picture was planned to show extra fleshing on the rear portion of the sale heifers.)

C.   Such animals should be provided on a free-choice basis a mineral supplement consisting of:

> 1 part trace-mineralized salt
> 2 parts defluorinated phosphate
> or dicalcium phosphate
> or steamed bonemeal

D.   Such animals may benefit from a supplement of 10,000 to 20,000 IU of vitamin A per head daily, or from a subcutaneous injection of 5 million IU of vitamin A in the fall.

E.   The daily roughage allowance of such animals is usually provided in either 1 or 2 feedings but any concentrate allowance is usually fed in a single feeding. At least half of the ration is best fed in the evening during cold weather.

F.   Should partial grazing at times be available for such animals, the above allowances of drylot feed should be reduced accordingly, but supplemental feeds will cause additional gain (Fig. 55–4).

V   Feeding mature beef breeding bulls in drylot.

A.   Feed mature bulls primarily on any good-quality hay along with a limited amount of haylage or silage, if available, with concentrates as in B on the following page.

  1.   Feed about 1.5% of the body weight as air-dry hay (haylage or silage may be substituted for up to about 50% of the hay on a DM basis).

FIGURE 55–4
Provision of supplemental feeds to stocker steers on winter pasture will cause them to weigh considerably more when spring pasture becomes available.

    2. The daily roughage allowance of such animals is usually provided in either one or two feedings.

B. A certain amount of supplementation of the above roughage allowance with concentrates is usually needed depending on the amount and quality of roughage fed and the amount of conditioning needed by the bull.

    1. Almost any combination of corn, oats, and/or barley will usually suffice; little if any high-protein meal is normally needed if the hay is of proper quality.

    2. Such animals should normally be fed from 5 to 10 lb of grain per head daily.

    3. The grain allowance is usually provided in a single feeding unless, for some reason, it is unusually high, in which case it might be divided into two feedings.

VI

Finishing calves and yearlings in drylot.

A. Upon arrival at the feedlot, all calves and yearlings should be held in a holding pen for 2 to 3 weeks for such routine procedures as castrating, dehorning, delousing, worming, implanting, vaccinations, or treatment of any existing disease.

B. While the above is taking place, such cattle should be conditioned for going on a full feed of the operation's finishing ration.

    1. They should be started on a good-quality roughage free choice (2.5% to 3% of body weight daily) with only 1 to 2 lb of grain per head daily. A low-level antibiotic supplement may be used in this connection if it is needed and authorized by law.

    2. For the first few days, it may be helpful to treat the drinking water of the cattle with 0.25% sodium bicarbonate (¼ lb/12 gal of water).

    3. Gradually increase the concentrate allowance for the cattle until by the end of the 2 to 3 week treatment period it will total about 2% of the animal's body weight daily with the roughage still being fed free choice at about 1% of the body weight daily.

    4. To place cattle on a full feed of the operation's finishing ration using either a self-feeding or a bunk feeding program. If the ration contains any high-moisture feeds, a bunk feeding program will have to be followed (Figs. 55–5, 55–6, 55–7).

C. Every precaution should be exercised if the cattle are put on self-feeding to avoid digestive disorders that will produce acidosis of the rumen, sometimes founder, and even death.

    1. If available, use a self-feeding mixture containing about 1/3 roughage, dry basis, for the first few days.

    2. If such a high roughage mixture is not available, gorge the animals with roughage before placing them on the self-feeding mixture.

FIGURE 55–5
A homemade self feeder suitable for feeding a complete
mixed ration to from 1 to 3 steers such as an FFA or 4-H
club member might have as a project. (Courtesy of the
University of Georgia College of Agriculture Experiment
Stations)

FIGURE 55–6
A self feeder for handling a complete mixed ration for up to
about 25 steers (or heifers) with refilling every 5–8 days.
(Courtesy of the University of Georgia College of Agriculture
Experiment Stations)

(a)

(b)

FIGURE 55–7
(a) An aerial view of steer feedlot. (b) A close-up view of steers
eating a ration mixture of corn silage and grain from a typical
feed bunk. (Courtesy of Dr. W. E. Wheeler, USDA, Roman L.
Hruska Meat Animal Research Center, Clay Center, Nebr.)

3. Use a self-feeding mixture that contains from 10%–20% roughage for the remainder of the feeding period. If a low-quality roughage is used it should be
   held to a minimum.

4. The treatment of the drinking water of such cattle with 0.25% sodium bicarbonate will help get them on self-feeding.

5. Individual animals suffering from acidosis will usually respond to treatment
   with supplemental dry roughage.

6. Make any changes in the ration gradually.

7. Never let the feeders run out of feed.

8. Calves and short yearlings fed in accordance with the above plan should on the average gain 3 to 3.5 lb per head daily over a feeding period of 100 to 150 days and require about 7–7.5 lb of air-dry feed per lb of gain.

D. If a bunk feeding program is necessary or preferred, the following suggestions may be helpful.

1. Keep the cattle just a little hungry the first few days on the finishing ration. Feed just under a full feed.

2. Arrange for feed bunks to be cleaned at frequent intervals—as often as necessary to keep the feed fresh and the bunks clean.

3. Try not to feed more than the cattle will eat by the next feeding but try not to allow them to be out of feed for too long a period of time—not more than about an hour at the most.

4. Mix feed as it is used—especially if the weather is hot and the ration contains one or more high-moisture feeds.

5. Treat sick cattle the same as in a self-feeding program.

6. Cattle on a good bunk feeding program should perform comparably to those on a self-feeding program.

VII Management practices for cattle in drylot.

A. **Additives—hormone-like.**   One of the greatest improvements in beef cattle feedlot performance came about by the discovery of the growth-stimulatory effect of diethylstilbestrol implants by Andrews and Dinusson at Purdue University in 1948. Following that announcement, many modifications of the practice have come into routine usage, whereas the use of diethylstilbestrol has become illegal. However, in general, the use of such products has resulted in a 7% to 10% increase in gain and an improvement in feed efficiency of a similar magnitude. A summary of the results from 37 trials with steers was summarized by researchers at Oklahoma State University (Duckett, S. K., et al. 1996. *The Prof. Anim. Scientist.* 12 [4]: 205), and presented in Table 55–1.

B. **Additives—ionophores.**   In the mid-70s, approval was obtained from the FDA for the introduction of the first ionophore for cattle feeding, namely, monensin. Since then, additional ionophores have become legal and are in use in cattle feeding. An oversimplified explanation of the function of ionophores is that, at the proper level, they cause increased production of propionic acid in the rumen resulting in greater energy production from the feedstuffs consumed by cattle. Growing cattle and finishing cattle respond differently to dietary ionophores. Finishing cattle fed effective ionophores tend to consume about 10% less feed and gain about the same—or slightly more—than cattle not fed ionophores, probably because a

Table 55–1
WEIGHTED LEAST SQUARE MEANS FOR CHANGE IN PERFORMANCE FROM IMPLANTING STEERS

First Implant	Reimplant	ADG[a]	FI[a]	FE[a] (feed:gain)
		—(kg)—		(kg:kg)
Compudose	None	0.15**	0.40**	−0.48**
Compudose + Finaplix	None	0.24**	0.47	−0.61*
Compudose + Finaplix	Finaplix	0.21**	0.18	−0.39**
Finaplix	None	0.16*	−1.74*	−1.02
Finaplix	Finaplix	0.03	−0.13	−0.24
Ralgro	None	0.12**	0.28	−0.41**
Ralgro	Ralgro	0.17**	0.90**	−0.29**
Ralgro	Synovex	0.26**	0.77*	−1.16**
Ralgro + Finaplix	None	0.10	−0.14	−0.46
Ralgro + Finaplix	Ralgro + Finaplix	0.11	0.77	−0.01
Revalor	None	0.19**	0.47**	−0.52**
Revalor	Revalor	0.32**	0.81**	−0.95**
Revalor	Synovex	0.22**	0.64	−0.73**
Synovex	None	0.20**	0.31*	−0.44**
Synovex	Finaplix	0.29**	0.53	−0.82**
Synovex	Revalor	0.22**	0.64	−0.75**
Synovex	Synovex	0.24**	0.67**	−0.56**
Synovex	Synovex + Finaplix	0.33**	0.70**	−0.66**
Synovex + Finaplix	None	0.33**	0.46**	−0.90**
Synovex + Finaplix	Finaplix	0.13*	0.07	−0.55**
Synovex + Finaplix	Synovex	0.30**	0.54	−0.45**
Synovex + Finaplix	Synovex + Finaplix	0.31**	0.65**	−0.74*
		0.23**	0.52**	−0.56*

[a]Change in ADG, FI, or FE: within-study comparisons calculated as ADG/FI/FE$_{(implant)}$−ADG/FI/FE$_{(non-implanted, control)}$.
*Change by implanting is unequal to 0 ($P<0.05$).
**Change by implanting is unequal to 0 ($P<0.01$).

finishing steer will eat to a certain energy satiety. A growing steer, that is one on pasture or a high-roughage diet, will consume the same amount of dry matter as one not fed the ionophore and tend to gain more rapidly. This is because the animal will eat until full, but will produce more NE$_{gain}$ from its ingested feed. A 1997 University of Minnesota presentation summarizing 67 finishing and 55 pasture or confinement (on principally forage diets) experiments was summarized in a *Feedstuffs* article (Discontanzo, A., et al. 1997. *Feedstuffs.* 69 [11]: 11. March 17, 1997), and presented in Table 55–2 and Table 55–3.

C. **Facilities.** Finishing cattle adapt well to outdoor feeding (Fig. 55–7) and to slatted floors (Fig. 55–8),

Table 55–2

EFFECTS OF APPROVED IONOPHORES ON PERFORMANCE OF FEEDLOT CATTLE UNDER VARIOUS DIETARY AND MANAGEMENT CONDITIONS

Item	None	Laidlomycin	Lasalocid	Monensin	Monensin Tylosin[a]	Tylosin	MSE[b]
Dose, mg/head/day	0	85.8	285.9	272.2	263.2	104.0	—
Does, g/ton	0	8	29	28	27	10	—
Means	49	37	22	29	33	20	—
Cattle	1,556	1,137	1,290	1,042	1,200	665	—
ADG, kg	1.39[c]	1.46[d]	1.38[c]	1.38[c]	1.39[c]	1.39[c]	.089
DMI, kg/day	9.34[c]	9.33[c]	8.91[d]	8.81[d]	8.73[d]	9.40[c]	2.238
FTG	6.81[c]	6.48[d]	6.52[d]	6.44[d]	6.35[d]	6.86[c]	1.783
FTG improvement, %	—	4.8	4.2	5.4	6.8	—	—

[a]Tylosin dose: 98 mg per head per day
[b]Mean square error of weighted (observations/mean) ANOVA
[c,d]Means differ (P<0.05)

Table 55–3

EFFECTS OF APPROVED IONOPHORES AND BAMBERMYCINS ON PERFORMANCE OF GRAZING CATTLE

Item	Control	Bambermycins	Lasalocid	Monensin	MSE[a]
Dose, mg/head/day	0	25	188	167	—
Means	70	10	14	47	—
Cattle	1,885	233	311	1,329	—
ADG, kg	0.64[b]	0.78[c]	0.78[c]	0.75[c]	0.068

[a]Mean square error of weighted (observations/mean) ANOVA
[b,c]Means differ (P<0.05)

FIGURE 55–8
An increasing number of cattle are being finished on slotted
floors such as the one shown, with provisions for waste accumulation and removal below. (Courtesy of the University
of Georgia College of Agriculture Coastal Plain Experiment
Station.)

56 Feeding Dairy Cattle on Drylot Rations

I General.

A. Most dairy cattle are fed a combination of roughage and concentrates.

B. Dairy cattle, when full fed, will consume up to about 3.5% of their body weight daily as air-dry feed, or the equivalent DM as high-moisture feed.

C. While alfalfa hay and corn silage are the most widely used roughages for dairy cattle, almost any combination of roughages available (hay, haylage, silage, or low-quality roughage) may be used in rations for dairy animals.

D. It is usually best for not more than about 2/3 of the roughage DM for dairy cattle to come from silage; haylage may be used more liberally than silage.

E. While corn, grain sorghum, oats, barley, and soybean meal are the concentrates most widely used for feeding dairy cattle, almost any other concentrate feed available may be used up to about 25% of the concentrates in balancing rations for dairy animals.

F. Unless salt has been included as needed in the ration mix for dairy cattle, it should be provided in a separate feeder on an *ad libitum* basis.

G. While the following guidelines will help one to approximate the feed needs of the different classes of dairy animals, they should not be used as a substitute for formulating a ration for each animal that will meet the nutritive requirements for that animal as shown in the NRC tables.

II **Feeding milking cows on drylot rations.**

A. A number of widely different plans are in use by the dairy industry for feeding the milking herd on drylot rations. The one that is probably the most practical and the most widely used might be referred to as the "group plan," or total-mixed-ration (TMR), Figure 56–1.

  1. The first step in carrying out the group plan is to divide the producing herd into groups according to the production of each cow. Usually 2 groups will suffice—the high producers in one group and the low producers in the other.

  2. Keep them separated for feeding purposes and group feed each group their respective allocation of hay or other dry roughage, and silage plus concentrate mixture.

  3. Feed the hay or other dry roughage once daily; divide the allocation of silage plus concentrate mixture into two equal feedings to be fed just following the morning and evening milkings.

FIGURE 56–1
Cows being group fed a total-mixed-ration (TMR) just after
going through the milking line. (Courtesy of the University of
Georgia College of Agriculture Experiment Stations)

4. Mixing the silage and concentrates prior to feeding is regarded as a desirable practice but if a mixing wagon is not available the concentrates may simply be poured over the silage with good results.

5. Determine the feed allocation for each group by balancing a daily ration for the average cow in each group based on body size and milk production, and then multiply the amount of each feed or feed mixture for the average cow by the number of cows in the group.

6. In balancing a daily ration for a milk cow, one should first decide on the roughage(s) to be fed and plan to feed from 1.5% to 2% of the cow's body weight as air-dry roughage daily or the equivalent in dry matter up to about 2/3 of the roughage DM from haylage and/or silage. The level of roughage used is usually determined by its availability, quality, and relative cost as a source of nutrients, but at least 1.5% of the cow's body weight in coarse roughage daily (air-dry basis) is required for normal rumen function and milk-fat synthesis.

7. The nutrients provided by the roughage(s) are deducted from the cow's total nutrient requirements and the remaining nutrients needed are supplied through the concentrate mixture.

8. If roughage is fed at the higher level (2% of body weight as air-dry roughage), then concentrates will be required about as follows to balance the ration: 1 lb concentrate for each 4 lb low-test milk (under 3.6% fat); 1 lb concentrate for each 3.5 lb, 3.6%–4.4% fat in milk; 1 lb concentrate for each 3 lb high-test milk (over 4.5% fat).

9. If less than the above amount of roughage is fed, then about 0.6 lb of additional concentrates should be fed for each lb of hay deleted per cow daily.

10. The lower the quality of roughage available, the less of it is usually required to provide the animal with its "roughage factor" needs.

11. Grinding and/or pelleting largely destroy the "roughage factor" value of feed for a dairy cow but otherwise do not significantly alter the feed's nutritive value.

12. Roughage analysis services are available in almost all areas of the country. These provide at nominal cost a more reliable measure of a roughage's composition and feeding value than would the use of visual estimates of quality and book values. These services will also, using available ingredients, provide computerized least-cost balanced rations for those requesting them. The use of such services is highly recommended.

B. **Formulations with neither alfalfa hay (or haylage) nor corn silage (a case history)**

1. Although Table 56–1 indicates 67% of the dairy herds that produce large amounts of milk receive corn silage and 55% receive alfalfa hay, it is possible to produce a rolling average of 20,000 lb of milk/cow/year using only grass hay as the total roughage. Undoubtedly, the use of high-quality alfalfa haylage

Table 56–1
CHARACTERIZATION OF MANAGEMENT PRACTICES OF TOP MILK PRODUCING HERDS IN THE UNITED STATES[a]

	Forage Levels Fed, %		
Forage	Lactating Herd	Dry Herd	Forage Analyses/Year
Corn silage	67.2	52.5	5.8
Legume hay	55.7	21.3	4.4
Legume haylage	49.2	13.1	7.2
Grass hay	24.6	57.4	1.8
Small grain silage	13.1	16.4	2.9
Grass haylage	11.5	11.5	6.5
Sorghum silage	8.2	11.5	2.7

Ration Additives		Alternative Feedstuffs	
Bicarbonate of soda	75.4	Whole cottonseed	72.1
Bypass protein (UIP)	68.9	Dried distillers grains	37.7
Magnesium oxide	65.6	Dried blood meal	37.7
Yeast culture	50.8	Meat and bonemeal	34.4
Zin Pro[b,1]	47.5	Molasses	29.5
Tallow (bovine fat)	45.9	Fish meal	19.7
Bypass fat	41.0	Wet brewer's grains	16.4
Niacin	37.7	Beet pulp	13.1
Buffer	32.7	Cottonseed hulls	9.8
Other fat source	21.3	Corn gluten meal	9.8
Dyna - K[b]	19.7	Peanut meal	8.2
Dynamate[b,2]	19.7	Soybean hulls	6.6
Trace minerals[3]	19.7	Dried brewer's grains	6.6
Beta carotene	11.5	Corn gluten feed	4.9
Methionine	9.8	Wet distillers grains	0
Choline chloride	6.6	Peanut hulls	0

[a]Jordan, E. R., and R. H. Fourdraine. 1993. A Symposium: Management for Herds to Produce 30,000 Pounds of Milk. *J. Dairy Sci.* 76:3247.
[b]Copyrighted names.
[1]Zinpro Corp., Chaska, MN.
[2]IMC, Mundelein, IL.
[3]Trace minerals other than trace-mineralized salt.

or alfalfa hay, or the use of corn silage would increase the rolling herd average and at the same time require less concentrates in the feeding program. For example, alfalfa hay (midbloom) contains 0.531 Mcal NE_{lac}/lb, and corn silage (90% dry matter equivalent) contains 0.64, whereas bermudagrass hay contains less (0.417). Alfalfa hay (midbloom) contains 15.3% total protein, whereas bermudagrass hay contains 42% less (8.9%), based on the appendix feed tables.

2. One such dairy farm in Crawford County, Arkansas, is owned by the father-son DeShazo team who maintain 200 cows in production at all times, using only grass hay as the roughage. Needless to say, the DeShazo's would be delighted to use high-quality alfalfa and/or corn silage in their dairy formulations, but in much of Crawford County, cultivated hay crops and row crops are most difficult to produce. Therefore, native grasses provide both grazing and harvested roughage for cattle feeding.

3. During pasture season, the cows graze between milkings, but there is also long-stemmed native grass hay available in their hay shed. The hay consists of bermudagrass, orchardgrass, and even some cheatgrass.

4. The DeShazo's control (lease or own) a total of 1500 acres to provide sufficient pasture and/or hay for this operation.

5. As indicated above, roughages (pasture and/or hay) are fed *ad libitum,* (Fig. 56.2) but the concentrate is controlled on an individual basis via computer based upon milk production during the lactation season.

6. The lactating herd is never confined to stalls except during the milking parlor phase (Fig. 56.3) when 16 cows (2 rows of 8) are being milked.

FIGURE 56–2
Only native grass hay is economically available for cows in western Arkansas; it is necessary to make up the difference between native grass hay and alfalfa hay by using amounts of concentrates in the DeShazo herd. (Photo by James Herndon)

7. The weight of the milk produced per day is fed into a computer that calculates how much grain concentrate each individual cow will receive during the next 24 hours. Twice each day in the milking phase, every cow receives five pounds of concentrate, except the cows with red ankle bands that indicate they are producing more than 90 pounds of milk per day. They receive 8 pounds of concentrate at each of 2 milkings. (Fig. 56.4).

8. Once a cow has completed the 2 milking phases, each day, she is free to go about eating roughage (pasture and/or grass hay).

9. Every cow has an individually coded neck piece containing a computer chip, suspended from a neck chain. There are 7 stalls that, when a coded neck piece comes in contact with the threshold of the feeding box, will begin distributing grain concentrate into the feeding box at the rate of four pounds per hour (Fig. 56.5).

10. Any given cow can quit feeding at the computerized feed box at any time and come back later for more feed. However, the computer keeps track of the total concentrate consumed by each cow and will not distribute any more grain when that cow has consumed her prescribed quota for that day.

FIGURE 56–3
One of 2 sides of a 16-cow herringbone milking accommodation. (Photo by James Herndon)

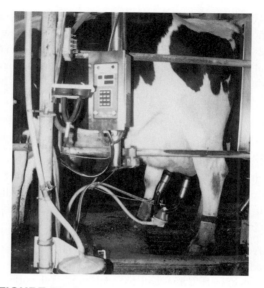

FIGURE 56–4
In the DeShazo herd, as in many good herds, milk production is weighed for each cow to the nearest one-tenth pound for calculating the concentrates she will receive during the next 24 hours. Note the band on the left rear leg, indicating that this cow produces more than 90 lb of milk/day. (Photo by James Herndon)

11. The cows are bred by artificial insemination, they are implanted with bovine somatotropin (BST), and they receive a combination of analyzed roughage and grain concentrate to meet their prescribed nutritional needs. In other words, the DeShazo father-son team is using fortified grain concentrate formulations as a substitute for what alfalfa and/or corn silage might supply their Holsteins.

12. However, one must use caution in recommending alfalfa in lactating cow diets. Table 56–2 demonstrates that the nutritional value of alfalfa can decline very rapidly unless it is harvested at the proper stage of maturity. Full-bloom alfalfa has drastically lower feeding value for the lactating cow than does pre-bloom alfalfa. In contrast, the digestibility of whole-plant corn silage does not change very much with advancing maturity (Table 56–3).

13. As illustrated in Table 56–1, managers of high-producing herds currently are using all sorts of ration additives, for example, 75% of them use bicarbonate of soda; 68.9% of them use bypass protein, etc. Maybe they derive benefits from most of these additives—and maybe not. However, if these managers are helping their dairy cows to produce 30,000 pounds of milk per year, as the old saying goes, "Don't knock success!" Erdman reported on combined data from 29 experiments in which 3 to 12 gm of niacin had been fed to lactating dairy

FIGURE 56–5
The DeShazo cows are never stanchioned except for milking.
They can eat long native grass hay, and measured quantities
of concentrates are metered to any of 7 individual feeding
stations, based upon data that evaluates their milk production for the past 24 hours. A computer chip in the neck chain
causes the automatic feeder to feed concentrate in accordance with the production of each individual cow. (Photo by
James Herndon)

Table 56–2

**STAGE OF ALFALFA MATURITY AND FORAGE—CONCENTRATE RATIOS
EFFECTS ON THE PRODUCTION OF 4% FCM MILK, POUNDS[a]**

	Forage: Concentrate Ratio			
Alfalfa Maturity	80:20	63:37	46:54	29:71
Prebloom	79.6	83.2	87.1	86.1
Early bloom	68.0	69.1	77.2	77.1
Midbloom	57.2	62.5	64.7	64.7
Full bloom	52.1	55.4	69.5	69.5

[a]Staples, D. R. 1992. "Large Dairy Herd Mgmt." *Am. Dairy Sci. Assn.* Savoy, IL. p. 347.

Table 56–3

**CORN PLANT MATURITY AT HARVEST EFFECT ON DRY MATTER
CONTENT, DIGESTION, AND INTAKE BY SHEEP[a]**

Stage of Maturity	Dry Matter, %	Dry matter Digestion, %	Dry Matter Intake, gm/kg Bodyweight
Early milk	21.0	67.3	46.0
Milk—early dough	23.4	71.0	53.4
Dough—dent	27.2	71.9	57.4
Glazed (mid milk line)	33.7	68.1	60.4
Flint	41.9	68.8	56.4

[a]Staples, C. R. 1992. "Large Dairy Herd Management." *Am. Dairy Sci.* Savoy, IL. p. 347.

---

cows daily. The summary reported no effect on milk fat or protein percent, but pounds of milk produced was increased by an average of 0.7 lb/hd/day. (Erdman, R. A., 1992, Large Dairy Herd Mgmt., *Amer. Dairy Sci. Ass'n,* Savoy, IL, p. 927)

14. Table 56–4 lists the normal level of some of the additives that top managers may add to lactation formulations.

15. Referring back to Table 56–1, note the list of alternate feedstuffs that top dairymen (rolling herd averages of 30,000 lb milk, or more) use. Obviously, greatest use is made of whole cottonseed (72.1%), whereas the survey indicates that very little use is made of peanut hulls by the top managers. Why do dairy herd managers use alternative feeds? Maybe substitutes come into the least-cost formula, or possibly many managers feel a specific feed substituted into the diet will help the cow do a better job.

16. Information about the top-producing dairy herds has been presented to offer dairymen who wish to keep abreast of the most modern techniques some ideas they might incorporate into their own program. Naturally, there is no one best program. With only grass hay as a viable roughage, the DeShazo's must build a balanced program around that. The Corn Belt dairyman, perhaps in Wisconsin, may use a high-quality roughage program incorporating alfalfa haylage and/or corn silage to start building a milk-producing diet.

III **Feeding dry cows on drylot rations.**

A. Milk cows are normally dried off for 6–8 weeks prior to freshening and are fed to meet the following nutritive needs during this period:

1. Maintenance needs.

Table 56–4
LEVELS OF ADDITIVES NORMALLY ADDED TO DAIRY DIETS[a]

Common Buffers	Amount/Day, lb
Sodium bicarbonate	0.25 to 0.50
Sodium sesquicarbonate	0.35 to 0.75
Magnesium oxide	0.10 to 0.20
Sodium bentonite	1 to 2
Calcium carbonate	0.25 to 0.33
Potassium carbonate	0.50 to 0.90

[a]Hudgens, M. F. 1992. "Selecting Feed Additives." "Large Dairy Herd Management." *Am. Dairy Sci. Assn.* Savoy, IL. p. 309.

2. Needs for fetal development.

3. Needs for growth up to 4 years of age.

4. Needs for replenishing the body's nutrient reserves; should gain about 200 lbs from the low point of previous lactation.

B. In order to accomplish the above feeding objectives, all dry cows should be fed as a separate group.

C. While there may be reasons on a particular farm for doing otherwise, dry cows may be given the same roughage allowance as is fed to the milking herd. However, if cows are fed heavily on corn silage during lactation, it may be advisable to feed little or no silage during the dry period.

D. Dry cows should be continued on limited grain feeding during the dry period in order to keep the rumen conditioned to handling concentrate feeds as well as for the nutrients it provides.

E. The level of grain feeding will vary depending on the circumstances but if about 2% of the body weight allowance of air-dry roughage is fed daily, then from 1/4 to 1/3 as much concentrates should be fed dry cows as are fed the milking herd. However, dry cows should not be fed so that they become excessively fat.

F. The same concentrate mixture as is fed to the low production group of milking cows would be hard to improve on for the dry cow group.

IV

A 365-day overview of feeding the higher milk producer. The dairy cow production year can be broken into about 4 or 5 logical programs reflecting changes in her nutritional needs:

A. **Calving to peak milk production.** Milk production in dairy cattle normally follows a curve in which it increases for 80 or 90 days after she gives birth to her calf.

This tendency needs to be aided by feeding higher-energy feedstuffs. It is at this time that the term "lead-feeding" comes into use. From a scientific point of view, one probably should calculate a high-producing dairy cow's needs once per week—admittedly, though, this may not be practical. Nevertheless, as the cow increases production during the first 3 months, it is beneficial to calculate her production needs and continue to add a little extra nutrients. If one were to calculate needs on the previous week's production, one would be behind in feeding for next week's needs during this period of increasing production. Feeding a bit more, then, is what is called lead feeding. Naturally, once her peak of production has been reached, she should be fed in accordance with her production—and later, pregnancy—needs.

In order to accomplish this, one should feed some of the best grade of hay possible and grain level should be increased. Attention should also be given to utilization of UIP (undegradable intake protein) to meet as much as 1/3 of the total protein requirements. If the diet gets too concentrated and there is danger of acidosis, sodium bicarbonate can be added as a rumen buffer.

B. "Flat-top" in production. This is the period when the cow has reached her peak of production and the manager hopes to be able to hold her as near this peak as possible. Naturally, there is going to be a gradual decline over this period from her peak of production at about 80 days on to day 210. Nutrient intake should be tied closely to production because lead-feeding during this period would not pay very good dividends. This is a period when she can start to build back some of the weight loss she has suffered during the earlier phase. By now she is pregnant and nutrients are beginning to be divided among 4 needs, namely (1) maintenance, (2) milk production, (3) forming her unborn calf, and (4) regaining parturition and early lactation weight losses.

C. "Easing-off" period. For the next 100 days, the cow is preparing seriously for her next calf and lactation. In doing this, she does quite well on much less-concentrated rations, therefore more forage and less concentrates are needed. Although one would not practice lead-feeding during this period of decline in milk production, it is a good idea to keep pretty close tabs on production and feed only sufficient concentrate to meet her needs for production, body condition, and age.

D. Rest period. During this nearly 2-month period the cow is not producing milk. She utilizes loose, long hay, and little or no concentrates are needed unless she needs to gain just a little more body condition.

E. Getting ready for next lactation. During this last 2 weeks of pregnancy she needs to be introduced again gradually to what her lactation diet will be like. In other words, start getting her back on concentrate feeding by feeding up to 0.5 percent of her body weight. Dry, long hay should be kept in the diet at this time to avoid potential digestive disturbances.

V Feeding dairy calves on drylot rations.

A. Provide a new-born calf with colostrum (milk) as its only feed through at least the first day of its life.

B. The colostrum may be made available to the calf through nursing or by hand feeding—if hand feeding is practiced, feed not more than 1 lb colostrum daily for each 10 lb of body weight in 2 feedings.

C. Milk feeding should be continued until the calf is about 4 to 5 weeks of age, using either fresh or whole milk or a good milk replacer at the rate of 1 lb of milk or an equivalent amount of milk replacer daily for each 10 lb of body weight.

D. If a milk replacer is used, it is usually best to use a commercially mixed product consisting primarily of milk products.

E. Thoroughly wash all milk-handling equipment between each feeding.

F. Early in the milk-feeding period, make available to the calves a palatable, nutritious mixture of concentrates so they can become accustomed to eating dry feed as soon as possible.

G. A satisfactory concentrate mixture may be prepared from almost any combination available of rolled corn, rolled grain sorghum, rolled oats, and rolled barley used along with about 10% wheat bran, 10% soybean meal, 10% cottonseed or linseed meal, 1% defluorinated phosphate, and 0.5% trace-mineralized salt.

H. Feed calves milk or milk replacer twice daily until they start eating the concentrate mixture, then once-a-day feeding of milk or milk replacer will suffice.

I. After the calves have become well established on concentrates, a limited amount of leafy, good-quality hay should be made available to them.

J. Milk feeding may be terminated at 4 to 5 weeks of age if the calves are consuming concentrates at the rate of 1.0–1.5 lb per 100 lbs of body weight daily plus hay free choice. If the hay is not of good quality or the calves do not consume as much as 1.5 lb per 100 lb body weight daily, more than the above amount of concentrates may be needed for normal growth.

K. After the calf is 1.5–2 months old, the protein level of the concentrate mix may be reduced to about 15%.

L. The calves should be continued on this feeding program until they are about 6 to 8 months old and then placed on a conventional stocker program.

M. If the objective of the calf feeding program is the production of veal calves rather than the production of replacement animals, then the calves should simply be full fed on whole milk or milk replacer with twice a day feeding until they are 6–8 weeks old.

VI   Feeding replacement dairy heifers on drylot rations.

A. While replacement dairy heifers are usually handled (from about 6 to 24 months of age) on pasture, it is sometimes necessary to feed them drylot rations during this period.

B. If handled in drylot, such animals should be fed primarily on fairly good-quality roughage (hay, haylage, or silage) plus supplemental concentrates as needed to meet their requirements (Fig. 56–6).

1. Feed almost any combination of good quality hay, haylage or silage available.

2. Feed about 2% or a little more of the body weight daily as air-dry hay (or the dry matter equivalent as haylage or silage).

3. Feed from 2–5 lbs of concentrates per head daily consisting of conventional grains and protein meals as needed to balance the ration.

4. For normal growth without fattening, heifers should be grouped by age and size for feeding purposes.

C. The daily roughage allowance of such animals is usually best divided into two feedings but the concentrate allowance is best fed with the evening feed.

D. Such animals should be provided, on a free-choice basis, a mineral supplement consisting of:

1 part trace-mineralized salt
2 parts defluorinated phosphate
  or dicalcium phosphate
  or steamed bonemeal

FIGURE 56–6
Developing heifers receive supplemental native grass hay
during winter pasturing season, in northwest Arkansas.
(Photo by James Herndon)

E.   Such animals may benefit from a supplement of 10,000 to 20,000 IU of vitamin A per head daily.

F.   If partial grazing is sometimes available for such animals, the above allowances of drylot feeds should be reduced according to the amount and quality of pasture available.

G.   There is quite a range in the average age of heifers at first calving. Crandall reported the average age at first calving among 2,865 herds in the Provo DHIA program was 27.3 months. (Crandall, B. 1996. *Secrets of the Top DHIA Performers.* DHIA. Provo, Utah. 1996) Data from the Nebraska Cooperative Extension indicated that a weight of 1,250 to 1,350 lb was the optimum weight for maximum milk production during the first lactation. (Keown. Nebguide G86-819. Univ. Neb. Lincoln. 1986) For heifers weighing 325 lb at 6 months of age, a gain of 1.5 lb per day for the next 21 months is necessary for her to reach 1250 lb by the age of 27 months. So, a good diet must be fed to heifers from 6 months of age to first calving in order to attain maximum lifetime milk production.

VII

Feeding mature dairy bulls in drylot.

A.   Feed mature dairy bulls primarily on any good-quality hay along with a limited amount of haylage or silage, if available, plus concentrates as in B below.

   1.   Feed about 1.5% of the body weight as air-dry hay (haylage or silage may be substituted for about 50% of the hay on a dry matter basis).

   2.   Divide each day's roughage allowance about equally between morning and evening feedings.

B.   A certain amount of concentrate supplementation of the above roughage allowance is usually needed, depending on the amount and quality of roughage fed and the amount of conditioning needed by the bull.

   1.   Almost any combination of corn, oats, barley, and/or grain sorghum will usually suffice; little if any high-protein meal is normally needed if the hay is of proper quality.

   2.   Such animals should normally receive from 5 to 10 lb of grain per head daily.

   3.   A high-level grain allowance is best divided between the morning and evening feedings; a low-level allowance may be fed in a single feeding, probably best with the evening feed.

C.   Dairy bulls may benefit from a supplement of 20,000 to 30,000 IU of vitamin A per head daily.

D.   If partial grazing is sometimes available for such animals, the above allowances of drylot feeds may be reduced according to the amount and quality of pasture available.

57 Veal Production

I General.

    A. The objective in veal production is to provide youthful white beef.

    B. Veal calves are primarily holstein bull calves produced as a byproduct of the dairy industry.

    C. The USDA definition of veal is, "Special-Fed calves fed a milk-based liquid diet throughout the feeding period of 15 to 20 weeks."

    D. Such calves are fed to a slaughter weight (finish weight) of 400 to 450 lb (180 to 200 kg). See Figure 57–1, and note the fleshing and filled-out hindquarters of this finished veal animal.

    E. Other names for this type of animal include "fancy," "formula-fed," "milk-fed," or "nature veal calves."

    F. The number of veal calves has declined by approximately 33% between 1980 and 1994 (1,106,000 vs 739,000).

    G. In the United States, veal production is centralized in 5 states, Wisconsin, Pennsylvania, Indiana, Ohio, and New York, which account for 85% of the total production.

Veal Consumption per Capita per Year (Boneless Equivalent) in the United States, 1960–1998

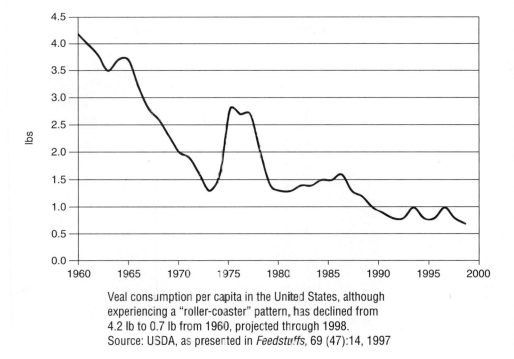

Veal consumption per capita in the United States, although
experiencing a "roller-coaster" pattern, has declined from
4.2 lb to 0.7 lb from 1960, projected through 1998.
Source: USDA, as presented in *Feedstuffs*, 69 (47):14, 1997

H. The remaining 15% is spread out over Michigan, Minnesota, Iowa, Maryland, Delaware, Kentucky, West Virginia, Vermont, Mississippi, and Missouri.

II

Principles of veal production.

A. Most veal calves are started on a milk replacer program at 5 to 7 days of age, at a weight of 100 lb.

B. The duration of the veal feeding program is 18 to 20 weeks (126 to 140 days.)

C. The feeding program is divided into two periods. All feed analyses that follow are based on an air-dry matter basis. The first period (starter) is 4 weeks in length during which a higher protein (20%) and a lower fat (16%) is fed. The second period (finisher) is 14 to 16 weeks long, at which time a lower protein (16%) and a higher fat (18%) is fed. However, in weeks 5 and 6, a blending of the 2 diets takes place to gradually accustom the calves to the change.

D. Dilution of the air-dry feed in water will be from 12 to 12.5% solids for the first 5 weeks. Concentration is calculated as follows: lb of feed per calf divided by total

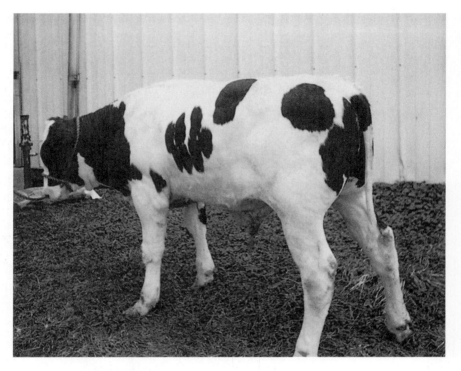

FIGURE 57–1
Holstein bull calves respond well to the veal nutrition program. (Photo by Dr. Dan Shields)

solution = % solids. During the last week or 2 of the 18 to 20 weeks, the solids content of the liquid replacer will be increased to about 20%.

E. The reason for the milk-product diet is that milk products contain very little iron; typically, veal starter formulations will contain about 50 ppm of iron. It is necessary to have iron for the formation of myoglobin, which in turn is necessary for the formation of the red color in meat. Fortunately for veal production, milk is about the only feedstuff that produces white meat, and it is a very nutritious food for young calves, when it is fortified properly.

F. As indicated by Figure 57–2, veal calves are individually fed and are maintained in individual stalls throughout the 18- to 20-week feeding period.

G. The liquid diets are fed in individual buckets (note buckets hanging above calves in Figure 57–2). The concentrated formulations supplied to the producer may be either in air-dry or liquid form to which water will be added in accordance with directions supplied by the manufacturer. Naturally, a justification for purchasing the concentrate in liquid form is that it has avoided the extra expense of drying the milk products to prepare the concentrate.

FIGURE 57–2
Veal calves are fed from individual drinking buckets and are maintained in individual stalls throughout the 18-week feeding period. (Photo by Dr. Dan Shields)

H. Since diets for veal calves must be regulated closely for iron levels and are quite complicated and highly fortified diets, only the most capable milk replacer manufacturer should formulate such products. Therefore, in discussing the actual feeding of veal calves, the next section will address such programs in terms of utilizing commercially prepared formulations.

I. Veal calves will gain an average of 2.5 to 2.8 lb (1.1 to 1.3 kg) per day in 18 to 20 weeks and will have a feed conversion of 1.7 to 1.9 units of feed/unit of gain.

J. Feed requirements will approximate 100 lb (45 kg), air-dry basis, of starter feed and 450 to 500 lb (200 to 225 kg) of finisher feed, for a total of 550 to 600 lb (250 to 275 kg.)

K. The amount of water prescribed to mix with the above amounts of air-dry formula is 2,850 lb, or approximately 5 times as much water as feed. In order for the fat to become emulsified in the mixture, hot water is required.

III

Characteristics of veal feeds and feeding.

A. Since much of veal milk replacer production uses whey portions of milk, a diagram of the partitioning milk is presented (Figure 57–3); milk products used in veal milk replacer formulations are enclosed in boxes in the diagram.

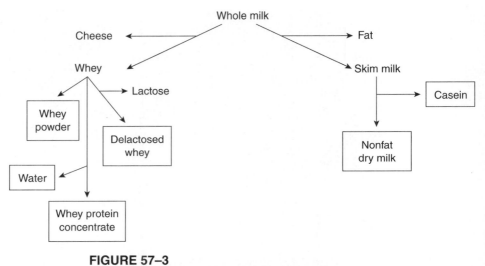

FIGURE 57–3
A diagram of the constituent parts of milk. The units enclosed
in boxes are used in producing veal milk-replacer diets.
(Source: Heinrichs. *Feedstuffs*. May 25, 1992)

B. Veal milk replacer starter is illustrated by accompanying feed tag for "Supreme Starter" in Figure 57–4.

    1. Uniform State Feed Laws require that guaranteed analyses of certain minimums and maximums for a limited number of characteristics must be included on feed tags, i.e., 20% protein, 16% fat, 0.90% to 1.10% calcium, 0.70% phosphorus, and 25,000 IU vitamin A/lb.

    2. The names of the ingredients contained in a feed mixture must also be included on the tag. Note the list of ingredients on the feed tag of the sample illustration.

C. Feeding directions are presented directing the veal manager on the routine which should be followed.

    1. The young calf, on day 1, is offered 0.40 lb of starter mixed in 2.60 lb of water twice daily. By day 5, small increases in starter and water are offered; by the end of day 14, each calf should be receiving 0.80 lb of starter in 5.70 lb of water, twice a day.

    2. By the end of week 4, the level of feeding should be 1.36 lb of starter mixed in 9.89 lb of water, twice daily.

D. Veal milk replacer finisher is introduced in week 5 by being blended with the starter. This transition should be completed during week 7. The finisher is offered, dissolved in water, until the 400-450 lb veals are ready for marketing. (A feed tag

---

SUPREME STARTER

VEAL MILK REPLACER—MILK REPLACER TO BE FED FOR VEAL PRODUCTION

GUARANTEED ANALYSIS

CRUDE PROTEIN.................................. .... MINIMUM .... 20.0 %
CRUDE FAT .... MINIMUM .... 16.0 %
CRUDE FIBER... MAXIMUM .... 0.15 %
CALCIUM .... MINIMUM .... 0.90 %
CALCIUM .... MAXIMUM .... 1.10 %
PHOSPHORUS.......... .... MINIMUM .... 0.70 %
VITAMIN A, IU PER LB .... MINIMUM................................25,000

INGREDIENTS

dried whey, dried whey protein concentrate, vegetable oil and animal fat (preserved with ethoxyquin), lecithin, dried whey products, DL-methionine, L-lysine, vitamin A acetate, D-activated animal sterol (source of vitamin D3), a-tocopherol acetate (source of vitamin E), ascorbic acid, choline chloride, niacin supplement, calcium pantothenate, riboflavin, pyridoxine hydrochloride, thiamine mononitrate, folic acid, menadione dimethypyrimidinol bisulfite (source of vitamin K3), biotin, cyanocobalamin (source of vitamin B12), dicalcium phosphate, calcium carbonate, ferrous sulfate, zinc sulfate, magnesium sulfate, manganese sulfate, copper sulfate, calcium iodate, cobalt sulfate, sodium selenite

DIRECTIONS FOR USE: SEE FEEDING SCHEDULE GUIDE BELOW.

50 LB NET WEIGHT (22.68 KG)

FIGURE 57–4
A typical feed tag for the starting nutrition program for veal calves. (Tag courtesy of Dr. Dan Shields, Vice-President, American Feeds and Livestock Co.)

is not presented for the finisher since it is so similar to the starter, except that it contains 16% protein and 18% fat, plus up to a maximum of 0.20% crude fiber.)

1.   By the beginning of week 8, each veal animal should be receiving 2.10 lb of finisher mixed in 11.90 lb of water, twice per day.

2.   By the end of week 12, veal consumption should consist of 2.91 lb finisher dissolved in 13.08 lb water, twice daily.

3.   On the last day of week 18, consumption should have increased to 3.25 lb of finisher in 13.15 lb of water, twice a day.

E.   Supplying water is especially critical in the feeding program for veal calves.

1.   Young calves should be offered water every day, between feeding, for the first 21 days.

2. After 21 days on the program, 8 lb of warm water should be offered to the calves at least 2 times per week. On hot days leave water in front of the calves—the older the calves, the more critical this becomes.

IV Typical blood picture of developing veal calves.

A. Blood profiles are only a guideline. However, a minimum of 10% of the calves should be sampled at least 3 times at weeks 3–4, 7–8, and 11–12, but all calves should be sampled week 11–12, in order that final adjustments can be made to move calves to desired market levels.

B. A weekly blood sampling might give a set of readings like those listed below:

Age (days)	WBC1	RBC2	HGB3	HCT4
14	9.7	8.4	10.7	31.7
28	8.3	8.0	9.5	27.7
42	8.7	8.1	9.3	26.8
56	9.1	8.3	9.2	26.3
70	8.6	8.3	9.1	25.1
84	8.7	8.2	8.7	24.1
98	8.6	8.0	8.5	23.1
112	8.7	7.7	8.2	21.6
126	8.6	7.3	7.9	20.4

[1]White blood cells (leukocytes)
[2]red blood cells (erythrocytes)
[3]hemoglobin
[4]hematrocrit

V Nutritional and disease problems.

A. Bloat may occur when the veal calf drinks liquids that are either too hot or too cold. Irregular feeding times, varying quantities of feeding liquids, or improper mixing temperature can also cause bloat.

B. Stress can precipitate health problems. Stresses include undue exposure in the sale barn; heat exhaustion; parasites, especially lice; and mild dehydration due to lack of adequate water.

C. Elevated WBC levels can indicate a bacterial infection, whereas lowered WBC levels can indicate a viral infection.

D. HGB and HCT are good tools to evaluate the iron status of the calf.

E. White muscle disease can result from a failure to treat with selenium and vitamin E upon arrival.

F. Scours (loose bowel movement) are normal in a veal barn, but they should be monitored and treated in accordance with veterinarian recommendations.

G. Lice and/or mange may appear on veal calves and should be treated to eliminate them.

H. Animal health products may be prescribed by your veterinarian and proper withdrawal procedures should be followed.

I. Most group medications are active in the GI tract of the calf and are quite effective in reducing both the good and bad bacterial population. In order to repopulate the GI tract with the desirable bacteria, it is good management to administer a good probiotic.

J. Common bacteria and bacterial problems include pneumonia, septicaemia, E. coli, salmonella, and pasteurella.

K. Common viruses are rota, corona, infectious bovine rhinotracheitis (IBR), bovine virus diarrhea (BVD), parainfluenza 3 (PI3), and bovine respiratory syncytical viruses (BRSV.)

L. Common protozoa parasites include cryptosporidium and giardia.

58 Feeding Horses in Drylot

I | **Digestive system of the horse is unique.**

- **A.** The upper part of the digestive system of the horse is much like that of the monogastric animal.

- **B.** At the lower end of the digestive system of the horse is its large fermentation organ, the cecum, that contains fermentation capability much like that of ruminant animals.

- **C.** Both the rumen and the cecum contain bacteria that can break down cellulose; however, in the ruminants the feed is fermented at the beginning of the digestive tract, whereas in the horse it is digested at the end of the digestive tract.

- **D.** There is a difference in eating rate between ruminants and horses. Whereas the ruminant, with its very large rumen, can eat feed rapidly and store it in its rumen for leisurely rumination later, the monogastric and the horse must eat more slowly. They must not force feed through the digestive tract too rapidly, before digestion is complete.

- **E.** Forcing feed too rapidly through the digestive tract of the horse causes undigested feed to enter the large intestine, with the result that starch residues are fermented too rapidly, causing excessive production of gases.

II **General nutrient requirements of the horse.**

A. Horses can survive very well on good pasture if minerals are supplied. In drylot feeding, consideration needs to be given to nutrient requirements so that a balanced diet can be provided.

B. Even though energy is a nutrient required by horses, it may become a problem. Obese horses may be more susceptible to stress and founder, or laminitis, which is characterized by excess heat and pain in the hooves. The hooves may grow grotesquely so that the animal does not want to move.

C. Depressed appetite is the usual indication of protein deficiency. Synthesis of microbial protein occurs in the cecum and colon of the horse, but it is not known how efficiently such protein is digested and absorbed.

D. Dietary minerals usually are not a problem. however, marked deficiencies of any of the essential minerals will naturally contribute to physiological problems. For example, deficiency of either calcium or phosphorus, or an improper ratio of the two can result in weakened bones. Even though salt is easy to provide, the horse owner may forget this simple need. Prolonged exercise, as in working horses or racing horses, will greatly increase the need for salt. The horse is one of the very few farm animals that sweats, and considerable salt is lost in sweat.

E. Vitamin A should be checked in a horse's diet. A deficiency of this vitamin can result in poor appetite, poor growth, nightblindness, excessive tearing, keratinization of the cornea of the eye and skin, respiratory problems, and reproductive problems. Vitamin D should not be a problem except for horses confined to the stable all day, since exposure to the sunlight will take care of their need for vitamin D. A word of caution should be given relative to vitamin E since it may have been overused with shy breeding mares, but probably to no avail.

III **Feedstuffs for horses.**

A. Horses are excellent grazers, as indicated earlier; in drylot, they like hay and will survive very well on high-quality hay. Table 58–1 is a guide to hay feeding for horses in most situations.

B. Of all the concentrates used for horses, oats are the first choice and shelled corn is second. Oats with the fibrous hull will form a loose mash in the horse's digestive tract; wheat, because of its bulky nature is used as part of the concentrate. Because of the tendency of wheat bran to absorb water, it's use in the diet will aid in bowel regulation. In areas where dryland farming is practiced, barley is a popular grain for horses.

C. There are many commercial horse feeds available.

1. **Complete feed.** Those who own only one or two horses may wish to purchase commercial feeds instead of working with several feeds. There are com-

Table 58–1

**HAY FEEDING GUIDE FOR HORSES NOT ON PASTURE,
IN POUNDS PER 100 POUNDS BODYWEIGHT**

Type of Horse	Alfalfa Hay		Legume-grass Hay
Stallions			
Breeding season	1 to 1.5	or	1 to 1.5
Nonbreeding season	0.7 to 1.5	or	0.7 to 1.5
Pregnant mares			0.7 to 1.5
Lactating mares	1 to 1.7		
Suckling foals	0.5 to 0.7		
Weanling foals			1.5 to 2.0
Yearling foals			0.5 to 0.7
Two year olds	1.0 to 1.5	or	1.0 to 1.7

plete feeds, usually in the form of large pellets, cubes, or wafers, that contain roughage, concentrate, and supplemental minerals and vitamins. They usually contain 12 to 14% protein and usually are used for horses 1 year or older.

2. **Concentrate.** Horsemen who grow their own hay may purchase the rest of the horse diet in a concentrate mixture containing ground and mixed grain, supplemental protein, and minerals. The protein in this will probably be 14 to 15%. It is fed along with their own hay.

3. **Protein supplement.** As the name implies, this is meant to supplement both hay and grain. The protein content is about 25%, with directions to add 0.5 to 0.75 lb to the grain and hay/day.

4. **Sweet feed.** This usually consists of a mixture of a low-grade roughage, such as oat hulls, and molasses—possibly 10%. It adds a pleasant aroma to the horse diet.

IV   Feeding management.

A. The part of a horse's ration over the 1% of its body weight as hay may consist of additional hay, or of concentrates if they are needed to meet the animal's energy and/or protein needs, or if concentrates are cheaper than hay as a source of nutrients. Some concentrates widely used in feeding horses are:

Oats	Wheat bran
Corn	Linseed meal
Barley	Soybean meal
Grain sorghum	Cottonseed meal

B. While the hay and concentrates of a horse's ration are sometimes ground and mixed together, because of the danger of overeating and the possible occurrence of colic, horses are not self-fed but are hand fed in either 2- or 3-times-a-day feedings on an individual basis.

C. While pelleting horse rations makes them less dusty, less bulky, and more palatable, it tends to destroy most of their roughage value and makes them more conducive to causing colic.

D. On the other hand, the coarse grinding of a horse's feed, without pelleting, does not greatly destroy its roughage value and tends to reduce the wear necessary on the horse's teeth to ready the feed for passage through the digestive tract.

E. Dustiness of a horse's ration may be controlled by adding about 3% to 5% molasses at mixing time or by dampening with water at feeding time.

F. While horses will eat haylage and silage, the pockets of mold frequently present in these feeds can cause colic in horses; consequently haylage and silage are not widely used as horse feeds.

G. Feed each horse according to its needs, feed regularly, and make any necessary changes in the ration gradually.

H. Reduce the grain allowance of horses on nonwork days by about 50% to avoid the occurrence of azoturia (Monday morning sickness).

I. Working or performing horses will need extra nutrients (Figure 58–1).

FIGURE 58–1
Working and performing horses will need extra nutrients—
especially energetic feeds.

V — Feeding open mares and geldings.

A. Feed open mares and geldings mainly in accordance with their size and activity, as outlined below:

	% of Body Wt, Air-Dry Basis		
	Minimum Hay	Additional Hay	Grain
Idle	1.0	0.75*	0.0
Light work	1.0	0.5*	0.5
Medium work	1.0	0.0	1.0

*This hay may be replaced with grain at the ratio of 0.6 lb of grain per lb of hay.

B. Such animals will require little, if any, protein supplement.

C. Provide such animals the following on a free-choice basis:

1. A mineral mixture consisting of:

> 1 part trace-mineralized salt
> 2 parts defluorinated phosphate
>   or dicalcium phosphate
>   or steamed bonemeal

2. A source of plain, loose salt (NaC1) to make it possible for a horse to replace salt lost from his body through perspiration without forcing the animal to consume unneeded calcium, phosphorus, and trace minerals as would be necessary if it had to rely on the overall mineral mixture for all of its salt needs.

VI — Feeding pregnant mares.

A. During the first 8 months of pregnancy, feed pregnant mares about the same as open mares, assuming they were fed an adequate ration while open.

B. The ration for mares during the last 3 months of pregnancy should be increased over that for open mares as follows:

1. Increase protein allowance by about 20% to 25%.

2. Increase energy allowance by about 5% to 10%.

3. Double the vitamin A allowance.

4. While the daily needs of the pregnant mare for Ca and P will increase during the last 90 days from the first 8 months from about 17% to as much as 33%,

these increased requirements should be provided by *ad libitum* consumption from a complete mineral mix.

VII   Feeding mares in lactation.

  A. In balancing a ration for a mare nursing a foal in drylot, the first step, as with balancing all horse rations, is to provide the minimum hay requirement—1% of the body weight daily as air-dry hay, preferably about 1/2 legume.

  B. As much grain (oats, corn, and/or barley) should then be provided as necessary to meet the mare's needs for energy. This will amount to a total grain allowance equal to approximately 1% of the mare's body weight daily.

  C. If it is then determined that additional protein is still needed, sufficient protein meal (soybean, linseed, or cottonseed meal) should be substituted for a like amount of grain already in the ration to make the protein, as well as the energy, adequate.

  D. Sufficient defluorinated phosphate, dicalcium phosphate, or steamed bonemeal should be added to make the phosphorus adequate and then sufficient ground limestone or oystershell flour added to provide any calcium still needed. If possible, these should be mixed into the grain mixture and force-fed to the lactating mare.

  E. From the standpoint of the health of both the mare and her foal, some supplemental vitamin A (10,000 to 30,000 IU, depending on the size of the mare) should be included in the mare's daily ration.

  F. Also include in each mare's grain mix daily, 0.1 lb of trace-mineralized salt, and, in addition, make plain salt available for free-choice consumption.

VIII   Feeding foals.

  A. If a mare, following foaling, milks normally, she will do an adequate job of meeting the foal's nutritional needs for the first 3 to 4 months.

  B. For up to 3–4 months of age, feeding the young foal is mostly a matter of acquainting it with feeds such as it will need to eat later on in order to meet its nutritional requirements.

  C. Starting when the foal is about 2 to 3 weeks old, it should be provided, within a creep, a small amount (less than a pound per head daily) of the concentrate mixture it will be receiving in larger amounts when it is 3 to 6 months old. Any uneaten feed should be removed and replaced daily with fresh feed.

  D. The grain mixture (E, below) should be provided in increasing amounts along with the mare's milk and a small allowance of good-quality hay until the foal is receiving 0.75 to 1.00 lb daily per 100 lb liveweight at weaning time at about six months of age.

E.   A concentrate mixture that would serve effectively for the above purpose is as follows:

Rolled oats	40.0
Cracked corn	20.0
Soybean meal	15.0
Wheat bran	10.0
Alfalfa meal	10.0
Molasses	3.5
Ground limestone	0.5
Defluorinated phosphate	0.5
Trace-mineralized salt	0.5

IX   Feeding weanlings, yearlings, and 2 year olds in drylot.

A.   Balancing rations for weanlings, yearlings, and 2 year olds is essentially the same in principle as for older classes of horses.

B.   Essentially the same feeds are used for feeding weanlings, yearlings, and 2 year olds as are used for feeding older horses; however, the better-quality feeds should be reserved for the younger animals.

C.   While circumstances sometimes make it necessary to feed young horses in drylot, it is much better for them to be grown out on pasture.

X   Table 58–2 presents a summary of programs for feeding horses.

Table 58–2
SUMMARY TABLE OF FEEDING HORSES

Category	Weight lb	Grain Mix/100 lb, Per Day, lb	Hay/100 lb/day, lb
Pregnant mares	900–1400	1.0	1.0 (grass-legume)
Stallions	900–1400	1.0	1.0 (grass-legume)
Foals, under 6 months	100–350	0.5–0.75	Free choice (legume)
Weanlings	350–450	1.0–1.75	Free choice (grass-legume)
Yearlings (May–Nov.)	450–700	—	Pasture
Yearlings (2nd winter)	700–1000	0.5–1.0	f.c. (grass-legume)
Light horse at work	900–1400	0.5–1.0	1.5–1.75
Mature, idle	900–1400	—	1.5–1.75 (winter, pasture, summer)

59 Feeding Chickens

<u>I</u>   **Essentials of a chicken diet.**

    A.   The first consideration is protein—more specifically, amino acids.

    B.   Even though the current NRC Bulletin (1994) lists a range of 23% protein for broiler diets down to 15% for growing chicks, the real emphasis is on needs for 11 essential amino acids, with the opportunity to substitute for a portion of 2 of the 11.

    C.   Energy requirement has a very high priority in setting up the essentials of a chick diet; in fact, energy needs will be met before protein needs, even if it means de-aminizing amino dietary protein to use it for energy.

    D.   Naturally, in nutrition, one must not assume that one nutrient is more important than another, because each is like a link in a chain—remove one nutrient and the entire chain is useless.

    E.   Minerals assume a very important role in all animal nutrition. However, because of the critical nature of the egg shell (largely calcium carbonate), calcium and other minerals related to egg shell formation assume an increasingly important role in poultry nutrition and physiology.

F.   Vitamins, even though required in such small quantities, are an extremely important consideration in any listing of poultry nutrient requirements. A good example is that vitamin D is critical for the absorption and utilization of calcium, which is so critical for egg shell formation.

II   Factors to consider in selecting feeds for poultry formulations.

A.   Even though we assume the chicken has a very dull sense of taste, still there seems to be some differentiation in their selection of feedstuffs—if they are given the opportunity to select.

1.   Wheat seems to be high on their preference list.

2.   The following whole grains, in decreasing order of preference, have been suggested: corn, kafir, barley, oats, sunflower seeds, peas, rye, and rough rice, with buckwheat not being eaten readily.

3.   Of the ground feeds, corn is first choice, followed by wheat middlings, wheat bran, meat products, and milk.

B.   Digestibility is critical, but fortunately this aspect seems to be tied to the order of acceptability listed above.

C.   The quality of chicken-marketable product is extremely critical, and it can be affected by feedstuffs consumed.

1.   Yellow coloring in body fat and legs and shanks of broilers is critical. This is derived from the xanthophyl in green and yellow feedstuffs such as grasses, dehydrated alfalfa meal, yellow corn, and corn gluten meal. Synthetic xanthophyl is available, also.

2.   Egg yolk color should be bright yellow. Highly undesirable is the olive green yolk in eggs taken from storage and caused by feeding gossypol (a cottonseed meal derivative) to the hens.

3.   Highly unsaturated fats in the diet can result in undesirable oily fat in the carcasses.

4.   Deficient dietary protein can cause smaller eggs; extra dietary protein will cause slightly larger eggs.

5.   Inadequate minerals and inadequate vitamin D can detract from egg shell breaking strength.

6.   Breaking strength and ash content of egg shells are related to adequacy of dietary manganese.

7.   Interior quality of eggs is not affected too greatly by diet except that rye and barley in the formulation might tend to cause thinner whites, and thus "older-looking" eggs.

III

Formulating for young chickens.

A. Such formulations usually consist of a combination of cereal grains and animal protein concentrates, plus lesser amounts of supplemental minerals and vitamins.

B. As indicated earlier, chicks do not have a protein requirement, *per se,* but rather a balance of 11 essential amino acids, plus additional amino acids to meet their total nitrogen needs.

C. Energy may be the limiting factor for the growth of egg-strain chicks reared under most environmental conditions.

D. Assuming no amino acid deficiency and an intake of 1 kg (2.2 lb) protein from hatch to 20 weeks, intake of 21 Mcal ME seems to be ideal.

E. Just before hatching, chicks absorb the unused yolk and come from the shell with adequate energy to last them for at least 3 days.

F. There is no danger to exposing newly hatched chicks to feed right after hatching—they probably will not eat any of it, anyway.

IV

Formula prototypes for egg-layer chick mashes. (Table 59–1) As indicated above, prototype diets are not meant to be rigid guides, but rather provide data that can assist in evaluating proposed diet formulations and in starting from scratch in making such formulations.

V

Feeding for top egg production.

A. Genetic selection has resulted in hens whose performance can be predicted rather closely.

B. The student will read many times in this text that the real requirement is not for protein, *per se,* but rather for essential amino acids plus minimal protein for adequate synthesis of nonessential amino acids. However, as rapidly as the hen's genetics are changing through selective breeding, there exists such gaps in the exact science of a laying hen's amino acid requirements that total protein must be included in dietary specifications.

C. Due to genetics and nutrition improvements, days of age at maturity have declined for several years at the rate of 1 day per year. Therefore, pullets that are at an anticipated mature weight 2 to 3 weeks earlier than used to be accepted, will be the most profitable birds.

D. An 18-week mature weight for pullets of 1380 g (3 lb) vs 1100 g (2.5 lb) resulted in 49.7 g early egg weights vs 46.9 g (Summers and Leeson. *Nutr. Rep. Int.* 40:645). Therefore, rapidly developing pullets cannot be emphasized too fully.

Table 59–1
FORMULA PROTOTYPES FOR CHICK MASHES
ONE METRIC TON BASIS (ENGLISH TON)

Feedstuff Classification	Starter (All-Mash)		Grower (All-Mash)	
	Mton	(Ton)	Mton	(Ton)
	kg	lb	kg	lb
High-energy grains, (corn, wheat, milo, oat groats)	450+	1,000+	365+	800+
Fibrous grains, (oats, wheat bran, wheat midds, barley)	0–135	0–300	0–275	0–600
Vegetable protein, (soybean meal, corn, gluten meal, peanut meal)	115–180	250–400	160–215	350–475
Animal protein, (fish meal, fish solubles, meat scraps)	35–55	75–125	35–55	75–125
Dehydrated alfalfa meal	0–45	0–100	0–45	0–100
B-vitamin carriers, (dist. solubles, milk by-products)	45	100	45	100
Calcium and phosphorus, (bonemeal, dicalcium phosphate)	14–28	30–60	28–40	60–90
Salt	4.5	10	4.5	10
$MnSO_4$ (64% feed grade)	0.18	0.4	0.18	0.4
Premix (B-vitamins, Vit A, D_3, K, trace minerals)	+	+	+	+

Analysis

Protein	18%	15.5
Calcium	0.85%	0.8%
Non-phytate phosphorus	0.38%	0.33%
Fiber	max. 4%	max. 5%
Metabolic energy	2,860 kcal/kg (1,300 kcal/lb)	2,860 kcal/kg (1,300 kcal/lb)

E. Length of day will influence growth of pullets and also quantity and quality of eggs.

F. Heat stress is a deterrent to egg production, probably caused by layers decreased desire for feed as well as using increased energy trying to keep cool.

G. Using enhancers such as molasses or increased levels of fats, to encourage increased energy and protein consumption is a possibility.

H. During periods of heat stress, it is critical to try to regulate calcium levels in the diet so that they are consuming at least 2.5 g of calcium/day.

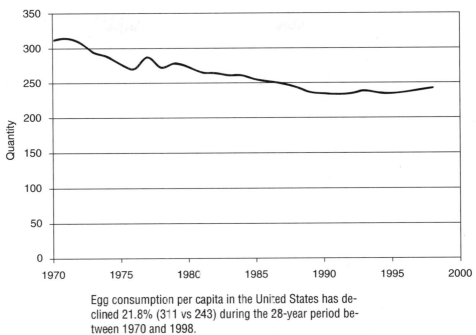

Egg Consumption per Capita per Year in the United States, 1970–1998

Egg consumption per capita in the United States has declined 21.8% (311 vs 243) during the 28-year period between 1970 and 1998.
Source: USDA, as presented in *Feedstuffs,* 69 (47):14, 1997

I. Energy is one of the hen's basic requirements. The cereal grains are excellent sources of energy and may make up 75% of the total formulation. Laying rations usually are built around 2 or more cereal grains of corn, oats, barley, wheat, and kafir. Energy levels can be boosted by raising the proportions of high-energy grains to fibrous grains, or by adding fat.

J. Protein quality should be high. A combination of 2 or more sources of supplemental protein, including animal sources of protein, such as fish meal, meat scraps, or poultry by-product meal are excellent as a portion of the supplemental protein (Table 59–2).

K. Although plant-source proteins (soybean meal) constitute the major portion of the supplemental protein, cottonseed meal should be used sparingly because of the effect of gossypol contained therein on egg yolk color after cold storage.

L. Crystalline methionine is economical enough to use as supplemental protein in typical corn-soy diets.

M. Naturally, vitamins should be in plentiful supply in the formulation, such as vitamin D_3 for optimal calcium utilization. Niacin may be borderline in corn-soy diets.

N. The critical need for minerals cannot be emphasized too strongly; calcium for egg shells; 0.5% to 1% salt; 0.4% manganese sulfate; and of course, hard grit so that the gizzard can conduct its grinding function.

Table 59–2
FORMULA PROTOTYPE FOR LAYER MASHES
(MTON AND ENGLISH TON, BASES)

Feedstuff Classification	Complete Mash kg	lb
High-energy grains (corn, wheat, milo, oat groats)	227+	500+
Fibrous grains (oats, wheat bran, wheat midds, barley)	0–365+	0–800
Vegetable protein (soybean meal, corn gluten meal, peanut meal)	113–136	250–300
Animal protein, (fish sols, poultry by-products, meat scraps)	23–45	50–100
Dehydrated alfalfa meal	45	100
Calcium and phosphorus, (bonemeal, dical, limestone)	45	100
B-vitamin carriers, (dist. sols., milk by-products)	23	50
Salt	9	20
$MnSO_4$ (64% feed grade)	0.23	0.5
Premix, (B-vitamins, vitamins A, D, K, and trace minerals)	+	+

Analysis	Grams feed intake/hen/day		
	80	100	120
Protein, %	18.8	15	12.5
Calcium, %	4.06	3.25	2.71
Non-phytate phosphorus, %	0.31	0.25	0.21
Metabolizable energy, kcal/kg	2900	2900	2900
(kcal/lb)	1300	1300	1300

O. The all-mash diet is an ideal method for providing a well-balanced diet. Its main drawback is that the all-mash method is more expensive than the combination of grain plus mash.

P. The combination system consists of keeping a 20% to 22% protein formulation in front of the hens at all times, and then hand feeding or hopper feeding a 10% protein grain mixture. However, the total protein of such method should not fall below 15%.

VI Broiler feeding.

A. Many of the innovations in broiler feeding occurred in the 1940s, for example high-energy diets, use of supplemental methionine, discovery of vitamin B_{12}, use of feed

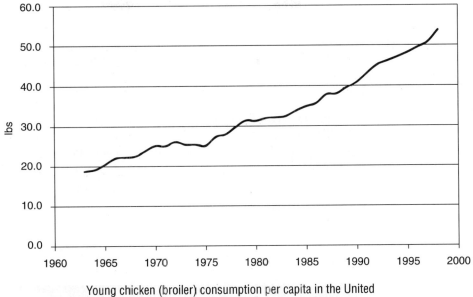

Young Chicken Consumption per Capita per Year
(Boneless Equivalent) in the United States, 1963–1998

Young chicken (broiler) consumption per capita in the United
States has had a steady increase of 189% between, 1963
and 1998 (18.7 vs. 54.1 lb).
Source: USDA, as presented in *Feedstuffs*, 69 (47):14, 1997

grade fats, calorie-to-protein ratios, use of linear programming to formulate least-cost formulations, discovery of nicarbizin, and the discovery of antibiotics. All of these contributed a great deal, which, when coupled with genetically superior growing birds, started the great growth in broiler production.

B. Protein sources for broilers are restricted primarily to soybean meal; fish meal is desirable but its cost limits its use. Poultry by-product meal has the added bonus of considerable fat for use in broiler diets.

C. Hydrolyzed feather meal is not too well used by broilers and its use should be limited to no more than 3% of the formulation.

D. Methionine, or a combination of methionine and cystine, constitutes the first limiting amino acid(s) in broiler formulations; lysine probably will be the second limiting amino acid.

E. Because of the rapidity of growth capable with broilers, energy becomes one of the limiting nutrients. Corn is the highest in energy of the common feed grains and will constitute the major feed grain to be utilized; corn also contributes a bonus in the form of xanthophyl to supply yellow coloring for body fat, and shank and leg color.

Current NRC recommendations suggest broiler diets should contain 3200 kcal ME_n/kg diet (1450/lb).

F. Fats are now an integral part of broiler diets. The usual added level is from 1% to 5%. Broilers do not use saturated fats as well as they do the less-saturated fats.

G. Producers plan to produce broilers that weigh, on average, more than 4 lb (1.8 kg) at 6 to 7 weeks of age, with a feed conversion of 2 lb (kg) feed/lb (kg) gain, and a mortality rate of less than 3%. With new advances, improvements of all these figures are becoming commonplace.

H. Some operators use a 2-stage diet program (starter and finisher) for broilers. There has been a marked trend toward at least 3 stages in recent years (starter, grower, and finisher) in order to make operations more profitable. Profits in broiler production are, at best, quite narrow.

I. In the 3-stage program, the starter program is used for 3 to 4 weeks, the grower for about 2 weeks, and the finisher diet for the remainder of the feeding period (Fig. 59–1).

FIGURE 59–1
Several thousand finished broiler chickens that were scheduled to be shipped to slaughter the evening this picture was made.

Table 59–3
EXAMPLE BROILER DIETS (2-PHASE PROGRAM)
ONE KILOGRAM MIX (1-TON MIX)

Ingredient	Starters (0–4 wks) kg	lb	Finishers (5–7 wks) kg	lb
Yellow corn	548	1,206	582	1,281
Soybean meal (48%)	219	389	177	389
Meat and bonemeal, 50%	64	141	56	123
Fish meal, 62%	23	50	23	50
Corn gluten meal, 60%	19	41	25	55
Fermentation by-product	2.3	5	2.3	5
Animal-vegetable fat	25	55	35	76
Salt	2.7	6	3.2	7
Vitamin mix[a]	4.5	10	4.5	10
Trace-mineral mix[b]	0.9	2	0.9	2
Methionine, DL	1.4	3	0.9	2
Analysis				
Metabolizable energy kcal/lb)	3,190 kcal/kg(1,450 kcal/lb)		3,300 kcal/kg	(1,500
Protein	23.0%		21.0%	
Fat	6.1%		7.2%	
Fiber	2.6%		2.6%	
Calcium	0.9%		0.8%	
Non-phytate phosphorus	0.54%		0.49%	

[a]Vitamin mix

[b]Trace mineral mix

Vitamin	Amt./kg of mix	Amt./lb of mix	Mineral	Percent/kg, lb of mix
Vitamin A	990,000 IU	450,000 IU	Manganese	6.5
Vitamin D$_3$	220,000 ICU	100,000 ICU	Zinc	6.5
Vitamin E	440 IU	200 IU	Iron	3.0
Choline	88,000 mg	40,000 mg	Copper	0.3
Pantothenic acid	1,760 mg	800 mg	Iodine	0.06
Riboflavin	880 mg	400 mg	Selenium	0.005
Niacin	6,600 mg	3,000 mg		
Folic acid	110 mg	50 mg		
Vitamin B$_{12}$	1,980 mcg	900 mcg		
Menadione sodium bisulfite (Vit. K)	660 mg	300 mg		

60 Feeding Turkeys

I   Starting and growing turkeys.

A. The nutrient requirements of turkeys are divided into the needs of birds that are used as a source of growth and those used for reproduction, in Nutrient Requirements of Poultry, ninth rev. ed., 1994. Nat'l. Acad. of Sci., Washington, D.C. These 2 categories differ largely in the proportion of nutrients devoted to productive use as opposed to those used for maintenance activities.

B. Substantial improvements in rates of gain and feed efficiencies of commercially available strains have occurred in the last decade.

C. These improvements have made it increasingly difficult to stay current with nutrient requirements because such recommendations must be based on feeding trials in order for them to be more than educated estimates.

D. The nutrient requirements published are "based on earlier research and the chronological age of experimental turkeys used at that time."

E. For the most part, these earlier recommended nutrient levels are still being employed by the turkey industry—they're just being initiated at an earlier age. Commercial experience has been sufficiently good that producers seem satisfied with that adjustment.

Turkey Consumption per Capita per Year (Boneless Equivalent) in the United States, 1960–1998

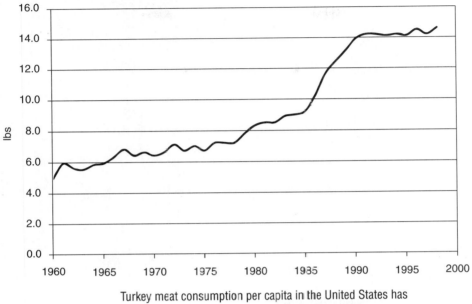

Turkey meat consumption per capita in the United States has experienced a 198% increase during the span of 38 years between 1960 and 1998 (4.9 vs 14.6 lb).
Source: USDA, as presented in *Feedstuffs*, 69 (47):14, 1997

F. The requirements are expressed concentrations in the feed; therefore it is critical that adequate intake of the ration occurs so that adequate intake of the nutrients is ensured.

G. Pelleting of the ration is used quite extensively and may increase nutrient digestibility in some constituent feedstuffs. Generally, pelleting facilitates feed intake, increases net energy of production from metabolizable energy, and reduces overall feed wastage. These benefits appear to be accentuated as feed nutrient level decreases and as birds become progressively older, provided the feed remains in pelleted form.

H. The 1994 NRC Bulletin recommends a 5-step protein level stepdown, starting at 28%, then going to 26% at 4 weeks, 22% at 8 weeks, 19% at 12 weeks, 16.5% at 15 weeks, and 14% from 20 to 24 weeks.

I. If the protein (amino acid) level is adequate, poults are most responsive to energy.

J. Metabolizable energy consumption per week will mirror the very rapid increase in size of growing poults; week 1—0.28 Mcal; week 4—2.0 Mcal; week 8—5.0 Mcal; week 12—9.0; week 16—12.3; week 20—15.2; and week 24—17.4 Mcal.

K. In corn-soy-based formulations, the sulfur amino acids, primarily methionine could be limiting, suggesting supplemental methionine.

L. Although the economic benefits for added fat in poult formulations appear to be questionable, fat fortification might increase total feed consumption slightly. How-

FIGURE 60–1
Seven thousand turkey poults in a 300-foot confinement
building. They have been on feed for 10 weeks.

ever, in hot weather, added fat may result in some growth rate stimulation. Optimal
weight gains for poults seem to occur at a temperature of 50° to 70°F (10° to 18°C),
whereas at elevated temperatures of 75° to 95°F (27° to 35°C), reduced rates of gain
may be anticipated.

M. Poults require vitamin A at a much greater level/kg diet than chickens (5000 IU vs
1500 IU for broilers).

N. Poults are grown to market size and weight in large confinement sheds (Fig. 60–1).

II

Turkey breeders, hens.

A. Excess bodyweight is less a problem in turkey hens than it is in toms because an
extensive loss in weight occurs in hens when egg laying progresses.

B. Inadequate weight gain prior to stimulatory lighting delays onset of the lay and reduces total egg production.

C. Size and condition at maturity are the most reliable indicators of successful reproductive performance.

D. In attempting to keep the hens from becoming obese by limiting feed intake, it is critical
that calcium and phosphorus do not become deficient or skeletal growth can be limited.

E. If the tendency toward obesity progresses, more desperate measures need to be taken, such as not feeding the hens either 1 or 2 days per week. This technique might be easier than trying to limit daily intake of feed to 90% of *ad lib.*

F. In planning for the first egg at 32 weeks of age, one should induce molting at 27 weeks of age by halving the day length down to about 6 hours; then come back with light stimulation at about 30 weeks of age.

G. As the hen progresses into egg laying, she declines in feed consumption to approximately 67% of what she was consuming before egg laying. Naturally, this decline in feed consumption, coupled with egg laying, results in marked body weight loss.

III Turkey breeders, toms.

A. Through the first 12 to 16 weeks of age, male and female turkeys being grown for reproductive purposes generally have been fed the same diet as birds intended for meat production. Thereafter, various efforts will be needed to avoid obesity in the breeders.

B. Naturally, since turkeys have been selected for rate of gain, the breeder males are going to become obese unless their energy intake is controlled. The most obvious way to do this is to feed a less energetic diet. However this practice must not be so severe that it will limit the quality of the semen.

C. Restriction of protein levels—while maintaining adequate amino acid balance—is another approach to controlling the tendency of toms to become obese.

BODY WEIGHTS AND FEED CONSUMPTION OF LARGE-TYPE TURKEYS DURING THE HOLDING AND BREEDING PERIODS

Age Weeks	Females			Males	
	Weight kg	Egg Production %	Feed Per Turkey/d g	Weight kg	Feed Per Turkey/d g
20	8.4	0	260	14.3	500
25	9.8	0	320	16.4	570
30	11.1	0*	310	19.1	630
35	11.1	68	280	20.7	620
40	10.8	64	280	21.8	570
45	10.5	58	280	22.5	550
50	10.5	52	290	23.2	560
55	10.5	45	290	23.9	570
60	10.6	38	290	24.5	580

Note: These values are based on experimental evidence between November and July; it is projected that summer breeders would produce 70% to 90% as many eggs and consume 60% to 80% as much feed as winter breeders. (Source: National Research Council. *Nutrient Requirements of Poultry.* ninth rev. ed., 1994.)
*Light stimulation is begun at this point.

61 Feeding Sheep in Drylot

I — General.

A. While a major portion of the feed for most classes of sheep normally comes from grazing, this particular section of the book deals only with the feeding of sheep in drylot.

B. Sheep, like cattle, are ruminants and should be fed accordingly.

1. Roughages make up from 50% to 100% of drylot sheep rations on a dry basis; the roughages may consist of almost any combination available of good-quality hay, haylage, and/or silage.

2. The remainder of a sheep's drylot ration may consist of almost any combination available of the feed grains, plant protein meals, cereal by-product feeds, and numerous other by-product concentrates that can contribute to a sheep's nutritive needs.

3. If concentrates are to make up more than about 10% of a sheep's ration, then the allowance beyond that point should be increased by about 0.1 lb/hd daily until the full allowance is reached.

Lamb and Mutton Consumption per Capita per Year
(Boneless Equivalent) in the United States, 1960–1998

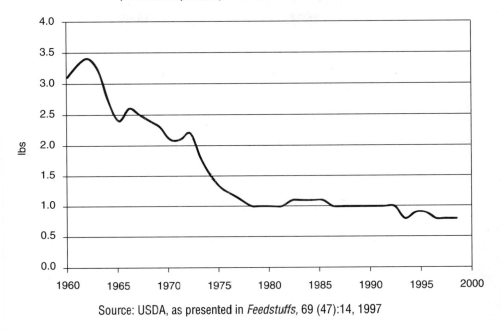

Source: USDA, as presented in *Feedstuffs,* 69 (47):14, 1997

C. Except for broken-mouth ewes (ewes with poor teeth), all classes of sheep chew their feed so thoroughly that there is little advantage realized from grinding it.

D. Sheep are much more sensitive than are cattle to high levels of copper intake so care must be exercised not to supply sheep with too much of this mineral. The requirement of sheep for copper is about 5 ppm of ration dry matter. However, less than 4 times this level has been found to be toxic under certain conditions. Since certain feeds and forages sometimes contain significant amounts of this mineral, then a trace mineral mix for sheep should probably contain no more than 0.1% copper, which should provide the minimum requirement.

E. Should circumstances seem to warrant, grain may be used to replace about half of the roughage in a conventional sheep ration (as described above) using 0.6 lb of grain DM to replace 1 lb of roughage DM. Such substitutions of grain for roughage should be made with extreme caution to prevent the overconsumption of grain and the occurrence of enterotoxemia and related digestive disorders.

II **Flushing ewes.**

A. Most sheepmen plan to have their breeding ewes in a gaining condition (flushing) just prior to and during the breeding season.

B. This practice seems to be conducive to a few more multiple births at lambing time.

C. Flushing can be accomplished during pasture season by turning the ewes to a luxuriant pasture that has been saved for this purpose.

D. If the ewes are in drylot at breeding time, the feeding of 0.5 lb of oats or shelled corn, in addition to their roughage, will accomplish the flushing effect.

III Feeding pregnant ewes—first 110 days.

A. The first 110 days of pregnancy usually are not too critical because the fetus or fetuses are in the first stages of development. Furthermore, if the ewe is in good condition, she need not gain weight during the early stage of pregnancy.

B. Throughout this period, ewes can be fed 2.5 lb of air-dry roughage equivalent, as hay or hay plus silage equivalent. If all the roughage is nonleguminous, then 0.5 lb of soybean meal or cottonseed meal should be fed per ewe daily.

C. In all ewe-feeding situations, be sure to offer a free-choice mineral mixture—protected from the weather—of 1 part dicalcium phosphate, 1 part ground limestone, and 1 part ruminant-type trace mineralized salt.

IV Feeding pregnant ewes—the last 6 weeks of pregnancy.

A. Pregnant ewes should gain 20 to 25 lbs during this period.

B. Pregnant ewes are prone to a condition called ketosis, which usually develops during the last week of pregnancy. This is the period when very rapid growth of the fetus occurs, calling for extra energy from its mother. If the ewe does not have sufficient energy in her ration, she will draw on the fat from her body to provide the extra energy needed for the developing fetus. Body fat is an excellent and available source of energy, but release of too much energy from body fat results in a buildup of 4-carbon fatty acid residues (beta hydroxy butyric acid, acetone, and keto butyric acid), all of which are acidic, in the bloodstream. Since all such compounds are acidic in nature, they overwhelm the acid-base balance of the body causing paralysis. Afflicted ewes have at least a 50:50 chance of dying unless they get relief. An intravenous injection of calcium gluconate may help. Also, feeding 0.5 lb of grain per day during the last 6 weeks of pregnancy would probably prevent the condition.

C. In the lamb-feeding section that follows, mention is made that so-called "stiff lamb disease" is caused by a deficiency of vitamin E and the trace mineral, selenium. Grains such as corn and wheat bran are excellent sources of vitamin E. Therefore, feeding a mixture of corn and wheat bran to the ewe at a level of 0.5 lb per day not only provides energy to avert ketosis, but also boosts the store of vitamin E in the unborn lamb. Based upon the legality of selenium, which has changed from time-to-time, supplemental selenium may be supplied in the "ruminant-type trace-mineralized salt" recommended for the free-choice mineral mixture discussed previously.

V — Feeding the lactating ewe.

A. Ewes may produce more than 4 quarts (3.8 liters) of milk per day at the height of their lactation. In addition, they are continuing to grow wool, so their post-parturient nutritional needs are demanding.

B. They should be given a light feeding the first 3 days post lambing, such as roughage.

C. Thereafter, lactation places about the greatest nutritive stress on an animal as does any body function, so ewes should be fed well during this period.

D. Ewes that are nursing twins should be encouraged to produce more milk than those nursing singles and should receive more nutrients.

E. Continue feeding roughage to lactating ewes in about the same amount as during gestation, plus grain and protein meal as needed to meet the nutrient needs for energy and protein.

    1. Ewes nursing single lambs need about 2 to 2.75 lb of total concentrate per head daily, with normal roughage feeding.

    2. Ewes nursing twins will require about 2.75 to 3.5 lb total concentrate per head daily with normal roughage feeding.

    3. How does one accomplish the feat of feeding ewes with twins more concentrate than is fed those with single lambs? Naturally, if the ewes were all in the same pen, this would be an impossible task. However, the serious shepherd will separate the two sets of ewes so that it is possible to differentiate in their feeding.

VI — Table 61–1 presents a summary feeding program for ewes.

VII — Milk replacers for lambs.

A. Often the sheepman finds himself with orphan lambs or a ewe with triplets, but with no capacity to handle more than two lambs.

B. Fortunately, ideal milk replacers are on the market that can be used for such lambs.

C. Such lambs should be separated from the main group and placed in a warmed (infra-red bulb?) and dry quarter.

D. Formulas are available that can be mixed with water at room temperature and stored in a refrigerator until fed.

E. Lambs do well on a cold milk replacer.

Table 61–1
DIETS SUITABLE FOR EWES IN VARIOUS STAGES OF PRODUCTION[1,2]

Production Stage	Pasture Hay lb	Corn Silage lb	Haylage lb	Oat Straw lb	Stalks lb	Grain lb	Soybean Meal lb
A. Maintenance							
1.	2.5						
2.		6.0					0.2
3.			6.0				
4.				ad lib			0.3
5.					grazing		
6.	grazing						
B. Gestation, first 110 days,							
1.	3.5						
2.	2.5					0.5	
3.	1.8					0.5	0.2
4.		8.0					0.2
5.			7.0			0.2	
6.	2.0				grazing		
7.	1.0				grazing	0.2	0.3
8.	grazing						
C. Gestation, last 6 weeks, add 0.5 lb grain to any of the diets listed above.							
D. Lactation							
1.	4.0					2.5	
2.		10.0				1.5	0.2
3.	1.0	8.0				1.5	0.2
4.				8.0		2.5	
5.	grazing						

[1]Per ewe, daily.
[2]Each feeding situation should be accompanied by free-choice mineral mixture consisting of 1 part dicalcium phosphate, 1 part ground limestone, and 1 part ruminant-type trace-mineralized salt.

F.   Table 61–2 shows the desirable characteristics of a lamb milk replacer (see label on bag.)

VIII   Lamb creep formulations.

A.   Solid feed in the form of a highly palatable concentrate mixture should be made available to all lambs—those on milk replacer as well as those nursing their mothers.

Table 61–2
CHARACTERISTICS OF A LAMB MILK REPLACER,
WITH PERCENTAGE RANGES, ON DRY BASIS

Ingredient	Unit
Fat	30–32%
Milk protein	22–24%
Fiber	less than 1%
Lactose (milk sugar)	22–25%
Ash	5–10%
Vitamin A	500 IU/lb
Vitamin D	50 IU/lb
Vitamin E	20 IU/lb
Niacin	10 mg/lb
Riboflavin	2 mg/lb
Calcium pantothenate	4 mg/lb
Vitamin B_{12}	10 mcg/lb
Aureomycin or terramycin	10 mg/lb
Iron (as ferrous sulfate)	8 g %
Copper (as cupric sulfate)	5 g %

B. University research has demonstrated that lambs as young as 21 days of age may be weaned off milk replacer quite satisfactorily. In fact, lambs weaned at this younger age adapt to a dry nonmilk diet more readily than lambs weaned at an older age, e.g., 35 days of age.

C. Table 61–3 lists formulas for simple and complex creep diets for nursing lambs.

IX   Post-weaning lamb feeding.

A. Nursing lambs that do especially well may have sufficient size and body finish to be marketed right at time of weaning. However, this is rather rare and usually occurs when ewes and lambs are quartered on excellent spring pasture containing lots of alfalfa.

B. Weaned lambs that have not reached market weight and finish should be placed in drylot for a finishing period.

C. The diet during the drylot finishing period consists largely of energy and protein with balancing minerals and just enough roughage to keep the feeder lambs from going off feed, or foundering. Lambs are especially sensitive to excessive energy in their diet and may develop a condition known as overeating disease. This con-

Table 61–3
CREEP FORMULATIONS FOR LAMBS—0.25 INCH PELLETS

Ingredient	Simple Formula %	Simple Formula %	Complex Formula %
Ground yellow corn	88	68	52.3
Dehydrated alfalfa meal	—	20	20
Cane sugar	—	—	10
Dried skim milk	—	—	5
Soybean meal	10	10	10
Dicalcium phosphate	1	1	1
Iodized salt	0.2	0.2	0.2
Premix	0.8[1]	0.8[1]	1.5[2]

[1]Premix per ton: 10 g mixed tocopherols, 5 million IU vitamin A, and 625 thousand IU vitamin D.

[2]Premix per ton: 40 g Aureomycin, 2 g thiamin, 4 g riboflavin, 32 g niacin, 20 g calcium pantothenate, 40 mg vitamin B_{12}, 100 g ascorbic acid, 10 g mixed tocopherols, 5 million IU vitamin A, 625 thousand IU vitamin D, 150 g ferrous sulfate, 12 g cupric sulfate, and 90 g manganese sulfate.

dition will usually result in the death of affected lambs. Fortunately, a vaccination is available from veterinarians that essentially eliminates the hazard of overeating disease.

D. Roughage to concentrate ratio should be close to 50:50. Experienced lamb feeders whose lambs have been vaccinated against overeating disease, can manage a 75% concentrate: 25% roughage feeding program with feeder lambs.

E. Table 61–4 presents some feeding programs for finishing out weanling lambs to market.

F. Modern day lambs on a full feed of a 50:50 roughage-to-concentrate finishing ration should consume feed and gain about as follows:

Wt of Lamb	% of Body Wt as Air-Dry Feed Daily	Total Air-dry Feed Daily	Daily Gain	Feed/Gain
lb	%	lb	lb	lb
66.0	4.78	3.22	0.44	7.32
77.0	4.44	3.44	0.48	7.17
88.0	4.44	3.89	0.55	7.07
99.0	4.22	4.11	0.55	7.47
110.0	4.00	4.44	0.48	9.25
121.0	3.89	4.67	0.44	10.61
Avg 93.5	4.30	3.96	0.49	8.15

Table 61–4
HANDY LAMB-FEEDING GUIDE

	lb/day
Legume hay	1.25 to 1.75
Grain	1.25 to 2
Grass-legume mixed hay	1.25 to 1.75
Grain	1.25 to 1.75
Protein meal	.20
Legume hay	1.25 to 1.75
Corn or sorghum silage	4 to 5
Protein meal	.20
Legume hay	1
Corn or sorghum silage	2 to 4
Grain	1.25 to 1.75
Protein meal	.20
Corn or sorghum silage	3 to 4
Grain	1.25 to 1.75
Protein meal	.20 to .30
Legume haylage	3 to 5
Grain	1.25 to 2
Protein meal	.10

G. Should circumstances seem to warrant, a concentrate to roughage ratio of 75:25 may be used for fattening lambs, but extreme caution will have to be exercised to prevent excessive losses from enterotoxemia.

62 Feeding Swine in Drylot

I   General.

A. Most hogs today are produced under drylot feeding; some producers still run pregnant gilts and sows on pasture.

B. Swine of all classes are fed on concentrate-type rations with very little if any roughage being fed. Some producers still use from 5% to 10% dehydrated alfalfa meal in their swine rations but where this is found, it is primarily with rations for sows and gilts with no access to pasture.

C. Corn is the most widely used high-energy feed for all classes of swine; however, grain sorghum, barley, oats, wheat, and various cereal by-product feeds are frequently used as a partial substitute for corn in swine rations.

D. Soybean meal is the most widely used high-protein feed for all classes of swine; however, peanut oil meal, cottonseed meal, meat meal, fish meal, and tankage are frequently used as a partial substitute for soybean meal in swine rations.

E. Commercial supplements are extensively used in the feeding of swine; there are three general types:

Pork Consumption per Capita per Year (Bone ess Equivalent) in the United States, 1960–1998

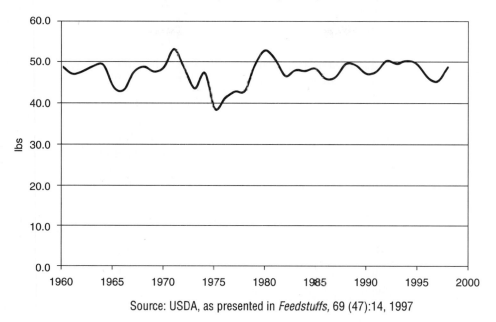

Source: USDA, as presented in *Feedstuffs,* 69 (47):14, 1997

1. Complete supplements. These include the needed protein, minerals, and vitamins and are used along with corn (or a suitable substitute) for a complete ration.

2. Mineral and vitamin supplements. These include the needed minerals and vitamins and are used along with corn and soybean meal (or suitable substitutes) for a complete ration.

3. Vitamin supplements. These are used along with corn, soybean meal (or suitable substitutes), and minerals as needed to produce a complete ration.

F. Hogs are nonruminants and, accordingly, may be self fed on concentrate-type rations with no adjustment period and no danger of causing digestive disorders. However, sows should be brought on to full feed gradually following farrowing to avoid certain metabolic disturbances.

G. Feed costs make up about 75% of the total cost of producing pork.

II

Feeding bred sows and gilts in drylot.

A. Prepare balanced rations for this class of swine, either by building the entire formulation or by purchasing a commercial supplement and using it with one or more of the feed grains to meet the nutrient requirements set forth in the Tables for Swine in the appendix. (Nutrient requirements for most swine feeding situations have been prepared by the Subcommittee on Swine Nutrition, National Academy of Science, and released in their 1998 publication, Nutrient Requirements of Swine. These have been included in the appendix of this text.)

1. In gestation rations, many swine managers like to include 10% oats or wheat bran, or both, to add bulk to the diet; others may not follow this practice.

2. Nutrient characteristics of the gestation diet will vary according to weight of the sow (125 to 200 kg [275 to 440 lb]) and how much she is expected to gain. This includes true body growth for growing gilts plus maternal tissue and products of conceptus (55 to 35 kg [121 to 77 lb]), with gilts gaining more and mature sows gaining less. The NRC Bulletin sets the anticipated number of pigs in the litter at 11 for the lightest gilt 14 for the 200 kg sow, and 12 for the rest. For an average 175 kg (385 lb) sow, the diet should contain 12.4% crude protein (specifically 0.54% total lysine since lysine is the first limiting amino acid in corn-soy type diets) 0.75% calcium, 0.60% phosphorus, and 3,265 kcal ME/kg (1484 kcal/lb). The sow should be fed 1.88 kg (4.14 lb) of this diet/day.

3. Although a well-fortified corn-soy diet should prove to be most adequate for pregnant gilt and sow diets for drylot feeding, some managers think the incorporation of either dehydrated alfalfa meal (10% to 15%) or corn silage is beneficial in terms of numbers of pigs farrowed and pigs weaned. This is an excellent modification of the pregnant sow feeding regimen, where such feeds are economically feasible.

B. The above specifications indicate "limit-feeding" will be needed to keep the sows from getting too fat during the gestation season. No one method for limiting feed intake is proposed, but such methods as hand feeding limited groups or including considerable roughage in the diet may help.

C. Where daily caloric intake is meant to be limited, feeding once per day may help.

D. Some managers like to add up to 5% fat to the gestation diets during the last 3 to 4 weeks prepartum for possible increased survival rate of the pigs and increased milk production of the dams.

III Feeding lactating sows and gilts in drylot.

A. Balanced rations should be formulated following the guidelines set up in the appendix for lactating gilts and sows.

1. Although a combination of corn and soybean meal should provide sufficient energy and protein (amino acids—especially lysine), variations are possible as to sources of these two nutrients.

2. The current NRC recommendations (1998) suggest the following dietary analyses should be observed for the lactation of sows whose post-farrowing weight is 175 kg (385 lb): a protein range from 16.3% to 19.2%; an energy con-

tent of 3,265 kcal/kg (1484 kcal/lb); total lysine ranging from 0.82 to 1.03%; and an air-dry feed intake of 3.56 to 6.40 kg (7.8 to 14 lb) per day, based on anticipated lactation weight change and daily gain of her litter of pigs. (Specific recommendations are found in the appendix tables for swine.)

B. During the first 24 hours after farrowing, feed the sow or gilt only a very small amount of her lactation diet (not more than a kilogram).

C. Over the next few days—about one week—gradually increase the allowance of the lactation ration until the sow or gilt is eating nearly everything offered to her. During the lactation period it is good to feed twice daily until the pigs are weaned.

D. During periods of heat stress, up to 5% fat is often included in lactation diets to ensure adequate energy intake.

E. It is important that the pigs have warm, dry quarters for resting (Fig. 62–1).

IV Feeding young and adult boars in drylot.

A. Use the same formulation for this class of swine as is used for pregnant sows and gilts.

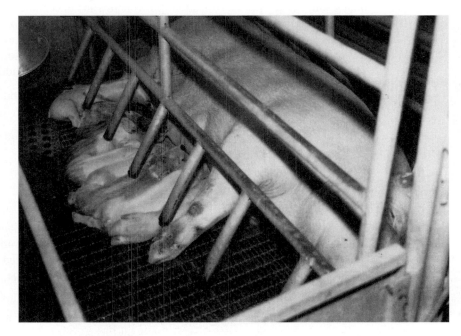

FIGURE 62–1
A dry slotted floor with plenty of room, a heat-providing infrared bulb, and a lactating mother in a stall combines optimal management practices for starting young pigs.

B. Feed young boars 2 to 2.7 kg (5 to 6 lb) once a day and adult boars, 4 to 5 lb once a day.

V Feeding early-weaned pigs.

A. Pigs weaned at 3 to 6 weeks of age are considered early weaned.

B. Pigs that are to be early weaned should have access to a starter type diet beginning at 1 week of age.

C. Mathematical equations have been used to estimate the amino acid requirements of weanling pigs weighing less than 20 kg (44 lb) and are presented as such in the NRC tables. The regression equation employed represents the best-fitting line through the following estimated requirements: 1.45% lysine at 5 kg (11 lb), 1.25% lysine at 10 kg (22 lb), 1.15% lysine at 15 kg (33 lb), and 1.05% lysine at 20 kg (44 lb).

D. Energy estimates have been determined in a similar manner. A 10 kg (22 lb) pig should consume about 0.7 kg (1.5 lb) feed, a 20 kg (44 lb) pig should consume about 1.3 kg (2.9 lb), etc. Therefore, exact lysine and energy requirements for the very young pig still need refinement. However, a well-balanced diet containing 21% protein, 1.15% lysine, 3,265 kcal ME/kg (1484 kcal ME/lb), 0.70% Ca, and 0.50% P, 0.25% salt, and adequate fortification with vitamins and trace minerals should be adequate for early-weaned pigs weighing 20 to 30 kg (44 to 66 lb). For lighter-weight pigs, the protein should be raised to about 23% and the total lysine to 1.35%. A typical corn-soy diet that meets these protein levels may not meet the total lysine level, so supplemental lysine in the form of lysine-HCL may need to be added.

E. Energy should not be a problem since the diet should be offered on an *ad libitum* basis.

F. Pigs are especially well-adapted to a slatted floor environment (Fig. 62–2).

VI Feeding grower-finisher pigs to market weight.

A. In Section 43 of this text, considerable space was devoted to balancing diets for swine using lysine—the first limiting amino acid in swine diets—rather than using total protein. Table 62–1 shows that when a corn-soy diet is balanced for the amino acid lysine, all of the 10 essential amino acids requirements are met, making corn-soy diets very convenient.

B. Section 43 also demonstrated that most of the macro-minerals, some micro-minerals, and several vitamins were deficient in a corn-soy diet and that additions must be made for them.

However, these deficiencies were corrected by incorporating "packages" with the total fortification equaling about 2.5% of the total formula.

FIGURE 62–2
Slatted floors provide a clean and dry environment for weanling pigs starting on grower-finisher diets.

C. Because of the availability, economy, amino acid adequacy, and general nutrient
adequacy of a corn-soy diet for both swine and poultry, corn was used for feeding livestock at a rate of more than 13 to 1 over the next most used energy
grain—sorghum grain—in 1994. Soybean meal was used at a rate of 15 to 1 over
the next most used oilseed meal—cottonseed meal—in 1994. (See Table 20-1 in
Section 20.)

D. Grower-finisher diets might be set up around 3 nutrient densities, as listed below:

	Weight of pig, kg (lb)		
Nutrient	*20–50 (44–110)*	*50–80 (110–176)*	*80–120 (176–264)*
Total lysine, %	0.95	0.75	0.60
Crude protein, %	18.0	15.5	13.2
Feed intake, kg (lb)	1.85 (4.08)	2.58 (5.67)	3.08 (6.77)
ME intake (kcal/day)	6,050	8,410	10,030
Calcium, %	0.60	0.50	0.45
Phosphorus, %	0.50	0.45	0.40
Available phosphorus, %	0.23	0.19	0.15

E. Diets with the above analyses may be offered on an *ad libitum* basis. One exception might be a diet that contains more than 9% cottonseed meal; cottonseed's

Table 62–1

NUTRIENTS IN CORN AND CORN + SOYBEAN MEAL (DEHULLED) COMPARED WITH THE NUTRIENT REQUIREMENTS OF A 40 KG (88 LB) GROWING PIG OF HIGH-MEDIUM LEAN GROWTH RATE (325 G [.71 LB] OF CARCASS FAT-FREE LEAN/DAY)[e]

Nutrient	Corn	Corn + Soybean Meal (74.1%:23.4%)	Requirement (40-kg pig)
Indispensable Amino Acids (%)			
Arginine	0.37	+1.09	0.35
Histidine	0.23	+0.47	0.29
Isoleucine	0.28	+0.71	0.49
Leucine	0.99	+1.59	0.86
Lysine	0.26	0.90	0.90
Methionine + cystine	0.36	+0.60	0.52
Phenylalanine + tyrosine	0.64	+1.46	0.83
Threonine	0.29	+0.65	0.59
Tryptophan	0.06	+0.20	0.16
Valine	0.39	+0.82	0.62
Mineral Elements			
Calcium (%)	0.03	−0.10	0.60
Phosphorus, total (%)	0.28	−0.37	0.50
Phosphorus, available (%)	0.04	−0.07	0.23
Sodium (%)	0.02	−0.02	0.10
Chlorine (%)	0.05	−0.05	0.08
Magnesium (%)	0.12	+0.16	0.04
Potassium (%)	0.33	+0.75	0.23
Sulfur (%)	0.13	0.20	—[a]
Copper (mg/kg)	3.0	+6.9	4.0
Iodine (mg/kg)	0.03	−0.04	0.14
Iron (mg/kg)	29	+63	60
Manganese (mg/kg)	7.0	+13.6	2.0
Selenium (mg/kg)	0.07	−0.12	0.15
Zinc (mg/kg)	18	−26	60
Vitamins			
Vitamin A (IU/kg)	213	−170	1,300
Vitamin D (IU/kg)	0	−0	150
Vitamin E (IU/kg)	8.3	−6.7	11
Vitamin K (mg/kg)	0	−0	0.50[b]
Biotin (mg/kg)	0.06	+0.11	0.05
Choline (g/kg)	0.62	+1.09	0.30
Folacin (mg/kg)	0.15	+0.43	0.30
Niacin, available (mg/kg)	0[c]	−5.2	10.0

Table 62–1 (Continued)

Nutrient	Corn	Corn + Soybean Meal (74.1%:23.4%)	Requirement (40-kg pig)
Pantothenic acid (mg/kg)	6.0	−8.0	8.0
Riboflavin (mg/kg)	1.2	−1.6	2.5
Thiamin (mg/kg)	3.5	+3.3	1.0
Vitamin B_6 (mg/kg)	5.0	−5.2	1.0
Vitamin B_{12} (μg/kg)	0	−0	10.0
Ascorbic acid	0	−0	—[d]
Linoleic acid (%)	1.9	+1.6	0.1

[a]The requirement is unknown but is met by the sulfur from methionine and cystine.
[b]The requirement is generally met by microbial synthesis.
[c]The niacin in cereal grain is unavailable.
[d]The requirement is met by metabolic synthesis.
[e]Nutrient Requirements of Swine, 1998. National Research Council.

gossypol content makes it unsatisfactory at a level of more than 9% of the total formulation.

F. Many swine feeders do not change diets for the last 2 formulations listed above, because it would not depress performance of pigs weighing 80 to 120 kg to continue receiving the formulation they received at the next lower weight bracket. However, the last diet listed contains less protein supplement, which would often decrease its cost, and specifications are included to conform to the needs of that weight pig, as set forth in the 1998 NRC recommendations.

G. Self-fed complete mixed rations are excellent for growing-finishing swine (Fig. 62–3).

FIGURE 62–3
Automatic feed tubes running along the ceiling drop feed
into feeders for these pigs on a slotted floor. Normal feeding
programs would have more pigs grouped together. The pigs
pictured above are on experimental diets at the University of
Arkansas Experiment Station.

63 Pastures-- General

I   **Definition of pasture.** A pasture is an area of land on which there is a growth of forage where designated livestock may graze at will. A good pasture is an area on which there is a growth of lush, green, nutritious, actively growing forage from which designated livestock can eat all they can consume in a relatively short period of time. Pastures vary greatly in quality depending on type, growing conditions, and stage of maturity.

II   **Classification of pastures.** Pastures are classified on the basis of various considerations.

A. **Legumes vs nonlegumes.**

     1. A *legume* is a crop that has the capacity to harbor nitrifying bacteria in nodules on its roots and so can meet a part, if not all, of its own nitrogen needs.

     2. A *nonlegume* is any crop that does not have the capacity to harbor nitrifying bacteria in nodules on its roots and so must depend on outside sources for its nitrogen supply.

B. **Annuals vs perennials.**

     1. An *annual* is a crop that must be propagated from seed each year it is grown.

2. A *perennial* is one that does not have to be reseeded each year but will reestablish itself from year to year from its roots.

C. **Summer vs winter.**

1. A *summer pasture* is one that starts growing with the onset of warm weather in the spring and continues growing until halted by frost in the fall.

2. A *winter pasture* is one that starts active growth in the fall, remains alive during the winter, and makes rapid growth during the late winter and spring, with little if any growth during the summer.

D. **Temporary vs permanent.**

1. A *temporary pasture* is one that is seeded on freshly cultivated soil for use through only one or a part of one grazing season. It will usually consist of an annual or a mixture of annuals.

2. A *permanent pasture* is one that, once established, remains as pasture for at least a period of years and in some instances continuously. It may consist of either perennials or reseeding annuals.

E. **Mixtures vs pure seedings.**

1. A *pasture mixture* is a combination of two or more pasture crops on the same area. They are usually crops that complement each other from the standpoint of growth characteristics and/or nutritive value.

2. A *pure seeding* is a pasture that supposedly consists of only one species, although absolutely pure stands are seldom found.

F. Information on the classification of the more commonly grown pasture crops is summarized in Table 63–1.

III

Effect of stage of maturity on nutrient content.

A. **Protein.**   Young plants are much richer in protein than maturing plants. Some of the leguminous plants (alfalfa, clover) may contain as much as 20% protein in the prebud stage, whereas 2 weeks later, in the full-bloom or post-full-bloom stage the level of protein in the same plant may drop precipitously. Mature, weathered plants of the same variety may drop to as low as half the protein contained in the young plant. So pasture plants should be kept in the growing stage, as much as possible.

B. **Fiber.**   On a dry matter basis, young, lush pasture may contain as much as 65% total digestible nutrients, whereas in the mature stage TDN in the same plant may be nearer 50%. Part of the maturing process is increased deposition of lignin (highly indigestible) in the growing plant as a means of giving the plant stem and branches rigidity. Weathered and bleached grass has little more feeding value than does straw.

Table 63–1
CLASSIFICATION SUMMARY OF MAJOR PASTURE CROPS

	Legume	Non-legume	Annual	Perennial	Summer	Winter	Temporary	Permanent
Bahiagrass		X		X	X			X
Bluegrass		X		X	X			X
Bermudagrass		X		X	X			X
Bromegrass		X		X	X			X
Coastal bermudagrass		X		X	X			X
Dallisgrass		X		X	X			X
Johnson grass		X		X	X			X
Orchardgrass		X		X	X*	X*		X
Redtop		X		X	X			X
Tall fescue		X		X		X		X
Timothy		X		X	X			X
Barley, winter		X	X			X	X	
Millet		X	X		X		X	
Oats, winter		X	X			X	X	
Rye		X	X			X	X	
Ryegrass		X	X			X	X	
Sudangrass		X	X		X		X	
Wheat, winter		X	X			X	X	
Alfalfa	X			X	X			X
Alsike clover	X			X	X		X	
Kudzu	X			X	X			X
Ladino clover	X			X	X*	X*		X
Red clover	X			X	X		X	
Sericea	X			X	X			X
White clover	X			X	X*	X*		X
Crimson clover	X		X			X	X	
Lespedeza	X		X		X			X
Arrowleaf clover	X		X			X	X	

*Summer in the North, winter in the South.

C. **Vitamins.** Lush, green growing plants are rich in the carotenes (beta carotene is the precursor from which most animals can synthesize their vitamin A needs), plus a host of B-vitamins, vitamin E, and ascorbic acid (vitamin C). Maturing plants gradually become depleted of most of these needed vitamins.

64

Some Important Facts about Some of the More Important Pasture Crops

I   Alfalfa.

A. Alfalfa is a perennial summer legume that is widely grown as a pasture crop over much of the country above the southern coastal states.

B. No grazing crop is more palatable, more nutritious, or more productive than is alfalfa when grown under favorable conditions.

C. As pointed out previously, alfalfa requires a fertile, well-drained and well-limed soil in order to thrive.

D. Pure stands of alfalfa are not ordinarily used as grazing for ruminants because of the danger of bloat.

E. To reduce the danger of bloat, alfalfa is usually seeded in combination with bromegrass, orchardgrass, timothy, and/or redtop when it is to be used for ruminant grazing.

F. Alfalfa sometimes suffers rather badly from winter kill under certain conditions.

G. Alfalfa also suffers rather severely in some areas, especially through the South, from insects and/or diseases, although genetically engineered strains are quite resistant.

FIGURE 64–1
Ewes and lambs grazing on orchardgrass pasture in mid-summer in Nebraska. (Courtesy of Dr. W. E. Wheeler, formerly with USDA, Roman L. Hruska Meat Animal Research Center, Clay Center, Nebr.)

II

Bromegrass, orchardgrass, timothy, and redtop.

A.   These are all perennial summer grasses that are found primarily in the Corn Belt and the northeastern states.

B.   As pasture forages, these grasses are widely used in combination with alfalfa and/or other legumes primarily to reduce the incidence of bloat among ruminant animals grazing the legumes.

C.   These grasses are not widely used for grazing purposes in either pure stands or combinations in the absence of legumes.

III

Red clover and alsike clover.

A.   Red clover and alsike clover are both short-lived summer perennial legumes found primarily through the Corn Belt and in the northeastern states.

B.   Both are widely used along with timothy and other grasses and legumes as components of rotational pastures where permanency is not a consideration.

C.   As components of such pastures, they are very palatable, nutritious, and productive.

D.   While not as bad as alfalfa about causing bloat, both red clover and alsike in pure stands may cause bloat, therefore both should be seeded with grasses when they are to be used for ruminant grazing.

E.   Since both are relatively short-lived as perennials, other longer lasting legumes such as alfalfa and ladino clover should be used along with or in the place of these two clovers in permanent pasture mixtures.

IV
_____ **Kentucky bluegrass.**

    A. Kentucky bluegrass is a rather short-growing perennial summer grass widely used for grazing throughout the Midwest and the northeastern states.

    B. It is found to a greater or lesser degree in most of the permanent pastures of these areas.

    C. Bluegrass, as it is commonly called, does best on fertile, well-drained soil that has been well limed or is naturally high in calcium content.

    D. It is very palatable and nutritious but is not a heavy producer except during the spring months.

    E. It is usually used in combination with certain short-growing legumes such as white or ladino clover to extend the grazing period.

V
_____ **Bermudagrass** (also called _common bermudagrass_).

    A. Bermudagrass is a close-growing perennial summer grass that, over the years, has been a widely prevalent weed crop in cotton and corn fields throughout the South. As these fields have been diverted from row crops to grazing, bermudagrass has in most instances taken over.

    B. There are probably more acres of bermudagrass pasture in the South than any other type.

    C. Young, immature bermudagrass makes excellent pasture, but it drops rapidly in nutritive value with advancing maturity.

    D. Bermudagrass will respond to fertilizer but not to the degree that coastal bermudagrass will.

    E. It is relatively shallow rooted and not very drought resistant.

    F. About 2–3 acres of bermudagrass with moderate fertilization will provide year-round feeding (pasture and hay) for one beef cow and her calf (Figs. 64–2, 64–3).

    G. Since bermudagrass is a nonlegume, it should be interseeded with legumes or receive liberal applications of nitrogen fertilizer for good results.

VI
_____ **Coastal bermudagrass.**

    A. Coastal bermudagrass is a hybrid bermudagrass developed by workers at the Georgia Coastal Plain Experiment Station during the 1940s.

    B. It ranks second in acreage only to bermudagrass as a grazing crop over most of the southern coastal states.

FIGURE 64–2
A brood cow herd running on a common bermudagrass pasture in North Georgia. (Courtesy of the University of Georgia College of Agriculture Experiment Station)

FIGURE 64–3
Brood cows and calves running on coastal bermudagrass pasture in South Georgia in August. (Courtesy of W. C. McCormick of the University of Georgia College of Agriculture Coastal Plain Experiment Station)

    C. Coastal bermudagrass will not survive the winters above the southern coastal states.

    D. It varies greatly in nutritive value during the course of the grazing season, depending on its degree of maturity, but it is very drouth resistant and has a high carrying capacity when properly fertilized.

    E. About 1 acre of coastal bermudagrass with moderately heavy fertilization will provide year-round feeding (grazing and hay) for one beef cow and her calf.

    F. A disadvantage of coastal bermudagrass is that it does not produce viable seed and must be propagated by the use of root stolons or sprigs. In a sense, however, this is an advantage since uncontrollable spread of the crop is thereby avoided.

    G. While very satisfactory as a grazing crop for beef brood herds, it is not regarded very highly as a grazing crop for fattening steers or producing dairy animals.

    H. Coastal bermudagrass, like bermudagrass, is a nonlegume and so is highly dependent on supplement sources of nitrogen for good results.

    I. Animals on coastal bermudagrass pasture will do best if it is kept closely grazed.

    J. Four new giant bermudagrass hybrids (coastcross 1, Midland, Tifton 78, and Tifton 44), which are reported to be more digestible and/or more winter hardy than Coastal, are now available as possible alternatives to Coastal as a grazing crop.

VII   Bahiagrass.

    A. Bahiagrass is a medium–tall-growing perennial summer grass; extensive acreages are grown for grazing over much of the southern Coastal Plain.

    B. It will not survive the winter very well above the lower Coastal Plain section.

    C. Bahiagrass will produce grazing essentially on a par with coastal Bermudagrass under Coastal Plain conditions.

    D. Its advantage over coastal bermudagrass is that it produces a viable seed and can be grown from seed.

    E. Like bermudagrass, it is a nonlegume and must be fertilized heavily with nitrogen for good results.

    F. Since bahiagrass does produce viable seed, a good portion of which is hard, its control sometimes poses a problem.

VIII   Fescue.

    A. Extensive acreages of various varieties of tall fescue, especially alta and Kentucky 31, are grown in many sections of the country above the Southern Coastal Plain for grazing purposes.

B. Fescue is a perennial winter grass, and stands of it are easily established from seed in areas where it is adapted.

C. Its major advantage is that it stays green throughout the winter in many sections and so is used extensively as a winter grazing crop, especially for beef brood cows and stocker animals (Fig. 64–4). It is also frequently used for other classes of livestock.

D. Several problems have been associated with the use of fescue for pasture and hay purposes. Beef cattle grazing on fescue or being fed on fescue hay sometime suffer from so-called fescue toxicity, fescue foot, and/or fat necrosis. Beef brood cows and milking dairy animals consuming significant amounts of fescue pasture and/or hay frequently are low milk producers, and young growing stock are slow gainers. Also, fescue has been suspected of being related to certain breeding problems of brood mares consuming this forage. These conditions are caused by a parasitic fungus known as endophyte.

E. The most common problem is fescue toxicity and the associated poor performance frequently experienced with beef cattle grazing on certain fields of fescue. Brood cows grazing such areas seem to be poor milk producers and their calves and young growing stock tend to be slow gainers. The condition is usually associated with a lingering rough hair coat and an elevated body temperature. Affected cattle usually display unusual rapid breathing and a more than normal tendency to remain in the shade and to stand in deep water when available. Such animals also tend to show nervousness and excessive salivation.

FIGURE 64–4
Beef brood cows on fescue pasture in midwinter in Georgia, supplemented with hay. (Courtesy of E. R. Beaty of the University of Georgia College of Agriculture Experiment Station)

F.   Workers at Auburn University have demonstrated that fescue toxicity is caused by a fungus present in some fescue plants. It has been shown also that a plant becomes infected with the fungus only through the individual seed from which it originates. Different stands of fescue vary in the degree of infestation. Certain stands of fescue that have been established either accidentally or on purpose using fungus-free seed are essentially fungus free.

G.   Livestock producers with fescue pastures should determine the degree of fungus infestation of such pastures and take appropriate action at the earliest possible date to cope with such contamination.

H.   Fungus contamination of existing fescue pastures may be dealt with in several ways.

   1.   One way is to supplement the fescue grazing with other feeds (preferably high in energy). This tends to reduce the fungus toxin intake and counteracts the harmful effects of the fungus toxin consumed.

   2.   A second approach to solving this problem, at least in part, is to overseed existing fescue stands with white or ladino clover or some other suitable legume. The presence of the legume in the pasture mix not only tends to reduce the amount of total fungus toxin consumed but also seems to counteract the harmful effects of the fungus toxin eaten. This seems to be over and above any improved nutrition that it may provide. This is probably the most practical approach to use with most moderately infected fescue stands since the cost is not excessive and most of the harmful effects of the fungus are avoided without any interruption in the use of the pasture. Also, the increased yields of more nutritious forage resulting from the presence of the legume in the forage mix is desirable in any case.

   3.   The third approach that might be used to cope with the fescue toxicity problem—especially with a highly infected pasture—would be to eradicate the fungus-infected sod and re-establish it using fungus-free seed.

      a.   A fungus-infected sod may be eradicated by growing a cultivated crop on the area for at least 1 year, making sure that all live fescue roots are destroyed and that fungus-infected plants do not produce seeds. The strategic use of an appropriate herbicide might also be used in hard-to-kill areas.

      b.   Once a fungus-infected sod has been eradicated from an area, a fungus-free sod may be established by using fungus-free seed and usual cultural practices. Fungus-free seed may be obtained by the following methods:

      •   Holding fungus-infected seed for a period of 1 year under normal storage conditions destroys the fungus in the seed.
      •   Using seed from fungus-free stands. An increasing supply of such seed of several strains and varieties is becoming available. Such seed certainly should be used not only in replacing old fungus-infected fescue sods with fungus-free stands of fescue but also in establishing fescue stands on new areas.

I. Results from Auburn University indicated that use of the wormer, Ivermectin, will greatly alleviate the symptoms of fescue toxicity. In a typical study, steers on high–endophyte-infected fescue that were treated with the worming medicine Ivermectin gained 116 lb in a 168 day trial, whereas untreated control animals on the same pasture gained only 80 lb in the same period of time. How Ivermectin protected the cattle exposed to endophyte fungus on fescue is not apparent. (Bransby, D. L. Large Animal Practice. 18 [No. 3]:16. 1997)

IX. Dallisgrass.

A. Dallisgrass is a perennial summer grass that prevails to varying degrees in many permanent pastures over much of the Southeast.

B. It thrives especially well in lowland pastures, but its growth is not restricted to such areas.

C. While it is sometimes included in seeding mixtures, its presence in the pasture sward is usually on a volunteer basis.

D. As a pasture forage, dallisgrass is regarded as about average in productivity and nutritive value.

E. A fungus known as *ergot* will sometimes develop on dallisgrass heads. If eaten by cattle, it may cause a condition known as *dallisgrass poisoning,* which is a nonfatal staggering condition from which animals afflicted normally recover when removed from the pasture.

X. Oats, rye, barley, and wheat.

A. Fall-seeded small grains are widely used in many sections of the country for grazing purposes.

B. Since they remain more or less green and growing throughout the fall, winter, and spring, they are used primarily for providing winter grazing (Fig. 64–5).

C. The farther south the location, generally speaking, the greater the amount of winter grazing provided by these crops.

D. To the extent that such grazing is available, it is generally very palatable and nutritious.

E. Rye is usually the most productive of these crops during the winter months and in the colder sections, but it is not as palatable and nutritious as oats, wheat, and barley.

F. When these crops are to be used strictly for grazing, they are sometimes seeded in combination with other crops such as ryegrass, crimson clover, and/or ladino clover.

G. Oats, rye, barley, and wheat are all subject to causing grass tetany, especially if highly fertilized.

FIGURE 64–5
Yearling steers on oat grazing in south Georgia in midwinter.
(Courtesy of W. C. McCormick of the University of Georgia
College of Agriculture Coastal Plan Experiment Station)

XI ____ Ryegrass.

A. Most ryegrass used for pasture is treated as a winter annual.

B. It is widely used throughout the South and in the Pacific Coast states as a component of mixtures for winter grazing.

C. It is usually used in combination with one or more of the small grains along with crimson and/or ladino clover in this connection.

D. It is palatable, nutritious, and a fairly heavy producer, and it helps extend the length of the grazing period.

E. It also adds strength to the pasture footing and helps to keep livestock from miring down during wet weather.

XII ____ Crimson clover.

A. Crimson clover is an annual winter legume used extensively in seeding mixtures for temporary winter grazing over much of the South.

B. Certain strains of crimson clover are reseeding and, if permitted to go to seed, will tend to reestablish themselves on permanent pasture areas each fall under southern conditions, thereby providing variable amounts of volunteer winter grazing.

C. Pure stands of lush crimson clover grazing will sometimes cause bloat in ruminant animals.

XIII White clover.

A. White clover is a close-growing, fine-stemmed perennial legume found in most of the more humid sections of the country.

B. It has a good capacity for reseeding itself where conditions are not favorable for perennial growth.

C. In the North, with ample moisture it continues active growth throughout the growing season but is killed back during the winter.

D. In the South, it is usually killed back by the summer heat and drouth but remains more or less in active growth during the late fall, winter, and spring.

E. White clover is very palatable and nutritious and is an important component of many permanent pasture swards.

F. It is usually most prevalent and most productive in the lower, wetter areas within a pasture.

G. White clover is not very drouth resistant and usually ceases active growth during prolonged periods of hot, dry weather.

H. Horses grazing white clover will salivate excessively (slobber).

XIV Ladino, Regal, and Tillman clovers.

A. Ladino clover is a type of giant white clover that was introduced into the United States from Italy several years ago.

B. Regal and Tillman are strains of the same giant white clover that have been selected and developed for their superior productivity.

C. As pasture forages, all 3 have characteristics similar to ordinary white clover except that all are taller growing and more productive.

D. Ladino clover has been widely used over the years in combination with bromegrass and orchardgrass in the North and with tall fescue above the Coastal Plain in the South in establishing improved permanent pasture.

E. Regal and Tillman clovers, because of their superior productivity, are currently being widely used in place of Ladino clover to establish such pastures.

F. Being taller growing, such clovers are able to compete better than the usual white clover in mixtures with tall-growing grasses.

G. Pure stands of these clovers are seldom used as grazing for ruminants because of the danger of bloat.

XV Annual lespedeza (commonly referred to as *lespedeza*).

A. Lespedeza is a summer annual legume that is widely grown over much of the central and southern sections of the country.

B. There are several different types and varieties, such as common, Korean, Kobe, and Tennessee 76.

C. Although it is an annual, lespedeza effectively reseeds itself under most conditions, and one type or another is found in more or lesser degrees in most of the permanent pastures of those areas where it is adapted.

D. Lespedeza is quite palatable and very nutritious, but it is not a heavy producer.

E. It has a relatively short growing period and consequently should be used in combination with other crops for best results.

F. Lespedeza seed is very nutritious, and cattle grazing on lespedeza that has gone to seed tend to fatten quite readily.

G. Lespedeza is not normally a bloat producer.

XVI Sericea lespedeza (commonly referred to as *sericea*).

A. Sericea is a perennial summer legume used extensively over the years in the South as a cover crop on land in the government soil bank.

B. Between that coming out of the soil bank and that seeded specifically for grazing purposes, there is a considerable acreage of sericea used for pasture, especially for beef cow herds, in the southeastern area.

C. Young, immature sericea is fairly palatable and nutritious, but as this crop matures, it becomes very unpalatable because of a buildup in its content of tannins.

D. For best results it should be grazed fairly closely or mowed frequently to keep down old growth and encourage new growth.

E. It is very drouth resistant and is frequently relied on by beef producers for grazing during periods of dry weather.

F. Sericea is one of the few legumes that does not ordinarily cause any bloat trouble when grazed in pure stand by ruminant animals.

XVII Kudzu.

A. Kudzu is a perennial summer leguminous vine sometimes used in the South for controlling erosion on steep land.

B. It is rather slow and tedious to establish since it is usually propagated from crowns.

C. Established areas of kudzu will provide excellent emergency grazing during periods of drouth.

D. It is quite palatable and nutritious but has a rather limited carrying capacity.

E. Heavy grazing during early spring is harmful to and may eliminate a stand of kudzu.

F. Kudzu is not normally a bloat producer.

XVIII Arrowleaf clover.

A. Arrowleaf clover is a winter annual legume that is gaining popularity as an early spring grazing crop over much of the Southeast (Figure 64–6).

B. Of the three strains available—Amclo, Yuchi, and Meechee—the first 2 are the most widely used.

C. It is ordinarily seeded in the fall—October, November, and December.

D. It requires a well-drained soil and does best on a fertile soil with a pH of 6.2 to 6.5.

E. The type of seedbed is not too important as long as the soil surface is fairly free of vegetative material.

FIGURE 64–6
Arrowleaf clover in late spring in Georgia. Arrowleaf clover is a good forage crop for the Southeast. (Courtesy of E. R. Beaty of the University of Georgia College of Agriculture Experiment Station)

F. Good stands of this crop have been obtained from top seeding closely grazed or burnt-over permanent pasture with 5–10 lbs of seed per acre.

G. Arrowleaf clover seed should be inoculated with type O inoculant for effective nitrogen fixation.

H. It produces a high percentage of hard seeds and tends to reseed itself under favorable conditions.

I. Arrowleaf clover makes its principal growth during March, April, May, and June.

J. The Amclo strain matures about 30 days later than crimson clover and the Yuchi strain about 15–20 days later than the Amclo.

K. It is a heavy producer of lush forage that is comparable to alfalfa in nutritive value.

L. Cattle seem to relish it, but not horses.

M. No trouble from bloat has been reported from grazing this group.

N. While grown primarily for grazing purposes, surplus late-growth arrowleaf clover is sometimes harvested for hay.

XIX Millet and sudangrass.

A. These are both rather coarse-growing summer annual grasses that are frequently planted for use as temporary summer grazing crops.

B. Several new varieties and/or crosses of both have been developed over the past several years, which are supposedly superior to the old varieties.

C. Such crops have, in the past, been planted most commonly for use with dairy herds, but their use for finishing steers on pasture is increasing in some sections.

D. They are both usually regarded as emergency crops to be used for grazing if needed but, if not needed for grazing, are harvested for hay or silage.

E. Second-growth sudangrass following a severe drought or an early frost will sometimes contain hydrocyanic acid, which is quite poisonous to livestock.

XX Johnsongrass.

A. Johnsongrass is a rank-growing perennial summer grass that occurs on a volunteer basis more or less throughout the southern states where it is one of the worst weeds of the row-crop farmer.

B. While seldom, if ever, planted as a pasture crop, it is sometimes used for grazing purposes when available (Fig. 64–7).

C. If grazed before becoming too mature, it makes a fairly satisfactory grazing crop, but it is very low in protein.

D. Continuous close grazing of Johnsongrass will tend to eliminate it from a pasture over a period of 2–3 years.

E. As with sudangrass, second-growth Johnsongrass following a severe drouth or an early frost sometimes contains hydrocyanic acid, which is very poisonous to livestock.

FIGURE 64–7
A brood cow herd grazing on Johnsongrass in north Georgia. Johnsongrass is ordinarily not seeded but is sometimes used for hay or pasture if available on a volunteer basis. (Courtesy of the University of Georgia College of Agriculture Experiment Station)

General Use of Pasture in Livestock Feeding

65

I   **It is impossible to balance rations precisely for grazing animals.**

A.   With a grazing program, the amount of forage consumed cannot be precisely controlled or measured.

B.   Under grazing conditions, it is impossible to know precisely the composition of the forage being consumed by the animal.

1.   Composition of forage varies during the course of the day, especially with respect to dry matter.

2.   Composition of forage varies from day to day, depending on weather conditions and stage of maturity.

3.   Composition of a forage species will vary, depending on the level of fertilization.

4.   Because of selective grazing on the part of the animal, it is impossible to know the exact nature of the forage being consumed by the animal.

C.   In general, pasture forage dry matter has a composition and feeding value comparable to that of hay and/or silage.

1.   Pasture is a roughage and should be so used in the feeding program.

2. In general, pasture may be used to replace the hay and/or silage in a balanced ration developed for drylot feeding.

3. Good lush pasture, in addition to replacing the hay and/or silage in a balanced ration for drylot feeding, may also reduce the amount of protein supplement required.

II Use of pasture in feeding beef cattle.

A. Most of the feed for beef brood cows and growing stock may come from pasture.

   1. If pasture is fair to good, only minerals will be required in addition to the pasture.

      a. Salt should always be provided. Trace-mineralized salt is probably best as insurance against a trace-mineral deficiency.

      b. In addition to salt, defluorinated phosphate, dicalcium phosphate, or steamed bonemeal, is usually provided on a free-choice basis (Figs. 65–1, 65–2).

FIGURE 65–1
Developing heifers on a lush spring pasture need few other nutrients than a free-choice mineral mixture (2 parts dicalcium phosphate and 1 part trace-mineralized salt). Loose plain salt can be provided in a second compartment.

FIGURE 65–2
A homemade type feeder such as is frequently used for providing a single mineral mix to cattle on pasture. (Courtesy of the University of Georgia College of Agriculture Experiment Stations)

    c. Some magnesium oxide might also be supplied during the late winter and early spring to prevent grass tetany if this is a problem.

2. If pasture is short or of poor quality, some supplemental protein and energy should be provided. This may be done by using one or more of the following in accordance with the animals' estimated nutrient requirements.

Good hay	"Hot meal"
Good silage	Range cubes
Grain	Protein blocks
Cottonseed meal	Liquid supplements
Soybean meal	

3. The adequacy of a pasture may be arrived at from the condition of the cows grazing it—if cows are losing flesh excessively, some supplemental feed should be provided.

4. Based on TDN provided, 1 pound of grain or other concentrates will replace 2 pounds of hay (or silage equivalent).

5. If the milk production of their mothers is inadequate, nursing calves on pasture are sometimes creep fed on a mixture of grain—almost any combination of corn, grain sorghum, oats, barley, and/or wheat is suitable (Fig. 65–3).

FIGURE 65–3
Suckling calves on dry winter forage will need additional energy and protein to supplement depressed milk production
by their mothers if normal growth is to be attained. One
should either give extra feed to the mothers to cause them to
produce more milk or creep-feed the calves energetic and
proteinaceous feeds during this period.

B. Good pasture will suffice as the primary feed for most classes of calves, yearlings,
and 2 year olds.

1. All such cattle will require supplemental minerals as outlined for brood cows
above.

2. Unless the pasture is of excellent quality, recently weaned calves should be fed
from 3 to 5 lbs of concentrates per head daily, along with the pasture and minerals.

3. The amount of supplemental feeding to be carried out along with pasture and
minerals for other classes of growing cattle will depend on the rate of growth
and degree of fattening desired, on the one hand, and the quality and abundance of the pasture, on the other.

4. As a general rule, pasture may be relied on to replace at least the roughage in
a balanced drylot ration for such cattle.

5. If the pasture for young cattle is short or of poor quality, it definitely should be
supplemented with hay, silage, and/or concentrates, as well as minerals.

6. If pasture of excellent quality is available over a period of several months,
some degree of fattening might be realized with such cattle on pasture with no
supplemental feeding other than minerals, depending on the following factors:

   a. Age of cattle. Older cattle will fatten on pasture more readily than
   young cattle. Calves will fatten on pasture only under the most favorable
   conditions.

b. **Quality of pasture.** Pastures decrease in nutritive value with advancing maturity, so cattle generally do best and fatten most readily on young, immature forage growth. To have such growth available over a period of time usually involves the use of temporary pastures.

c. **Freedom from diseases and parasites.** Animals afflicted with diseases and/or parasites (either internal or external) are inclined to be slow gainers and usually are unable to gain sufficiently on pasture to fatten.

d. **Environmental conditions.** The English breeds of cattle do not like to graze in the hot midday sun of the summer months. Also, pastures become mature during the summer, and parasites are worst during this period. Hence, if pasture is adequate, grazing cattle usually do best during the fall, winter, and spring months.

e. **Genetic potential of cattle to perform.** Cattle that have been bred to perform well, especially those selected to perform well on pasture, will fatten most effectively on pasture.

7. If the above-mentioned factors are all favorable, cattle, including calves, may attain a low-choice finish on pasture. However, at least some of the above factors are usually not favorable, and so pastures are not ordinarily relied on for producing choice grade slaughter cattle.

III Pastures for feeding dairy cattle.

A. *Good pasture* is satisfactory as the only feed, other than minerals, for feeding dry cows and growing dairy animals.

B. For milking cows, good pasture may be used to replace all of the roughage and possibly part of the concentrates, depending on the following:

1. **Quality of pasture.** Good pasture may be used as a greater portion of the ration than poor pasture without hurting production.

2. **Price of concentrates and milk.** The lower the milk prices are and the higher the concentrate prices are, the more economical it might be to use pasture in the feeding program for milking cows.

3. **Production potential of cows.** Cows with a low production potential may produce up to their potential on good pasture and limited grain, whereas cows with high production potential will respond favorably to less pasture and more concentrates.

C. Cows on pasture are inclined to produce off-flavored milk.

1. This is especially true for pastures with wild onions or bitterweed. Such pastures should not be used for milking cows.

2. Most lush green pasture tends to impart varying degrees of off flavor to milk. Most such flavors can be avoided by keeping cows off pasture for several hours prior to milking.

FIGURE 65–4
A dairy cow grazing on lush pasture. While pastures were formerly used widely for feeding dairy herds, this is no longer true. Land prices in the vicinity of dairies are too high to permit its use for pasture. Also, pastures are too unreliable, too variable in quality, and too often a source of off flavors in the milk to be looked upon with favor by many dairymen. (Courtesy of Ralston Purina Co., St. Louis, Mo.)

D. Lush green pasture will supply more protein than corn or sorghum silage and usually more than even good hay—hence, good pasture may reduce the amount of protein supplement that would otherwise be required.

E. Modern dairies are using pasture less and less in their feeding program (Figure 65–4).

    1. The price of the land on which most dairies are located is too high to justify its use for pasture.

    2. The control of off flavors is too great a problem.

    3. Quality of pasture varies too much from week to week.

IV

Pasture for sheep.

A. Pasture may be used as the only feed, other than minerals, for all classes of sheep.

    1. Sheep will come closer to producing a top-quality product (wool and fat market lambs) on pasture alone than will any other class of livestock.

2. If pasture is used as the only feed in a sheep program, it will need to be extra good during the following periods:

    a. For the flushing period just prior to and during the breeding season to help bring ewes into heat and increase the number of multiple births.

    b. For 2–3 weeks just prior to lambing to prevent a condition known as *pregnancy disease,* which is caused by a too-low level of digestible carbohydrates.

    c. During the nursing period—best for both ewes and lambs, but if the supply is short, good pasture may be used as a creep feed for lambs only.

3. All sheep and especially those on pasture must be subjected to a rigid program of internal parasite control.

4. Sheep are most likely to bloat on legume pastures.

V

Pasture for horses.

A. Pasture may be used as the only feed, other than minerals, for all classes of non-working horses.

B. Pasture may be used for working horses during nonworking hours if supplemented with grain at the rate of 0.75–1.50 lb of grain per hundred lb of body weight, depending on the amount of work being done.

C. Hay is better than pasture for working horses. Hay will give a horse better wind and better endurance.

D. Lush pasture is not a desirable feed for saddle horses to be ridden on bridle paths or in the showring. It tends to produce a very loose, messy feces.

VI

Pasture for hogs.

A. Up to about the early 1950s, pasture was regarded as an essential factor in swine production.

    1. Until that time, it was regarded as the most economical source of certain unidentified nutritive factors.

    2. Until then, the use of freshly planted temporary pasture for sows with young pigs was regarded as the best way to keep internal parasite infestation of young pigs at a minimum.

    3. Good pasture may be used to replace a varying proportion of the drylot concentrate mixture for swine depending on the class of animal.

        a. 10–20% of the concentrate feed for growing-fattening pigs.

    b.   20–50% of the concentrate feed for sows nursing pigs.

    c.   50–75% of the concentrate feed for pregnant sows.

4.   When pasture is used to replace a portion of the concentrate feed for swine, it is necessary to restrict the level of concentrates fed to the level desired by hand feeding. In other words, swine will not voluntarily reduce concentrate consumption on pasture if concentrates are self fed.

5.   Most present-day, up-to-date swine operations do not use pastures for feeding lactating sows or growing-fattening pigs.

    a.   With present-day knowledge of nutrition, pastures are not needed as a source of unidentified factors.

    b.   It is becoming increasingly more difficult to produce parasite-free pigs on pasture. It can be done more effectively on concrete floors.

    c.   At present-day prices, nutrients can be provided more cheaply with concentrates than with pasture.

6.   Pasture may or may not be used for pregnant sows in modern-day swine operations. Some producers consider pasture desirable for this class of animals, but others believe that they can be handled more cheaply and just as effectively in drylot on dry feed.

66

Use of Performance Stimulants

I

In this section, the authors wish to discuss several groups of substances that are not recognized as feed nutrients but that, under certain circumstances, stimulate or promote the more effective use of feed nutrients by an animal in a way that results in a more rapid gain, a higher production, and/or greater feed efficiency. We will not discuss all of the products that over the years have been proposed, studied, and/or used in this connection. Instead it is the purpose of this section to simply recognize the existence of such products and consider briefly some of the thinking behind their use, and to recognize and discuss briefly most of those in general use by the nation's livestock industry today.

A. At one time or another during the past 25 years, such products have been developed for all the different species of livestock. More have probably been developed for beef cattle than for any other species, but to a greater or lesser degree all species have been involved.

B. In most instances, the claims of such products have been increased rate of growth. In some instances, it has been increased milk production, and in nearly all instances increased efficiency of feed utilization has been the primary goal.

C. Some stimulate appetite, some improve digestion, others alter metabolism, and still others control diseases or parasites.

D. These products are available in the form of a subcutaneous implant, a drench or bolus, or a feed additive. Some are available in more than one form.

E. Performance stimulants involve several different types of products.

    1. Several are antibiotics. An antibiotic is a substance produced by one organism that has an inhibiting effect on the growth of another. It may function as a performance stimulant by controlling subclinical levels of certain diseases or it may serve simply to bring about a more favorable microflora in the rumen or lower digestive tract. As used in some feeding programs, antibiotics might be more properly classed as a medicine than as a performance stimulant. A list (not necessarily complete) of antibiotic products that at one time or another have been used as performance stimulants during the past 25 years follows:

        a. Bacitracin

        b. Chlortetracycline (Aureomycin)

        c. Erythromycin

        d. Neomycin

        e. Oleandomycin

        f. Oxytetracycline (Terramycin)

        g. Penicillin

        h. Streptomycin

        i. Tylosin (Tylan)

        j. Lincomycin

        k. Virginiamycin

        l. Bambermycin (Flavomycin)

        m. Zinc bacitracin

        n. Rumensin (Monensin)

        o. Bovatec (Lasalocid)

        p. Carbodox

    2. Some are chemoantibacterial or chemotherapeutic compounds. These are compounds that act similarly to antibiotics but are of purely chemical origin as contrasted to being of biological origin. Included in this group are compounds such as the following:

        a. Arsanilic acid or sodium arsanilate

        b. 3-nitro-4-hydroxyphenyl arsonic acid

        c. Ethylenediamine dehydroiodide (EDDI)

    d. The nitrofurans

- Furazolidone (NF 180)
- Furaltadone
- Nitrofurazone
- Nihydrazone

    e. Sulfamethazine

    f. Other sulfa compounds

3. Some are referred to as chemobiotics. A chemobiotic is a combination of a chemoantibacterial compound and an antibiotic. Chemobiotics (sometimes referred to as chembiotics) act similarly to antibiotics but have been formulated to do jobs that neither antibiotics nor chemoantibacterial compounds can do alone. Included in this group are products such as the following:

    a. Dynafac

    b. Hygromycin

4. Some are hormone or hormone-like products. A hormone is a chemical substance secreted into the body fluids by an endocrine gland. Several hormone or hormone-like products are available on the market for which performance stimulating effects are claimed. Most of these are recommended for use in beef cattle feeding. Some of the more common ones are as follows:

    a. Stilbestrol (Diethylstilbestrol or DES) (No longer approved for use in livestock feeding)

    b. MGA (Melangestrol acetate)

    c. Synovex S (Progesterone + estradiol benzoate) Implant Steer-oid also contains these hormones.

    d. Synovex H (Testosterone propionate + estradiol benzoate)

    e. Rapigain (Testosterone + stilbestrol)

    f. Ralgro (Zeranol or zearalanol)

    g. Compudose (Estradiol 17B)

5. Others are anthelmintics. An anthelmintic is a remedy for worms, or worming agent. While such materials sometimes stimulate performance, they do so more as a medicine than as a performance stimulant. Included in this group are materials such as the following:

    a. Hygromycin

    b. Phenothiazine

    c. Piperazine

    d. Thiabendazole (Thibenzole)

    e. Tramisol

    f. Loxon

    g. Ivomec (Ivermectin)

    h. Paratect (Morantel tartrate)

6. A few function as pH regulators. The proper functioning of the different parts of an animal's digestive tract is pH related and animal performance has sometimes been improved through the use of pH regulators in the feeding program. Included among such products are these:

    a. Sodium bicarbonate

    b. Ground limestone

7. One group functions as thyroid regulators. Thyroxin(e) produced by the thyroid gland controls an animal's rate of metabolism. Animals sometimes show a performance response to a change in rate of metabolism as the result of altering the amount of thyroxin present. Included in this group are such products as the following:

    a. Thyroprotein (Iodinated casein)—A synthetic thyroxin product (sold under several trade names) sometimes used to supplement the thyroid's production of this hormone.

    b. Thiourea—A thyroid inhibitor.

    c. Thiouracil—A thyroid inhibitor.

    d. Methimazole (Tapazole)—A thyroid inhibitor.

8. Some work as a bloat preventive. Bloat, even when it does not cause death, is a deterrent to the efficient performance of ruminants on many legume pastures. A relatively new product, poloxaline, is being marketed under several trade names to help prevent trouble from bloat.

9. Some act as a surfactant. The addition of certain surface-active agents to feed has been proposed as a means of exposing the feed particles more completely to the digestive juices thereby effecting more complete digestion. Several products have been studied in this connection, such as the following:

    a. Lecithin

    b. Silicone

10. Some are a feed flavor. A high level of feed consumption is generally recognized as essential for efficient animal performance. Various feed flavors have been used in an effort to increase feed consumption. Among such products are these:

    a. Anise oil.

    b. Miscellaneous synthetic flavors.

11. Some act as a tranquilizer. Some of the tranquilizers used over the past several years in human medicine have been used by livestock researchers in an effort to calm animals in the feedlot and in turn improve feedlot performance. Examples of tranquilizers studied include the following:

    a. Hydroxyzine

    b. Trifluomeprazine

    c. Reserpine

12. Some are used as an enzyme preparation. Feed digestion is basic to feed utilization, and an essential element of efficient feed digestion is an adequate amount of the proper enzymes. The digestion of most feed nutrients is dependent on the presence of specific enzymes. An absence or shortage of any of these enzymes can result in inefficient feed digestion and in turn inefficient feed utilization. There are enzyme products on the market that have been designed to supplement the enzymes naturally present in an animal's digestive tract in an effort to effect more efficient digestion. Benefits from the use of such products in a feeding program seem to vary according to the circumstances. Their continued use in any feeding operation would have to be decided on the basis of benefits realized vs cost. An enzyme feed additive presently on the market is Zyme-All.

II

Some of the performance stimulants that have had or are having a significant influence on the nation's livestock production are as follows:

A. Antibiotics. Over the past 25 years, many different antibiotic products have been studied from the standpoint of their use in livestock (and poultry) rations at subtherapeutic levels to increase rate of growth and feed efficiency. Most were not without beneficial effects under certain circumstances. Some, however, showed greater and/or more frequent beneficial effects at a lower cost than others. Accordingly, penicillin and the tetracyclines (Chlortetracycline and Oxytetracycline as Aureomycin and Terramycin, respectively) have gained widespread use as feed additives. However, within the past few years the Food and Drug Administration (FDA) has become concerned about the following:

1. The use of these antibiotics in feeds at subtherapeutic levels may give rise to populations of bacteria that are resistant to penicillin, the tetracyclines, and other antibiotics.

2. This resistance might interfere with subsequent treatment of sick livestock.

3. This resistance might be transferred to bacteria in humans, thus leading to the possibility of untreatable disease in humans.

As a result, FDA has proposed in separate proposals (8/30/77 and 10/21/77) that the use of both penicillin and the tetracyclines at subtherapeutic levels in animal feeds be discontinued and that the use of these materials otherwise for the treatment of livestock (and poultry) diseases be carried out under the supervision of a veterinarian. Regulations in keeping with either or both of these proposals may very well be implemented sometime in the future. Other antibiotics may in time be added to the above. In the meantime, livestock (and poultry) researchers will no doubt be busy identifying other antibacterial materials that might be effectively substituted for penicillin and the tetracyclines in animal feeding.

A relatively new antibiotic feed additive, which has entered the cattle feedlot picture with wide acclaim, is Rumensin (Monensin). This is a biologically active compound produced by a strain of *Streptomyces cinnamonensis*. Rumensin, one of the so-called ionophores, when added to cattle rations, tends to increase the molar proportion of propionic acid and decrease the proportion of acetic and butyric acids produced in the rumen. As a result, there is an improvement in the efficiency of energy utilization by the animal. While rate of gain is normally improved little, if any, by this product, total feed consumption is sufficiently reduced to effect an improvement on the average of 10.6% in feed efficiency, according to the manufacturer. It seems to work in either steers or heifers and with all types of rations. Recommended level of use of Rumensin is 15 mg per lb of total air-dry feed whether fed as a supplement or mixed in the ration. There is no withdrawal period required.

Following the introduction of Rumensin into the cattle feeding picture was another ionophore, lasalocid, marketed as Bovatec. Bovatec, when used at recommended levels in cattle rations, functions very much as Rumensin does to improve feed efficiency. The manufacturer of Bovatec also claims more rapid gains from the use of this product. As with Rumensin, Bovatec requires no withdrawal period. Both products are toxic to horses and should be kept away from this class of livestock. Other ionophores are under study as feed additives but none have as yet received FDA approval for such use.

B. The status of certain chemoantibacterial drugs that have been widely used in swine and poultry feeds is also cloudy. FDA has proposed to withdraw approval of the four nitrofurans (furazolidone, furaltadone, nitrofurazone, and nihydrazone) for use in feed because they are suspected of producing cancer in humans. Also, the high percentage of pork carcasses, which continue to show violative levels of sulfa (more than 0.1 ppm in the tissue), continues to cause FDA concern about the use of sulfa drugs in swine feeds in the future, at least without a carefully controlled feed mixing program and an adequate withdrawal period.

C. While certain combinations of antibiotic and chemoantibacterial compounds (chemobiotics) have been used rather extensively as feed additives over the past few years, most of these have involved the use of penicillin and/or one or both of the tetracyclines. In view of FDA's proposed action to ban further use of the latter materials in feeds without veterinary authorization, the use of such preparations in mixed feeds in the future looks doubtful.

D. The use of hormone and hormone-like performance stimulants has become a standard practice in beef production, especially in the finishing phase. Their use is frequently the difference between profit and loss from cattle feeding. There are several such products and a brief discussion of the use of each follows:

1. Diethylstilbestrol (stilbestrol, DES) is generally recognized as the original performance stimulant, having been introduced to the industry in 1948 by two Purdue University researchers (F. N. Andrews and W. E. Dinusson). It was researched as a subcutaneous implant (Fig. 66–1) at levels of 60 mg (effective level needed to be only 36 mg). Increased rate of gain (10% to 15%) and improvements in feed efficiency (10% to 15%) caused the beef cattle industry to take notice and to inquire where they could obtain the drug. In the early 50s, Iowa State University researchers (Wise Burroughs, et al.) found that oral DES was just as effective as implanted DES, and patented the usage of oral DES in beef cattle diets. DES was found to be effective for young calves and for growing cattle—both in the feedlot and on pasture. It was not recommended that DES be used with replacement heifers because of the possible future effect on breeding performance. Even though opening the field of hormone-like substances to stimulate rates of gain and improve feed efficiency represented a new era in cattle feeding, the use of DES in the category was doomed by the passage of the Delaney Amendment, which prohibited use of any substance that had been shown to cause cancer in humans or animals.

It was subsequently demonstrated that diethylstilbestrol would, at least at certain levels, induce cancer in mice. Also, traces of stilbestrol were found in livers from cattle that had received this product as an implant or a feed additive. Consequently, even though the level of DES used to produce cancer in mice

FIGURE 66–1
Legal hormone implants are best implanted in the ear. Avoid puncturing an ear vein. The ear is discarded in the packing plant, so there is no danger of its getting into the human food chain. (Photo by James Herndon)

was many thousand times that found in the beef livers, and the detection of DES in livers was possible only as the result of new, sophisticated analytical equipment, the presence of this product in the livers at any level was illegal based on the Delaney amendment.

As a result, the Food and Drug Administration, which is responsible for administering this law, ruled in 1973 that further use of diethylstilbestrol with feedlot cattle and sheep must be discontinued. The ruling was unpopular with cattle and sheep feeders, as well as with manufacturers of DES, and the legality of the ruling was challenged in the courts for 5–6 years. During this time, the ban on the use of stilbestrol was temporarily lifted, subject to certain restrictions, until its legality could be established by the courts. In the meantime, cattle and sheep feeders continued to use DES in their finishing operations.

However, in 1979, following prolonged court proceedings, the FDA was authorized to proceed with plans to terminate further use of DES for livestock finishing purposes over several months. While there have been problems in obtaining the compliance of certain cattle feeders with this final ruling, recent action by the FDA is directed toward gaining 100% compliance and the complete elimination of stilbestrol from the nation's cattle and sheep feeding programs.

2. A performance stimulant product that many cattle feeders are using with good results is zeranol (sometimes referred to as zearalanol), which is sold commercially under the trade name "Ralgro." Zeranol is produced by a mold, *Giberella zeae,* grown under special conditions. Although not really a hormone, it does tend to have a hormone-like effect in the animal's body. Ralgro implants are manufactured to contain 12 mg of zeranol each. Three pellets (36 mg) are used for finishing steers and heifers, and 2 pellets (24 mg) for growing cattle. Ralgro implants are placed just beneath the skin near the base of the ear. Each implant is effective for 3 to 4 months and Ralgro implants may be repeated every 3 to 4 months from birth to market provided the last is no closer than 65 days to slaughter. From 6%–10% improvement in rate of gain and 6%–8% improvement in feed efficiency are claimed for this product. It is effective with both steers and heifers.

3. Another performance stimulant for which some rather significant improvements in both rate of gain and feed efficiency have been reported is synovex— Synovex-S for steers and Synovex-H for heifers. Synovex-S implant is a combination of 200 mg of progesterone and 20 mg of estradiol benzoate and is used in steers weighing 400 to 1000 lb. Synovex-H is a combination of 200 mg of testosterone propionate and 20 mg of estradiol benzoate and is used in heifers weighing 400 to 800 lb. The Synovex implant consists of a cartridge containing 8 small pellets that are implanted beneath the skin of the back of the ear. Synovex implants are effective for 100 to 120 days and should be made no closer than 60 days before slaughter.

4. Another performance stimulant compound that is recommended for use with feedlot heifers only is Melangestrol acetate (MGA). This material is approved by the FDA for feeding at levels of 0.25 to 0.50 mg per head daily, but must be

withdrawn from the feed at least 48 hours prior to slaughter. Claims are made by the manufacturer of 6.5% greater feed efficiency and 10.3% faster gain from the use of this product. In conjunction with these benefits, there is a suppression of heat in feedlot animals.

5. Another new implant, called Compudose, has recently been introduced to the steer implant market. Like certain other performance stimulants, it embraces the use of estradiol to improve rate and efficiency of gain and is designed for use with both young growing stock and finishing animals. The implant supposedly effects a controlled release of the estradiol over a 200-day period. There is no required period of time between implantation with compudose and slaughter.

6. Another implant that has recently been made available to the steer feeder is Steer-oid. It embraces the same two hormones (progesterone and estradiol benzoate) in identical amounts (200 mg and 20 mg) as that used in Synovex-S implants discussed earlier. Improved rates of gain and feed efficiency are claimed by the manufacturer. Steer-oid implant requires no withdrawal period.

7. There has been considerable interest lately in the possibility of using combinations of performance stimulants. As far as is known, such a practice would be legal provided all withdrawal requirements and implant restrictions were met. Which combinations might be most profitable is not yet known. It would seem that combining Rumensin with one of the hormone-like implants might be effective since they work in entirely different ways in the animal.

8. The final decision regarding the use of a performance stimulant should, of course, be made by the producer, after he has obtained all the information he can on the subject. Such factors as class of animal, days until slaughter, withdrawal time required, benefits, possible adverse side effects, and cost per animal should all be considered. In all instances, the manufacturer's recommendations and advice concerning the use of a particular stimulant, as well as all government regulations in that regard, should be obtained and closely followed.

III

Parts I and II of this section dealt with the general principles of feed "performance stimulants," along with some of the background and history of such. Part III is meant to acquaint the student with the names of most of the compounds that are recognized as feed additives and are controlled by the Food and Drug Administration (FDA) division of the U.S. Government. In addition to the individual compounds, many combinations of compounds have been cleared by the FDA. The feed additives have been divided into Class I drugs (those with no required withdrawal period) and Class II drugs (those with a specified withdrawal period). Such specifications will not be included in this text since the purpose of this listing is merely to acquaint the student with such drugs. Bear in mind that without FDA clearance, no one is permitted to use such drugs. Therefore, it is up to the registered feed manufacturer/dealer to know all the rules and ramifications of using

FDA-controlled drugs, and dealers must inform purchasers of these rules and ramifications. As indicated above, it is beneficial for the student to know about and to be able to discuss the general principles of most of these additives.

Generic names will be given but not chemical names. Why? The first additive, known as Amprolium, has a chemical name that requires 77 spaces on the word processor, i.e. 1-(4-amino-2-n-propyl-5-pyrimidinylmethyl)-2-picolinium chloride hydrochloride.

Disclaimer: No single drug has been omitted intentionally. No combinations of drugs have been included. Any misspelled names or lack of Registered Trademarks is unintentional. The following presentation is an attempt to interpret material presented in the 1997 Feed Additive Compendium published by the Miller Publishing Co., 1240 Whitewater Drive, Minnetonka, MN 55343. Descriptions have been abbreviated so that the student might get a general appreciation of feed additives that are available as of 1997. This presentation is in no way meant to be a recommendation. The authors recognize that the products represent valuable sale items for the manufacturers and for distributors who have the right to sell them and there is no intent to either discourage or encourage the sale of such items. Descriptions of proposed functions may have been abbreviated to save space, but not to omit any valuable information. Some products may have been inadvertently omitted.

Amprolium, chickens, for the control of coccidiosis.

Amprovine, (see Amprolium)

Ampralan, (see Apramycin)

Apramycin, pigs, for control of weanling scours.

Arsanilic Acid, chickens and turkeys, growth promotion, improved feed efficiency, improving pigmentation, control of coccidiosis; swine, increased rate of gain, improved feed efficiency, control of dysentary (bloody type).

Atgard, (see Dichlorvos)

Aureomycin, (see Chlortetracycline)

Avatec, (see Lasalocid)

Aviax, (see Semduramicin)

Baciferm®, (see Bacitracin Zinc)

Bacitracin Methylene Disalicylate, chickens and turkeys, increased rate of gain and improved feed efficiency; laying hens, increased egg production; broilers, control of necrotic enteritis, an aid in preventing coccidiosis, growth promotion; cattle, reduction in number of liver condemnations due to abscesses; swine, control of dysentary, increased rate of gain, and feed efficiency.

Bacitracin Zinc, chickens and turkeys, swine, and cattle increased rate of gain and improved feed efficiency; laying hens, increased egg production and improved feed efficiency.

Bambermycins, broilers, turkeys, swine, and cattle increased rate of gain and improved feed efficiency.

Banminth Premix-48, (see Pyrantel Tartrate)

Bio-Cox®, (see Salinomycin)

Bloat Guard®, (see Poloxalene)

BMD, (see Bacitracin Methylene Disalicylate)

Bovatec, (see Lasalocid)

Carbodox, swine, control of swine dysentary and bacterial swine enteritis, increased rate of gain, improved feed efficiency.

Cattlyst, (see Laidlomycin Propionate)

Chlortetracycline, chickens and turkeys, control of several infectious situations that are susceptible to chlortetracycline; swine, increased rate of gain and improved feed efficiency, control of certain infectious-type conditions that are susceptible to chlortetracycline; calves, increased gain, improved feed efficiency, control of certain strains of the "shipping fever" complex; older cattle, control of certain organisms that manifest as a type of pneumonia; sheep, increased rate of gain and improved feed efficiency; reduced incidence of vibrionic abortion.

CLTC, (see Chlortetracycline)

Coban®, (see Monensin)

Coyden®, (see Clopidol)

CTC, (see Chlortetracycline)

Cygro®, (see Maduramicin Ammonium)

Deccox®, (see Decoquinate)

Decoquinate, chickens (broilers), cattle, and sheep, for prevention of coccidiosis.

Dichlorvos, swine, for control and removal of certain intestinal parasites, an aid in improving litter production.

Erythromycin, chickens (layers-breeders), and turkeys, an aid in prevention and control of respiratory disease.

Fenbendazole, swine, for removal of lungworms and several gastrointestinal worms.

Flavomycin, (see Bambermycins)

Fortracin, (see Bacitracin Methylene Disalicylate)

Gainpro, (see Bambermycins)

Halofuginone Hydrobromide, chickens, for the prevention of certain types of coccidiosis (based on source of cause), improved feed efficiency in broilers; turkeys, increased rate of gain and prevention of coccidiosis.

Histostat, (see Nitarsone)

Hygromix, (see Hygromycin B)

Hygromycin B, chickens, control infestation of large roundworms and capillary worms; swine, control of large roundworms and whipworms.

Iveremectin, swine, treatment and control of gastrointestinal roundworms, kidney worms, lungworms, lice, and mange mites.

Laidlomycin Propionate, beef cattle (confined), improved feed efficiency and increased rate of gain.

Larvadex, chickens, control of soldier fly and housefly in and around caged or slatted floor laying hens and breeder chickens.

Lasalocid, chickens (broilers) and turkeys, prevention of coccidiosis; cattle, improved feed efficiency and increased rate of gain for cattle fed in confinement for slaughter; increased rate of gain for pasture cattle (slaughter, stocker, feeder cattle, and dairy and beef replacement heifers), control of coccidiosis; sheep, control of coccidiosis.

Levamisole Hydrochloride, cattle, controlling gastrointestinal worms, lungworms, intestinal worms, stomach worms; swine, controlling following nematode infections, large roundworms, nodular worms, lungworms, intestinal threadworms, kidney worms.

Lincomix, (see Lincomycin)

Lincomycin, chickens (broilers), control of necrotic enteritis, increased rate of gain, improved feed efficiency; swine, control of dysentary, increased rate of gain, improved feed efficiency.

Maduramicin Ammonium, chicken (broilers), prevention of coccidiosis.

Maxiban, (see Narasin/Nicarbizin)

Mecadox, (see Carbadox)

Melengesterol Acetate, cattle (beef heifers), increased gain, improved feed efficiency, suppression of estrus in heifers meant for slaughter.

Methoprene, cattle (beef and dairy), insect growth regulator during fly season to prevent the breeding of the horse fly in the manure of treated cattle.

MGA, (see Melengesterol Acetate)

Monensin, chickens, prevention of certain types of coccidiosis; turkeys, control of coccidiosis; cattle, improved feed efficiency, control of coccidiosis, increased rate of gain on pasture.

Monteban, (see Narasin)

Morantel Tartrate, cattle, control and removal of mature gastrointestinal nematodes including stomach worms of the small intestine and worms of the large intestine.

Mycostatin-20, (see Nystatin)

Narasin, chickens (broilers), control of coccidiosis.

Nicarb, (see Nicarbazin)

Nicarbazin, chickens, aids in the prevention of cecal and intestinal coccidiosis.

Nitarsone, chickens and turkeys as an aid in prevention of blackhead.

Novobiocin, chickens and turkeys, an aid in prevention of breast blisters associated with staphylococcal infections, aids in prevention of outbreak of fowl cholera.

Nystatin, chickens and turkeys, an aid in control of crop mycosis and mycotic diarrhea.

OTC, OXTC®, (see Oxytetracycline)

Oxytetracycline, chicken, increased rate of gain, improved feed efficiency, control of types of synovitis, chronic respiratory disease, air sac infection; turkeys, increased rate of gain, improved feed efficiency, control of hexamitiasis, synovitis; swine, increased rate of gain, improved feed efficiency, for treatment of bacterial enteritis, control of bacterial pneumonia; breeding swine, reducing abortions and spread of leptospirea; cattle, aid in control of shipping fever complex, increased rate of gain, improved feed efficiency, for treatment of bacterial enteritis, reduction of liver condemnations due to abscesses; sheep, increased rate of gain, improved feed efficiency.

Penicillin, chickens and turkeys, increased rate of gain, improved feed efficiency, prevention of chronic respiratory disease in chickens, prevention of bluecomb disease, prevention of sinusitis in turkeys; swine, increased rate of gain, improved feed efficiency.

Poloxalene, cattle, prevention of legume (alfalfa, clover) and wheat pasture bloat.

Pro-Gen, (see Arsanilic Acid)

Propylene Glycol, cattle, as an aid in the prevention of acetonemia (ketosis).

Pyrantel Tartrate, swine, aid in controlling large roundworm and nodular worm.

Rabon, cattle, horses, and swine prevents development of face flies, horn flies, house flies, and stable flies in the manure of treated animals.

Robenidine Hydrochloride, chickens (broilers), as an aid in prevention of coccidiosis.

Robenz, (see Robenidine Hydrochloride)

Roxarsone, chickens and turkeys, increased rate of gain, improved feed efficiency, improved pigmentation; swine, increased rate of gain, improved feed efficiency.

Rumatel, (see Morantel Tartrate)

Rumensin, (see Monensin)

Sacox, (see Salinomycin)

Safe-Guard, (see Fenbendazole)

Salinomycin, chickens, prevention of coccidiosis, improved feed efficiency.

Semduramicin, chickens (broilers), prevention of coccidiosis.

Sirlene, (see Propylene Glycol)

Stafac, (see Virginiamycin)

Stenerol, (see Halofuginone Hydrobromide)

Sulfamethazine, (see Chlortetracycline plus Sulfamethazine and Penicillin)

Sulfathiazole, (see Sulfamethazine)

Swine Oral Larvicide, (see Rabon)

TBZ, (see Thiabendazole)

Terramycin, (see Oxytetracycline)

Thiabendazole, cattle, sheep, and goats, control of infections of gastrointestinal roundworms; swine, aid in prevention of infections of large roundworms.

Tiamulin, swine, increased weight gain, improved feed efficiency, control of swine dysentary associated with *Serpulina hyodysenteriae.*

Tilmicosin, swine, control of respiratory disease.

Tramisol, (see Levamisole Hydrochloride)

Tylan, (see Tylosin)

Tylosin, chickens (growers, layers), increased weight gain, improved feed efficiency, aid in the control of chronic respiratory disease; swine, increased rate of gain, improved feed efficiency; cattle (beef), reduced incidence of liver abscesses.

V-Max, (see Virginiamycin)

Virginiamycin, chickens, increased weight gain of broilers, for prevention of necrotic enteritis caused by *Clostridium perfringens;* turkeys, increased weight gain and improved feed efficiency; swine, increased weight, improved feed efficiency, treatment and control of dysentary; cattle (fed in confinement and for slaughter), increased weight gain, improved feed efficiency, reduction in incidence of liver abscesses.

Zinc Bacitracin, (see Bacitracin Zinc)

Zoalene, chickens and turkeys (grown for meat purposes only), aid in prevention and control of coccidiosis.

Zoamix, (see Zoalene)

67 Feeding Animal Wastes

I. ___ Every livestock and poultry operation is confronted with the problem of waste disposal.

A. Livestock and poultry wastes have, over the years, been disposed of principally as a soil fertilizer. As livestock and poultry production have been intensified on certain farms and in certain areas, the limited land acreage available nearby and the restricted rate of waste application that can be tolerated on crop land has caused the matter of livestock and poultry waste disposal to become a major problem.

B. On many farms, pastures have been so heavily covered with waste that they have become unsuited for grazing purposes. Problems such as nitrate poisoning, grass tetany, and fat necrosis have been associated with the excessive application of animal wastes to pasture land.

C. In many instances of heavy waste production and disposal, available runoff systems have proven to be inadequate for preventing stream contamination. This is especially true in view of the greater emphasis currently being placed on pollution control.

D. The growth of residential developments near areas of intensive livestock and/or poultry production has also complicated the matter of waste disposal.

482

II _____ As a result of the above developments, the need for an alternate method for the economical disposal of livestock and poultry waste has been recognized. Not only must the method, to be effective, be able to handle sizeable amounts of the waste, but it must do so at a cost that would make it practical to use in production operations. It has been proposed by certain workers in the field that the disposal of animal wastes might be facilitated through their use as a feed material.

A. It has been pointed out that waste products consist primarily of feces and urine plus any bedding that has been used, along with any water added either inadvertently or on purpose during cleaning and/or accumulation.

1. The feces dry matter consists mostly of undigested feed plus a small but variable amount of metabolic excretions.

2. The urine dry matter consists entirely of metabolic excretions.

3. Bedding materials usually consist of any of several different low-moisture, high-fiber products of varying feeding value.

B. While feces consist to a large degree of undigested feed and for all practical purposes are considered to be indigestible, this is not to say that fed again in proper combinations with other feed materials they would not be subject to further digestion.

C. While urine dry matter is all water soluble and, for the most part, is in a highly absorbable form, it is quantitatively a very small part of the overall waste dry matter.

1. Other than nonprotein nitrogen, it contains little if any nutritive materials that could be of practical feeding value.

2. Even the nonprotein nitrogen (NPN) must be converted into protein by microbial synthesis before it is in a form that an animal can use.

3. However, properly handled, animal urine can be a valuable source of NPN and, in turn, protein for ruminant animals.

4. The metabolic excretions of the kidneys, which in most animals are eliminated in the urine, are secreted into the cloaca of poultry and eliminated along with the fecal material through the anus.

D. The bedding portion of animal wastes, as already indicated, usually consists of dry, high-fiber materials such as straws, low-quality hays, hulls, cobs, wood shavings, sawdust, and possibly others.

1. Broiler (poultry) operations commonly use wood shavings or sawdust, when available, but are by no means restricted to these materials. Other materials are sometimes used for bedding purposes by broiler operators but these are usually limited to moderately small particled products or coarsely ground materials.

2. Egg producers, most of whom presently have caged layer or similar operations, usually use no bedding and consequently normally deal with a wetter but less fibrous product than where a litter base has been used.

3. Swine operations use, when needed, bedding material of the same general type as is used by poultry producers; however, many modern swine producers do not use any bedding.

   a. They either use a solid concrete floor, which is hosed off at frequent intervals, or a slotted floor with all excreta being collected in a pit below.

   b. It seems that bedding is not necessary for a hog's warmth, except in extremely cold weather, nor is it otherwise necessary for the animal's comfort and well-being.

   c. Since sanitation can be more adequately maintained through the use of solid concrete or slotted floors, bedding has been eliminated in many swine operations.

   d. Bedding-free swine waste is usually quite high in water but relatively low in fibrous material; however, its dry matter is normally quite high in other nutrients.

4. Beef cattle, dairy cattle, horses, and sheep are most commonly bedded, when bedding is needed, with straw and low-quality hay. This is not to say, however, that other materials are not sometimes used as bedding for these livestock classes. The need for bedding is dispensed with in some instances through the use of slotted floors, especially with beef cattle on feed.

III

The chemical composition of the waste from different livestock species differs greatly, as does the composition of the waste from different livestock operations within the same species.

A. One of the major factors influencing the composition of waste dry matter is the composition of the ration being fed. Since poultry, dairy cattle, and swine are usually fed rations higher in protein than are beef cattle, horses, and sheep, waste from the former is usually considerably higher in total nitrogen than that from the latter.

B. Since most bedding materials are quite high in crude fiber and relatively low in other nutritive components, their presence in waste material tends to contribute a similar composition to the overall material; consequently, bedding-free waste is usually higher in the more valuable nutrients, on a dry matter basis.

C. On the other hand, bedding-free waste is usually characterized by a great deal of moisture, making it difficult and unpleasant to store and use. While the storage and use of wastes might be greatly facilitated by drying, drying is rapidly becoming an unpopular practice in view of the prevailing energy shortage and the high cost of labor.

D. Another factor that can have a major influence on the composition of waste—especially its nitrogen content—is the conditions under which it is stored and the length of time held in storage. Moisture is especially conducive to the loss of nitrogen as ammonia from waste in storage, particularly during hot weather. Losses of one-half or more of the nitrogen from waste during accumulation are not uncommon.

IV Of the various types of livestock and poultry waste available, that from poultry is presently being used the most widely for feeding purposes.

A. Poultry wastes at the present are, for the most part, of two general types, that from broiler houses and that from layer hen operations.

  1. Most broilers are produced in bedded houses.

   a. Wood shavings is the most commonly used bedding material for broiler houses in most areas; however, such products as sawdust, peanut hulls, cottonseed hulls, oat hulls, rice hulls, ground corn cobs, and ground sugarcane bagasse have, on occasion, been used for this purpose.

   b. Sometimes a broiler house is cleaned out between each batch of birds with fresh bedding being provided for each batch.

   c. In other instances, the house is cleaned only after every 2–5 batches of birds with only a limited amount of new bedding being added with each new batch.

  2. Layer hens, on the other hand, are usually housed in cages or on wire.

   a. With either arrangement, the hen's excreta falls through and accumulates on the ground or floor below.

   b. The excreta sometimes is collected on a belt that moves it in relatively fresh form into some type of dehydration unit for drying.

   c. Otherwise, the accumulated droppings are removed at the operator's convenience with whatever equipment is available.

   d. Very little if any bedding is normally involved.

B. Both broiler and layer wastes are currently being widely used in one form or another as a feed for beef cattle.

  1. One of the major factors influencing the use of poultry waste for feeding purposes is that of water content. Both fresh broiler excreta and fresh layer excreta are quite high in moisture content (71%–76%), making both unsuitable for use as a feed ingredient in the fresh form.

   a. Excreta accumulations under hens dry very slowly by natural processes.

    • Sometimes they are artificially dried to make them suitable for storage and use as a feed material.

- Drying also prevents their undue decomposition and loss of ammonia.

- Layer waste is also sometimes used in a semifresh or partially dried form as a feed material.

- In the above form, it is sometimes used as a silage component or as a component of mixed rations.

- The levels used in either instance will depend on the circumstances.

- While pure poultry waste does not noticeably mold at any moisture level, feed mixtures containing wet waste do tend to mold and must be mixed at frequent intervals to avoid mold damage.

- The frequency of mixing necessary depends on the prevailing environmental temperature, the amount of moisture in the waste material being used, and the percentage of the product being included in the mixture.

- In the absence of established data, the necessary frequency of mixing can be determined only by trial and error. Daily mixing is usually best.

- If the waste is used in making silage, it is usually in conjunction with a highly fermentable, high-fiber crop such as corn, grain sorghum, or sweet sorghum forage and makes up not more than about 1/3 of the silage dry matter.

b. Broiler excreta, on the other hand, which is normally accumulated on a litter base, undergoes considerable drying during accumulation; also, the litter material that is normally quite dry tends to lower the average moisture content of the overall waste material.

- Consequently, litter-based broiler waste, which is quite variable in moisture content, under average management will usually run from 20% to 30% moisture at the time it is removed from the house.

- At cleaning-out time, it is sometimes placed in an open shed in small piles (about 4–5 feet deep) and allowed to dry further (Fig. 67–1).

- Under such conditions, the waste normally goes through a heat that tends to have a self-drying effect.

- Broiler waste dries slowly under such an arrangement and tends to lose a certain amount of nitrogen and energy.

- Broiler waste handled in this manner normally does not suffer from mold damage.

- It is usually suitable for use in a dry mixed ration in about 6 to 8 week's time, or less.

- However, if the moisture content of the waste at this point is still considerable, it will be necessary to mix the ration at frequent intervals to avoid mold damage.

FIGURE 67–1
Sawdust based broiler waste stockpiled under polyethylene
cover for use as needed for feeding purposes. (Courtesy of the
University of Georgia College of Agriculture Experiment Stations)

- As with the layer waste, the frequency of mixing will depend on the
  prevailing environmental temperatures, the percentage of moisture in
  the waste, and the amount of the material being used in the ration.

- While daily mixing is usually not necessary with such rations, in the
  absence of established data one can be guided as to length of mixing
  interval only by trial and error

c. In piling broiler waste in an open shed for drying and/or storage, care must
   be exercised not to create a set of conditions that can produce spontaneous
   combustion.

- While these conditions are not definitely established, they are known
  to involve a moisture level comparable to that of the waste from many
  broiler houses, a considerable depth of pile, entrance of air into the pile
  at a certain unknown rate, and possibly other requisites.

- About the only practical precaution that one can take against spontaneous combustion of broiler waste is not to pile it too deeply.

- While the authors have personally used 4-foot-deep piles without
  complications, they cannot guarantee that this depth of pile will be
  safe under all conditions.

- Neither can they state that some greater depth might not be safe to use; however a pile approximately 10 ft high is known to have burned on one occasion.

- During the past few years, in view of the moisture content of broiler waste and the trouble and expense of drying it, there has been considerable interest in the ensiling of this material for feeding purposes (discussed subsequently).

C. Nutrient composition of poultry manures.

    1. Crude protein.

        a. Broiler excreta produced by birds on a conventional 24% crude protein diet can be expected to have about 35% crude protein on a dry basis if collected under favorable conditions.

        b. It appears that well-preserved layer hen excreta will run this high or higher in crude protein content.

        c. Layer hen excreta, when accumulated in an undried state, may end up with only a fraction of its original nitrogen content, depending on the degree of decomposition and ammonia loss it has undergone.

        d. Litter-based broiler waste will usually run from 20% to 30% crude protein, dry basis, although individual samples of such waste will frequently run on either side of this range.

        e. The proportion of excreta to litter base is a major factor in determining the crude protein level in the overall waste material, as is the amount of decomposition and resulting loss of ammonia.

- Since most litter bases are low in the various nutritive fractions other than crude fiber, the more of it present in relation to the excreta, the higher the fiber will be and the lower will be the other nutritive fractions in the overall waste product.

- The proportion of excreta to litter base will depend largely on the amount of litter base used initially, the number of groups of birds produced in a house between cleanings, and the amount of fresh bedding added with each new group.

- The prevailing temperature and moisture conditions under which the manure was accumulated are major factors in determining the amount of decomposition and ammonia loss, and, in turn, its crude protein content at cleaning-out time.

- Cool, dry conditions seem to be conducive to holding crude protein losses down.

        f. About half of the total crude protein of poultry wastes is true protein.

- The actual amount seems to range from about 1/3 to 2/3 depending on the circumstances.

- This is probably related to the degree of decomposition.

- The percentage of the crude protein of poultry waste that is true protein does not seem to be greatly influenced by whether it is broiler or layer waste.

- In any event, a relatively high percentage of the crude protein of poultry waste is present in the form of nonprotein nitrogen materials.

2. Crude fiber.

   a. On the basis of available data, it would appear that both broiler excreta and layer excreta in general will run from about 10% to 15% crude fiber, dry basis.

      - The actual level of fiber in the excreta will be determined primarily by the level of the fiber in the feed fed.
      - Since layers are ordinarily fed more fiber than are broilers, layer excreta should run somewhat higher than broiler excreta in fiber content.

   b. Since most commonly used litter bases are quite high in crude fiber, litter-based poultry wastes are also normally quite high in this nutrient, usually running between 15% and 30% depending on the litter base used and the percentage of it in the overall waste material.

3. Ether extract.

   a. Poultry wastes in general are quite low in ether extract, usually running from 1% to 4% in this nutritive fraction.

   b. Figures on either side of the above range are not at all uncommon.

4. Ash.

   a. Poultry wastes vary greatly in ash content but in general are relatively high in this chemical fraction.

   b. Based on available data, it would appear that broiler excreta will run about 20% ash on a dry basis while layer excreta will run from 25% to 40%.

   c. Since most bedding materials will run much lower than the above in ash content, the respective litter-based wastes should run below the above figures in percentage of ash.

   d. A considerable amount of soil from the floor of the house is sometimes removed with the waste as a house is cleaned, and waste samples containing more than 50% ash on a dry basis are not uncommon.

   e. Except for the calcium and phosphorus present in the ash, this fraction can contribute little of a positive nature to the feeding value of the waste.

f. The higher the level of ash in a feed sample, other things being equal, the lower will be its energy value.

5. Nitrogen-free extract.

a. On the basis of available data, it appears that poultry wastes in general usually run from about 25% to 35% nitrogen-free extract, but individual samples will frequently show a nitrogen-free extract (NFE) level outside of this range.

b. The above level seems to prevail whether it is broiler or layer waste and whether it is litter-based waste or pure excreta, which would seem to indicate that litter base materials and pure poultry excreta are similar in NFE content.

c. While we usually think of NFE as consisting mostly of sugars and starch and while 25%–35% of sugars and starch would be a significant level of these materials in any product, it is doubtful that poultry waste NFE is made up mostly of these two fractions.

d. It would appear that poultry waste NFE contains a high percentage of lignins and hemicelluloses of low digestibility.

6. Calcium and phosphorus.

a. While available data on the calcium and phosphorus content of broiler and layer excreta is quite limited, that which is available indicates that both excreta are relatively high in these nutrients.

b. It appears that layer waste in general will run considerably higher than broiler waste in calcium content, no doubt as a result of the higher level of calcium normally included in layer diets.

c. Litter-based broiler waste seems to run about 2% to 3% in calcium and a little lower than this in phosphorus, on a dry basis.

d. Layer excreta, while normally running about 2% to 3% phosphorus, will usually contain about 7% to 9% calcium.

7. Total digestible nutrients.

a. In digestion trials with sheep, Fontenot and co-workers at Virginia in 1966 found litter-based broiler waste to have an average calculated TDN value of 59.8%, dry basis, when fed at levels of either 25% or 50% of the ration.

b. In digestion trials with cows, the senior author realized an average calculated TDN value, dry basis, of 31.8% on pine-shavings-based broiler waste when fed at a 75% level in the ration.

c. A possible explanation for the difference in the TDN values realized is that the broiler waste could have a positive associative effect on the digestibility of the remainder of the ration at lower levels of waste feeding (up to about 50% waste) but a negative effect on the digestibility of the remainder of the ration, as well as on its own digestibility, at higher levels of this material.

   d. It would appear that litter-based poultry wastes are, at best, low in energy value, and at high levels of feeding this material may be an extremely poor source of feed energy.

V **As a feed material, poultry wastes are used primarily as feed for beef cattle.**

A. The use of poultry wastes as a cattle feed has been primarily as a component of rations for finishing steers (Fig. 67–2).

   1. While litter-based broiler wastes have been most widely used, litter-free layer waste is receiving increasing attention in this connection.

   2. The broiler wastes are usually used in the semi-air-dry state directly from the broiler house or from a stock pile.

   3. Frequently broiler waste is stored and used in the form of silage.

FIGURE 67–2
These steer calves had an average daily gain of 2.45 lb (1.11 kg) over a 143-day feeding period on a ration that was 70% ground shelled corn and 30% broiler waste on a dry weight basis. (Courtesy of the University of Georgia College of Agriculture Experiment Stations)

4. Litter-free layer waste has been used in both the "as-is" and the dehydrated forms for steer-feeding purposes.

   a. Dehydrated layer waste, however, is costly to produce and has not given as good a performance as expected in feeding trials.

   b. Certain producers, however, claim satisfactory results from using layer waste in the "as-is" form in rations for finishing steers.

5. The nutritional value of poultry wastes for the most part lies in their content of crude protein and in their content of calcium and phosphorus.

6. Litter-based wastes are also useful for meeting an animal's needs for roughage.

7. Since about 35% to 65% of the crude protein in poultry waste is in the form of nonprotein nitrogen, it is most effectively used as a source of protein by ruminant animals, and particularly by those on grain feeding; the finishing steer particularly meets these requirements.

8. If litter-based broiler waste is used to make up approximately 25% of the ration for a fattening steer (dry basis), the protein needs of the animal will usually be more than met; however, this might not be true when the waste has undergone considerable ammonia loss.

9. The litter-based portion of this amount of waste will usually meet the steer's need for roughage.

10. It will also usually more than meet the animal's need for calcium and phosphorus.

11. Either semi–air-dried or ensiled waste can be used in this connection with about equal effectiveness.

12. Top performance can be expected from steers where a ration of litter-based broiler waste and ground shelled corn in the ratio of 25:75 is used, dry basis.

13. While litter-free layer waste may be used as a source of crude protein, calcium, and phosphorus for finishing steers, it would not provide the needed roughage.

14. Under circumstances where poultry waste is abundant and cheap, feeders sometimes use more than is needed to meet the protein, calcium, phosphorus, and roughage requirements.

   a. The above is especially true for broiler waste and as much as 40% to 50% of it is sometimes included in steer finishing rations with satisfactory results.

   b. While rate of gain usually goes down as the level of waste increases to more than about 25%, the feeding of this much waste is in some instances a profitable practice.

   c. However, the use of more than about 50% broiler waste in a steer finishing ration is of doubtful merit and as much as 75% broiler waste along with 25% corn in a finishing ration for calves is known to produce unsatisfactory results.

- Gains are low from the start and after the first 50 to 60 days on such a ration, some of the animals will start losing weight and taking on the appearance of severe unthriftiness.

- The unthrifty appearance can be reversed by transferring these animals to a conventional ration.

15. There have been no reports of infectious diseases attributed to the feeding of waste-containing rations.

16. Meat from steers produced on waste-containing rations has been just as acceptable as that of control animals.

17. No excessive levels of harmful residues have been reported in the meat from steers finished on waste-containing rations.

B. Use of poultry wastes in beef cattle wintering and stocker rations.

1. Poultry waste has proven to be a satisfactory source of crude protein, calcium, and phosphorus for wintering cows and heifers, and for stocker animals.

2. Litter-based wastes will also meet the "roughage" needs of such animals.

3. Under circumstances where poultry waste is abundant and cheap it has been used as a major part (more than 50%) of the wintering ration for cows and heifers with no ill effects.

4. However, infectious diseases may be transmitted just as readily through waste-containing rations as through conventional rations.

C. Use of poultry waste as a feed for dairy cattle.

1. Published research on the use of poultry waste in rations for dairy cattle to date is quite limited and the authors are not aware that the dairy industry widely uses this material for feeding purposes.

2. There is no reason to think, however, that the nutrients in poultry waste would not be used as effectively by dairy cattle as by beef animals.

3. Before poultry waste is used, as a general practice, for dairy cattle feeding, more research should be conducted on the effect, if any, of this material on the milk at all practical levels of feeding.

D. Feeding poultry waste to other classes of livestock.

1. Sheep, like cattle, are ruminants and from that standpoint should be able to handle the NPN and fiber of poultry waste without difficulty.

a. However, sheep are more sensitive to copper than are cattle and can be killed by the copper present in some poultry wastes.

b. For this and perhaps other reasons poultry waste is not widely used as a sheep feed.

2.   There are isolated reports of horses having eaten poultry waste but these are not many. As subject as horses are to digestive disorders from eating low-quality feed, it would not seem advisable to use poultry waste in horse rations.

3.   Since hogs are nonruminants and have no means for utilizing either NPN or fiber, poultry wastes in their existing form are not regarded as suitable for swine-feeding purposes.

VI

Use of cattle waste as a feed. Other than poultry waste, about the only type of animal waste that has been used for feeding purposes is cattle waste.

A.   Floor scrapings from cattle feedlots have been used effectively as components of cattle rations. Anthony at Auburn University in 1974 (unpublished data) reported 9.4% increased gain and 24.1% reduction in amount of basal ration required per pound of gain from the inclusion of 40% wet waste, collected from the feeding floor, with the basal mix. A reduction of 32.5% in amount of basal ration required per pound of gain was realized when 60% wet waste was mixed with the basal ration but there was an 18.4% lower daily gain on this treatment.

B.   While sometimes used as a ration component in the unprocessed form, such waste is more frequently mixed with a fermentable forage and made into silage for use in cattle rations.

C.   The more nutritious the ration is, the greater the feeding value of the waste.

1.   Waste produced on a high-grain finishing ration will have a feeding value about equal to fair-quality hay.

2.   The waste produced on a low-quality roughage ration will have essentially no feeding value other than to serve as a source of bulk.

D.   The proportion of cattle waste to be used in a ration or a silage mixture should be gauged accordingly.

E.   About half of the waste from a cattle finishing operation is as much as one can hope to recycle through the cattle producing it.

F.   The rest, if it is to be fed, would have to be used for feeding other cattle such as stocker or breeding animals.

G.   In fact, cattle waste is better suited as a feed for stocker or breeding cattle than it is for finishing animals.

VII

Use of swine waste as a feed.

A.   Research on the use of swine waste for feeding purposes has, for the most part, taken 2 approaches.

1. One has been to dry the swine waste and recycle it as a swine feed ingredient.

2. The other has been to subject the wet swine feces to biological degradation and then use the resulting microorganismal growth in either the wet or dried form for swine-feeding purposes.

B. While a certain amount of favorable results have been obtained from using both approaches, as the authors are aware, the recycling of swine waste as a swine feed has not yet been accepted by the swine industry.

C. The use of swine waste for feeding other classes of livestock has not been investigated.

VIII   Ensiling poultry wastes.

A. During the past few years, there has been considerable interest in the ensiling of broiler waste for feeding purposes.

1. Some producers are sprinkling down their broiler houses prior to cleaning and then placing the dampened waste material containing around 35% to 40% moisture in a silo.

2. If the silo is an oxygen-limiting structure and the waste is in good condition containing 35% to 40% moisture, the resulting silage is well preserved and comes out as a slightly acid, pleasant smelling product.

3. If the silo is not an oxygen-limiting structure, there will be a considerable loss of ammonia during storage and the product comes out smelling of ammonia with a pH of around 8 to 10.

   a. In spite of its high pH, the product is, for the most part, free of mold and is readily eaten by cattle after they have been accustomed to it and it has been mixed with other feed.

   b. Adding to the waste at storage time about 10% of a highly fermentable carbohydrate such as ground shelled corn or cane molasses will provide a more typical silage fermentation and produce a more typical silage product.

B. Because of its high moisture content and low fiber content, layer waste does not make a satisfactory product for ensiling purposes if used alone.

C. Poultry wastes are sometimes added to forage that is being used for silage to raise the silage's protein content.

1. Either layer waste or broiler waste may be used for this purpose.

2. If the forage is of good quality, such as corn fodder, as much as 1/3 to 1/2 of the silage dry matter may come from the waste without interfering with the quality of the fermentation, under good ensiling conditions.

IX

One of the major concerns that has prevailed in connection with using waste as a livestock feed has been whether or not such products would be approved by Federal and State officials for livestock feeding purposes.

A. To date, the Food and Drug Administration has taken no official position toward wastes as a feed ingredient.

B. Several states have given approval to the distribution and sale of such products, mainly poultry wastes in one form or another, for feeding purposes. In all such instances, however, regulations have been established that attempt to protect the consumer against potential health hazards and prevent fraudulent practices in this connection.

C. To some people the thought of eating meat produced with rations containing waste is repulsive. Consider the following facts, however:

1. What a steer uses from the waste in its ration undergoes thorough digestion before it is deposited in the animal as meat.

2. Poultry waste or any other waste used in a steer's ration is never in any closer contact with the meat of the steer than it was with the meat of the animal that produced it.

3. Poultry waste or any other waste used in a steer's ration is never in any closer contact with the meat of the steer than is the waste resulting from the rest of the steer's ration.

4. No objectional features have ever been associated with the meat from animals fed rations containing animal waste.

68 Aflatoxin in Feeds

I

Most people are familiar with mold in one form or another. Mold is the wooly-like growth that frequently develops on the surface of damp organic matter held under aerobic conditions at temperatures between 45°F and 100°F. There are many different types and strains.

II

Molds produce *metabolites,* many of which are toxic to other organisms. Such toxic metabolites are referred to as *mycotoxins.* Certain mycotoxins are harmful to large animals, including livestock. Such mycotoxins are sometimes present in feed that has been held under mold-producing conditions.

III

Of the many different mycotoxins that sometimes occur in feeds, the aflatoxin group seems to pose the greatest threat to animal health and efficient livestock production.

A. The aflatoxin group embraces several different types. Aflatoxin B_1, B_2, G_1, and G_2 are the most common.

B. Of the different aflatoxins, aflatoxin B_1 is usually the most prevalent and the most toxic. The other types rank as follows, in descending order of toxicity: G_1, B_2, G_2.

C. Aflatoxin is produced primarily by the mold *Aspergillus flavus* although other *Aspergillus* molds are sometimes present and aflatoxin-like mycotoxins are produced by other molds.

D. The aflatoxin-producing mold *Aspergillus flavus* is most likely to occur at harmful levels in grains and seeds that have been stored too wet for an extended period but may, under certain conditions, develop prior to harvest.

    1. Corn has received the greatest attention as a possible aflatoxin source, but barley, grain sorghum, wheat, cottonseed, peanuts, and soybeans as well as processed feeds produced from these grains and seeds have sometimes shown a high level of aflatoxin.

    2. Grains and seeds that have had insect damage in the field are especially susceptible to preharvest mold infestation. Mechanical damage to grains and seeds at the time of harvest is also conducive to mold development.

    3. A proper state of dryness probably offers the surest protection against mold damage to grains, seeds, and other feeds in storage.

        a. Moisture levels of more than 13% in grains and seeds are generally favorable for mold growth in storage, although this varies depending on factors such as the following:

- Type of grain or seed
- Temperature
- Humidity
- Amount of aeration
- Time in storage

        b. To avoid mold formation in processed feeds, the moisture level should not exceed about 11%, depending on the circumstances.

        c. Leaky bins and bins in which sweating occurs will tend to produce pockets of grain, seed, or feed that contain mold.

        d. The treatment of high-moisture feed grains with propionic acid or some similar preservative is sometimes used to keep down mold formation.

    4. Aflatoxin-producing molds are most active at 75°F to 90°F. Cold weather, of course, tends to check mold growth.

E. Should grains, seeds, or other products that are to be fed to livestock or used in preparing livestock feeds show any evidence of mold, they should be checked for aflatoxin before being used and then be used in a manner that is commensurate with their aflatoxin content.

    1. Grains and seeds that show mold damage may be given a preliminary check for the presence of aflatoxin by exposing such products to black light and observing them for a bright-greenish-yellow (BGY) fluorescence, which would indicate aflatoxin contamination. However, this is basically a test for the presence of aflatoxin and does not provide a measure of the amount present. It is effective with whole grains only.

2. Should the presence of aflatoxin be indicated in a feed material by black light examination, then it should be subjected to a quantitative aflatoxin determination before being used for feeding purposes.

   a. This must be done in a laboratory that is especially equipped to carry out the aflatoxin determination.

   b. About 3–4 lbs of a representative sample should be submitted in a container for this analysis.

   c. Aflatoxin determinations are ordinarily made by either thin layer or column chromatography.

   d. The different aflatoxin types are normally quantitated separately and then totaled together.

   e. Aflatoxin levels are normally reported in parts per billion (ppb), which is the same as micrograms per kilogram (mcg/kg).

   f. The level of aflatoxin present in a feed material should dictate its use for feeding purposes, as discussed subsequently.

3. Should a group of animals be performing abnormally poorly on a particular ration, and there is no obvious explanation for such performance, then their feed should be analyzed for aflatoxin and the feeder be guided accordingly.

F. Aflatoxin is thought to have its harmful effects on livestock in several possible ways.

1. It may disrupt the normal fermentation processes of the rumen, causing inefficient digestion.

2. A disruption of the rumen fermentation might cause a breakdown in the normal biosynthesis of critical amino acids and vitamins, thereby resulting in a deficiency of certain nutrients.

3. The aflatoxin may be absorbed from the digestive tract into the body's system where it disrupts certain metabolic processes.

4. It has been suggested that aflatoxin harms animals by interfering with proper antibody production, thereby making them more susceptible to disease.

5. Aflatoxin is considered to be a potent carcinogen.

G. Following are some of the symptoms that have been associated with aflatoxicosis in one or more classes of livestock.

1. Off feed
2. Reduced feed consumption
3. Reduced gains
4. Lowered milk production
5. Inefficient feed utilization
6. Diarrhea
7. Scours
8. Bloody diarrhea
9. Enteritis
10. GI tract lesions
11. Prolapsed rectum
12. Enlarged gall bladder
13. Enlarged liver
14. Greyish-colored livers
15. Fibrosis of the liver
16. Liver lesions
17. Liver carcinomas
18. Polydipsea
19. Polyuria
20. Dehydration
21. Enlargement of kidneys
22. Kidney lesions
23. Necrosis in renal tubes
24. Breeding problems
25. Abortion
26. Metritis
27. Enlarged vulva
28. Respiratory problems
29. Nasal discharge
30. Runny eyes
31. Droopy ears
32. Dry muzzles
33. Rough haircoat
34. Loss of hair
35. Internal hemorrhaging
36. Increase in mastitis
37. Aflatoxin in milk
38. Sluggishness
39. Depression
40. Lethargy
41. Death

H. Table 68–1 shows approximate maximum levels of aflatoxin that may be present in the ration of different classes of livestock without doing harm to the animal or contaminating food produced by (milk) or from (meat) the animal.

1. Since aflatoxin is a recognized carcinogen, there is a zero tolerance of it in all food products, including meat and milk.

2. Aflatoxin can be detected in milk from cows on a ration containing as little as 15–20 ppb of this contaminant on an air-dry basis.

3. Milk will be free of aflatoxin after 4 days of feeding noncontaminated feed.

4. Young animals are more susceptible to harm from aflatoxin than are older animals; consequently it is necessary to keep the aflatoxin at relatively low levels in rations for females nursing young and in creep feeds for young animals.

5. Swine are somewhat more subject to aflatoxin poisoning than are cattle, and rations should be formulated with respect to their content of this material.

6. Aflatoxin can affect an animal's breeding performance; hence, rations for breeding animals should be held at a lower level of aflatoxin than those for feedlot animals.

7. There is some evidence that aflatoxin is less harmful to sheep than to cattle; however, until more results are available, their tolerance of aflatoxin will be shown to be comparable to that of cattle in this text.

Table 68–1
SUGGESTED MAXIMUM LEVELS OF AFLATOXIN FOR DIFFERENT CLASSES OF LIVESTOCK

Class of Livestock	Suggested Maximum Level of Aflatoxin in Air-Dry Ration ppb (mcg/kg)
Lactating dairy cows	20
Young pigs	20 (In starter or creep ration)
Young colts	20 (In creep ration)
Cows nursing young	100
Ewes nursing young	100
Mares nursing young	100
Sows nursing young	100
Young calves	100 (In creep ration)
Young lambs	100 (In creep ration)
Dry sows and boars	200
Dry mares, geldings, and stallions	200
Growing-finishing pigs	200
Dry cows and bulls	400
Dry ewes and rams	400
Feedlot cattle	500
Feedlot lambs	500
None	More than 500

8. There has been very little research done on the aflatoxin tolerance of horses; it has been assumed, however, that they would be more like the hog than the cow in this regard since they are also nonruminants.

9. No aflatoxin has been found in the tissues of animals fed rations containing no more aflatoxin than that suggested as a maximum for a particular class of livestock, as shown in Table 68–1.

I. Since the aflatoxin level of the total ration determines the acceptability of the ration for feeding purposes from an aflatoxin standpoint, many farmers follow the practice of mixing an aflatoxin-contaminated feed with aflatoxin-free feeds to lower the average aflatoxin level of the overall mixture to a point that will permit its use for the livestock being fed.

1. In order to carry out such a dilution procedure, it is necessary to know the aflatoxin level in the contaminated feed or feeds as determined by laboratory analysis.

2. Usually only one aflatoxin-contaminated feed is involved.

3. The other feeds in the ration are then considered as a single quantity in calculating the average level of aflatoxin in the overall ration.

4. The contaminated feed does not have to be physically mixed with the rest of the ration to realize the diluting effect of the noncontaminated ration components as long as the latter are received by the animal as a part of its daily ration.

5. Where the aflatoxin level in the individual feed components is known the level of aflatoxin in an overall ration may be calculated by using the following format:

Ration Components	% in Ration (kg/100 kg)	Aflatoxin Level ppb (mcg/kg)	Total Aflatoxin mcg
Contaminated corn	70	200	14,000
Other feeds	30	0	0
Total ration	100	x	14,000

$$x = \frac{14,000}{100} = 140 \text{ mcg/kg or 140 ppb}$$

6. Let us assume that 2 of the ration components contained aflatoxin at different levels and one desires to know the aflatoxin level of the overall ration. The calculation would be made as follows:

Ration Components	% in Ration (kg/100 kg)	Aflatoxin Level ppb (mcg/kg)	Total Aflatoxin mcg
Contaminated corn	70	200	14,000
Contaminated CSM	10	800	8,000
Other feeds	20	0	0
Total ration	100	x	22,000

$$x = \frac{22,000}{100} = 220 \text{ mcg/kg or 220 ppb}$$

7. It is frequently necessary to determine how much of an aflatoxin-contaminated feed can be used in a ration without exceeding the tolerance level for the animals being fed. The same basic format used in making the preceding calculations would be used for making this determination.

8. Let us assume that we wanted to use a lot of corn that contains 487 ppb of aflatoxin to formulate a ration for a class of livestock with a maximum tolerance level of 200 ppb aflatoxin. In using the preceding format to make this calculation, we simply start out knowing how much aflatoxin can be tolerated in the overall ration and then work backwards to arrive at how much of the contaminated feed can be used without exceeding the maximum tolerance level, which in this instance is 200 ppb. Using the preceding format, then, the calculation would be made as follows:

Ration Components	% in Ration (kg/100 kg)	Aflatoxin Level ppb (mcg/kg)	Total Aflatoxin mcg
Contaminated corn	x	447	20,000
Other feeds	100 − x	0	0
Total ration	100	299	20,000

$$x = \frac{20,000}{447} = 41.1\% \qquad 100 - x = 58.9\%$$

9. The "other feeds" in this calculation would include all the aflatoxin-free feeds used in the ration. These feeds would embrace the protein, mineral, and vitamin supplements as well as other sources of energy, as long as they contained no aflatoxin. Even additional corn may be included in this group as long as it is aflatoxin-free.

10. Should one desire to use more than 1 contaminated feed in a ration that is not to exceed a certain aflatoxin level, it is necessary to use them in a certain ratio. Once the aflatoxin level of the combined feeds is calculated, the maximum level of these combined feeds in the ration should be determined. The ratio of the 2 feeds to use would have to be an arbitrary decision of the feeder depending on the following:

    a. The feeds involved.

    b. The aflatoxin level of each feed.

    c. The class of livestock to be fed.

    d. The amount of the different feeds on hand in relation to the need for each in the ration.

11. To illustrate this calculation, let us assume that we had a quantity of corn containing 200 ppb aflatoxin and some cottonseed meal carrying 800 ppb that we wanted to work into a ration for growing-finishing shotes. We have about 10 times as much corn as CSM. Also, it is known that we cannot use more than about 9% CSM in the ration for growing-finishing swine without danger of gossypol poisoning. Consequently, it was decided to use the corn and CSM in the ratio of 10 parts corn to 1 part CSM. With this ratio of corn, CSM would use the 2 feeds up simultaneously and would not exceed the limit of 9% in the use of CSM in the ration for growing-finishing swine. Both feeds would serve a worthwhile purpose in the nutrition of this class of livestock.

12. The first step, then, is to calculate the aflatoxin level of the 10 corn:1 CSM combination as follows:

Ration Components	% in Combination (kg/100 kg)	Aflatoxin Level ppb (mcg/kg)	Total Aflatoxin mcg
Corn	91	200	18,200
CSM	9	800	7,200
Corn-CSM Comb.	100	x	25,400

$$x = \frac{25,400}{100} = 254 \text{ mcg/kg or } 254 \text{ ppb}$$

13. Using the above figure (254 ppb), it is possible to calculate the amount of the contaminated corn-CSM combination that can be used in the ration without exceeding the maximum permissible level as follows:

Ration Components	% in Ration (kg/100 kg)	Aflatoxin Level ppb (mcg/kg)	Total Aflatoxin mcg
Corn + CSM Comb.	x	254	20,000
Other feeds	100 – x	0	0
Total ration	100	200	20,000

$$x = \frac{20,000}{254} = 78.7 \text{ kg of corn-CSM comb. per 100 kg of ration (\%)}$$

$$100 - x = 100 - 78.7 = 21.3 \text{ kg of other feeds per 100 kg of ration (\%)}$$

14. From the above, it is apparent that if the corn and CSM are used in a 10:1 ratio, a total of 78.7 kg of the combination could be used per 100 kg of ration without exceeding the maximum permissible level of 200 ppb of aflatoxin in rations for growing-finishing pigs. The figure for "other feeds" is the difference between 78.7 and 100, or 21.7%. These feeds may be any additional protein and/or energy feeds, along with mineral and vitamin supplements as needed to balance the ration. It has already been assumed that these feeds contain no aflatoxin. If there is any reason to suspect that they might contain aflatoxin, they should be analyzed.

15. In some instances, it is not possible to feed a complete mixed ration and the aflatoxin-containing feed or feeds must be mixed with only a part of the overall ration. The rest of the ration, usually consisting of such feeds as aflatoxin-free hay and/or silage, would be fed separately. In such cases the ppb permissible level for aflatoxin still applies to the overall ration and not just to the part carrying the contaminated feed. Consequently the level of aflatoxin in the portion carrying the contaminated feed may have an aflatoxin level in ppb that is higher than the maximum level permissible in the overall ration.

16. Let us assume that a dairyman has a quantity of corn containing 113 ppb of aflatoxin that he wants to use in the ration for his milking herd. His ration formula calls for 25% corn, 20% other concentrates and 55% hay and silage, air-dry basis. The concentrates are to be fed as a mix—the roughages to be fed separately. The maximum level of aflatoxin to be fed to milking cows is 20 ppb in the overall ration. However, under the circumstances it is possible to add the aflatoxin-contaminated corn to the concentrate portion of the ration only. The question is how much of the contaminated corn should be added to the concentrate mix. We know that corn is not to be more than 55.55% (25% of the total ration) of the total concentrates, but it is necessary to know what part of this may be contaminated corn without exceeding the maximum permissible level of 20 ppb for the overall ration. In order to obtain this figure, the calculation must first be made on the basis of the total ration as follows:

Ration Components	% in Ration (kg/100 kg)	Aflatoxin Level ppb (mcg/kg)	Total Aflatoxin mcg
Contaminated corn	x	113	2,000
Aflatoxin-free corn	25	0	0
Other concentrates	20	0	0
Hay and silage (air-dry basis)	55	0	0
Total ration	100	20	2,000

$$x = \frac{2,000}{113} = 17.7 \text{ kg/100 kg } (\%)$$

$$25 - x = 7.3 \text{ kg/100 kg } (\%)$$

17. The proportion that the various concentrate components would be of the overall concentrate mix are then calculated as follows:

Concentrate Mix Components	Amount Per 45 kg of Mix kg	Amount Per 100 kg of Mix kg (%)
Contaminated corn	17.7	37.78
Aflatoxin-free corn	7.3	17.78
Other concentrates	20.0	44.44
Overall conc. mix	45.0	100.00

18. The concentrate portion of this ration contains 42.7 ppb of aflatoxin. While this is more than double the maximum permissible aflatoxin level for milking cows (20 ppb), when the aflatoxin level is calculated on the basis of the complete ration, roughage included, it is only 20 ppb.

J. Sometimes, because of an extra-high level of aflatoxin in a feed or because of a low tolerance level for the class of animals being fed, it is not possible or practical to make the feed suitable for feeding by diluting it with other feeds. In such cases, it may be necessary to detoxify the contaminated feed before attempting to use it in the feeding program. The method offering the most promise to date for the detoxification of aflatoxin in feed is treating the feed with ammonia.

1. It has definitely been shown that ammoniation is an effective way to detoxify aflatoxin in feeds.

2. Thus far there have been no ill effects to livestock reported from the use of ammoniated feeds.

3. It now appears that it may be possible to effectively detoxify aflatoxin-contaminated grains for only a few cents per bushel.

4. Unfortunately, equipment for carrying out the ammoniation process has not yet been developed for general farm use.

5. Also, grain ammoniation has not yet been approved by the Food and Drug Administration and consequently feeds containing ammoniated ingredients cannot be sold in interstate commerce.

6. There is still much research to be conducted before the ammoniation process can become standard operating procedure in the industry.

    a. There is need for information on the best moisture level for carrying out the process.

    b. Figures are needed on the amount of ammonia necessary and the treatment time to produce the desired results.

    c. Information is needed on the amount of drying and/or aeration necessary to remove the ammonia odor to a point where it does not affect feed acceptance.

    d. The cost of applying the ammoniation treatment and the effect of such treatment on the nutritional value of the feed must be determined.

7. If further work continues to provide results that are as encouraging as those to date, it would seem that ammoniation is at least one workable solution to the aflatoxin detoxification problem.

K. In 1997, the FDA approved irradiation for detoxification of meat products.

69 Study Questions and Problems

I. Calculate the percentage of crude protein in the following feed mixture:

Ground shelled corn	400 lb
Ground barley grain	300 lb
Cane molasses	100 lb
44% soybean meal	200 lb
Defluorinated phosphate	10 lb
Trace-mineralized salt	20 lb
Urea (45% N)	30 lb

SOLUTION:

	% Prot (from Appendix Except for Urea)	lb Prot Per lb Feed		lb Feed in Mix		Total lb Prot from Each Feed
Corn	9.6	0.096	×	400	=	38.4
Barley	11.9	0.119	×	300	=	35.7

continued

	% Prot (from Appendix Except for Urea)	lb Prot Per lb Feed		lb Feed in Mix		Total lb Prot from Each Feed
Molasses	4.4	0.044	×	100	=	4.4
SBM	44.6	0.446	×	200	=	89.2
Defluorinated phosphate	0.0	0.000	×	10	=	0.0
TM salt	0.0	0.000	×	20	=	0.0
Urea	281.2	2.812	×	30	=	84.4
TOTAL				1,060		252.1

$$\frac{252.1}{1,060} \times 100 = 23.78\% \text{ (answer)}$$

II. Calculate the percentage crude fiber in the feed mixture in I.

ANSWER: 3.57%

III. Calculate the percentage of calcium in the following feed mixture:

Ground shelled corn	550 lb
41% cottonseed meal	50 lb
Ground oats	400 lb
Steamed bonemeal	5 lb
Ground limestone	5 lb
Trace-mineralized salt	5 lb
Urea (45% N)	25 lb

SOLUTION:

	% Ca (from Appendix Except for Urea)	lb Ca Per lb of Feed	lb Feed in Mix	Total lb Ca from Each Feed
Corn	0.03	0.0003	550	0.165
Cottonseed meal	0.19	0.0019	50	0.095
Oats	0.07	0.0007	400	0.280
Bonemeal	30.92	0.3092	5	1.546
Limestone	34.00	0.3400	5	1.700
TM salt	0.0	0.0000	5	0.000
Urea	0.0	0.0000	25	0.000
TOTAL			1,040	3.786

$$\frac{3.786}{1,040} \times 100 = 0.36\% \text{ (answer)}$$

IV. **Calculate the percentage of phosphorus in the feed mixture in III.**

ANSWER: 0.38%

V. **Calculate the percentage of crude protein in a mixture of 1,900 lb cane molasses and 100 lb urea (45% N).**

ANSWER: 18.24%

VI. **A farmer has a silo 20 ft in diameter and 60 ft high.**

1. How many tons of silage can be stored in this silo?

2. How many pounds of shelled corn could be stored in this silo?

SOLUTION:

1. Tons of silage that can be stored in silo.

$$\text{Volume} = \text{Area of base} \times \text{height}$$
$$V = \pi r^2 \times h$$
$$V = 3.1416 \times 10^2 \times 60$$
$$V = 18{,}849.6 \text{ ft}^3$$

Approximate weight of silage: 45 lb/ft^3

$$\frac{(18{,}849.6)\,(45)}{2{,}000} = 424.1 \text{ tons } (answer)$$

2. Pounds of shelled corn that can be stored in silo.

$$\frac{\text{Volume in ft}^3}{\text{Ft}^3 \text{ per bu}} \times \text{Wt per bu shelled corn} = \text{Capacity of silo in lb of shelled corn}$$

$$\frac{18{,}849.6}{1.25} \times 56 = 844{,}462 \text{ lb } (answer)$$

VII. **A farmer has a silo 14 ft in diameter and 30 ft high.**

1. What is the volume of the silo in cubic feet?

2. What is the capacity of the silo in lb of barley grain?

ANSWERS:

1. 4,618.2 ft^3
2. 177,337 lb

VIII. A metal drum is 22 in. in diameter and 34 in. high.

     1. What is the capacity of the drum in gallons?

     2. What is the capacity of the drum in bushels?

SOLUTION:

     1. Capacity of drum in gallons.

$$\text{Cap in gal} = \frac{\pi r^2 \times h}{231 \ (\text{in}^3 \text{ per gal})}$$

$$= \frac{3.1416 \times 11^2 \times 34}{231}$$

$$= 55.95 \text{ gal } (answer)$$

     2. Capacity of drum in bushels.

$$\text{Cap in bu} = \frac{\pi r^2 \times h}{2{,}150 \ (\text{in}^3 \text{ per bu})}$$

$$= \frac{3.1416 \times 11^2 \times 34}{2{,}150}$$

$$= 6.01 \text{ bu } (answer)$$

IX. Calculate the pounds of cane molasses that can be stored in a horizontal tank 16 ft long and 6 ft in diameter.

SOLUTION:

$$\begin{aligned} \text{Volume} &= \text{Area of end} \times \text{length} \\ &= \pi r^2 \times \text{length} \\ &= 3.1416 \times 3^2 \times 16 \\ &= 452.4 \text{ ft}^3 \end{aligned}$$

$$\text{Ft}^3 \times \text{gal per ft}^3 \times \text{Wt per gal of molasses} = \text{Total lb of molasses}$$
$$452.4 \times 7.48 \times 11.75 = 39{,}761 \text{ lb } (answer)$$

or

$$\frac{\text{Ft}^3 \times \text{in}^3 \text{ per ft}^3}{\text{In}^3 \text{ per gal}} \times \text{Wt per gal of molasses} = \text{Total lb of molasses}$$

$$\frac{452.4 \times 1,728}{231} \times 11.75 = 39,764 \text{ lb } (\textit{answer})$$

X. A farmer has a cylindrical feed bin with a cone-shaped bottom. The bin is 69 in. in diameter and has an overall depth of 88 in. The cone section has a depth of 56 in. What is the capacity of the bin in bushels?

SOLUTION:

Volume of cylindrical section.

$$V = \pi r^2 \times \text{Depth}$$
$$= 3.1416 \times 34.5^2 \times (88-56)$$
$$= 3.1416 \times 1,190.25 \times 32$$
$$= 119,657.3 \text{ in}^3$$

Volume of cone section.

$$V = \pi r^2 \times \text{Depth} \times 1/3$$
$$= 3.1416 \times 1,190.25 \times 56 \times 1/3$$
$$= 69,800.1 \text{ in}^3$$

Capacity of bin in bu.

$$C = \frac{\text{Total volume in in}^3}{\text{In}^3 \text{ in a bu}}$$
$$= \frac{119,657.3 + 69,800.1}{2,150.42}$$
$$= 88.1 \text{ bu } (\textit{answer})$$

XI. A farmer has a cylindrical grain storage bin with a cone-shaped roof. The bin is 18 ft in diameter and has a height to the eave of 128 in and to the peak of the cone of 164 in. What is the capacity of the bin in bu when filled to the peak of the cone?

ANSWER: 2375.8 bu.

XII. A farmer has a cylindrical bin 20 ft in diameter filled to a height of 30 ft with snapped corn. Calculate the lb of grain, cobs, and shucks in the bin.

SOLUTION:

First, calculate the volume of corn in ft^3.

$$V = \pi r^2 \times h$$
$$= 3.1416 \times 10^2 \times 30$$
$$= 9{,}424.8 \text{ ft}^3$$

Next, calculate the bu of snapped corn in bin.

$$\frac{\text{Ft}^3 \text{ in bin}}{\text{Ft}^3 \text{ per bu snapped corn}} = \text{bu snapped corn}$$

$$\frac{9{,}424.8}{3.5} = 2{,}692.8 \text{ bu}$$

Next, calculate the lb of snapped corn in the bin.

$$2{,}692.8 \text{ bu} \times 80 = 215{,}424 \text{ lb snapped corn}$$

Next, calculate the lb of grain, cobs, and shucks.

$$70\% \text{ of } 215{,}424 = 150{,}798 \text{ lb grain}$$
$$17.5\% \text{ of } 215{,}424 = 37{,}699 \text{ lb cobs}$$
$$12.5\% \text{ of } 215{,}324 = 26{,}928 \text{ lb shucks}$$

XIII. **A farmer has a watering tank that is 2 ft high, 3 ft wide, and 10 ft long, and has circular ends. What is the capacity of the tank in gallons?**

SOLUTION:

To solve, divide the tank into 2 sections. Consider the 2 ends as 1 section—a cylinder with a depth of 2 ft and a radius 1/2 the width of the tank (1/2 × 3 ft = 1.5 ft).

Consider the remainder of the tank as the other section—a rectangular-shaped section that is 2 ft high, 3 ft wide, and 7 ft (10–3) long.

Combine the two sections to arrive at total capacity.

ANSWER: 420 gal.

XIV. **A farmer has a feed tub that is 11 in deep and has a diameter of 16.5 in at the bottom and 20 in at the top. What is the capacity of the tub in gallons? In bushels?**

SOLUTION:

To solve, consider the tub to be the top section of a cone from which the bottom section (a smaller cone) has been removed.

First, calculate the height of the two cones by using a ratio.

$$x{:}20 = (x-11){:}16.5$$
$$16.5x = 20x - 220$$
$$3.5x = 220$$
$$x = 62.86 \text{ in (height of large cone)}$$
$$x - 11 = 51.86 \text{ in (height of smaller cone)}$$

Next, calculate the volume of the 2 cones.

$$3.1416 \times 8.25^2 \times 51.86 \times 1/3 = 3,696.3 \text{ in}^3 \text{ in smaller cone}$$
$$3.1416 \times 10^2 \times 62.86 \times 1/3 = 6,582.7 \text{ in}^3 \text{ in larger cone}$$

The difference in the volume of the 2 cones equals the volume of the tub:

$$6582.7 - 3696.3 = 2,886.4 \text{ in}^3$$
$$= 12.5 \text{ gal}$$
$$\hspace{3cm} (\textit{answers})$$
$$= 1.34 \text{ bu}$$

XV. **Using ground shelled corn and 44% soybean meal, formulate a 1,000 lb, 16% crude protein mix.**

SOLUTION:

USE OF THE SQUARE METHOD

			Percent	Per 1,000 lb
Corn 9.6	*from* 16	28.6	81.71%	817.1 lb corn
SBM 44.6	*minus*	6.4	18.29	182.9 lb SBM
TOTALS		35.0	100.00%	

or

USE OF ALGEBRAIC EQUATIONS

$$x = \text{Percent of corn in mix}$$

$$y = \text{Percent of SBM in mix}$$

$$0.096x + 0.446y = 16.0$$

$$0.096 + 0.096y = 9.6$$

$$0.35y = 6.4$$

$$y = 18.29\% \text{ SBM in mix}$$

$$x = 100 - y = 81.71\% \text{ corn mix}$$

XVI. **Calculate on a ton basis what combination of cane molasses and urea (45% N) would provide a 20% crude protein product.**

SOLUTION:

USE OF THE SQUARE METHOD

				Per 100 lb	*Per 2,000 lb*
Molasses 4.4	*from*		261.2	94.4	1,888 lb molasses
	20				
Urea 281.2	*minus*		15.6	5.6	112 lb urea
			276.8		

USE OF ALGEBRAIC EQUATIONS

ANSWER: Same as above.

XVII. **A farmer wants to add enough urea (45% N) to some sorghum forage (1.9% crude protein) as the silo is filled to produce a silage containing 3.5% crude protein. How many lb of the urea should be added per ton of fresh forage?**

SOLUTION:

USE OF ALGEBRAIC EQUATIONS

Let $x = \%$ urea

Let $y = \%$ forage

$$2.812x + 0.019y = 3.5$$
$$0.019x + 0.019y = 1.9$$
$$2.793x + 0.0y = 1.5$$
$$x = 0.573 \ (\% \text{ urea})$$
$$y = 100 - x = 99.427 \ (\% \text{ forage})$$

If 0.573 lb of urea is added per 99.427 lb of forage, then 11.53 lb (*answer*) of urea would be added per 2,000 lb of forage.

USE OF THE SQUARE METHOD

ANSWER: Same as above.

XVIII. **A farmer is making silage from overripe corn containing 47% dry matter. He wants to add sufficient water to the corn forage as it enters the silo to bring the dry matter content of the silage down to 33%. How much water should be added per ton of corn forage? (Use both the Square Method and equations.)**

ANSWER: 848.5 lb or 101.7 gal.

XIX. **Using a mixture of 2 parts shelled corn and 1 part oats in conjunction with 44% soybean meal, formulate a 1,200 lb, 15% crude protein mix for lactating sows.**

SOLUTION:

First, calculate the percentage of protein in the corn-oats mixture.

$$2 \times 9.6 = 19.2$$
$$1 \times 11.8 = \underline{11.8}$$
$$31.0 \div 3 = 10.33\%$$

Then proceed with equations as follows:

Let x = % of corn-oats mixture in overall mix

Let y = % SBM in overall mix

$$0.1033x + 0.4460y = 15.0$$
$$0.1033x + 0.1033y = 10.33$$
$$0.3427y = 4.67$$

$$y = 13.63\% \text{ SBM}$$

$$x = 100 - y = 86.37\% \text{ corn-oats mixture}$$

Per 1,200 lb

$$13.63 \times 12 = 163.6 \text{ lb SBM}$$

$$86.37 \times 12 = 1036.4 \text{ lb corn and oats} \quad \begin{cases} 690.9 \text{ lb of corn} \\ 345.5 \text{ lb of oats} \end{cases}$$

Calculate using the Square Method.

ANSWER: Same as above.

XX. Using the following feeds:

    1. Ground grain sorghum grain.

    2. Brewers dried grains.

    3. 41% cottonseed meal.

    4. Ground oats grain.

    5. 44% soybean meal.

Calculate a satisfactory 2,000 lb, 16% crude protein concentrate mix for a dairy cow.

SOLUTION:

Use either Square Method or equations to calculate. An innumerable number of satisfactory combinations are possible.

XXI. The daily protein and metabolizable energy requirements of a 300 kg yearling finishing steer are 0.92 kg and 21.7 Mcal, respectively. Upon subtracting the protein and energy of 1.5 kg of coastal bermudagrass hay to be fed to the steer, there still remains a need for 0.840 kg of protein and 19.04 Mcal ME to be derived from the concentrates—ground shelled corn and 41% cottonseed meal. Using simultaneous equations, calculate the amount of each of these 2 feeds that should be fed to meet the protein and ME needs.

    SOLUTION:

$$\text{Let } x = \text{Amount of corn}$$

$$\text{Let } y = \text{Amount of CSM}$$

Then

$$0.096x + 0.410y = 0.840$$

$$3.03x + 2.80y = 19.04$$

$$3.03x + 12.94y = 26.52$$

$$-10.14y = -7.48$$

$$y = 0.738 \text{ kg } 41\% \text{ CSM}$$

$$0.096x + (0.410 \times 0.738) = 0.840$$

$$0.096x + 0.303 = 0.840$$

$$0.096x = 0.537$$

$$x = 5.59 \text{ kg sh corn}$$

XXII. **If the following feeds can be bought at the prices indicated, which would be the best buy as a source of TDN for fattening cattle?**

Shelled corn	$1.96 per bu
Barley grain	1.68 per bu
Grain sorghum	2.96 per cwt
Oats grain	1.05 per bu
Cane molasses	0.36 per gal

SOLUTION:

	Cost Per lb	Cost Per cwt	TDN Per cwt	Cost per lb TDN
Corn	3.50¢	$3.50	77.0	4.55¢
Barley	3.50	3.50	74.0	4.73
Sorghum	2.96	2.96	78.0	3.79
Oats	3.28	3.28	68.0	4.82
Molasses	3.06	3.06	54.0	5.67

ANSWER:

Grain sorghum is the cheapest source of TDN.

XXIII. **The following feeds are available at the prices indicated. Which would be the cheapest satisfactory source of protein for growing-fattening pigs?**

44% soybean meal	$183.00 per ton
49% soybean meal	205.00 per ton
36% cottonseed meal	158.00 per ton
41% cottonseed meal	173.00 per ton
Peanut oil meal	200.00 per ton
Meat scrap	220.00 per ton

SOLUTION:

	Cost Per lb	Cost Per cwt	Protein Per cwt	Cost Per lb Protein
44% SBM	9.15¢	$ 9.15	44.6	20.52¢
49% SBM	10.25	10.25	49.7	20.62
41% CSM	8.65	8.65	41.0	21.10
POM	10.00	10.00	48.1	20.79
Scrap	11.00	11.00	51.4	21.40

ANSWER:

From the above, it is apparent that 44% SBM is the cheapest source of protein under the prevailing prices. Since soybean meal is a very satisfactory protein supplement for growing-fattening pigs, it would be the best buy.

XXIV. **Using the Peterson method, calculate the nutritive worth of ground cottonseed when shelled corn is $2.00 per bu and 44% soybean meal is $100.00 per ton.**

SOLUTION:

$$\text{Price of corn per ton} = \frac{\$2.00}{56} \times 2,000 = \$71.43$$

Corn and SBM Constants from Peterson Feed Tables		Price of Corn and SBM Per Ton		
0.904	×	$ 71.43	=	$64.57
0.224	×	100.00	=	22.40
				$86.97 per ton
		(answer)		

XXV. **A farmer plans to feed out 60 head of steer calves from an average initial weight of 500 lb to an average finished weight of 1,000 lb using ground earcorn, 41% cottonseed meal, and minerals. Calculate the amount of each feed he will need to carry out this operation.**

SOLUTION:

Approximate ration mix to be fed.

Ground earcorn	90.0%
41% cottonseed meal	9.0%
Defluorinated phosphate	0.33%
Ground limestone	0.33%
Trace-mineralized salt	0.33%

Total gain to be produced.

$$60 \times 500 = 30,000$$

Total feed to be needed.

$$\text{lb gain} \times \text{feed/lb gain} = \text{total feed needed}$$
$$30,000 \times 7.5 = 225,000 \text{ lb}$$

Amounts of individual feeds needed.

$$90\% \text{ of } 225,000 = 202,500 \text{ lb earcorn}$$
$$9\% \text{ of } 225,000 = 20,250 \text{ lb } 41\% \text{ CSM}$$
$$0.33\% \text{ of } 225,000 = 750 \text{ lb defluorinated phosphate}$$
$$= 750 \text{ lb ground limestone}$$
$$= 750 \text{ lb TM salt}$$

XXVI. **A farmer has 50 head of dry, pregnant beef cows averaging approximately 1,000 lb in weight that he plans to carry through a 100-day wintering period in drylot. Calculate the amount of feed he will need for this purpose.**

SOLUTION:

First, balance a daily ration for a 1,000 lb dry, pregnant beef cow, and then multiply each item in the ration by the number of cows (50) and then by the number of days (100).

XXVII. A farmer is using a swine mixture consisting of the following ingredients in the amounts indicated:

Ingredient	%
Ground shelled corn	79.55
Soybean meal, 44%	17.81
Defluorinated Phosphate	.87
Ground limestone	.77
TM salt	.50
Vitamin supplement	.50
Total	100.00

A. The above mix contains 16% crude protein while only 13% crude protein is required for the animals being fed. How might this mix be altered to provide only 13% crude protein using the same ingredients as above?

SOLUTION: Refer to Section 51, paragraph III.

B. Upon checking the mix formulated in A for phosphorus content it is found to provide .474% phosphorus, while the phosphorus requirement for the animals being fed is only .40%. How might this mix be altered to provide the proper level of phosphorus without disturbing its protein content?

SOLUTION: Refer to Section 51, paragraph IV.

C. Upon checking the mix formulated in B for calcium content it is found to contain .62% Ca whereas the NRC requirement for the class of pigs being fed is only .50%. How might this formula be altered to lower its calcium content from .62% to .50% without changing the mix significantly in its content of other critical nutrients?

SOLUTION: Refer to Section 51, paragraph V.

XXVIII. A farmer desires to feed a group of pigs on the following feed mixture:

Ingredient	%
Ground shelled corn	86.37
49% soybean meal	11.38
Ground limestone	0.49
Defluorinated phosphate	0.76
Trace-mineralized salt	0.50
Vitamin supplement	0.30
Antibiotic supplement	0.20
TOTAL	100.00

A. In order to simplify the ration mixing process he plans to combine all of the ration ingredients other than the corn into a premix to use as a single ingredient along with the corn at ration mixing time. Calculate the composition of the premix.

SOLUTION:

In order to calculate the composition of the premix, one first lists the premix ingredients and the amount of each per 100 lb of the overall mix. Each of these amounts is then converted to a percentage of the premix as follows:

Ingredient	% in Overall Mix	To Convert to Percentage	% in Premix
49% soybean meal	11.38	11.38/13.63 × 100 = 83.49	
Ground limestone	0.49	0.49/13.63 × 100 = 3.59	
Defluorinated phosphate	0.76	0.76/13.63 × 100 = 5.58	
Trace-mineralized salt	0.50	0.50/13.63 × 100 = 3.67	
Vitamin supplement	0.30	0.30/13.63 × 100 = 2.20	
Antibiotic supplement	0.20	0.20/13.62 × 100 = 1.47	
TOTAL	13.63		100.00

B. Calculate the percentage composition of a premix that embraces all of the ingredients in the above mix other than the corn and soybean meal.

SOLUTION: Follow same procedure as in A above.

C. How much of the premix should be used in mixing a ton of the overall ration?

ANSWER: 45 lb.

XXIX. **Select from the choices offered the most appropriate answer for each of the following:**

A. Weight in lb of total air-dry ration for

1.	A 1000-lb dry beef brood cow	6	12	18	24
2.	A 1000-lb lactating beef brood cow	10	15	20	25
3.	A 500-lb stocker heifer	6	12	18	24
4.	A 600-lb finishing yearling steer	6	12	18	24
5.	A 400-lb dairy heifer	10	15	20	25
6.	An 1800-lb beef breeding bull	12	22	32	42
7.	An 1800-lb dairy breeding bull	12	22	32	42
8.	A 500-lb fattening steer calf	5	10	15	20
9.	A 1000-lb idle mature gelding	18	24	30	36
10.	An 1100-lb cattle horse used regularly	22	26	30	34

11.	A 1200-lb lactating mare	18	26	30	36
12.	A 1200-lb mule at hard work	20	27	32	38
13.	A 150-lb lactating ewe	6	8	10	12
14.	An 80-lb fattening lamb	3	6	9	16
15.	A 110-lb growing-finishing pig	6	9	12	15
16.	A 400-lb lactating sow/12 pigs	5	10	15	20
17.	A 400-lb gestating sow	4	8	12	16
18.	An 1100-lb high-producing dairy cow	20	24	28	33

B. Percent concentrates in the ration for

1.	A 1000-lb dry beef brood cow	<10	25	40	>60
2.	A 1000-lb lactating beef brood cow	<25	40	60	>80
3.	A 500-lb stocker heifer	<20	35	50	>65
4.	A 600-lb fattening yearling steer	<25	40	60	>80
5.	A 400-lb dairy heifer	<20	35	50	>65
6.	An 1800-lb beef breeding bull	<25	40	60	>80
7.	An 1800-lb dairy breeding bull	<25	40	60	>80
8.	A 500-lb fattening steer calf	<20	40	60	>80
9.	A 1000-lb idle mature gelding	<10	30	60	>90
10.	An 1100-lb cattle horse used regularly	<25	50	75	>90
11.	A 1200-lb lactating mare	<25	50	75	>90
12.	A 1200-lb mule at hard work	<20	40	60	>80
13.	A 150-lb lactating ewe	<25	40	60	>80
14.	An 80-lb fattening lamb	<25	50	75	>90
15.	A 110-lb growing-finishing pig	<25	40	60	>95
16.	A 400-lb lactating sow/12 pigs	<25	50	75	>90
17.	A 400-lb gestating sow	<25	50	75	>90
18.	An 1100-lb high-producing dairy cow	<25	50	75	>95

XXX. Provide correct answers for each of the following

1. Pounds per bushel of shelled corn
2. Pounds per bushel of oats grain
3. Pounds per bushel of barley grain
4. Pounds per bushel of grain sorghum grain
5. Pounds per bushel of wheat grain
6. Pounds per bushel of soybeans
7. Pounds per bushel of light oats
8. Pounds per bushel of heavy oats
9. Pounds per bushel of ear corn
10. Pounds per bushel of snapped corn

11. Pounds per gallon of water
12. Pounds per gallon of vegetable oil
13. Pounds per gallon of milk
14. Pounds per gallon of cane molasses
15. Pounds per cubic foot of silage in a horizontal silo
16. Pounds per cubic foot of silage in an upright silo
17. Pounds per cubic foot of water
18. Pounds per cubic foot of cane molasses
19. Cubic inches per gallon of water
20. Cubic inches per gallon of cane molasses
21. Cubic feet per bushel of shelled corn
22. Cubic feet per bushel of soybeans
23. Cubic inches per bushel of cane molasses
24. Grams per pound
25. Grams per ounce
26. Ounces per pound
27. Pounds per kilogram
28. Grams per kilogram
29. Milligrams per gram
30. Milligrams per kilogram
31. Micrograms per gram
32. Micrograms per kilogram
33. Parts per million is the same as what per kilogram
34. Parts per billion is the same as what per kilogram
35. Pounds per ton
36. Kilograms per ton
37. Kilograms per metric ton
38. Pounds per metric ton
39. Centimeters per inch
40. Millimeters per inch
41. Miles in a kilometer
42. Kilometers in a mile
43. Liters per quart
44. Quarts per liter
45. Square feet in an acre
46. Square meters in a hectare
47. Acres in a hectare
48. Hectares in an acre
49. Formula for converting degrees Fahrenheit to degrees Celsius
50. Formula for converting degrees Celsius to degrees Fahrenheit
51. Formula for calculating area of a circle
52. Formula for calculating the volume of a cylinder
53. Formula for calculating the volume of a cone

XXXI. The following forage crops fall into one of the following categories. Please indicate which one, by number.

1. Summer annual legume
2. Summer annual nonlegume
3. Winter annual legume
4. Winter annual nonlegume
5. Summer perennial legume
6. Summer perennial nonlegume
7. Winter perennial legume
8. Winter perennial nonlegume

___ 1. Common bermudagrass
___ 2. Rye
___ 3. Ladino clover
___ 4. Bahiagrass
___ 5. Sericea
___ 6. Ryegrass
___ 7. Oats
___ 8. Alfalfa
___ 9. Coastal bermudagrass
___ 10. Crimson clover
___ 11. Barley
___ 12. Kudzu
___ 13. Fescue
___ 14. Lespedeza
___ 15. Millet
___ 16. Orchardgrass
___ 17. Sudangrass
___ 18. Arrowleaf clover
___ 19. Dallisgrass
___ 20. Wheat

XXXII. Following is a list of products that have been used or considered for use as performance stimulants in the feeding of livestock. Into which of the following categories would each fall?

A. Antibiotic
B. Chemoantibacterial
C. Chemobiotic
D. Hormone or hormone-like
E. Anthelmintic
F. pH regulator
G. Thyroid regulator
H. Bloat preventive

I.   Surfactant

J.   Feed flavor

K.   Tranquilizer

L.   Enzyme preparation

M.   None of the above

1.   Bacitracin

2.   Arsanilic acid

3.   Dynafac

4.   Stilbestrol

5.   Hygromycin

6.   Sodium bicarbonate

7.   Chlortetracycline

8.   3-nitro-4-hydroxyphenyl arsonic acid

9.   Melangestrol acetate

10.   Phenothiazine

11.   Thyroprotein

12.   Lecithin

13.   Erythromycin

14.   Ethylenediamine dehydroiodide

15.   Synovex

16.   Neomycin

17.   Piperazine

18.   Thiourea

19.   Silicone

20.   Oleandomycin

21.   Nitrofurans

22.   Thiabendazole

23.   Anise oil

24.   Oxytetracycline

25.   Furazolidone

26.   Rapigain

27.   Hydroxyzine

28.   Penicillin

29.   Furaltadone

30.   Ground limestone

31.   Trifluomeprazine

32.   Zyme—all

33.   Streptomycin

34.   Nitrofurazone

35.   Ralgro

36.   Tramisol

37.   Thiouracil

38.   Tylosin

39.   Nihydrazone

40. Reserpine
41. Lincomycin
42. Sulfamethazine
43. Virginiamycin
44. Loxon
45. Methimazole
46. Poloxaline
47. Flavomycin
48. Zeranol
49. Zinc bacitracin
50. Rumensin
51. Bovatec
52. Ivomec
53. Paratect
54. Steer-oid

XXXIII. **Certain performance stimulants are sometimes referred to with different names or initials. Correlate those names in the first column with those in the second.**

___ *1.* Tylosin	*1.* Aureomycin
___ *2.* Arsanilic acid	*2.* Terramycin
___ *3.* Stilbestrol	*3.* Tylan
___ *4.* Melangestrol acetate	*4.* Chemotherapeutic
___ *5.* Rumensin	*5.* Sodium arsanilate
___ *6.* Ralgro	*6.* EDDI
___ *7.* Chlortetracycline	*7.* NF 180
___ *8.* Thiabendazole	*8.* DES
___ *9.* Chemoantibacterial	*9.* MGA
___ *10.* Thyroprotein	*10.* Zeranol
___ *11.* Methimazole	*11.* Thibenzole
___ *12.* Furazolidone	*12.* Iodinated casein
___ *13.* Oxytetracycline	*13.* Tapazole
___ *14.* Ethylenediamine dehydroiodide	*14.* Monensin
___ *15.* Bovatec	*15.* Lasalocid

XXXIV. **A farmer has 2 lots of corn—one lot contains 267 ppb of aflatoxin; the other is aflatoxin-free.**

    A. What would the aflatoxin concentration be in a 60:40 mixture of the 2 lots of corn?

SOLUTION:

Feed Components	kg Per 100 kg of Mixture	Aflatoxin Level ppb or mcg/kg	Aflatoxin in 100 kg of Mix mcg
Contam. corn	60	267	16,020
A-free corn	40	0	0
TOTAL MIX	100	x	16,020

$$x = \frac{16,020}{100} = 160.2 \text{ ppb}$$

B. What combination of the two lots of corn would give you a mixture which contains only 80 ppb aflatoxin?

SOLUTION:

Feed Components	kg Per 100 kg of Mixture	Aflatoxin Level ppb or mcg/kg	Aflatoxin in 100 kg of Mix mcg
Contam. corn	x	267	8,000
A-free corn	100 − x	0	0
TOTAL MIX	100	80	8,000

$$x = \frac{8,000}{267} = 30.0\%$$

$$100 - 30 = 70\%$$

C. Let us assume the 2 lots of corn contained 267 and 20 ppb aflatoxin, respectively. What combination of these 2 lots of corn would give a mixture that contains 80 ppb aflatoxin.

Feed Components	kg Per 100 kg of Mixture	Aflatoxin Level ppb of mcg/kg	Aflatoxin in 100 kg of Mix
267 ppb corn	x	267	267x
20 ppb corn	100 − x	20	2,000 − 20x
TOTAL MIX	100	80	2,000 + 247x or 8,000

$$x = 24.29 \% \text{ (267 ppb corn)}$$
$$100 - x = 75.71\% \text{ (20 ppb corn)}$$

XXXV. Indicate for each of the following classes of livestock the maximum level of aflatoxin that each can tolerate in its ration, expressed as mcg of this toxin per kg of air-dry feed. (Possible answers: 20, 100, 200, 300, 400, 500, 600)

Feedlot cattle
Lactating dairy cows
Dry sows and boars
Creep ration for lambs
Mares nursing young
Growing-finishing pigs
Dry mares, geldings, and stallions
Feedlot lambs
Creep ration for pigs
Cows nursing young
Ewes nursing young
Creep ration for colts
Dry cows and bulls
Sows nursing young
Creep ration for calves
Dry ewes and rams

70 Tables on Nutrient Requirements

Following are several tables which list the nutrient requirements of various classes of beef cattle, dairy cattle, horses, sheep, poultry, and swine, respectively. These requirements are those published by the National Research Council of the National Academy of Sciences and are presented herein with the permission of the academy. The specific publications from which the respective sets of requirements were obtained are as listed below:

- *NUTRIENT REQUIREMENTS OF BEEF CATTLE,* Seventh Revised Edition,1996. Publication ISBN 0–309–05426–5.

- *NUTRIENT REQUIREMENTS OF DAIRY CATTLE,* Sixth Revised Edition, 1989. Publication ISBN 0–309–03826–X.

- *NUTRIENT REQUIREMENTS OF HORSES,* Fifth Revised Edition, 1989. Publication ISBN 0–309–30989–4.

- *NUTRIENT REQUIREMENTS OF SHEEP,* Sixth Revised Edition, 1985. Publication ISBN 0–309–03596–I.

- *NUTRIENT REQUIREMENTS OF POULTRY,* Ninth Revised Edition, 1994. Publication ISBN 0–309–04892–3.

- *NUTRIENT REQUIREMENTS OF SWINE,* Tenth Revised Edition, 1998. Publication ISBN 0–309–05993–3.

The requirements for the different livestock species as presented in the following tables are basically the same as those given in the above publications. In a few instances, however, to facilitate their use in this book, some of the figures have been transposed from one base to another, such as kilograms to pounds, grams to kilograms, vitamin A to carotene, and digestible energy to TDN. In each case, standard conversion values have been used.

Table Beef-1
NUTRIENT REQUIREMENTS FOR GROWING AND FINISHING CATTLE

Wt @ Small Marbling	533 kg
Weight Range	200–450 kg
ADG Range	0.50–2.50 kg
Breed Code	1 Angus

Body Weight, kg		200	250	300	350	400	450
Maintenance Requirements							
NE$_m$	Mcal/d	4.1	4.84	5.55	6.23	6.89	7.52
MP	g/d	202	239	274	307	340	371
Ca	g/d	6	8	9	11	12	14
P	g/d	5	6	7	8	10	11

Growth Requirements
  (ADG)

NE$_g$ Required for Gain, Mcal/d

		200	250	300	350	400	450
0.5	kg/d	1.27	1.50	1.72	1.93	2.14	2.33
1.0	kg/d	2.72	3.21	3.68	4.13	4.57	4.99
1.5	kg/d	4.24	5.01	5.74	6.45	7.13	7.79
2.0	kg/d	5.81	6.87	7.88	8.84	9.77	10.68
2.5	kg/d	7.42	8.78	10.06	11.29	12.48	13.64

MP Required for Gain, g/d

		200	250	300	350	400	450
0.5	kg/d	154	155	158	157	145	133
1.0	kg/d	299	300	303	298	272	246
1.5	kg/d	441	440	442	432	391	352
2.0	kg/d	580	577	577	561	505	451
2.5	kg/d	718	712	710	687	616	547

Calcium Required for Gain, g/d

		200	250	300	350	400	450
0.5	kg/d	14	13	12	11	10	9
1.0	kg/d	27	25	23	21	19	17
1.5	kg/d	39	36	33	30	27	25
2.0	kg/d	52	47	43	39	35	32
2.5	kg/d	64	59	53	48	43	38

Phosphorus Required for Gain, g/d

		200	250	300	350	400	450
0.5	kg/d	6	5	5	4	4	4
1.0	kg/d	11	10	9	8	8	7
1.5	kg/d	16	15	13	12	11	10
2.0	kg/d	21	19	18	16	14	13
2.5	kg/d	26	24	22	19	17	15

Table Beef-2
DIET EVALUATION FOR GROWING AND FINISHING CATTLE

| Wt @ Small Marbling | | 533 kg | | | | | |
| Breed Code | | 1 Angus | | | | | |

Ration	eNDF % DM	TDN % DM	NE_m Mcal/kg	NE_g Mcal/kg	CP % DM	DIP % CP	Weight Class	NE Adjuster
A	57	50	1.00	0.45	7.4	88	325	100%
B	43	60	1.35	0.77	10.0	78	350	100%
C	30	70	1.67	1.06	12.6	72.4	375	100%
D	5	80	1.99	1.33	14.4	48.5	400	100%
E	3	90	2.29	1.59	16.6	44.2	425	100%

Body Weight, kg	DMI Adjuster	DMI kg/d	ADG kg/d	DIP	UIP	MP	Ca	P
				--------	balances, g/d	--------	-- requirements, % of DM --	
300—A	100%	7.9	0.32	1	0	0	0.22%	0.13%
—B	100%	8.4	0.89	0	0	0	0.35%	0.18%
—C	100%	8.2	1.36	2	0	0	0.48%	0.24%
—D	100%	7.7	1.69	1	2	1	0.60%	0.29%
—E	100%	7.1	1.90	1	2	1	0.71%	0.34%
325—A	100%	8.4	0.32	1	14	11	0.21%	0.13%
—B	100%	8.9	0.89	0	38	30	0.33%	0.18%
—C	100%	8.7	1.36	2	57	46	0.45%	0.22%
—D	100%	8.2	1.69	1	73	58	0.55%	0.27%
—E	100%	7.6	1.90	1	82	66	0.65%	0.31%
350—A	100%	8.9	0.32	1	27	22	0.20%	0.13%
—B	100%	9.4	0.89	0	75	60	0.31%	0.17%
—C	100%	9.2	1.36	2	114	91	0.42%	0.21%
—D	100%	8.7	1.69	1	143	114	0.51%	0.25%
—E	100%	8.0	1.90	1	160	128	0.60%	0.29%
375—A	100%	9.4	0.32	1	40	32	0.20%	0.13%
—B	100%	9.9	0.89	0	111	89	0.30%	0.16%
—C	100%	9.7	1.36	2	169	135	0.39%	0.20%
—D	100%	9.1	1.69	1	212	169	0.48%	0.24%
—E	100%	8.4	1.90	1	238	190	0.56%	0.28%
400—A	100%	9.8	0.32	1	53	43	0.19%	0.12%
—B	100%	10.4	0.89	0	147	118	0.28%	0.16%
—C	100%	10.2	1.36	2	223	178	0.37%	0.19%
—D	100%	9.6	1.69	2	279	223	0.44%	0.23%
—E	100%	8.8	1.90	1	314	251	0.52%	0.26%
425—A	100%	10.3	0.32	1	66	53	0.19%	0.12%
—B	100%	10.9	0.89	0	182	146	0.27%	0.15%
—C	100%	10.6	1.36	2	276	221	0.35%	0.19%
—D	100%	10.0	1.69	2	346	277	0.42%	0.22%
—E	100%	9.3	1.90	1	388	311	0.48%	0.25%

Table Beef-3
NUTRIENT REQUIREMENTS FOR GROWING BULLS

Wt @ Maturity	890 kg
Weight Range	300–800 kg
ADG Range	0.50–2.50 kg
Breed Code	1 Angus

Body Weight, kg		300	400	500	600	700	800
Maintenance Requirements							
NE_m	Mcal/day	6.38	7.92	9.36	10.73	12.05	13.32
MP	g/d	274	340	402	461	517	572
Ca	g/d	9	12	15	19	22	25
P	g/d	7	10	12	14	17	19
Growth Requirements							
ADG							
NE_g Required for Gain, Mcal/d							
0.5	kg/d	1.72	2.13	2.52	2.89	3.25	3.59
1.0	kg/d	3.68	4.56	5.39	6.18	6.94	7.67
1.5	kg/d	5.74	7.12	8.42	9.65	10.83	11.97
2.0	kg/d	7.87	9.76	11.54	13.23	14.85	16.41
2.5	kg/d	10.05	12.47	14.74	16.90	18.97	20.97
MP Required for Gain, g/d							
0.5	kg/d	158	145	122	100	78	58
1.0	kg/d	303	272	222	175	130	86
1.5	kg/d	442	392	314	241	170	102
2.0	kg/d	577	506	400	299	202	109
2.5	kg/d	710	617	481	352	228	109
Calcium Required for Gain, g/d							
0.5	kg/d	12	10	9	7	6	4
1.0	kg/d	23	19	16	12	9	6
1.5	kg/d	33	27	22	17	12	7
2.0	kg/d	43	35	28	21	14	8
2.5	kg/d	53	43	34	25	16	8
Phosphorus Required for Gain, g/d							
0.5	kg/d	5	4	3	3	2	2
1.0	kg/d	9	8	6	5	4	2
1.5	kg/d	13	11	9	7	5	3
2.0	kg/d	18	14	11	8	6	3
2.5	kg/d	22	17	14	10	6	3

Table Beef-4
DIET EVALUATION FOR GROWING BULLS

| Wt @ Maturity | 890 kg |
| Breed Code | 1 Angus |

Ration	eNDF % DM	TDN % DM	NE_m Mcal/kg	NE_g Mcal/kg	CP % DM	DIP % CP	Weight Class	NE Adjuster
A	43	50	1.00	0.45	8.2	80	325	100%
B	37	65	1.51	0.92	10.9	78	350	100%
C	30	70	1.67	1.06	12.0	76	375	100%
D	20	75	1.83	1.20	13.4	73	400	100%
E	5	80	1.99	1.33	13.8	51	425	100%

Body Weight, kg	DMI Adjuster	DMI kg/d	ADG kg/d	DIP	UIP	MP	Ca	P
				--------	balances, g/d	--------	-- requirements, % of DM --	
300—A	100%	7.9	0.22	5	103	83	0.18%	0.12%
—B	100%	8.3	1.02	4	8	6	0.39%	0.20%
—C	100%	8.2	1.23	2	−3	−2	0.45%	0.23%
—D	100%	8.0	1.41	3	10	8	0.51%	0.25%
—E	100%	7.7	1.56	5	−2	−2	0.56%	0.27%
325—A	100%	8.4	0.22	5	119	95	0.18%	0.12%
—B	100%	8.8	1.02	5	51	41	0.36%	0.19%
—C	100%	8.7	1.23	2	49	39	0.42%	0.21%
—D	100%	8.5	1.41	3	70	56	0.47%	0.24%
—E	100%	8.2	1.56	6	63	51	0.52%	0.26%
350—A	100%	8.9	0.22	5	134	107	0.18%	0.12%
—B	100%	9.4	1.02	5	94	75	0.34%	0.18%
—C	100%	9.2	1.23	2	100	80	0.39%	0.20%
—D	100%	9.0	1.41	3	129	103	0.44%	0.22%
—E	100%	8.7	1.56	6	128	102	0.48%	0.24%
375—A	100%	9.4	0.22	6	149	119	0.18%	0.12%
—B	100%	9.8	1.02	5	136	109	0.32%	0.17%
—C	100%	9.7	1.23	2	150	125	0.37%	0.19%
—D	100%	9.4	1.41	3	187	149	0.41%	0.21%
—E	100%	9.1	1.56	6	191	153	0.45%	0.23%
400—A	100%	9.8	0.22	6	161	131	0.17%	0.12%
—B	100%	10.3	1.02	5	177	142	0.31%	0.17%
—C	100%	10.2	1.23	2	199	159	0.35%	0.19%
—D	100%	9.9	1.41	3	244	195	0.39%	0.20%
—E	100%	9.6	1.56	7	253	202	0.42%	0.22%
425—A	100%	10.3	0.22	6	169	143	0.17%	0.12%
—B	100%	10.8	1.02	6	218	174	0.29%	0.16%
—C	100%	10.6	1.23	2	247	198	0.33%	0.18%
—D	100%	10.4	1.41	3	300	240	0.36%	0.19%
—E	100%	10.0	1.56	7	314	251	0.40%	0.21%

Table Beef-5
NUTRIENT REQUIREMENTS OF PREGNANT REPLACEMENT HEIFERS

Mature Weight	*533 kg*
Calf Birth Weight	*40 kg*
Age @ Breeding	*15 Months*
Breed Code	*1 Angus*

	Months Since Conception								
	1	*2*	*3*	*4*	*5*	*6*	*7*	*8*	*9*
NE_m required, Mcal/d									
Maintenance	5.98	6.14	6.30	6.46	6.61	6.77	6.92	7.07	7.23
Growth	2.29	2.36	2.42	2.48	2.54	2.59	2.65	2.71	2.77
Pregnancy	0.03	0.07	0.16	0.32	0.64	1.18	2.08	3.44	5.37
Total	8.31	8.57	8.87	9.26	9.79	10.55	11.65	13.23	15.37
MP required, g/d									
Maintenance	295	303	311	319	326	334	342	349	357
Growth	118	119	119	119	119	117	115	113	110
Pregnancy	2	4	7	18	27	50	88	151	251
Total	415	425	437	457	472	501	545	613	718
Minerals									
Calcium required, g/d									
Maintenance	10	11	11	11	12	12	12	13	13
Growth	9	9	9	8	8	8	8	8	8
Pregnancy	0	0	0	0	0	0	12	12	12
Total	19	19	20	20	20	20	33	33	33
Phosphorus required, g/d									
Maintenance	8	8	8	9	9	9	10	10	10
Growth	4	4	3	3	3	3	3	3	3
Pregnancy	0	0	0	0	0	0	7	7	7
Total	12	12	12	12	12	13	20	20	20
ADG, kg/d									
Growth	0.39	0.39	0.39	0.39	0.39	0.39	0.39	0.39	0.39
Pregnancy	0.03	0.05	0.08	0.12	0.19	0.28	0.40	0.57	0.77
Total	0.42	0.44	0.47	0.51	0.58	0.67	0.79	0.96	1.16
Body weight, kg									
Shrunk body	332	343	355	367	379	391	403	415	426
Gravid uterus mass	1	3	4	7	12	19	29	44	64
Total	333	346	360	375	391	410	432	459	491

Table Beef-6
BREED MAINTENANCE REQUIREMENT MULTIPLIERS,
BIRTH WEIGHTS, AND PEAK MILK PRODUCTION

Breed	Code	NE_m (BE)	Birth Wt. kg (CBW)	Peak Milk Yield, kg/day (PKYD)
Angus	1	1.00	31	8.0
Braford	2	0.95	36	7.0
Brahman	3	0.90	31	8.0
Brangus	4	0.95	33	8.0
Braunvieh	5	1.20	39	12.0
Charolais	6	1.00	39	9.0
Chianina	7	1.00	41	6.0
Devon	8	1.00	32	8.0
Galloway	9	1.00	36	8.0
Gelbvieh	10	1.00	39	11.5
Hereford	11	1.00	36	7.0
Holstein	12	1.20	43	15.0
Jersey	13	1.20	31	12.0
Limousin	14	1.00	37	9.0
Longhorn	15	1.00	33	5.0
Maine Anjou	16	1.00	40	9.0
Nellore	17	0.90	40	7.0
Piedmontese	18	1.00	38	7.0
Pinzgauer	19	1.00	40	11.0
Polled Hereford	20	1.00	33	7.0
Red Poll	21	1.00	36	10.0
Sahiwal	22	0.90	38	8.0
Salers	23	1.00	35	9.0
Santa Gertrudis	24	0.90	33	8.0
Shorthorn	25	1.00	37	8.5
Simmental	26	1.20	39	12.0
South Devon	27	1.00	33	8.0
Tarentaise	28	1.00	33	9.0

Table Beef-7
DIET EVALUATION FOR PREGNANT REPLACEMENT HEIFERS

Mature Weight	533 kg
Calf Birth Weight	40 kg
Age @ Breeding	15 Months
Breed Code	1 Angus

Ration	TDN % DM	NE$_m$ Mcal/kg	NE$_g$ Mcal/kg	CP % DM	DIP % DM	DMI Factor
A	50	1.00	0.45	8.2	80	100%
B	60	1.35	0.77	9.8	80	100%
C	70	1.67	1.06	11.4	80	100%

		Months Since Conception								
		1	2	3	4	5	6	7	8	9
A	NE$_m$ req. factor	100%	100%	100%	100%	100%	100%	100%	100%	100%
	DM, kg	8.5	8.8	9.0	9.2	9.4	9.7	9.9	10.1	10.3
	NE allowed ADG	0.35	0.34	0.33	0.31	0.28	0.22	0.12	0.00	0.00
	DIP balance, g/d	5	5	5	6	6	6	6	6	6
	UIP balance, g/d	75	79	83	87	90	92	90	66	−53
	MP balance, g/d	60	63	67	69	72	74	72	52	−42
	Ca % DM	0.22%	0.21%	0.21%	0.20%	0.19%	0.18%	0.28%	0.25%	0.25%
	P % DM	0.17%	0.17%	0.16%	0.16%	0.15%	0.14%	0.19%	0.16%	0.16%
B	DM, kg	9.0	9.3	9.5	9.7	10.0	10.2	10.4	10.7	10.9
	NE allowed ADG	0.96	0.96	0.95	0.92	0.88	0.82	0.71	0.54	0.30
	DIP balance, g/d	4	4	4	4	4	4	4	4	4
	UIP balance, g/d	5	14	22	30	38	49	54	46	18
	MP balance, g/d	4	11	18	24	31	40	43	37	14
	Ca % DM	0.36%	0.35%	0.33%	0.32%	0.31%	0.29%	0.38%	0.34%	0.29%
	P % DM	0.27%	0.27%	0.26%	0.26%	0.25%	0.23%	0.27%	0.24%	0.20%
C	DM, kg	8.8	9.1	9.3	9.5	9.8	10.0	10.2	10.4	10.7
	NE allowed ADG	1.47	1.46	1.45	1.42	1.38	1.31	1.19	1.02	0.77
	DIP balance, g/d	2	2	2	2	2	2	2	2	2
	UIP balance, g/d	−66	−54	−43	−32	−19	−1	10	8	−18
	MP balance, g/d	−53	−43	−34	−26	−15	−1	8	6	−14
	Ca % DM	0.48%	0.47%	0.45%	0.43%	0.41%	0.39%	0.48%	0.43%	0.38%
	P % DM	0.37%	0.36%	0.35%	0.35%	0.33%	0.31%	0.35%	0.32%	0.28%

Note: Requirements are for NE allowed ADG and target weight. NE allowed ADG is ADG independent of conceptus gain.

Table Beef-8
NUTRIENT REQUIREMENTS OF BEEF COWS

Mature Weight	533 kg	Milk Fat	4.0%
Calf Birth Weight	40 kg	Milk Protein	3.4%
Age @ Calving	60 Months	Calving Interval	12 Months
Age @ Weaning	30 Weeks	Time Peak	8.5 Weeks
Peak Milk	8 kg	Milk SNF	8.3%
Breed Code	1 Angus		

	Month Since Calving											
	1	2	3	4	5	6	7	8	9	10	11	12
NE_m Req. Factor	100%	100%	100%	100%	100%	100%	100%	100%	100%	100%	100%	100%
NE_m required, Mcal/d												
Maintenance	10.25	10.25	10.25	10.25	10.25	10.25	8.54	8.54	8.54	8.54	8.54	8.54
Growth	0.00	0.00	0.00	0.00	0.00	0.00	0.00	0.00	0.00	0.00	0.00	0.00
Lactation	4.78	5.74	5.17	4.13	3.10	2.23	0.00	0.00	0.00	0.00	0.00	0.00
Pregnancy	0.00	0.00	0.01	0.03	0.07	0.16	0.32	0.64	1.18	2.08	3.44	5.37
Total	15.03	15.99	15.43	14.41	13.42	12.64	8.87	9.18	9.72	10.62	11.98	13.91
MP required, g/d												
Maintenance	422	422	422	422	422	422	422	422	422	422	422	422
Growth	0	0	0	0	0	0	0	0	0	0	0	0
Lactation	349	418	376	301	226	163	0	0	0	0	0	0
Pregnancy	0	0	1	2	4	7	14	27	50	88	151	251
Total	770	840	799	724	651	591	436	449	471	510	573	672
Calcium required, g/d												
Maintenance	16	16	16	16	16	16	16	16	16	16	16	16
Growth	0	0	0	0	0	0	0	0	0	0	0	0
Lactation	16	20	18	14	11	8	0	0	0	0	0	0
Pregnancy	0	0	0	0	0	0	0	0	0	12	12	12
Total	33	36	34	31	27	24	16	16	16	29	29	29

Table Beef-8 (Continued)

							Months Since Calving					
	1	2	3	4	5	6	7	8	9	10	11	12
NE_m Req. Factor	100%	100%	100%	100%	100%	100%	100%	100%	100%	100%	100%	100%
Phosphorus required, g/d												
Maintenance	13	13	13	13	13	13	13	13	13	13	13	13
Growth	0	0	0	0	0	0	0	0	0	0	0	0
Lactation	9	11	10	8	6	4	0	0	0	0	0	0
Pregnancy	0	0	0	0	0	0	0	0	0	5	5	5
Total	22	24	23	21	19	17	13	13	13	18	18	18
ADG, kg/d												
Growth	0.00	0.00	0.00	0.00	0.00	0.00	0.00	0.00	0.00	0.00	0.00	0.00
Pregnancy	0.00	0.00	0.02	0.03	0.05	0.08	0.12	0.19	0.28	0.40	0.57	0.77
Total	0.00	0.00	0.02	0.03	0.05	0.08	0.12	0.19	0.28	0.40	0.57	0.77
Milk kg/d	6.7	8.0	7.2	5.8	4.3	3.1	0.0	0.0	0.0	0.0	0.0	0.0
Body weight, kg												
Shrunk body	533	533	533	533	533	533	533	533	533	533	533	533
Conceptus	0	0	1	1	3	4	7	12	19	29	44	64
Total	533	533	534	534	536	537	540	545	552	562	577	597

Table Beef-9
DIET EVALUATION FOR BEEF COWS

Mature Weight	533 kg	Milk Fat	4.0%
Calf Birth Weight	40 kg	Milk Protein	3.4%
Age @ Calving	60 Months	Calving Interval	12 Months
Age @ Weaning	30 Weeks	Time Peak	8.5 Weeks
Peak Milk	8 kg	Milk SNF	8.3%
Breed Code	1 Angus		

Ration	TDN % DM	ME Mcal/kg	NE_m Mcal/kg	CP % DM	DIP % CP	DMI Factor
A	50	1.84	1.00	7.9	82.5	100%
B	60	2.21	1.35	7.8	100.0	100%
C	70	2.58	1.67	9.1	100.0	100%

Months Since Calving

	1	2	3	4	5	6	7	8	9	10	11	12
NE_m Req. Factor	100%	100%	100%	100%	100%	100%	100%	100%	100%	100%	100%	100%
A Milk kg/d	6.7	8.0	7.2	5.8	4.3	3.1	0.0	0.0	0.0	0.0	0.0	0.0
DM, kg	11.14	11.40	12.12	11.83	11.54	11.30	10.68	10.68	10.68	10.68	10.68	10.68
Energy balance, Mcal/d	−3.90	−4.59	−3.31	−2.58	−1.88	−1.34	1.81	1.50	0.95	0.06	−1.30	−3.24
DIP balance, g/d	7	7	7	7	7	7	6	6	6	6	6	6
UIP balance, g/d	−201	−270	−169	−96	−24	34	175	170	142	93	14	−110
MP balance, g/d	−161	−216	−136	−77	−19	27	149	136	113	75	11	−88
Ca % DM	0.65%	0.70%	0.62%	0.57%	0.52%	0.47%	0.34%	0.34%	0.34%	0.59%	0.59%	0.59%
P % DM	0.20%	0.21%	0.19%	0.18%	0.16%	0.15%	0.12%	0.12%	0.12%	0.17%	0.17%	0.17%
Reserves flux/mo, Mcal	−148	−174	−126	−98	−71	−51	55	46	29	2	−50	−123

Table Beef-9 (Continued)

	Months Since Calving											
	1	2	3	4	5	6	7	8	9	10	11	12
NE_m Req. Factor	100%	100%	100%	100%	100%	100%	100%	100%	100%	100%	100%	100%
B DM, kg	11.96	12.23	12.72	12.43	12.14	11.90	11.28	11.28	11.28	11.28	11.28	11.28
Energy balance, Mcal	1.07	0.47	1.69	2.32	2.92	3.38	6.32	6.00	5.46	4.56	3.20	1.27
DIP balance, g/d	5	5	5	5	5	5	5	5	5	5	5	5
UIP balance, g/d	18	−47	44	114	182	233	221	221	221	221	209	85
MP balance, g/d	14	−38	35	91	146	189	304	291	269	230	167	68
Ca % DM	0.27%	0.30%	0.27%	0.25%	0.22%	0.20%	0.15%	0.15%	0.15%	0.25%	0.25%	0.25%
P % DM	0.19%	0.20%	0.18%	0.17%	0.16%	0.14%	0.11%	0.11%	0.11%	0.16%	0.16%	0.16%
Reserves flux/mo, Mcal	32	14	51	71	89	103	192	183	166	139	97	39
C DM, kg	13.16	13.42	13.79	13.50	13.21	12.97	12.35	12.35	12.35	12.35	12.35	12.35
Energy balance, Mcal/d	6.99	6.48	7.65	8.18	8.69	9.07	11.80	11.49	10.95	10.05	8.69	6.76
DIP balance, g/d	3	3	3	3	3	3	2	2	2	2	2	2
UIP balance, g/d	295	233	314	308	301	296	282	282	282	282	282	282
MP balance, g/d	236	187	256	308	360	401	509	496	473	435	371	272
Ca % DM	0.25%	0.27%	0.25%	0.23%	0.20%	0.19%	0.13%	0.13%	0.13%	0.23%	0.23%	0.23%
P % DM	0.17%	0.18%	0.17%	0.15%	0.14%	0.13%	0.10%	0.10%	0.10%	0.14%	0.14%	0.14%
Reserves flux/mo, Mcal	212	197	233	249	264	276	359	349	333	306	261	205

Table Beef-10
APPROXIMATE TOTAL DAILY WATER INTAKE OF BEEF CATTLE[a]

Weight		Temperature in °F (°C)[b]										
		40 (4.4)		50 (10.0)		60 (14.4)		70 (21.1)		80 (26.6)		90 (32.2)	
kg	lb	Liter	Gal	Liter	Gal	Liter	Gal	Liter	Gal	Liter	Gal	Liter	Gal
Growing Heifers, Steers, and Bulls													
182	400	15.1	4.0	16.3	4.3	18.9	5.0	22.0	5.8	25.4	6.7	36.0	9.5
273	600	20.1	5.3	22.0	5.8	25.0	6.6	29.5	7.8	33.7	8.9	48.1	12.7
364	800	23.0	6.3	25.7	6.8	29.9	7.9	34.8	9.2	40.1	10.6	56.8	15.0
Finishing Cattle													
273	600	22.7	6.0	24.6	6.5	28.0	7.4	32.9	8.7	37.9	10.0	54.1	14.3
364	800	27.6	7.3	29.9	7.9	34.4	9.1	40.5	10.7	46.6	12.3	65.9	17.4
454	1,000	32.9	8.7	35.6	9.4	40.9	10.8	47.7	12.6	54.9	14.5	78.0	20.6
Wintering Pregnant Cows[c]													
409	900	25.4	6.7	27.3	7.2	31.4	8.3	36.7	9.7	—	—	—	—
500	1,100	22.7	6.0	24.6	6.5	28.0	7.4	32.9	8.7	—	—	—	—
Lactating Cows[d]													
409	900	43.1	11.4	47.7	12.6	54.9	14.5	64.0	16.9	67.8	17.9	61.3	16.2
Mature bulls													
636	1,400	30.3	8.0	32.6	8.6	37.5	9.9	44.3	11.7	50.7	13.4	71.9	19.0
727	1,600+	32.9	8.7	35.6	9.4	40.9	10.8	47.7	12.6	54.9	14.5	78.0	20.6

[a]Winchester and Morris (1956).

[b]Water intake of a given class of cattle in a specific management regime is a function of dry matter intake and ambient temperature. Water intake is quite constant up to 40°F (4.4°C).

[c]Dry matter intake has a major influence on water intake. Heavier cows are assumed to be higher in body condition and to require less dry matter and, thus, less water intake.

[d]Cows larger than 409 kg (900 lbs) are included in this recommendation.

Table Beef-11
ESTIMATED BIRTH WEIGHT OF CALVES OF DIFFERENT BREEDS OR BREED CROSSES, kg

Breed	BIF	AFRC	MARC
Angus	31	26	35
Brahman	31	—	41
Braford	36	—	—
Brangus	33	—	—
Braunvieh	—	—	39
Charolais	39	43	40
Chianina	—	—	41
Devon	32	34	—
Galloway	—	—	36
Gelbvieh	39	—	39
Hereford	36	35	37
Holstein	—	43	—
Jersey	—	25	31
Limousin	37	38	39
Longhorn	—	—	33
Maine-Anjou	40	—	41
Nellore	—	—	40
Piedmontese	—	—	38
Pinzgauer	33	—	40
Polled Hereford	33	—	36
Red Poll	—	—	36
Sahiwal	—	—	38
Santa Gertrudis	33	—	—
Salers	35	—	38
Shorthorn	37	32	39
Simmental	39	43	40
South Devon	33	42	38
Tarentaise	33	—	38

Note: BIF, Beef Improvement Federation; AFRC, Agricultural and Food Research Council; MARC, Roman L. Hruska U.S. Meat Animal Research Center (USDA/ARS).

Sources: Beef Improvement Federation (1990), AFRC (1990), MARC, from data reported by Cundiff et al. (1988), and Gregory et al. (1982), which are from a particular sire breed on mature Angus and Hereford cows.

Table Dairy-1
DRY MATTER INTAKE REQUIREMENTS TO FULFILL NUTRIENT ALLOWANCES FOR MAINTENANCE, MILK PRODUCTION, AND NORMAL LIVE WEIGHT GAIN DURING MID AND LATE LACTATION

Live Wt (lb)	*800*	*900*	*1,000*	*1,100*	*1,200*	*1,300*	*1,400*	*1,500*	*1,600*	*1,700*	*1,800*
FCM (4%)[a] (lb)						% of Live Weight[b,c]					
20	2.8	2.6	2.5	2.3	2.2	2.1	2.1	2.0	1.9	1.9	1.8
30	3.2	3.0	2.9	2.7	2.6	2.5	2.4	2.3	2.2	2.1	2.1
40	3.6	3.4	3.2	3.1	2.9	2.8	2.7	2.5	2.4	2.4	2.3
50	4.0	3.8	3.6	3.4	3.2	3.1	3.0	2.8	2.7	2.6	2.5
60	4.4	4.1	3.9	3.7	3.5	3.4	3.2	3.1	3.0	2.9	2.7
70	4.8	4.6	4.3	4.0	3.8	3.6	3.5	3.3	3.2	3.1	2.9
80	5.4	5.1	4.7	4.3	4.1	3.8	3.7	3.5	3.4	3.2	3.1
90	—	5.5	5.1	4.7	4.4	4.1	3.9	3.7	3.6	3.4	3.3
100	—	—	5.5	5.0	4.7	4.4	4.2	3.9	3.8	3.6	3.5
110	—	—	—	5.4	5.1	4.8	4.5	4.2	4.0	3.8	3.7
120	—	—	—	—	5.4	5.0	4.8	4.5	4.3	4.1	3.9
130	—	—	—	—	—	5.4	5.1	4.8	4.6	4.4	4.2

Note: The following assumptions were made in calculating the dry matter intake requirements shown in Table Dairy-1:

1. The basic or reference cow used for the calculations weighed 1,320 lb and produced milk with 4% milk fat. Other live weights in the table and corresponding fat percentages were 881 lb and 5% fat; 1,100 lb and 4.5% fat; and 1,540 and 1,760 lb and 3.5% fat.

2. The concentration of energy in the diet for the reference cow was 0.65 Mcal of NEL/lb of DM for milk yields equal to or less than 22 lb/day. It increased linearly to 0.78 Mcal of NEL/lb for milk yields equal to or greater than 88 lb/day.

3. The energy concentrations of the diets for all other cows were assumed to change linearly as their energy requirements for milk production, relative to maintenance, changed in a manner identical to that of the 1,320-lb cow as she increased in milk yield from 22 to 88 lb/day.

4. Enough DM to provide sufficient energy for cows to gain 0.055% of their body weight daily was also included in the total. If cows do not consume as much DM as they require, as calculated from Table Dairy-1, their energy intake will be less than their requirements. The result will be a loss of body weight, reduced milk yields, or both. If cows consume more DM than what is projected as required from Table Dairy-1, the energy concentration of their diet should be reduced or they may become overly fat.

[a]4% fat-corrected milk (lb) = (0.4) (lb of milk) + (15) (lb of milk fat).
[b]Probable DMI may be up to 18% less in early lactation.
[c]DMI as a percentage of live weight may be 0.02% less per 1% increase in diet moisture content above 50% if fermented feeds constitute a major portion of the diet.

Table Dairy-2
DAILY NUTRIENT REQUIREMENTS OF GROWING DAIRY CATTLE AND MATURE BULLS[a]

Live Weight (kg)	Gain (g)	Dry Matter Intake[a] (kg)	Energy					Protein			Minerals		Vitamins	
			NEM (Mcal)	NEG (Mcal)	ME (Mcal)	DE (Mcal)	TDN (kg)	UIP (g)	DIP (g)	CP (g)	Ca (g)	P (g)	A (1,000 IU)	D (1,000 IU)
Growing Large-Breed Calves Fed Only Milk or Milk Replacer														
40	200	0.48	1.37	0.41	2.54	2.73	0.62	—	—	105	7	4	1.70	0.26
45	300	0.54	1.49	0.56	2.86	3.07	0.70	—	—	120	8	5	1.94	0.30
Growing Large-Breed Calves Fed Milk Plus Starter Mix														
50	500	1.30	1.62	0.72	5.90	6.42	1.46	—	—	290	9	6	2.10	0.33
75	800	1.98	2.19	1.30	8.98	9.78	2.22	—	—	435	16	8	3.20	0.50
Growing Small-Breed Calves Fed Only Milk or Milk Replacer														
25	200	0.38	0.96	0.37	2.01	2.16	0.49	—	—	84	6	4	1.10	0.16
30	300	0.51	1.10	0.52	2.70	2.90	0.66	—	—	112	7	4	1.30	0.20
Growing Small-Breed Calves Fed Milk Plus Starter Mix														
50	500	1.43	1.62	0.72	6.49	7.06	1.60	—	—	315	10	6	2.10	0.33
75	600	1.76	2.19	0.96	7.98	8.69	1.97	—	—	387	14	8	3.20	0.50
Growing Veal Calves Fed Only Milk or Milk Replacer														
40	200	0.45	1.37	0.55	1.89	2.07	0.47	—	—	100	7	4	1.70	0.26
50	400	0.57	1.62	0.57	2.39	2.63	0.59	—	—	125	9	5	2.10	0.33
60	540	0.80	1.85	0.81	2.84	3.17	0.71	—	—	176	13	8	2.60	0.40
75	900	1.36	2.19	1.47	4.82	5.39	1.21	—	—	300	16	9	3.20	0.50
100	1,250	2.00	2.72	2.26	6.22	7.06	1.58	—	—	440	20	11	4.20	0.66
125	1,250	2.38	3.21	2.44	7.40	8.40	1.88	—	—	524	22	13	5.30	0.82
150	1,100	2.72	3.69	2.29	8.46	9.60	2.15	—	—	598	24	15	6.40	0.99

continued

Large-Breed Growing Females

100	600	2.63	2.72	1.22	7.03	8.13	1.84	317	57	421	17	9	4.24	0.66
100	700	2.82	2.72	1.44	7.54	8.72	1.98	346	75	452	18	9	4.24	0.66
100	800	3.02	2.72	1.66	8.06	9.32	2.11	374	92	483	18	10	4.24	0.66
150	600	3.51	3.69	1.45	9.14	10.61	2.41	283	150	562	19	11	6.36	0.99
150	700	3.75	3.69	1.71	9.76	11.33	2.57	307	173	600	19	12	6.36	0.99
150	800	3.99	3.69	1.97	10.39	12.07	2.74	331	196	639	20	12	6.36	0.99
200	600	4.39	4.57	1.65	11.14	12.99	2.95	254	239	631	20	14	8.48	1.32
200	700	4.68	4.57	1.95	11.87	13.84	3.14	274	267	686	21	14	8.48	1.32
200	800	4.97	4.57	2.25	12.62	14.71	3.34	294	295	741	22	15	8.48	1.32
250	600	5.31	5.41	1.84	13.10	15.33	3.48	229	326	637	22	16	10.60	1.65
250	700	5.65	5.41	2.18	13.94	16.32	3.70	246	359	678	23	17	10.60	1.65
250	800	5.99	5.41	2.51	14.79	17.32	3.93	263	393	726	24	17	10.60	1.65
300	600	6.26	6.20	2.02	15.05	17.69	4.01	209	413	752	23	17	12.72	1.98
300	700	6.66	6.20	2.39	16.00	18.81	4.27	223	452	799	24	18	12.72	1.98
300	800	7.06	6.20	2.77	16.97	19.95	4.52	236	490	848	25	19	12.72	1.98
350	600	7.29	6.96	2.20	17.01	20.09	4.56	193	501	874	24	18	14.84	2.31
350	700	7.75	6.96	2.60	18.09	21.36	4.84	204	545	930	25	19	14.84	2.31
350	800	8.21	6.96	3.01	19.18	22.64	5.14	214	590	985	26	20	14.84	2.31
400	600	8.39	7.69	2.37	19.03	22.58	5.12	182	592	1,007	25	19	16.96	2.64
400	700	8.92	7.69	2.80	20.23	24.00	5.44	190	641	1,070	26	20	16.96	2.64
400	800	9.46	7.69	3.24	21.44	25.44	5.77	198	692	1,135	26	21	16.96	2.64
450	600	9.59	8.40	2.53	21.12	25.18	5.71	176	686	1,151	26	19	19.08	2.97
450	700	10.20	8.40	2.99	22.46	26.78	6.07	182	742	1,224	28	20	19.08	2.97
450	800	10.82	8.40	3.46	23.81	28.40	6.44	187	799	1,298	29	21	19.08	2.97
500	600	10.93	9.09	2.69	23.32	27.96	6.34	175	785	1,311	28	20	21.20	3.30
500	700	11.63	9.09	3.18	24.81	29.74	6.75	179	848	1,395	28	20	21.20	3.30
500	800	12.33	9.09	3.68	26.32	31.55	7.16	182	913	1,480	29	21	21.20	3.30
550	600	12.42	9.77	2.84	25.67	30.95	7.02	180	891	1,490	28	20	23.32	3.63
550	700	13.22	9.77	3.37	27.33	32.95	7.47	183	963	1,587	28	20	23.32	3.63
550	800	14.04	9.77	3.90	29.02	34.99	7.94	185	1,035	1,685	29	21	23.32	3.63
600	600	14.11	10.43	3.00	28.23	34.24	7.77	193	1,007	1,694	28	20	25.44	3.96
600	700	15.05	10.43	3.55	30.09	36.50	8.28	194	1,088	1,805	28	21	25.44	3.96
600	800	15.99	10.43	4.11	31.98	38.79	8.80	195	1,170	1,919	29	21	25.44	3.96

Table Dairy-2 (Continued)

Live Weight (kg)	Gain (g)	Dry Matter Intake[a] (kg)	Energy NEM (Mcal)	NEG (Mcal)	ME (Mcal)	DE (Mcal)	TDN (kg)	Protein UIP (g)	DIP (g)	CP (g)	Minerals Ca (g)	P (g)	Vitamins A (1,000 IU)	D (1,000 IU)
			Small-Breed Growing Females											
100	400	2.41	2.72	0.91	6.34	7.35	1.67	249	38	386	15	8	4.24	0.66
100	500	2.64	2.72	1.16	6.92	8.03	1.82	275	59	422	16	8	4.24	0.66
100	600	2.86	2.72	1.40	7.51	8.71	1.98	300	80	458	17	9	4.24	0.66
150	400	3.31	3.69	1.09	8.39	9.78	2.22	222	129	512	17	10	6.36	0.99
150	500	3.60	3.69	1.39	9.12	10.63	2.41	243	156	567	18	11	6.36	0.99
150	600	3.89	3.69	1.69	9.86	11.50	2.61	263	185	622	19	11	6.36	0.99
200	400	4.24	4.57	1.26	10.38	12.16	2.76	201	217	513	19	13	8.48	1.32
200	500	4.60	4.57	1.60	11.25	13.19	2.99	217	251	562	20	13	8.48	1.32
200	600	4.96	4.57	1.95	12.14	14.23	3.23	232	286	611	20	14	8.48	1.32
250	400	5.24	5.41	1.41	12.36	14.57	3.30	185	305	629	21	15	10.60	1.65
250	500	5.68	5.41	1.80	13.38	15.78	3.58	197	346	681	21	16	10.60	1.65
250	600	6.12	5.41	2.20	14.43	17.01	3.86	209	389	735	22	16	10.60	1.65
300	400	6.34	6.20	1.56	14.38	17.06	3.87	176	395	761	22	16	12.72	1.98
300	500	6.87	6.20	1.99	15.57	18.48	4.19	184	445	824	23	17	12.72	1.98
300	600	7.40	6.20	2.43	16.79	19.92	4.52	192	495	888	23	17	12.72	1.98
350	400	7.57	6.96	1.71	16.50	19.71	4.47	173	490	909	23	17	14.84	2.31
350	500	8.20	6.96	2.18	17.87	21.35	4.84	178	548	985	23	18	14.84	2.31
350	600	8.85	6.96	2.66	19.28	23.03	5.22	183	608	1,062	24	18	14.84	2.31
400	400	8.98	7.69	1.84	18.77	22.58	5.12	177	592	1,078	24	18	16.96	2.64
400	500	9.74	7.69	2.35	20.36	24.50	5.56	181	661	1,169	24	19	16.96	2.64
400	600	10.52	7.69	2.87	21.98	26.45	6.00	183	730	1,263	25	19	16.96	2.64
450	400	10.64	8.40	1.98	21.27	25.80	5.85	191	706	1,276	27	18	19.08	2.97
450	500	11.56	8.40	2.52	23.12	28.04	6.36	193	786	1,387	28	19	19.08	2.97
450	600	12.50	8.40	3.08	25.01	30.33	6.88	194	867	1,500	28	19	19.08	2.97
			Small-Breed Growing Males											
100	800	2.80	2.72	1.42	7.48	8.66	1.96	401	65	448	18	10	4.24	0.66

continued

100	900	2.97	2.72	1.60	7.92	9.16	2.08	433	79	475	19	10	4.24	0.66
100	1,000	3.13	2.72	1.79	8.36	9.67	2.19	465	93	501	20	11	4.24	0.66
150	800	3.60	3.69	1.64	9.52	11.03	2.50	364	155	576	20	12	6.36	0.99
150	900	3.80	3.69	1.85	10.03	11.63	2.64	393	172	607	21	13	6.36	0.99
150	1,000	3.99	3.69	2.07	10.55	12.22	2.77	422	190	639	22	13	6.36	0.99
200	800	4.43	4.57	1.84	11.48	13.34	3.03	333	241	709	22	15	8.48	1.32
200	900	4.66	4.57	2.08	12.06	14.02	3.18	359	262	745	23	15	8.48	1.32
200	1,000	4.89	4.57	2.33	12.66	14.71	3.34	385	284	782	24	16	8.48	1.32
250	800	5.27	5.41	2.03	13.37	15.58	3.53	305	325	778	24	17	10.60	1.65
250	900	5.53	5.41	2.30	14.03	16.35	3.71	329	350	837	25	18	10.60	1.65
250	1,000	5.80	5.41	2.57	14.70	17.13	3.89	352	375	897	26	18	10.60	1.65
300	800	6.13	6.20	2.21	15.22	17.80	4.04	281	408	771	25	19	12.72	1.98
300	900	6.43	6.20	2.51	15.96	18.66	4.23	302	436	827	25	19	12.72	1.98
300	1,000	6.73	6.20	2.80	16.70	19.53	4.43	323	464	884	26	20	12.72	1.98
350	800	7.02	6.96	2.38	17.06	20.02	4.54	261	490	843	26	20	14.84	2.31
350	900	7.36	6.96	2.70	17.88	20.98	4.76	280	522	883	26	20	14.84	2.31
350	1,000	7.70	6.96	3.02	18.70	21.94	4.98	298	554	924	27	21	14.84	2.31
400	800	7.96	7.69	2.55	18.91	22.27	5.05	244	572	955	26	21	16.96	2.64
400	900	8.34	7.69	2.89	19.80	23.32	5.29	260	608	1,001	27	21	16.96	2.64
400	1,000	8.72	7.69	3.24	20.71	24.39	5.53	277	644	1,046	28	22	16.96	2.64
450	800	8.95	8.40	2.71	20.78	24.56	5.57	230	656	1,074	29	21	19.08	2.97
450	900	9.37	8.40	3.08	21.76	25.72	5.83	245	696	1,125	29	22	19.08	2.97
450	1,000	9.80	8.40	3.44	22.75	26.89	6.10	259	736	1,176	29	23	19.08	2.97
500	800	10.00	9.09	2.87	22.69	26.92	6.11	220	742	1,201	29	21	21.20	3.30
500	900	10.48	9.09	3.25	23.76	28.19	6.39	233	786	1,257	29	22	21.20	3.30
500	1,000	10.95	9.09	3.64	24.84	29.47	6.68	246	830	1,314	29	23	21.20	3.30
550	800	11.14	9.77	3.02	24.66	29.38	6.66	213	831	1,336	29	21	23.32	3.63
550	900	11.66	9.77	3.43	25.82	30.76	6.98	225	879	1,399	29	22	23.32	3.63
550	1,000	12.19	9.77	3.84	27.00	32.16	7.29	236	927	1,463	30	23	23.32	3.63
600	800	12.36	10.43	3.17	26.71	31.95	7.25	211	923	1,483	29	21	25.44	3.96
600	900	12.95	10.43	3.60	27.97	33.47	7.59	221	976	1,554	29	22	25.44	3.96
600	1,000	13.54	10.43	4.03	29.25	34.99	7.94	231	1,029	1,624	30	23	25.44	3.96
650	800	13.69	11.07	3.32	28.86	34.67	7.86	212	1,020	1,643	29	21	27.56	4.29
650	900	14.35	11.07	3.77	30.24	36.33	8.24	222	1,078	1,722	29	22	27.56	4.29

Table Dairy-2 (Continued)

Live Weight (kg)	Gain (g)	Dry Matter Intake[a] (kg)	NEM (Mcal)	NEG (Mcal)	ME (Mcal)	DE (Mcal)	TDN (kg)	UIP (g)	DIP (g)	CP (g)	Ca (g)	P (g)	A (1,000 IU)	D (1,000 IU)
							Large-Breed Growing Males (Continued)							
650	1,000	15.01	11.07	4.22	31.63	38.00	8.62	230	1,137	1,801	30	23	27.56	4.29
700	800	15.16	11.70	3.46	31.14	37.59	8.52	219	1,124	1,820	29	22	29.68	4.62
700	900	15.90	11.70	3.93	32.64	39.40	8.94	227	1,187	1,907	29	22	29.68	4.62
700	1,000	16.63	11.70	4.40	34.16	41.23	9.35	235	1,252	1,996	30	23	29.68	4.62
750	800	16.79	12.33	3.60	33.59	40.73	9.24	232	1,235	2,015	29	22	31.80	4.95
750	900	17.62	12.33	4.09	35.23	42.73	9.60	239	1,305	2,114	29	23	31.80	4.95
750	1,000	18.45	12.33	4.58	36.89	44.74	10.15	246	1,376	2,213	30	23	31.80	4.95
800	800	17.56	12.94	3.74	35.12	42.59	9.66	216	1,303	2,107	29	22	33.92	5.28
800	900	18.41	12.94	4.25	36.83	44.67	10.13	221	1,377	2,210	29	23	33.92	5.28
800	1,000	19.28	12.94	4.76	38.55	46.76	10.61	227	1,451	2,313	30	23	33.92	5.28
							Small-Breed Growing Males							
100	500	2.45	2.72	1.02	6.54	7.56	1.72	287	41	392	16	8	4.24	0.66
100	600	2.64	2.72	1.23	7.04	8.15	1.85	316	58	422	17	9	4.24	0.66
100	700	2.83	2.72	1.45	7.55	8.74	1.98	345	75	453	18	9	4.24	0.66
150	500	3.28	3.69	1.20	8.55	9.92	2.25	257	129	525	18	11	6.36	0.99
150	600	3.52	3.69	1.46	9.16	10.64	2.41	282	151	563	19	11	6.36	0.99
150	700	3.76	3.69	1.71	9.78	11.36	2.58	306	174	601	19	12	6.36	0.99
200	500	4.12	4.57	1.37	10.45	12.18	2.76	232	213	573	20	13	8.48	1.32
200	600	4.40	4.57	1.66	11.17	13.02	2.95	252	241	629	20	14	8.48	1.32
200	700	4.69	4.57	1.96	11.90	13.87	3.15	273	268	684	21	14	8.48	1.32
250	500	4.99	5.41	1.53	12.31	14.41	3.27	210	296	598	21	16	10.60	1.65
250	600	5.32	5.41	1.86	13.14	15.38	3.49	228	328	638	22	16	10.60	1.65
250	700	5.66	5.41	2.19	13.97	16.35	3.71	245	361	679	23	17	10.60	1.65
300	500	5.89	6.20	1.68	14.15	16.64	3.77	193	378	707	23	17	12.72	1.98
300	600	6.28	6.20	2.04	15.09	17.74	4.02	207	415	754	23	17	12.72	1.98
300	700	6.68	6.20	2.41	16.04	18.85	4.28	221	453	801	24	18	12.72	1.98
350	500	6.86	6.96	1.82	16.01	18.91	4.29	180	461	823	23	18	14.84	2.31

continued

350	600	7.31	6.96	2.22	17.06	20.15	4.57	191	503	877	24	18	14.84	2.31
350	700	7.76	6.96	2.62	18.13	21.41	4.86	203	547	932	25	19	14.84	2.31
400	500	7.90	7.69	1.96	17.91	21.25	4.82	171	545	947	24	19	16.96	2.64
400	600	8.41	7.69	2.39	19.08	22.64	5.14	180	594	1,010	25	19	16.96	2.64
400	700	8.94	7.69	2.82	20.27	24.06	5.46	189	644	1,073	26	20	16.96	2.64
450	500	9.03	8.40	2.10	19.87	23.70	5.37	166	634	1,083	28	19	19.08	2.97
450	600	9.62	8.40	2.55	21.18	25.26	5.73	174	689	1,155	28	19	19.08	2.97
450	700	10.23	8.40	3.01	22.51	26.84	6.09	180	744	1,227	28	20	19.08	2.97
500	500	10.28	9.09	2.23	21.93	26.29	5.96	167	726	1,233	28	19	21.20	3.30
500	600	10.96	9.09	2.71	23.39	28.04	6.36	173	788	1,315	28	20	21.20	3.30
500	700	11.65	9.09	3.20	24.87	29.81	6.76	177	851	1,398	28	20	21.20	3.30
550	500	11.67	9.77	2.36	24.12	29.08	6.60	174	825	1,400	28	19	23.32	3.63
550	600	12.46	9.77	2.87	25.75	31.05	7.04	178	895	1,495	28	20	23.32	3.63
550	700	13.26	9.77	3.39	27.40	33.03	7.49	181	966	1,591	28	20	23.32	3.63
600	500	13.25	10.43	2.48	26.50	32.14	7.29	187	933	1,590	28	19	25.44	3.96
600	600	14.16	10.43	3.02	28.32	34.35	7.79	190	1,012	1,699	28	20	25.44	3.96
600	700	15.08	10.43	3.57	30.17	36.59	8.30	192	1,091	1,810	28	21	25.44	3.96

Maintenance of Mature Breeding Bulls

500	—	7.89	9.09	—	15.79	19.15	4.34	161	472	789	20	12	21.20	3.30
600	—	9.05	10.43	—	18.10	21.95	4.98	155	573	905	24	15	25.44	3.96
700	—	10.16	11.70	—	20.32	24.64	5.59	148	670	1,016	28	18	29.68	4.62
800	—	11.23	12.94	—	22.46	27.24	6.18	142	764	1,123	32	20	33.92	5.28
900	—	12.27	14.13	—	24.53	29.76	6.75	135	854	1,227	36	22	38.16	5.94
1,000	—	13.28	15.29	—	26.55	32.20	7.30	129	943	1,328	41	25	42.40	6.60
1,100	—	14.26	16.43	—	28.52	34.59	7.85	122	1,029	1,426	45	28	46.64	7.26
1,200	—	15.22	17.53	—	30.44	36.92	8.37	115	1,113	1,522	49	30	50.88	7.92
1,300	—	16.16	18.62	—	32.32	39.21	8.89	108	1,196	1,616	53	32	55.12	8.58
1,400	—	17.09	19.68	—	34.17	41.45	9.40	102	1,277	1,709	57	35	59.36	9.24

Note: The following abbreviations were used: NEM, net energy for maintenance; NEG, neg energy for gain; ME, metabolizable energy; DE, digestible energy; TDN, total digestible nutrients; UIP, undegraded intake protein; DIP, degraded intake protein; CP, crude protein.
ᵃThe data for DMI are not requirements per se, unlike the requirements for net energy maintenance, net energy gain, and absorbed protein. They are not intended to be estimates of voluntary intake but are consistent with the specified dietary energy concentrations. The use of diets with decreased energy concentrations will increase dry matter intake needs; metabolizable energy, digestible energy, and total digestible nutrient needs; and crude protein needs. The use of diets with increased energy concentrations will have opposite effects on these needs.

Table Dairy-3
DAILY NUTRIENT REQUIREMENTS OF LACTATING AND PREGNANT COWS

Live Weight (kg)	Energy				Total Crude Protein (g)	Minerals		Vitamins	
	NEL (Mcal)	ME (Mcal)	DE (Mcal)	TDN (kg)		Ca (g)	P (g)	A	D (1,000 IU)
Maintenance of Mature Lactating Cows[a]									
400	7.16	12.01	13.80	3.13	318	16	11	30	12
450	7.82	13.12	15.08	3.42	341	18	13	34	14
500	8.46	14.20	16.32	3.70	364	20	14	38	15
550	9.09	15.25	17.53	3.97	386	22	16	42	17
600	9.70	16.28	18.71	4.24	406	24	17	46	18
650	10.30	17.29	19.86	4.51	428	26	19	49	20
700	10.89	18.28	21.00	4.76	449	28	20	53	21
750	11.47	19.25	22.12	5.02	468	30	21	57	23
800	12.03	20.20	23.21	5.26	486	32	23	61	24
Maintenance Plus Last 2 Months of Gestation of Mature Dry Cows[b]									
400	9.30	15.26	18.23	4.15	890	26	16	30	12
450	10.16	16.66	19.91	4.53	973	30	18	34	14
500	11.00	18.04	21.55	4.90	1,053	33	20	38	15
550	11.81	19.37	23.14	5.27	1,131	36	22	42	17
600	12.61	20.68	24.71	5.62	1,207	39	24	46	18
650	13.39	21.96	26.23	5.97	1,281	43	26	49	20
700	14.15	23.21	27.73	6.31	1,355	46	28	53	21
750	14.90	24.44	29.21	6.65	1,427	49	30	57	23
800	15.64	25.66	30.65	6.98	1,497	53	32	61	24

continued

Milk Production—Nutrients/kg of Milk of Different Fat Percentages

(Fat %)									
3.0	0.64	1.07	1.23	0.280	78	2.73	1.68	—	—
3.5	0.69	1.15	1.33	0.301	84	2.97	1.83	—	—
4.0	0.74	1.24	1.42	0.322	90	3.21	1.98	—	—
4.5	0.78	1.32	1.51	0.343	96	3.45	2.13	—	—
5.0	0.83	1.40	1.61	0.364	101	3.69	2.28	—	—
5.5	0.88	1.48	1.70	0.385	107	3.93	2.43	—	—

Live Weight Change During Lactation—Nutrients/kg of Weight Change[c]

Weight loss	−4.92	−8.25	−9.55	−2.17	−320	—	—	—	—
Weight gain	5.12	8.55	9.96	2.26	320	—	—	—	—

Note: The following abbreviations were used: NEL, net energy for lactation; ME, metabolizable energy; DE, digestible energy; TDN, total digestible nutrients.

[a]To allow for growth of young lactating cows, increase the maintenance allowances for all nutrients except vitamins A and D by 20% during the first lactation and 10% during the second lactation.

[b]Values for calcium assume that the cow is in calcium balance at the beginning of the last 2 months of gestation. If the cow is not in balance, then the calcium requirement can be increased from 25% to 33%.

[c]No allowance is made for mobilized calcium and phosphorus associated with live weight loss or with live weight gain. The maximum daily nitrogen available from weight loss is assumed to be 30 g or 234 g of crude protein.

Table Dairy-4
DAILY NUTRIENT REQUIREMENTS OF LACTATING COWS USING ABSORBABLE PROTEIN

Live Weight (lb)	Fat (%)	Milk (lb)	Live Weight Change (lb)	Dry Matter Intake (lb)	Energy			Protein		Minerals	
					NELDM (Mcal/lb)	NEL (Mcal)	TDN (lb)	UIP (lb)	DIP (lb)	Ca (lb)	P (lb)
				Intake at 100% of the Requirement for Maintenance, Lactation, and Weight Gain							
900	4.5	14.0	0.495	20.79	0.64	13.39	13.07	0.98	1.50	0.085	0.056
900	4.5	29.0	0.495	27.42	0.68	18.73	18.20	1.51	2.25	0.137	0.087
900	4.5	43.0	0.495	32.76	0.72	23.71	22.94	1.92	2.95	0.185	0.117
900	4.5	58.0	0.495	37.86	0.77	29.04	28.00	2.29	3.72	0.237	0.149
900	4.5	74.0	0.495	44.54	0.78	34.73	33.45	2.81	4.52	0.292	0.183
900	5.0	14.0	0.495	21.26	0.64	13.70	13.37	1.01	1.54	0.088	0.058
900	5.0	29.0	0.495	28.13	0.69	19.36	18.80	1.55	2.34	0.144	0.092
900	5.0	43.0	0.495	33.70	0.73	24.65	23.84	1.97	3.09	0.195	0.124
900	5.0	58.0	0.495	38.98	0.78	30.31	29.19	2.34	3.90	0.251	0.158
900	5.0	74.0	0.495	46.61	0.78	36.35	35.00	2.94	4.74	0.310	0.194
900	5.5	14.0	0.495	21.73	0.64	14.01	13.66	1.05	1.59	0.092	0.060
900	5.5	29.0	0.495	28.84	0.69	20.00	19.41	1.59	2.43	0.151	0.096
900	5.5	43.0	0.495	34.62	0.74	25.59	24.73	2.01	3.23	0.206	0.130
900	5.5	58.0	0.495	40.49	0.78	31.57	30.41	2.43	4.08	0.265	0.166
900	5.5	74.0	0.495	48.68	0.78	37.96	36.56	3.08	4.97	0.328	0.205
1,100	4.0	18.0	0.605	24.62	0.64	15.86	15.47	1.11	1.86	0.102	0.067
1,100	4.0	36.0	0.605	32.01	0.68	21.87	21.25	1.71	2.70	0.160	1.103
1,100	4.0	55.0	0.605	38.77	0.73	28.21	27.29	2.23	3.60	0.221	0.140
1,100	4.0	73.0	0.605	44.45	0.77	34.22	32.98	2.65	4.46	0.279	0.176
1,100	4.5	91.0	0.605	54.13	0.78	42.22	40.65	3.39	5.58	0.359	0.225
1,100	4.5	18.0	0.605	25.23	0.64	16.25	15.86	1.16	1.91	0.107	0.070
1,100	4.5	36.0	0.605	32.89	0.69	22.65	22.00	1.76	2.81	0.169	0.108
1,100	4.5	55.0	0.605	39.95	0.74	29.41	28.43	2.29	3.77	0.234	0.149
1,100	4.5	73.0	0.605	45.92	0.78	35.81	34.49	2.72	4.69	0.297	0.187
1,100	4.5	91.0	0.605	54.13	0.78	42.22	40.65	3.39	5.58	0.359	0.225
1,100	5.0	18.0	0.605	25.74	0.65	16.65	16.24	1.19	1.97	0.111	0.072

continued

1,100	5.0	36.0	0.605	33.76	0.69	23.44	22.75	1.81	2.92	0.177	0.113
1,100	5.0	55.0	0.605	41.11	0.74	30.61	29.57	2.35	3.95	0.248	0.157
1,100	5.0	73.0	0.605	47.97	0.78	37.41	36.02	2.85	4.92	0.314	0.198
1,100	5.0	91.0	0.605	56.68	0.78	44.20	42.56	3.55	5.86	0.381	0.239
1,300	3.0	23.0	0.715	27.80	0.64	17.91	17.47	1.21	2.16	0.115	0.076
1,300	3.0	47.0	0.715	36.40	0.68	24.87	24.17	1.94	3.13	0.181	0.116
1,300	3.0	70.0	0.715	43.53	0.72	31.55	30.53	2.51	4.07	0.244	0.115
1,300	3.0	93.0	0.715	49.90	0.77	38.22	36.85	3.01	5.02	0.306	0.194
1,300	3.0	117.0	0.715	57.94	0.78	45.19	43.52	3.67	6.00	0.372	0.234
1,300	3.5	23.0	0.715	28.58	0.64	18.41	17.96	1.27	2.23	0.121	0.079
1,300	3.5	47.0	0.715	37.55	0.69	25.90	25.15	2.00	3.27	0.192	0.123
1,300	3.5	70.0	0.715	45.05	0.73	33.08	31.98	2.59	4.29	0.261	0.165
1,300	3.5	93.0	0.715	51.70	0.78	40.25	38.77	3.09	5.32	0.329	0.208
1,300	3.5	117.0	0.715	61.21	0.78	47.74	45.97	3.88	6.36	0.400	0.251
1,300	4.0	23.0	0.715	29.25	0.65	18.91	18.45	1.31	2.30	0.127	0.083
1,300	4.0	47.0	0.715	38.68	0.70	26.92	26.13	2.07	3.42	0.204	0.130
1,300	4.0	70.0	0.715	46.53	0.74	34.60	33.43	2.67	4.52	0.277	0.176
1,300	4.0	93.0	0.715	54.22	0.78	42.28	40.72	3.24	5.61	0.351	0.221
1,300	4.0	117.0	0.715	64.49	0.78	50.29	48.43	4.09	6.72	0.428	0.269
1,500	3.0	26.0	0.825	31.23	0.64	20.12	19.63	1.32	2.48	0.132	0.087
1,500	3.0	52.0	0.825	40.51	0.68	27.66	26.88	2.10	3.53	0.203	0.130
1,500	3.0	78.0	0.825	48.58	0.73	35.21	34.07	2.75	4.60	0.274	0.174
1,500	3.0	104.0	0.825	55.76	0.77	42.76	41.22	3.31	5.67	0.345	0.218
1,500	3.0	130.0	0.825	64.50	0.78	50.30	48.44	4.03	6.73	0.416	0.262
1,500	3.5	26.0	0.825	32.11	0.64	20.69	20.18	1.39	2.56	0.138	0.091
1,500	3.5	52.0	0.825	41.79	0.69	28.80	27.97	2.18	3.69	0.215	0.138
1,500	3.5	78.0	0.825	50.27	0.73	36.91	35.69	2.84	4.84	0.292	0.186
1,500	3.5	104.0	0.825	57.77	0.78	45.02	43.36	3.40	6.01	0.370	0.233
1,500	3.5	130.0	0.825	68.14	0.78	53.14	51.17	4.26	7.13	0.447	0.281
1,500	4.0	26.0	0.825	32.83	0.65	21.25	20.73	1.43	2.64	0.144	0.094
1,500	4.0	52.0	0.825	43.04	0.70	29.93	29.05	2.25	3.86	0.228	0.146
1,500	4.0	78.0	0.825	51.92	0.74	38.61	37.30	2.92	5.09	0.311	0.197
1,500	4.0	104.0	0.825	60.65	0.78	47.29	45.55	3.58	6.33	0.395	0.249
1,500	4.0	130.0	0.825	71.77	0.78	55.97	53.90	4.50	7.53	0.478	0.300
1,700	3.0	29.0	0.935	34.60	0.64	22.29	21.75	1.43	2.80	0.148	0.097
1,700	3.0	57.0	0.935	44.56	0.68	30.42	29.56	2.27	3.93	0.224	0.145

Table Dairy-4 (Continued)

Live Weight (lb)	Fat (%)	Milk (lb)	Live Weight Change (lb)	Dry Matter Intake (lb)	NELDM (Mcal/lb)	Energy NEL (Mcal)	TDN (lb)	Protein UIP (lb)	DIP (lb)	Minerals Ca (lb)	P (lb)
Intake at 100% of the Requirement for Maintenance, Lactation, and Weight Gain (Continued)											
1,700	3.0	86.0	0.935	53.55	0.73	38.84	37.58	2.99	5.12	0.304	0.193
1,700	3.0	114.0	0.935	61.28	0.77	46.96	45.28	3.59	6.28	0.380	0.240
1,700	3.0	143.0	0.935	71.01	0.78	55.38	53.33	4.39	7.46	0.459	0.289
1,700	3.5	29.0	0.935	35.55	0.64	22.93	22.37	1.50	2.89	0.155	0.102
1,700	3.5	57.0	0.935	45.96	0.69	31.66	30.75	2.35	4.11	0.238	0.153
1,700	3.5	86.0	0.935	55.41	0.74	40.71	39.36	3.09	5.39	0.324	0.206
1,700	3.5	114.0	0.935	63.48	0.78	49.45	47.63	3.69	6.64	0.407	0.257
1,700	3.5	143.0	0.935	75.01	0.78	58.50	56.33	4.65	7.90	0.494	0.311
1,700	4.0	29.0	0.935	36.35	0.65	23.56	22.98	1.55	2.98	0.162	0.106
1,700	4.0	57.0	0.935	47.33	0.70	32.91	31.94	2.43	4.29	0.252	0.161
1,700	4.0	86.0	0.935	57.22	0.74	42.59	41.14	3.18	5.67	0.345	0.219
1,700	4.0	114.0	0.935	66.60	0.78	51.94	50.02	3.88	7.00	0.435	0.274
1,700	4.0	143.0	0.935	79.01	0.78	61.62	59.34	4.91	8.34	0.528	0.332
Intake at 85% of the Requirement for Maintenance and Lactation											
900	4.5	43.0	-1.516	25.31	0.76	19.18	18.49	1.47	2.32	0.185	0.117
900	4.5	58.0	-1.875	31.30	0.76	23.71	22.87	2.08	2.93	0.237	0.149
900	4.5	74.0	-2.257	37.69	0.76	28.55	27.53	2.76	3.58	0.292	0.183
900	5.0	43.0	-1.579	26.37	0.76	19.97	19.26	1.54	2.43	0.195	0.124
900	5.0	58.0	-1.960	32.72	0.76	24.79	23.91	2.20	3.08	0.251	0.158
900	5.0	74.0	-2.366	39.50	0.76	29.92	28.86	2.92	3.77	0.310	0.194
900	5.5	43.0	-1.642	27.42	0.76	20.77	20.03	1.62	2.54	0.206	0.130
900	5.5	58.0	-2.045	34.14	0.76	25.86	24.94	2.32	3.23	0.265	0.166
900	5.5	74.0	-2.474	41.31	0.76	31.29	30.18	3.08	3.96	0.328	0.205
1,100	4.0	55.0	-1.802	30.08	0.76	22.79	21.98	1.78	2.83	0.221	0.140
1,100	4.0	73.0	-2.206	36.82	0.76	27.89	26.90	2.51	3.51	0.279	0.176
1,100	4.5	91.0	-2.743	45.79	0.76	34.69	33.46	3.43	4.43	0.359	0.225

continued

1,100	4.5	55.0	-1.882	31.43	0.76	23.81	22.96	1.90	2.97	0.234	0.149
1,100	4.5	73.0	-2.313	38.61	0.76	29.25	28.21	2.67	3.70	0.297	0.187
1,100	4.5	91.0	-2.743	45.79	0.76	34.69	33.46	3.43	4.43	0.359	0.225
1,100	5.0	55.0	-1.963	32.77	0.76	24.83	23.94	2.02	3.11	0.248	0.157
1,100	5.0	73.0	-2.420	40.40	0.76	30.60	29.52	2.82	3.88	0.314	0.198
1,100	5.0	91.0	-2.877	48.02	0.76	36.38	35.09	3.63	4.66	0.381	0.239
1,300	3.0	70.0	-2.009	33.54	0.76	25.40	24.50	2.04	3.19	0.244	0.155
1,300	3.0	93.0	-2.458	41.03	0.76	31.08	29.97	2.88	3.95	0.306	0.194
1,300	3.0	117.0	-2.926	48.84	0.76	37.00	35.68	3.75	4.74	0.372	0.234
1,300	3.5	70.0	-2.112	35.25	0.76	26.70	25.75	2.19	3.37	0.261	0.165
1,300	3.5	93.0	-2.594	43.30	0.76	32.80	31.64	3.07	4.19	0.329	0.208
1,300	3.5	117.0	-3.097	51.71	0.76	39.17	37.78	3.99	5.04	0.400	0.251
1,300	4.0	70.0	-2.214	36.96	0.76	28.00	27.01	2.34	3.55	0.277	0.176
1,300	4.0	93.0	-2.730	45.58	0.76	34.53	33.30	3.27	4.42	0.351	0.221
1,300	4.0	117.0	-3.269	54.57	0.76	41.34	39.87	4.24	5.34	0.428	0.269
1,500	3.0	78.0	-2.238	37.36	0.76	28.30	27.30	2.28	3.60	0.274	0.174
1,500	3.0	104.0	-2.745	45.83	0.76	34.71	33.48	3.23	4.46	0.345	0.218
1,500	3.0	130.0	-3.252	54.29	0.76	41.13	39.67	4.17	5.31	0.416	0.262
1,500	3.5	78.0	-2.352	39.27	0.76	29.75	28.69	2.45	3.80	0.292	0.186
1,500	3.5	104.0	-2.897	48.37	0.76	36.64	35.34	3.45	4.72	0.370	0.233
1,500	3.5	130.0	-3.443	57.48	0.76	43.54	41.99	4.45	5.64	0.447	0.281
1,500	4.0	78.0	-2.467	41.18	0.76	31.19	30.09	2.62	4.00	0.311	0.197
1,500	4.0	104.0	-3.050	50.92	0.76	38.57	37.20	3.67	4.99	0.395	0.249
1,500	4.0	130.0	-3.633	60.66	0.76	45.95	44.32	4.72	5.98	0.478	0.300
1,700	3.0	86.0	-2.464	41.14	0.76	31.17	30.06	2.53	4.01	0.304	0.193
1,700	3.0	114.0	-3.011	50.26	0.76	38.07	36.72	3.54	4.93	0.380	0.240
1,700	3.0	143.0	-3.576	59.70	0.76	45.23	43.62	4.59	5.88	0.459	0.289
1,700	3.5	86.0	-2.590	43.25	0.76	32.76	31.60	2.71	4.23	0.324	0.206
1,700	3.5	114.0	-3.178	53.05	0.76	40.19	38.76	3.79	5.22	0.407	0.257
1,700	3.5	143.0	-3.786	63.20	0.76	47.88	46.18	4.90	6.25	0.494	0.311
1,700	4.0	86.0	-2.717	45.35	0.76	34.36	33.14	2.89	4.45	0.345	0.219
1,700	4.0	114.0	-3.345	55.84	0.76	42.30	40.80	4.03	5.51	0.435	0.274
1,700	4.0	143.0	-3.996	66.71	0.76	50.53	48.74	5.20	6.61	0.528	0.332

NOTE: The following abbreviations were used: NELDM, net energy for lactation/lb of dry matter; NEL, net energy for lactation; TDN, total digestible nutrients; UIP, undegraded intake protein; DIP, degraded intake protein.

Table Dairy-5
RECOMMENDED NUTRIENT CONTENT OF DIETS FOR DAIRY CATTLE

			Lactating Cow Diets					Early Lactation (wks 0–3)	Dry Pregnant Cows	Calf Milk Replacer	Calf Starter Mix	Growing Heifers and Bulls[a]			Mature Bulls	Maximum Tolerable Levels[b,c]
Cow Wt (lb)	Fat (%)	Wt Grain (lb/d)	——— Milk Yield (lb/d) ———									3–6 Mos	6–12 Mos	>12 Mos		
900	5.0	0.50	14	29	43	58	74									
1,100	4.5	0.60	18	36	55	73	91									
1,300	4.0	0.72	23	47	70	93	117									
1,500	3.5	0.82	26	52	78	104	130									
1,700	3.5	0.94	29	57	86	114	143									

Nutrient											
Energy											
NEL, Mcal/lb	0.65	0.69	0.73	0.78	0.78	0.76	0.57	—	—	—	—	—	—	—
NEm, Mcal/lb	—	—	—	—	—	—	—	—	—	0.77	0.72	0.63	0.52	—
NEg, Mcal/lb	—	—	—	—	—	—	—	—	—	0.49	0.44	0.37	—	—
ME, Mcal/lb	1.07	1.16	1.25	1.31	1.31	1.27	0.93	1.71	1.41	1.18	1.12	1.03	0.91	—
DE, Mcal/lb	1.26	1.35	1.44	1.50	1.50	1.46	1.12	1.90	1.60	1.37	1.31	1.22	1.10	—
TDN, % of DM	63	67	71	75	75	73	56	95	80	69	66	61	55	—
Protein equivalent														
Crude protein, %	12	15	16	17	18	19	12	22	18	16	12	12	10	—
UIP, %	4.5	5.4	5.7	6.0	6.3	7.2	—	—	—	8.2	4.3	2.1	—	—
DIP, %	7.9	8.8	9.7	10.4	10.4	9.7	—	—	—	4.6	6.4	7.2	—	—
Fiber content (min.)[d]														
Crude fiber, %	17	17	15	15	17	17	22	—	—	13	15	15	15	—
Acid detergent fiber, %	21	21	19	19	21	21	27	—	—	16	19	19	19	—
Neutral detergent fiber, %	28	28	25	25	28	28	35	—	—	23	25	25	25	—
Ether extract (min.), %	3	3	3	3	3	3	3	10	3	3	3	3	3	—
Minerals														
Calcium, %	0.43	0.53	0.60	0.65	0.66	0.77	0.39[c]	0.70	0.60	0.52	0.41	0.29	0.30	2.00
Phosphorus, %	0.28	0.34	0.38	0.42	0.41	0.49	0.24	0.60	0.40	0.31	0.30	0.23	0.19	1.00
Magnesium, %[f]	0.20	0.20	0.20	0.25	0.25	0.25	0.16	0.07	0.10	0.16	0.16	0.16	0.16	0.50
Potassium, %[g]	0.90	0.90	0.90	1.00	1.00	1.00	0.65	0.65	0.65	0.65	0.65	0.65	0.65	3.00
Sodium, %	0.18	0.18	0.18	0.18	0.18	0.18	0.10	0.10	0.10	0.10	0.10	0.10	0.10	—
Chlorine, %	0.25	0.25	0.25	0.25	0.25	0.25	0.20	0.20	0.20	0.20	0.20	0.20	0.20	—
Sulfur, %	0.20	0.20	0.20	0.20	0.20	0.25	0.16	0.29	0.20	0.16	0.16	0.16	0.16	0.40
Iron, ppm	50	50	50	50	50	50	50	100	50	50	50	50	50	1,000
Cobalt, ppm	0.10	0.10	0.10	0.10	0.10	0.10	0.10	0.10	0.10	0.10	0.10	0.10	0.10	10.00
Copper, ppm[h]	10	10	10	10	10	10	10	10	10	10	10	10	10	100
Manganese, ppm	40	40	40	40	40	40	40	40	40	40	40	40	40	1,000
Zinc, ppm	40	40	40	40	40	40	40	40	40	40	40	40	40	500
Iodine, ppm[i]	0.60	0.60	0.60	0.60	0.60	0.60	0.25	0.25	0.25	0.25	0.25	0.25	0.25	50.00[j]
Selenium, ppm	0.30	0.30	0.30	0.30	0.30	0.30	0.30	0.30	0.30	0.30	0.30	0.30	0.30	2.00

continued

Vitamins[k]														
A, IU/lb	1,450	1,450	1,450	1,450	1,450	1,800	1,800	1,700	1,000	1,000	1,000	1,000	1,450	30,000
D, IU/lb	450	450	450	450	450	450	540	270	140	140	140	140	140	4,500
E, IU/lb	7	7	7	7	7	7	7	18	11	11	11	11	7	900

NOTE: The values presented in this table are intended as guidelines for the use of professionals in diet formulation. Because of the many factors affecting such values, they are not intended and should not be used as a legal or regulatory base.

[a]The approximate weight for growing heifers and bulls at 3–6 mos is 331 lb; at 6–12 mos, it is 559 lb; and at more than 12 mos, it is 881 lb. The approximate average daily gain is 1.543 lb/day.

[b]The maximum safe levels for many of the mineral elements are not well defined and may be substantially affected by specific feeding conditions. Additional information is available in *Mineral Tolerance of Domestic Animals* (NRC, 1980).

[c]Vitamin tolerances are discussed in detail in *Vitamin Tolerance of Animals* (NRC, 1987b).

[d]It is recommended that 75% of the NDF in lactating cow diets be provided as forage. If this recommendation is not followed, a depression in milk fat may occur.

[e]The value for calcium assumes that the cow is in calcium balance at the beginning of the dry period. If the cow is not in balance, then the dietary calcium requirement should be increased by 25% to 33%.

[f]Under conditions conducive to grass tetany (see text), magnesium should be increased to 0.25% or 0.30%.

[g]Under conditions of heat stress, potassium should be increased to 1.2% (see text).

[h]The cow's copper requirement is influenced by molybdenum and sulfur in the diet (see text).

[i]If the diet contains as much as 25% strongly goitrogenic feed on a dry basis, the iodine provided should be increased two times or more.

[j]Although cattle can tolerate this level of iodine, lower levels may be desirable to reduce the iodine content of milk.

[k]The following minimum quantities of B-complex vitamins are suggested per unit of milk replacer: niacin, 2.6 ppm; pantothenic acid, 13 ppm; riboflavin, 6.5 ppm; pyridoxine, 6.5 ppm; folic acid, 0.5 ppm; biotin, 0.1 ppm; vitamin B_{12}, 0.07 ppm; thiamin, 6.5 ppm; and choline, 0.26%. It appears that adequate amounts of these vitamins are furnished when calves have functional rumens (usually at 6 weeks of age) by a combination of rumen synthesis and natural feedstuffs.

Table Dairy-6
MAXIMUM TOLERABLE DIETARY LEVELS
OF CERTAIN ELEMENTS

Element	Maximum Tolerable Level (ppm)
Aluminum	1,000[a]
Arsenic	
Inorganic	50
Organic	100
Bromine	200
Cadmium	0.5[b]
Fluorine	40[c]
Lead	30[b]
Mercury	2[b]
Molybdenum	10[d]
Nickel	50
Vanadium	50

SOURCE: NRC (1980).

[a]As soluble salts of high bioavailability. Higher levels of less soluble forms found in natural substances can be tolerated.

[b]Levels are based on human food residue considerations.

[c]As sodium fluoride or fluorides of similar toxicity. The maximum safe level of fluorine for growing heifers and bulls is lower than for other dairy cattle. Somewhat higher levels are tolerated when fluorine is from less available sources such as phosphates (see text). Morphological lesions in cattle teeth may be seen when dietary fluoride for the young exceeds 20 ppm, but a relationship between the lesions caused by fluoride levels below the maximum tolerable levels and animal performance has not been established.

[d]Toxicity related to the dietary level of copper.

Table Horse-1
DAILY NUTRIENT REQUIREMENTS OF PONIES (200-KG MATURE WEIGHT)

Animal	Weight (kg)	Daily Gain (kg)	DE (Mcal)	Crude Protein (g)	Lysine (g)	Calcium (g)	Phosphorus (g)	Magnesium (g)	Potassium (g)	Vitamin A (10³IU)
Mature Horses										
Maintenance	200		7.4	296	10	8	6	3.0	10.0	6
Stallions (breeding season)	200		9.3	370	13	11	8	4.3	14.1	9
Pregnant mares										
9 months	200		8.2	361	13	16	12	3.9	13.1	12
10 months			8.4	368	13	16	12	4.0	13.4	12
11 months			8.9	391	14	17	13	4.3	14.2	12
Lactating mares										
Foaling to 3 months	200		13.7	688	24	27	18	4.8	21.2	12
3 months to weaning	200		12.2	528	18	18	11	3.7	14.8	12
Working horses										
Light work[a]	200		9.3	370	13	11	8	4.3	14.1	9
Moderate work[b]	200		11.1	444	16	14	10	5.1	16.9	9
Intense work[c]	200		14.8	592	21	18	13	6.8	22.5	9
Growing Horses										
Weanling, 4 months	75	0.40	7.3	365	15	16	9	1.6	5.0	3
Weanling, 6 months										
Moderate growth	95	0.30	7.6	378	16	13	7	1.8	5.7	4
Rapid growth	95	0.40	8.7	433	18	17	9	1.9	6.0	4
Yearling, 12 months										
Moderate growth	140	0.20	8.7	392	17	12	7	2.4	7.6	6
Rapid growth	140	0.30	10.3	462	19	15	8	2.5	7.9	6
Long yearling, 18 months										
Not in training	170	0.10	8.3	375	16	10	6	2.7	8.8	8
In training	170	0.10	11.6	522	22	14	8	3.7	12.2	8
Two year old, 24 months										
Not in training	185	0.05	7.9	337	13	9	5	2.8	9.4	8
In training	185	0.05	11.4	485	19	13	7	4.1	13.5	8

Table Horse-2
DAILY NUTRIENT REQUIREMENTS OF HORSES (400-KG MATURE WEIGHT)

Animal	Weight (kg)	Daily Gain (kg)	DE (Mcal)	Crude Protein (g)	Lysine (g)	Calcium (g)	Phosphorus (g)	Magnesium (g)	Potassium (g)	Vitamin A (10^3 IU)
Mature Horses										
Maintenance	400		13.4	536	19	16	11	6.0	20.0	12
Stallions (breeding season)	400		16.8	670	23	20	15	7.7	25.5	18
Pregnant mares										
9 months	400		14.9	654	23	28	21	7.1	23.8	24
10 months			15.1	666	23	29	22	7.3	24.2	24
11 months			16.1	708	25	31	23	7.7	25.7	24
Lactating mares										
Foaling to 3 months	400		22.9	1,141	40	45	29	8.7	36.8	24
3 months to weaning	400		19.7	839	29	29	18	6.9	26.4	24
Working horses										
Light work[a]	400		16.8	670	23	20	15	7.7	25.5	18
Moderate work[b]	400		20.1	804	28	25	17	9.2	30.6	18
Intense work[c]	400		26.8	1,072	38	33	23	12.3	40.7	18
Growing Horses										
Weanling, 4 months	145	0.85	13.5	675	28	33	18	3.2	9.8	7
Weanling, 6 months										
Moderate growth	180	0.55	12.9	643	27	25	14	3.4	10.7	8
Rapid growth	180	0.70	14.5	725	30	30	16	3.6	11.1	8
Yearling, 12 months										
Moderate growth	265	0.40	15.6	700	30	23	13	4.5	14.5	12
Rapid growth	265	0.50	17.1	770	33	27	15	4.6	14.8	12
Long yearling, 18 months										
Not in training	330	0.25	15.9	716	30	21	12	5.3	17.3	15
In training	330	0.25	21.6	970	41	29	16	7.1	23.4	15
Two year old, 24 months										
Not in training	365	0.15	15.3	650	26	19	11	5.7	18.7	16
In training	365	0.15	21.5	913	37	27	15	7.9	26.2	16

Table Horse-3
DAILY NUTRIENT REQUIREMENTS OF HORSES (500-KG MATURE WEIGHT)

Animal	Weight (kg)	Daily Gain (kg)	DE (Mcal)	Crude Protein (g)	Lysine (g)	Calcium (g)	Phosphorus (g)	Magnesium (g)	Potassium (g)	Vitamin A (10^3IU)
Mature Horses										
Maintenance	500		16.4	656	23	20	14	7.5	25.0	15
Stallions (breeding season)	500		20.5	820	29	25	18	9.4	31.2	22
Pregnant mares										
9 months	500		18.2	801	28	35	26	8.7	29.1	30
10 months			18.5	815	29	35	27	8.9	29.7	30
11 months			19.7	866	30	37	28	9.4	31.5	30
Lactating mares										
Foaling to 3 months	500		28.3	1,427	50	56	36	10.9	46.0	30
3 months to weaning	500		24.3	1,048	37	36	22	8.6	33.0	30
Working horses										
Light work[a]	500		20.5	820	29	25	18	9.4	31.2	22
Moderate work[b]	500		24.6	984	34	30	21	11.3	37.4	22
Intense work[c]	500		32.8	1,312	46	40	29	15.1	49.9	22
Growing Horses										
Weanling, 4 months	175	0.85	14.4	720	30	34	19	3.7	11.3	8
Weanling, 6 months										
Moderate growth	215	0.65	15.0	750	32	29	16	4.0	12.7	10
Rapid growth	215	0.85	17.2	860	36	36	20	4.3	13.3	10
Yearling, 12 months										
Moderate growth	325	0.50	18.9	851	36	29	16	5.5	17.8	15
Rapid growth	325	0.65	21.3	956	40	34	19	5.7	18.2	15
Long yearling, 18 months										
Not in training	400	0.35	19.8	893	38	27	15	6.4	21.1	18
In training	400	0.35	26.5	1,195	50	36	20	8.6	28.2	18
Two year old, 24 months										
Not in training	450	0.20	18.8	800	32	24	13	7.0	23.1	20
In training	450	0.20	26.3	1,117	45	34	19	9.8	32.2	20

Table Horse-4
DAILY NUTRIENT REQUIREMENTS OF HORSES (600-KG MATURE WEIGHT)

Animal	Weight (kg)	Daily Gain (kg)	DE (Mcal)	Crude Protein (g)	Lysine (g)	Calcium (g)	Phosphorus (g)	Magnesium (g)	Potassium (g)	Vitamin A (10³IU)
Mature Horses										
Maintenance	600		19.4	776	27	24	17	9.0	30.0	18
Stallions (breeding season)	600		24.3	970	34	30	21	11.2	36.9	27
Pregnant mares										
9 months	600		21.5	947	33	41	31	10.3	34.5	36
10 months			21.9	965	34	42	32	10.5	35.1	36
11 months			23.3	1,024	36	44	34	11.2	37.2	36
Lactating mares										
Foaling to 3 months	600		33.7	1,711	60	67	43	13.1	55.2	36
3 months to weaning	600		28.9	1,258	44	43	27	10.4	39.6	36
Working horses										
Light work[a]	600		24.3	970	34	30	21	11.2	36.9	27
Moderate work[b]	600		29.1	1,164	41	36	25	13.4	44.2	27
Intense work[c]	600		38.8	1,552	54	47	34	17.8	59.0	27
Growing Horses										
Weanling, 4 months	200	1.00	16.5	825	35	40	22	4.3	13.0	9
Weanling, 6 months										
Moderate growth	245	0.75	17.0	850	36	34	19	4.6	14.5	11
Rapid growth	245	0.95	19.2	960	40	40	22	4.9	15.1	11
Yearling, 12 months										
Moderate growth	375	0.65	22.7	1,023	43	36	20	6.4	20.7	17
Rapid growth	375	0.80	25.1	1,127	48	41	22	6.6	21.2	17
Long yearling, 18 months										
Not in training	475	0.45	23.9	1,077	45	33	18	7.7	25.1	21
In training	475	0.45	32.0	1,429	60	44	24	10.2	33.3	21
Two year old, 24 months										
Not in training	540	0.30	23.5	998	40	31	17	8.5	27.9	24
In training	540	0.30	32.3	1,372	55	43	24	11.6	38.4	24

Table Horse-5
DAILY NUTRIENT REQUIREMENTS OF HORSES (700-KG MATURE WEIGHT)

Animal	Weight (kg)	Daily Gain (kg)	DE (Mcal)	Crude Protein (g)	Lysine (g)	Calcium (g)	Phosphorus (g)	Magnesium (g)	Potassium (g)	Vitamin A (10³ IU)
Mature Horses										
Maintenance	700		21.3	851	30	28	20	10.5	35.0	21
Stallions (breeding season)	700		26.6	1,064	37	32	23	12.2	40.4	32
Pregnant mares										
9 months	700		23.6	1,039	36	45	34	11.3	37.8	42
10 months			24.0	1,058	37	46	35	11.5	38.5	42
11 months			25.5	1,124	39	49	37	12.3	40.9	42
Lactating mares										
Foaling to 3 months	700		37.9	1,997	70	78	51	15.2	64.4	42
3 months to weaning	700		32.4	1,468	51	50	31	12.1	46.2	42
Working horses										
Light work[a]	700		26.6	1,064	37	32	23	12.2	40.4	32
Moderate work[b]	700		31.9	1,277	45	39	28	14.7	48.5	32
Intense work[c]	700		42.6	1,702	60	52	37	19.6	64.7	32
Growing Horses										
Weanling, 4 months	225	1.10	19.7	986	41	44	25	4.8	14.6	10
Weanling, 6 months										
Moderate growth	275	0.80	20.0	1,001	42	37	20	5.1	16.2	12
Rapid growth	275	1.00	22.2	1,111	47	43	24	5.4	16.8	12
Yearling, 12 months										
Moderate growth	420	0.70	26.1	1,176	50	39	22	7.2	23.1	19
Rapid growth	420	0.85	28.5	1,281	54	44	24	7.4	23.6	19
Long yearling, 18 months										
Not in training	525	0.50	27.0	1,215	51	37	20	8.5	27.8	24
In training	525	0.50	36.0	1,615	68	49	27	11.3	36.9	24
Two year old, 24 months										
Not in training	600	0.35	26.3	1,117	45	35	19	9.4	31.1	27
In training	600	0.35	36.0	1,529	61	48	27	12.9	42.5	27

Table Horse-6
DAILY NUTRIENT REQUIREMENTS OF HORSES (800-KG MATURE WEIGHT)

Animal	Weight (kg)	Daily Gain (kg)	DE (Mcal)	Crude Protein (g)	Lysine (g)	Calcium (g)	Phosphorus (g)	Magnesium (g)	Potassium (g)	Vitamin A (10^3 IU)
Mature Horses										
Maintenance	800		22.9	914	32	32	22	12.0	40.0	24
Stallions (breeding season)	800		28.6	1,143	40	35	25	13.1	43.4	36
Pregnant mares										
9 months	800		25.4	1,116	39	48	37	12.2	40.6	48
10 months			25.8	1,137	40	49	37	12.4	41.3	48
11 months			27.4	1,207	42	52	40	13.2	43.9	48
Lactating mares										
Foaling to 3 months	800		41.9	2,282	81	90	58	17.4	73.6	48
3 months to weaning	800		35.5	1,678	60	58	36	13.8	52.8	48
Working horses										
Light work[a]	800		28.6	1,143	40	35	25	13.1	43.4	36
Moderate work[b]	800		34.3	1,372	48	42	30	15.8	52.1	36
Intense work[c]	800		45.7	1,829	64	56	40	21.0	69.5	36
Growing Horses										
Weanling, 4 months	250	1.20	21.4	1,070	45	48	27	5.3	16.1	11
Weanling, 6 months										
Moderate growth	305	0.90	22.0	1,100	46	41	23	5.7	18.0	14
Rapid growth	305	1.10	24.2	1,210	51	47	26	6.0	18.6	14
Yearling, 12 months										
Moderate growth	460	0.80	28.7	1,291	55	44	24	7.9	25.4	21
Rapid growth	460	0.95	31.0	1,396	59	49	27	8.1	25.9	21
Long yearling, 18 months										
Not in training	590	0.60	30.2	1,361	57	43	24	9.6	31.3	27
In training	590	0.60	39.8	1,793	76	56	31	12.6	41.2	27
Two year old, 24 months										
Not in training	675	0.40	28.7	1,220	49	40	22	10.6	35.0	30
In training	675	0.40	39.1	1,662	66	54	30	14.5	47.6	30

Table Horse-7
DAILY NUTRIENT REQUIREMENTS OF HORSES (900-KG MATURE WEIGHT)

Animal	Weight (kg)	Daily Gain (kg)	DE (Mcal)	Crude Protein (g)	Lysine (g)	Calcium (g)	Phosphorus (g)	Magnesium (g)	Potassium (g)	Vitamin A (10^3 IU)
Mature Horses										
Maintenance	900		24.1	966	34	36	25	13.5	45.0	27
Stallions (breeding season)	900		30.2	1,207	42	37	26	13.9	45.9	40
Pregnant mares	900									
9 months			26.8	1,179	41	51	39	12.9	42.9	54
10 months			27.3	1,200	42	52	39	13.1	43.6	54
11 months			29.0	1,275	45	55	42	13.9	46.3	54
Lactating mares										
Foaling to 3 months	900		45.5	2,567	89	101	65	19.6	82.8	54
3 months to weaning	900		38.4	1,887	66	65	40	15.5	59.4	54
Working horses										
Light work[a]	900		30.2	1,207	42	37	26	13.9	45.9	40
Moderate work[b]	900		36.2	1,448	51	44	32	16.7	55.0	40
Intense work[c]	900		48.3	1,931	68	59	42	22.2	73.4	40
Growing Horses										
Weanling, 4 months	275	1.30	23.1	1,154	48	53	29	5.8	17.7	12
Weanling, 6 months										
Moderate growth	335	0.95	23.4	1,171	49	44	24	6.2	19.6	15
Rapid growth	335	1.15	25.6	1,281	54	50	28	6.5	20.2	15
Yearling, 12 months										
Moderate growth	500	0.90	31.2	1,404	59	49	27	8.6	27.7	22
Rapid growth	500	1.05	33.5	1,509	64	54	30	8.8	28.2	22
Long yearling, 18 months										
Not in training	665	0.70	33.6	1,510	64	49	27	10.9	35.4	30
In training	665	0.70	43.9	1,975	83	64	35	14.2	46.2	30
Two year old, 24 months										
Not in training	760	0.45	31.1	1,322	53	45	25	12.0	39.4	34
In training	760	0.45	42.2	1,795	72	61	34	16.2	53.4	34

Note: Mares should gain weight during late gestation to compensate for tissue deposition. However, nutrient requirements are based on maintenance body weight.

[a]Examples are horses used in Western and English pleasure, bridle path hack, equitation, etc.
[b]Examples are horses used in ranch work, roping, cutting, barrel racing, jumping, etc.
[c]Examples are horses used in race training, polo, etc.

Table Horse-8
NUTRIENT CONCENTRATIONS IN TOTAL DIETS FOR HORSES AND PONIES (DRY MATTER BASIS)

	Digestible Energy[a]		Diet Proportions		Crude Protein (%)	Lysine (%)	Calcium (%)	Phosphorus (%)	Magnesium (%)	Potassium (%)	Vitamin A	
	(Mcal/kg)	(Mcal/lb)	Conc. (%)	Hay (%)							(IU/kg)	(IU/lb)
Mature Horses												
Maintenance	2.00	0.90	0	100	8.0	0.28	0.24	0.17	0.09	0.30	1830	830
Stallions	2.40	1.10	30	70	9.6	0.34	0.29	0.21	0.11	0.36	2640	1200
Pregnant mares												
9 months	2.25	1.00	20	80	10.0	0.35	0.43	0.32	0.10	0.35	3710	1680
10 months	2.25	1.00	20	80	10.0	0.35	0.43	0.32	0.10	0.36	3650	1660
11 months	2.40	1.10	30	70	10.6	0.37	0.45	0.34	0.11	0.38	3650	1660
Lactating mares												
Foaling to 3 months	2.60	1.20	50	50	13.2	0.46	0.52	0.34	0.10	0.42	2750	1250
3 months to weaning	2.45	1.15	35	65	11.0	0.37	0.36	0.22	0.09	0.33	3020	1370
Working horses												
Light work[b]	2.45	1.15	35	65	9.8	0.35	0.30	0.22	0.11	0.37	2690	1220
Moderate work[c]	2.65	1.20	50	50	10.4	0.37	0.31	0.23	0.12	0.39	2420	1100
Intense work[d]	2.85	1.30	65	35	11.4	0.40	0.35	0.25	0.13	0.43	1950	890
Growing Horses												
Weanling, 4 months	2.90	1.40	70	30	14.5	0.60	0.68	0.38	0.08	0.30	1580	720
Weanling, 6 months												
Moderate growth	2.90	1.40	70	30	14.5	0.61	0.56	0.31	0.08	0.30	1870	850
Rapid growth	2.90	1.40	70	30	14.5	0.61	0.61	0.34	0.08	0.30	1630	740
Yearling, 12 months												
Moderate growth	2.80	1.30	60	40	12.6	0.53	0.43	0.24	0.08	0.30	2160	980
Rapid growth	2.80	1.30	60	40	12.6	0.53	0.45	0.25	0.08	0.30	1920	870
Long yearling, 18 months												
Not in training	2.50	1.15	45	55	11.3	0.48	0.34	0.19	0.08	0.30	2270	1030
In training	2.65	1.20	50	50	12.0	0.50	0.36	0.20	0.09	0.30	1800	820
Two year old, 24 months												
Not in training	2.45	1.15	35	65	10.4	0.42	0.31	0.17	0.09	0.30	2640	1200
In training	2.65	1.20	50	50	11.3	0.45	0.34	0.20	0.10	0.32	2040	930

[a]Values assume a concentrate feed containing 3.3 Mcal/kg and hay containing 2.0 Mcal/kg of dry matter.
[b]Examples are horses used in Western and English pleasure, bridle path hack, equitation, etc.
[c]Examples are horses used in ranch work, roping, cutting, barrel racing, jumping, etc.
[d]Examples are race training, polo, etc.

Table Horse-9

NUTRIENT CONCENTRATIONS IN TOTAL DIETS FOR HORSES AND PONIES (90% DRY MATTER BASIS)

	Digestible Energy[a]		Diet Proportions		Crude Protein (%)	Lysine (%)	Calcium (%)	Phosphorus (%)	Magnesium (%)	Potassium (%)	Vitamin A	
	(Mcal/kg)	(Mcal/lb)	Conc. (%)	Hay (%)							(IU/kg)	(IU/lb)
Mature Horses												
Maintenance	1.80	0.80	0	100	7.2	0.25	0.21	0.15	0.08	0.27	1650	750
Stallions	2.15	1.00	30	70	8.6	0.30	0.26	0.19	0.10	0.33	2370	1080
Pregnant mares												
9 months	2.00	0.90	20	80	8.9	0.31	0.39	0.29	0.10	0.32	3330	1510
10 months	2.00	0.90	20	80	9.0	0.32	0.39	0.30	0.10	0.33	3280	1490
11 months	2.15	1.00	30	70	9.5	0.33	0.41	0.31	0.10	0.35	3280	1490
Lactating mares												
Foaling to 3 months	2.35	1.10	50	50	12.0	0.41	0.47	0.30	0.09	0.38	2480	1130
3 months to weaning	2.20	1.05	35	65	10.0	0.34	0.33	0.20	0.08	0.30	2720	1240
Working horses												
Light work[b]	2.20	1.05	35	65	8.8	0.32	0.27	0.19	0.10	0.34	2420	1100
Moderate work[c]	2.40	1.10	50	50	9.4	0.35	0.28	0.22	0.11	0.36	2140	970
Intense work[d]	2.55	1.20	65	35	10.3	0.36	0.31	0.23	0.12	0.39	1760	800
Growing Horses												
Weanling, 4 months	2.60	1.25	70	30	13.1	0.54	0.62	0.34	0.07	0.27	1420	650
Weanling, 6 months												
Moderate growth	2.60	1.25	70	30	13.0	0.55	0.50	0.28	0.07	0.27	1680	760
Rapid growth	2.60	1.25	70	30	13.1	0.55	0.55	0.30	0.07	0.27	1470	670
Yearling, 12 months												
Moderate growth	2.50	1.15	60	40	11.3	0.48	0.39	0.21	0.07	0.27	1950	890
Rapid growth	2.50	1.15	60	40	11.3	0.48	0.40	0.22	0.07	0.27	1730	790
Long yearling, 18 months												
Not in training	2.30	1.05	45	55	10.1	0.43	0.31	0.17	0.07	0.27	2050	930
In training	2.40	1.10	50	50	10.8	0.45	0.32	0.18	0.08	0.27	1620	740
Two year old, 24 months												
Not in training	2.20	1.00	35	65	9.4	0.38	0.28	0.15	0.08	0.27	2380	1080
In training	2.40	1.10	50	50	10.1	0.41	0.31	0.17	0.09	0.29	1840	840

[a]Values assume a concentrate feed containing 3.3 Mcal/kg and hay containing 2.00 Mcal/kg of dry matter.
[b]Examples are horses used in Western and English pleasure, bridle path hack, equitation, etc.
[c]Examples are horses used in ranch work, roping, cutting, barrel racing, jumping, etc.
[d]Examples are race training, polo, etc.

Table Horse-10
OTHER MINERALS AND VITAMINS FOR HORSES AND PONIES
(DRY MATTER BASIS)

	Adequate Concentrations in Total Rations				
	Maintenance	*Pregnant and Lactating Mares*	*Growing Horses*	*Working Horses*	*Maximum Tolerance Levels*
Minerals					
Sodium (%)	0.10	0.10	0.10	0.30	3[a]
Sulfur (%)	0.15	0.15	0.15	0.15	1.25
Iron (mg/kg)	40	50	50	40	1,000
Manganese (mg/kg)	40	40	40	40	1,000
Copper (mg/kg)	10	10	10	10	800
Zinc (mg/kg)	40	40	40	40	500
Selenium (mg/kg)	0.1	0.1	0.1	0.1	2.0
Iodine (mg/kg)	0.1	0.1	0.1	0.1	5.0
Cobalt (mg/kg)	0.1	0.1	0.1	0.1	10
Vitamins					
Vitamin A (IU/kg)	2,000	3,000	2,000	2,000	16,000
Vitamin D (IU/kg)[b]	300	600	800	300	2,200
Vitamin E (IU/kg)	50	80	80	80	1,000
Vitamin K (mg/kg)	c				
Thiamin (mg/kg)	3	3	3	5	3,000
Riboflavin (mg/kg)	2	2	2	2	
Niacin (mg/kg)					
Pantothenic acid (mg/kg)					
Pyridoxine (mg/kg)					
Biotin (mg/kg)					
Folacin (mg/kg)					
Vitamin B_{12} (μg/kg)					
Ascorbic acid (mg/kg)					
Choline (mg/kg)					

[a]As sodium chloride.
[b]Recommendations for horses not exposed to sunlight or to artificial light with an emission spectrum of 280–315 nm.
[c]Blank space indicates that data are insufficient to determine a requirement or maximum tolerable level.

Table Horse-11
EXPECTED FEED CONSUMPTION BY HORSES (% BODY WEIGHT)[a]

	Forage	Concentrate	Total
Mature Horses			
Maintenance	1.5–2.0	0–0.5	1.5–2.0
Mares, late gestation	1.0–1.5	0.5–1.0	1.5–2.0
Mares, early lactation	1.0–2.0	1.0–2.0	2.0–3.0
Mares, late lactation	1.0–2.0	0.5–1.5	2.0–2.5
Working Horses			
Light work	1.0–2.0	0.5–1.0	1.5–2.5
Moderate work	1.0–2.0	0.75–1.5	1.75–2.5
Intense work	0.75–1.5	1.0–2.0	2.0–3.0
Young horses			
Nursing foal, 3 months	0	1.0–2.0	2.5–3.5
Weanling foal, 6 months	0.5–1.0	1.5–3.0	2.0–3.5
Yearling foal, 12 months	1.0–1.5	1.0–2.0	2.0–3.0
Long yearling, 18 months	1.0–1.5	1.0–1.5	2.0–2.5
Two year old, 24 months	1.0–1.5	0.0–1.5	1.75–2.5

[a]Air-dry feed (about 90% DM).

Table Sheep-1
NUTRIENT REQUIREMENTS OF SHEEP
(DAILY NUTRIENTS PER ANIMAL)

Body Weight kg(lb)	Dry Matter[a] kg(lb)	Roughage DM basis kg(lb)	Energy[b] TDN kg(lb)	Energy[b] ME Mcal	Total Protein kg(lb)	Digestible Protein kg(lb)	Ca kg(lh)	P kg(lh)	Carotene (mg)	Vitamin A (IU)
Ewes[e]										
Maintenance										
50(110)	1.0(2.2)	1.0 (2.2)	0.55(1.21)	1.98	.089(.196)	.048(.106)	.0030(.0066)	.0028(.0062)	1.9	1275
60(132)	1.1(2.4)	1.1 (2.4)	0.61(1.35)	2.20	.098(.216)	.053(.117)	.0031(.0068)	.0029(.0064)	2.2	1530
70(154)	1.2(2.6)	1.2 (2.6)	0.66(1.46)	2.38	.107(.236)	.058(.128)	.0032(.0070)	.0030(.0066)	2.6	1785
80(176)	1.3(2.9)	1.3 (2.9)	0.72(1.59)	2.60	.116(.256)	.063(.139)	.0033(.0073)	.0031(.0068)	3.0	2040
Nonlactating and first 15 weeks of gestation										
50(110)	1.1(2.4)	1.1 (2.4)	0.60(1.32)	2.16	.099(.218)	.054(.119)	.0030(.0066)	.0028(.0062)	1.9	1275
60(132)	1.3(2.9)	1.3 (2.9)	0.72(1.59)	2.60	.117(.258)	.064(.141)	.0031(.0068)	.0029(.0064)	2.2	1530
70(154)	1.4(3.1)	1.4 (3.1)	0.77(1.70)	2.78	.126(.278)	.069(.152)	.0032(.0070)	.0030(.0066)	2.6	1785
80(176)	1.5(3.3)	1.5 (3.3)	0.82(1.81)	2.96	.135(.298)	.074(.163)	.0033(.0073)	.0031(.0068)	3.0	2040
Last 6 weeks of gestation or last 8 weeks of lactation suckling singles										
50(110)	1.7(3.7)	1.56(3.44)	0.99(2.18)	3.58	.158(.348)	.088(.194)	.0041(.0090)	.0039(.0086)	6.2	4250
60(132)	1.9(4.2)	1.75(3.86)	1.10(2.43)	3.97	.177(.390)	.099(.218)	.0044(.0097)	.0041(.0090)	7.5	5100
70(154)	2.1(4.6)	1.93(4.26)	1.22(2.69)	4.40	.195(.430)	.190(.240)	.0045(.0099)	.0043(.0095)	8.8	5950
80(176)	2.2(4.8)	2.02(4.45)	1.28(2.82)	4.62	.205(.452)	.114(.251)	.0048(.0106)	.0045(.0099)	10.0	6800

continued

First 8 weeks of lactation suckling singles or last 8 weeks of lactation suckling twins

50(110)	2.1(4.6)	1.41(3.11)	1.36(3.00)	4.90	.218(.481)	.130(.287)	.0109(.0240)	.0078(.0172)	6.2	4250
60(132)	2.3(5.1)	1.54(3.40)	1.50(3.31)	5.41	.239(.527)	.143(.315)	.0115(.0254)	.0082(.0181)	7.5	5100
70(154)	2.5(5.5)	1.68(3.70)	1.63(3.59)	5.88	.260(.573)	.155(.342)	.0120(.0265)	.0086(.0190)	8.8	5950
80(176)	2.6(5.7)	1.74(3.84)	1.69(3.73)	6.10	.270(.595)	.161(.355)	.0126(.0278)	.0090(.0198)	10.0	6800

First 8 weeks of lactation suckling twins

50(110)	2.4(5.3)	1.61(3.55)	1.56(3.44)	5.63	.276(.609)	.173(.381)	.0125(.0276)	.0089(.0196)	6.2	4250
60(132)	2.6(5.7)	1.74(3.84)	1.69(3.73)	6.10	.299(.659)	.187(.412)	.0130(.0287)	.0094(.0207)	7.5	5100
70(154)	2.8(6.2)	1.88(4.15)	1.82(4.01)	6.57	.322(.710)	.202(.455)	.0134(.0295)	.0095(.0209)	8.8	5950
80(176)	3.0(6.6)	2.01(4.43)	1.95(4.30)	7.04	.345(.761)	.216(.476)	.0144(.0318)	.0102(.0225)	10.0	6800

Replacement lambs and yearlings[d]

30(66)	1.3(2.9)	1.08(2.38)	0.81(1.79)	2.92	.130(.287)	.075(.165)	.0059(.0130)	.0033(.0073)	1.9	1275
40(88)	1.4(3.1)	1.29(2.84)	0.82(1.81)	2.96	.133(.293)	.074(.163)	.0061(.0135)	.0034(.0075)	2.5	1700
50(110)	1.5(3.3)	1.50(3.31)	0.83(1.83)	2.99	.133(.293)	.073(.161)	.0063(.0139)	.0035(.0077)	3.1	2125
60(132)	1.5(3.3)	1.50(3.31)	0.82(1.81)	2.96	.133(.293)	.072(.159)	.0065(.0143)	.0036(.0079)	3.8	2250

Rams

Replacement lambs and yearlings[d]

40(88)	1.8(4.0)	1.21(2.67)	1.17(2.58)	4.22	.184(.406)	.108(.238)	.0063(.0139)	.0035(.0077)	2.5	1700
60(132)	2.3(5.1)	2.12(4.67)	1.38(3.04)	4.98	.219(.483)	.122(.269)	.0072(.0159)	.0040(.0088)	3.8	2550
80(176)	2.8(6.2)	2.80(6.17)	1.54(3.40)	5.56	.249(.549)	.134(.295)	.0079(.0174)	.0044(.0097)	5.0	3400
100(220)	2.8(6.2)	2.80(6.17)	1.54(3.40)	5.56	.249(.549)	.134(.295)	.0083(.0183)	.0046(.0101)	6.2	4250
120(265)	2.6(5.7)	2.60(5.73)	1.43(3.15)	5.16	.231(.509)	.125(.276)	.0085(.0187)	.0047(.0104)	7.5	5100

Table Sheep-1 (Continued)

Body Weight kg(lb)	Dry Matter[a] kg(lb)	Roughage DM basis kg(lb)	Energy[b] TDN kg(lb)	Energy[b] ME Mcal	Total Protein kg(lb)	Digestible Protein kg(lb)	Ca kg(lb)	P kg(lb)	Carotene (mg)	Vitamin A (IU)
Lambs										
Finishing (Gaining .20–.25 kg [.44–.55 lb] per head per day)[e]										
30(66)	1.3(2.9)	.98(2.16)	0.83(1.83)	2.99	.143(.315)	.087(.192)	.0048(.0106)	.0030(.0066)	1.1	765
35(77)	1.4(3.1)	.94(2.07)	0.94(2.07)	3.39	.154(.340)	.094(.207)	.0048(.0106)	.0030(.0066)	1.3	892
40(88)	1.6(3.5)	.93(2.05)	1.12(2.47)	4.04	.176(.388)	.107(.236)	.0050(.0110)	.0031(.0068)	1.5	1020
45(99)	1.7(3.7)	.99(2.18)	1.19(2.62)	4.30	.187(.412)	.114(.251)	.0050(.0110)	.0031(.0068)	1.7	1148
50(110)	1.8(4.0)	1.04(2.29)	1.26(2.78)	4.54	.198(.437)	.121(.267)	.0050(.0110)	.0031(.0068)	1.9	1275
55(121)	1.9(4.2)	1.10(2.43)	1.33(2.93)	4.80	.209(.461)	.127(.280)	.0050(.0110)	.0031(.0068)	2.1	1402
Early-weaned (Gaining .25–.30 kg [.55–.66 lb] per head per day)[f]										
10(22)	0.6(1.3)	.30(.66)	0.44(.97)	1.59	.096(.212)	.069(.152)	.0024(.0053)	.0016(.0035)	1.2	850
20(44)	1.0(2.2)	.50(1.10)	0.73(1.61)	2.63	.160(.353)	.115(.254)	.0036(.0079)	.0024(.0053)	2.5	1700
30(66)	1.4(3.1)	.70(1.54)	1.02(2.25)	3.68	.196(.432)	.133(.293)	.0050(.0110)	.0033(.0073)	3.8	2250

[a] To convert dry matter to an as-fed basis, divide dry matter by percentage of dry matter as fed and multiply by 100.

[b] 1 kg TDN = 4.4 Mcal DE (digestible energy). DE may be converted to ME (metabolizable energy) by multiplying by 82%.

[c] Values are for ewes in moderate condition, not excessively fat or thin. Fat ewes should be fed at the next lower weight, thin ewes at the next higher weight. Once maintenance weight is established, such weight would follow through all production phases.

[d] Requirements for replacement lambs (ewe and ram) start when the lambs are weaned.

[e] Maximum gains expected. If lambs are held for later market, they should be fed as replacement ewe lambs are fed. Lambs capable of gaining faster than indicated should be fed at higher level. Lambs finish at the maximum rate if they are self-fed.

[f] A 40-kg early-weaned lamb should be fed the same as a finishing lamb of the same weight.

Table Sheep-2
NUTRIENT CONTENT OF DIETS FOR SHEEP
(NUTRIENT CONCENTRATION IN DIET DRY MATTER)

| Body Weight kg(lb) | Daily Dry Matter[a] Per Animal kg(lb) | Energy[b] | | | Total Protein % | Digestible Protein % | Ca % | P % | Carotene mg/kg(lb) | Vitamin A IU/kg(lb) |
		Rough-age %	TDN %	ME Mcal/kg (lb)						
Ewes[d]										
Maintenance										
50(110)	1.0(2.2)	100	55	2.0(.91)	8.9	4.8	.30	.28	1.9(.86)	1275(578)
60(132)	1.1(2.4)	100	55	2.0(.91)	8.9	4.8	.28	.26	2.0(.91)	1391(631)
70(154)	1.2(2.6)	100	55	2.0(.91)	8.9	4.8	.27	.25	2.2(1.00)	1488(675)
80(176)	1.3(2.9)	100	55	2.0(.91)	8.9	4.8	.25	.24	2.3(1.04)	1569(712)
Nonlactating and first 15 weeks of gestation										
50(110)	1.1(2.4)	100	55	2.0(.91)	9.0	4.9	.27	.25	1.7(.77)	1159(526)
60(132)	1.3(2.9)	100	55	2.0(.91)	9.0	4.9	.24	.22	1.7(.77)	1177(534)
70(154)	1.4(3.1)	100	55	2.0(.91)	9.0	4.9	.23	.21	1.9(.86)	1275(578)
80(176)	1.5(3.3)	100	55	2.0(.91)	9.0	4.9	.22	.21	2.0(.91)	1360(617)
Last 6 weeks of gestation or last 8 weeks of lactation suckling singles										
50(110)	1.7(3.7)	92	58	2.1(.95)	9.3	5.2	.24	.23	3.6(1.63)	2500(1134)
60(132)	1.9(4.2)	92	58	2.1(.95)	9.3	5.2	.23	.22	3.9(1.77)	2684(1217)
70(154)	2.1(4.6)	92	58	2.1(.95)	9.3	5.2	.21	.20	4.2(1.90)	2833(1285)
80(176)	2.2(4.8)	92	58	2.1(.95)	9.3	5.2	.21	.20	4.5(2.04)	3091(1402)

(Continued)

Table Sheep-2 (Continued)

| Body Weight kg(lb) | Daily Dry Matter[a] Per Animal kg(lb) | Rough- age % | Energy[b] | | Total Protein % | Digest- ible Protein % | Ca % | P % | Carotene mg/kg(lb) | Vitamin A IU/kg(lb) |
			TDN %	ME Mcal/kg (lb)						
First 8 weeks of lactation suckling singles or last 8 weeks of lactation suckling twins										
50(110)	2.1(4.6)	67	65	2.4(1.09)	10.4	6.2	.52	.37	3.0(1.36)	2024(918)
60(132)	2.3(5.1)	67	65	2.4(1.09)	10.4	6.2	.50	.36	3.3(1.50)	2217(1005)
70(154)	2.5(5.5)	67	65	2.4(1.09)	10.4	6.2	.48	.34	3.5(1.59)	2380(1079)
80(176)	2.6(5.7)	67	65	2.4(1.09)	10.4	6.2	.48	.34	3.8(1.72)	2615(1186)
First 8 weeks of lactation suckling twins										
50(110)	2.4(5.3)	67	65	2.4(1.09)	11.5	7.2	.52	.37	2.6(1.18)	1771(803)
60(132)	2.6(5.7)	67	65	2.4(1.09)	11.5	7.2	.50	.36	2.9(1.32)	1962(890)
70(154)	2.8(6.2)	67	65	2.4(1.09)	11.5	7.2	.48	.34	3.1(1.41)	2125(964)
80(176)	3.0(6.6)	67	65	2.4(1.09)	11.5	7.2	.48	.34	3.3(1.50)	2267(1028)
Replacement lambs and yearlings[d]										
30(66)	1.3(2.9)	83	62	2.2(1.00)	10.0	5.8	.45	.25	1.5(.68)	981(445)
40(88)	1.4(3.1)	92	60	2.1(.95)	9.5	5.3	.44	.24	1.8(.82)	1214(551)
50(1.10)	1.5(3.3)	100	55	2.0(.91)	8.9	4.8	.42	.23	2.1(.95)	1417(643)
60(132)	1.5(3.3)	100	55	2.0(.91)	8.9	4.8	.43	.24	2.5(1.13)	1700(771)

continued

Rams

Replacement lambs and yearlings[d]

40(88)	1.8(4.0)	67	65	2.4(1.09)	10.2	6.0	.35	.19	1.4(.63)	944(428)
60(132)	2.3(5.1)	92	60	2.1(.95)	9.5	5.3	.31	.17	1.7(.77)	1109(503)
80(176)	2.8(6.2)	100	55	2.0(.91)	8.9	4.8	.28	.16	1.8(.82)	1214(551)
100(220)	2.8(6.2)	100	55	2.0(.91)	8.9	4.8	.30	.17	2.2(1.00)	1518(688)
120(265)	2.6(5.7)	100	55	2.0(.91)	8.9	4.8	.33	.18	2.9(1.32)	1962(890)

Lambs

Finishing (Gaining .20–.25 kg [.44–.55 lb] per head per day)[e]

30(66)	1.3(2.9)	75	64	2.3(1.04)	11.0	6.7	.37	.23	0.8(.36)	588(263)
35(77)	1.4(3.1)	67	67	2.4(1.09)	11.0	6.7	.34	.21	0.9(.41)	637(289)
40(88)	1.6(3.5)	58	70	2.5(1.13)	11.0	6.7	.31	.19	0.9(.41)	638(289)
45(99)	1.7(3.7)	58	70	2.5(1.13)	11.0	6.7	.29	.18	1.0(.45)	675(306)
50(110)	1.8(4.0)	58	70	2.5(1.13)	11.0	6.7	.28	.17	1.1(.50)	708(321)
55(121)	1.9(4.2)	58	70	2.5(1.13)	11.0	6.7	.26	.16	1.1(.50)	738(335)

Early-weaned (Gaining .25–.30 kg [.55–.66 lb] per head per day)[f]

10(22)	0.6(1.3)	50	73	2.6(1.18)	16.0	11.5	.40	.27	2.0(.91)	1417(643)
20(44)	1.0(2.2)	50	73	2.6(1.18)	16.0	11.5	.36	.24	2.5(1.13)	1700(771)
30(66)	1.4(3.1)	50	73	2.6(1.18)	14.0	9.5	.36	.24	2.7(1.22)	1821(826)

a To convert dry matter to an as-fed basis, divide dry matter by percentage of dry matter as fed and multiply by 100.

b 1 kg TDN = 4.4 Mcal DE (digestible energy). DE may be converted to ME (metabolizable energy) by multiplying by 82%. Because of rounding errors, calculations between Table Sheep-1 and Table Sheep-2 may not give the same values.

c Values are for ewes in moderate condition, not excessively fat or thin. Fat ewes should be fed at the next lower weight, thin ewes at the next higher weight. Once maintenance weight is established, such weight would follow through all production phases.

d Requirements for replacement lambs (ewe and ram) start when the lambs are weaned.

e Maximum gains expected. If lambs are held for later market, they should be fed as replacement ewe lambs are fed. Lambs gaining faster than indicated should be fed a higher level. Lambs finish at the maximum rate if they are self-fed.

f A 40-kg early-weaned lamb should be fed the same as a finishing lamb of the same weight.

Table Poultry-1
NUTRIENT REQUIREMENTS OF IMMATURE LEGHORN-TYPE CHICKENS AS PERCENTAGES OR UNITS PER KILOGRAM OF DIET

Nutrient	Unit	White-Egg-Laying Strains				Brown-Egg-Laying Strains			
		0 to 6 Weeks; 450 g^a; 2,850^b	*6 to 12 Weeks;* 980 g^a; 2,850^b	*12 to 18 Weeks;* 1,375 g^a; 2,900^b	*18 Weeks to First Egg;* 1,475 g^a; 2,900^b	*0 to 6 Weeks;* 500 g^a; 2,800^b	*6 to 12 Weeks;* 1,100 g^a; 2,800^b	*12 to 18 Weeks;* 1,500 g^a; 2,850^b	*18 Weeks to First Egg;* 1,600 g^a; 2,850^b
Protein and Amino Acids									
Crude protein[c]	%	18.00	16.00	15.00	17.00	17.00	15.00	14.00	16.00
Arginine	%	1.00	0.83	0.67	0.75	0.94	0.78	0.62	0.72
Glycine + serine	%	0.70	0.58	0.47	0.53	0.66	0.54	0.44	0.50
Histidine	%	0.26	0.22	0.17	0.20	0.25	0.21	0.16	0.18
Isoleucine	%	0.60	0.50	0.40	0.45	0.57	0.47	0.37	0.42
Leucine	%	1.10	0.85	0.70	0.80	1.00	0.80	0.65	0.75
Lysine	%	0.85	0.60	0.45	0.52	0.80	0.56	0.42	0.49
Methionine	%	0.30	0.25	0.20	0.22	0.28	0.23	0.19	0.21
Methionine + cystine	%	0.62	0.52	0.42	0.47	0.59	0.49	0.39	0.44
Phenylalanine	%	0.54	0.45	0.36	0.40	0.51	0.42	0.34	0.38
Phenylalanine + tyrosine	%	1.00	0.83	0.67	0.75	0.94	0.78	0.63	0.70
Threonine	%	0.68	0.57	0.37	0.47	0.64	0.53	0.35	0.44
Tryptophan	%	0.17	0.14	0.11	0.12	0.16	0.13	0.10	0.11
Valine	%	0.62	0.52	0.41	0.46	0.59	0.49	0.38	0.43
Fat									
Linoleic acid	%	1.00	1.00	1.00	1.00	1.00	1.00	1.00	1.00
Macrominerals									
Calcium[d]	%	0.90	0.80	0.80	2.00	0.90	0.80	0.80	1.80
Nonphytate phosphorus	%	0.40	0.35	0.30	0.32	0.40	0.35	0.30	0.35
Potassium	%	0.25	0.25	0.25	0.25	0.25	0.25	0.25	0.25
Sodium	%	0.15	0.15	0.15	0.15	0.15	0.15	0.15	0.15
Chlorine	%	0.15	0.12	0.12	0.15	0.12	0.11	0.11	0.11
Magnesium	mg	600.00	500.0	400.0	400.0	570.0	470.0	370.0	370.0

continued

Trace Minerals									
Manganese	mg	60.0	30.0	30.0	30.0	56.0	28.0	28.0	28.0
Zinc	mg	40.0	35.0	35.0	35.0	38.0	33.0	33.0	33.0
Iron	mg	80.0	60.0	60.0	60.0	75.0	56.0	56.0	56.0
Copper	mg	*5.0*	4.0	4.0	4.0	5.0	4.0	4.0	4.0
Iodine	mg	0.35	0.35	0.35	0.35	0.33	0.33	0.33	0.33
Selenium	mg	0.15	*0.10*	0.10	0.10	0.14	0.10	0.10	0.10
Fat-Soluble Vitamins									
A	IU	1,500.0	1,500.0	1,500.0	1,500.0	1,420.0	1,420.0	1,420.0	1,420.0
D_3	ICU	200.0	200.0	200.0	300.0	190.0	190.0	190.0	280.0
E	IU	*10.0*	5.0	5.0	5.0	9.5	4.7	4.7	4.7
K	mg	0.5	0.5	0.5	0.5	0.47	0.47	0.47	0.47
Water-Soluble Vitamins									
Riboflavin	mg	3.6	1.8	1.8	2.2	3.4	1.7	1.7	1.7
Pantothenic acid	mg	*10.0*	*10.0*	*10.0*	10.0	9.4	9.4	9.4	9.4
Niacin	mg	27.0	11.0	11.0	11.0	26.0	10.3	10.3	10.3
B_{12}	mg	0.009	0.003	0.003	0.004	0.009	0.003	0.003	0.003
Choline	mg	1,300.0	900.0	500.0	500.0	1,225.0	850.0	470.0	470.0
Biotin	mg	*0.15*	*0.10*	*0.10*	0.10	0.14	0.09	0.09	0.09
Folic acid	mg	*0.55*	*0.25*	*0.25*	0.25	0.52	0.23	0.23	0.23
Thiamin	mg	*1.0*	*0.8*	*0.8*	0.8	1.0	0.8	0.8	0.8
Pyridoxine	mg	3.0	3.0	3.0	3.0	2.8	2.8	2.8	2.8

Note: Where experimental data are lacking, values typeset in bold italics represent an estimate based on values obtained for other ages or related species.

[a] Final body weight.

[b] These are typical dietary energy concentrations for diets based mainly on corn and soybean meal, expressed in kcal ME_n/kg diet.

[c] Chickens do not have a requirement for crude protein *per se*. There, however, should be sufficient crude protein to ensure an adequate nitrogen supply for synthesis of nonessential amino acids. Suggested requirements for crude protein are typical of those derived with corn-soybean meal diets, and levels can be reduced somewhat when synthetic amino acids are used.

[d] The calcium requirement may be increased when diets contain high levels of phytate phosphorus (Nelson, 1984).

Table Poultry-2
NUTRIENT REQUIREMENTS OF LEGHORN-TYPE LAYING HENS AS PERCENTAGES OR UNITS PER KILOGRAM OF DIET (90% DRY MATTER)

Nutrient	Unit	Dietary Concentrations Required by White-Egg Layers at Different Feed Intakes			Amounts Required per Hen Daily (mg or IU)		
		80[a,b]	100[a,b]	120[a,b]	White-Egg Breeders at 100 g of Feed per Hen Daily[b]	White-Egg Layers at 100 g of Feed per Hen Daily	Brown-Egg Layers at 110 g of Feed per Hen Daily[c]
Protein and Amino Acids							
Crude protein[d]	%	18.8	15.0	12.5	15,000	15,000	16,500
Arginine[e]	%	0.88	0.70	0.58	700	700	770
Histidine	%	0.21	0.17	0.14	170	170	190
Isoleucine	%	0.81	0.65	0.54	650	650	715
Leucine	%	1.03	0.82	0.68	820	820	900
Lysine	%	0.86	0.69	0.58	690	690	760
Methionine	%	0.38	0.30	0.25	300	300	330
Methionine + cystine	%	0.73	0.58	0.48	580	580	645
Phenylalanine	%	0.59	0.47	0.39	470	470	520
Phenylalanine + tyrosine	%	1.04	0.83	0.69	830	830	910
Threonine	%	0.59	0.47	0.39	470	470	520
Tryptophan	%	0.20	0.16	0.13	160	160	175
Valine	%	0.88	0.70	0.58	700	700	770
Fat							
Linoleic acid	%	1.25	1.0	0.83	1,000	1,000	1,100
Macrominerals							
Calcium[f]	%	4.06	3.25	2.71	3,250	3,250	3,600
Chloride	%	0.16	0.13	0.11	130	130	145
Magnesium	mg	625	500	420	50	50	55
Nonphytate phosphorus[g]	%	0.31	0.25	0.21	250	250	275
Potassium	%	0.19	0.15	0.13	150	150	165
Sodium	%	0.19	0.15	0.13	150	150	165

continued

Trace Minerals

Nutrient	Unit						
Copper	mg	?	?	?	?	?	?
Iodine	mg	0.044	0.035	0.029	0.010	0.004	***0.004***
Iron	mg	56	45	38	6.0	4.5	***5.0***
Manganese	mg	25	20	17	2.0	2.0	***2.2***
Selenium	mg	0.08	0.06	0.05	0.006	0.006	***0.006***
Zinc	mg	44	35	29	4.5	3.5	***3.9***

Fat-Soluble Vitamins

Nutrient	Unit						
A	IU	3,750	3,000	2,500	300	300	***330***
D$_3$	IU	375	300	250	30	30	***33***
E	IU	***6***	***5***	4	***1.0***	0.5	***0.55***
K	mg	0.6	0.5	0.4	0.1	***0.05***	***0.055***

Water-Soluble Vitamins

Nutrient	Unit						
B$_{12}$	mg	***0.004***	***0.004***	***0.004***	0.008	***0.0004***	***0.0004***
Biotin	mg	***0.13***	***0.10***	***0.08***	***0.01***	***0.01***	***0.011***
Choline	mg	1,310	1,050	875	105	105	***115***
Folacin	mg	0.31	0.25	0.21	0.035	0.025	***0.028***
Niacin	mg	12.5	10.0	8.3	1.0	1.0	***1.1***
Pantothenic acid	mg	2.5	2.0	1.7	0.7	0.20	***0.22***
Pyridoxine	mg	3.1	2.5	2.1	0.45	0.25	***0.28***
Riboflavin	mg	3.1	2.5	2.1	0.36	0.25	0.28
Thiamin	mg	0.88	0.70	0.60	0.07	0.07	***0.08***

Note: Where experimental data are lacking, values typeset in bold italics represent an estimate based on values obtained for other ages or related species.

[a] Grams feed intake per hen daily.

[b] Based on dietary ME_n concentrations of approximately 2,900 kcal/kg and an assumed rate of egg production of 90% (90 eggs per 100 hens daily).

[c] Italicized values are based on those from white-egg layers but were increased 10% because of larger body weight and possibly more egg mass per day.

[d] Laying hens do not have a requirement for crude protein *per se*. However, there should be sufficient crude protein to ensure an adequate supply of nonessential amino acids. Suggested requirements for crude protein are typical of those derived with corn-soybean meal diets, and levels can be reduced somewhat when synthetic amino acids are used.

[e] Italicized amino acid values for white-egg-laying chickens were estimated by using Model B (Hurwitz and Bornstein, 1973), assuming a body weight of 1,800 g and 47 g of egg mass per day.

[f] The requirement may be higher for maximum eggshell thickness.

[g] The requirement may be higher in very hot temperatures.

Table Poultry-3
NUTRIENT REQUIREMENTS OF BROILERS AS PERCENTAGES OR UNITS PER KILOGRAM OF DIET (90% DRY MATTER)

Nutrient	Unit	0 to 3 Weeks[a] 3,200[b]	3 to 6 Weeks[a] 3,200[b]	6 to 8 Weeks[a] 3,200[b]
Protein and Amino Acids				
Crude protein[c]	%	23.00	20.00	18.00
Arginine	%	1.25	1.10	1.00
Glycine + serine	%	1.25	*1.14*	*0.97*
Histidine	%	0.35	*0.32*	*0.27*
Isoleucine	%	0.80	*0.73*	*0.62*
Leucine	%	1.20	*1.09*	*0.93*
Lysine	%	1.10	1.00	0.85
Methionine	%	0.50	0.38	0.32
Methionine + cystine	%	0.90	0.72	0.60
Phenylalanine	%	0.72	*0.65*	*0.56*
Phenylalanine + tyrosine	%	1.34	*1.22*	*1.04*
Proline	%	0.60	*0.55*	*0.46*
Threonine	%	0.80	0.74	0.68
Tryptophan	%	0.20	0.18	0.16
Valine	%	0.90	*0.82*	0.70
Fat				
Linoleic acid	%	1.00	1.00	1.00
Macrominerals				
Calcium[d]	%	1.00	0.90	0.80
Chlorine	%	0.20	0.15	0.12
Magnesium	mg	600	*600*	*600*
Nonphytate phosphorus	%	0.45	0.35	0.30
Potassium	%	0.30	0.30	0.30
Sodium	%	0.20	0.15	0.12
Trace Minerals				
Copper	mg	8	*8*	*8*
Iodine	mg	*0.35*	0.35	0.35
Iron	mg	80	*80*	*80*
Manganese	mg	*60*	60	60
Selenium	mg	0.15	0.15	0.15
Zinc	mg	40	*40*	*40*
Fat-Soluble Vitamins				
A	IU	*1,500*	*1,500*	*1,500*
D₃	ICU	200	200	200
E	IU	*10*	*10*	*10*
K	mg	0.50	0.50	0.50

continued

Table Poultry-3 (Continued)

Nutrient	Unit	0 to 3 Weeks[a] 3,200[b]	3 to 6 Weeks[a] 3,200[b]	6 to 8 Weeks[a] 3,200[b]
Water-Soluble Vitamins				
B$_{12}$	mg	0.01	*0.01*	*0.007*
Biotin	mg	0.15	0.15	*0.12*
Choline	mg	1,300	*1,000*	750
Folacin	mg	0.55	0.55	*0.50*
Niacin	mg	35	30	25
Pantothenic acid	mg	*10*	*10*	*10*
Pyridoxine	mg	3.5	3.5	*3.0*
Riboflavin	mg	3.6	3.6	3
Thiamin	mg	*1.80*	*1.80*	*1.80*

Note: Where experimental data are lacking, values typeset in bold italics represent an estimate based on values obtained for other ages or related species.

[a]The 0 to 3, 3 to 6, and 6 to 8 week intervals for nutrient requirements are based on chronology for which research data were available; however, these nutrient requirements are often implemented at younger age intervals or on a weight-of-feed consumed basis.

[b]These are typical dietary energy concentrations, expressed in kcalME_n/kg diet. Different energy values may be appropriate depending on local ingredient prices and availability.

[c]Broiler chickens do not have a requirement for crude protein *per se*. There, however, should be sufficient crude protein to ensure an adequate nitrogen supply for synthesis of nonessential amino acids. Suggested requirements for crude protein are typical of those derived with corn-soybean meal diets, and levels can be reduced when synthetic amino acids are used.

[d]The calcium requirement may be increased when diets contain high levels of phytate phosphorus (Nelson, 1984).

Table Poultry-4
NUTRIENT REQUIREMENTS OF TURKEYS AS PERCENTAGES OR UNITS PER KILOGRAM OF DIET
(90 % DRY MATTER)

Nutrient	Unit	Growing Turkeys, Males and Females						Breeders	
		0 to 4 Weeks[a]; 0 to 4 Weeks[b]; 2,800[c]	4 to 8 Weeks[a]; 4 to 8 Weeks[b]; 2,900[c]	8 to 12 Weeks[a]; 8 to 11 Weeks[b]; 3,000[c]	12 to 16 Weeks[a]; 11 to 14 Weeks[b]; 3,100[c]	16 to 20 Weeks[a]; 14 to 17 Weeks[b]; 3,200[c]	20 to 24 Weeks[a]; 17 to 20 Weeks[b]; 3,300[c]	Holding; 2,900[c]	Laying Hens; 2,900[c]
Protein and Amino Acids									
Protein[d]	%	28.0	26	22	19	16.5	14	12	14
Arginine	%	1.6	1.4	1.1	0.9	0.75	0.6	0.5	0.6
Glycine + serine	%	1.0	0.9	0.8	0.7	0.6	0.5	0.4	0.5
Histidine	%	0.58	0.5	0.4	0.3	0.25	0.2	0.2	0.3
Isoleucine	%	1.1	1.0	0.8	0.6	0.5	0.45	0.4	0.5
Leucine	%	1.9	1.75	1.5	1.25	1.0	0.8	0.5	0.5
Lysine	%	1.6	1.5	1.3	1.0	0.8	0.65	0.5	0.6
Methionine	%	0.55	0.45	0.4	0.35	0.25	0.25	0.2	0.2
Methionine + cystine	%	1.05	0.95	0.8	0.65	0.55	0.45	0.4	0.4
Phenylalanine	%	1.0	0.9	0.8	0.7	0.6	0.5	0.4	0.55
Phenylalanine + tyrosine	%	1.8	1.6	1.2	1.0	0.9	0.9	0.8	1.0
Threonine	%	1.0	0.95	0.8	0.75	0.6	0.5	0.4	0.45
Tryptophan	%	0.26	0.24	0.2	0.18	0.15	0.13	0.1	0.13
Valine	%	1.2	1.1	0.9	0.8	0.7	0.6	0.5	0.58
Fat									
Linoleic acid	%	1.0	1.0	0.8	0.8	0.8	0.8	0.8	1.1
Macrominerals									
Calcium[e]	%	1.2	1.0	0.85	0.75	0.65	0.55	0.5	2.25
Nonphytate phosphorus[f]	%	0.6	0.5	0.42	0.38	0.32	0.28	0.25	0.35
Potassium	%	0.7	0.6	0.5	0.5	0.4	0.4	0.4	0.6
Sodium	%	0.17	0.15	0.12	0.12	0.12	0.12	0.12	0.12
Chlorine	%	0.15	0.14	0.14	0.12	0.12	0.12	0.12	0.12
Magnesium	mg	500	500	500	500	500	500	500	500

continued

Trace Minerals									
Manganese	mg	60	60	60	60	60	60	60	60
Zinc	mg	70	65	50	40	40	40	40	65
Iron	mg	80	60	60	60	50	50	50	60
Copper	mg	8	8	6	6	6	6	6	8
Iodine	mg	0.4	0.4	0.4	0.4	0.4	0.4	0.4	0.4
Selenium	mg	0.2	0.2	0.2	0.2	0.2	0.2	0.2	0.2
Fat-Soluble Vitamins									
A	IU	5,000	5,000	5,000	5,000	5,000	5,000	5,000	5,000
D_3[g]	ICU	1,100	1,100	1,100	1,100	1,100	1,100	1,100	1,100
E	IU	12	12	10	10	10	10	10	25
K	mg	1.75	1.5	1.0	0.75	0.75	0.50	0.5	1.0
Water-Soluble Vitamins									
B_{12}	mg	0.003	0.003	0.003	0.003	0.003	0.003	0.003	0.003
Biotin[h]	mg	0.25	0.2	0.125	0.125	0.100	0.100	0.100	0.20
Choline	mg	1,600	1,400	1,100	1,100	950	800	800	1,000
Folacin	mg	1.0	1.0	0.8	0.8	0.7	0.7	0.7	1.0
Niacin	mg	60.0	60.0	50.0	50.0	40.0	40.0	40.0	40.0
Pantothenic acid	mg	10.0	9.0	9.0	9.0	9.0	9.0	9.0	16.0
Pyridoxine	mg	4.5	4.5	3.5	3.5	3.0	3.0	3.0	4.0
Riboflavin	mg	4.0	3.6	3.0	3.0	2.5	2.5	2.5	4.0
Thiamin	mg	2.0	2.0	2.0	2.0	2.0	2.0	2.0	2.0

NOTE: Where experimental data are lacking, values typeset in bold italics represent estimates based on values obtained from other ages or related species or from modeling experiments.

[a] The age intervals for nutrient requirements of males are based on actual chronology from previous research. Genetic improvements in body weight gain have led to an earlier implementation of these levels, at 0 to 3, 3 to 6, 6 to 9, 9 to 12, 12 to 15, and 15 to 18 weeks, respectively, by the industry at large.

[b] The age intervals for nutrient requirements of females are based on actual chronology from previous research. Genetic improvements in body weight gain have led to an earlier implementation of these levels, at 0 to 3, 3 to 6, 6 to 9, 9 to 12, 12 to 14, and 14 to 16 weeks, respectively, by the industry at large.

[c] These are approximate metabolizable energy (ME) values provided with typical corn-soybean-meal-based feeds, expressed in kcal ME_n/kg diet. Such energy, when accompanied by the nutrient levels suggested, is expected to provide near-maximum growth, particularly with pelleted feed.

[d] Turkeys do not have a requirement for crude protein per se. There, however, should be sufficient crude protein to ensure an adequate nitrogen supply for synthesis of nonessential amino acids. Suggested requirements for crude protein are typical of those derived with corn-soybean meal diets, and levels can be reduced when synthetic amino acids are used.

[e] The calcium requirement may be increased when diets contain high levels of phytate phosphorus (Nelson, 1984).

[f] Organic phosphorus is generally considered to be associated with phytin and of limited availability.

[g] These concentrations of vitamin D are considered satisfactory when the associated calcium and phosphorus levels are used.

[h] Requirement may increase with wheat-based diets.

Table Poultry-5
TYPICAL BODY WEIGHTS, FEED REQUIREMENTS, AND ENERGY CONSUMPTION OF BROILERS

Age (weeks)	Body Weight (g)		Weekly Feed Consumption (g)		Cumulative Feed Consumption (g)		Weekly Energy Consumption (kcal ME/bird)		Cumulative Energy Consumption (kcal ME/bird)	
	Male	Female	Male	Female	Male	Female	Male	Female	Male	Female
1	152	144	135	131	135	131	432	419	432	419
2	376	344	290	273	425	404	928	874	1,360	1,293
3	686	617	487	444	912	848	1,558	1,422	2,918	2,715
4	1,085	965	704	642	1,616	1,490	2,256	2,056	5,174	4,771
5	1,576	1,344	960	738	2,576	2,228	3,075	2,519	8,249	7,290
6	2,088	1,741	1,141	1,001	3,717	3,229	3,651	3,045	11,900	10,335
7	2,590	2,134	1,281	1,081	4,998	4,310	4,102	3,459	16,002	13,794
8	3,077	2,506	1,432	1,165	6,430	5,475	4,585	3,728	20,587	17,522
9	3,551	2,842	1,577	1,246	8,007	6,721	5,049	3,986	25,636	21,508

Note: Values are typical for broilers fed well-balanced diets providing 3,200 kcal ME/kg.

Table Poultry-6
GROWTH RATE AND FEED AND ENERGY CONSUMPTION OF LARGE-TYPE TURKEYS

Age (weeks)	Body Weight (kg)		Feed Consumption per Week (kg)		Cumulative Feed Consumption (kg)		ME Consumption per Week (Mcal)	
	Male	Female	Male	Female	Male	Female	Male	Female
1	0.12	0.12	0.10	0.10	0.10	0.10	0.28	0.28
2	0.25	0.24	0.19	0.18	0.29	0.28	0.53	0.5
3	0.50	0.46	0.37	0.34	0.66	0.62	1.0	1.0
4	1.0	0.9	0.70	0.59	1.36	1.21	2.0	1.7
5	1.6	1.4	0.85	0.64	2.21	1.85	2.5	1.9
6	2.2	1.8	1.10	0.80	3.31	2.65	3.2	2.3
7	3.1	2.3	1.40	0.98	4.71	3.63	4.1	2.8
8	4.0	3.0	1.73	1.21	6.44	4.84	5.0	3.5
9	5.0	3.7	2.00	1.42	8.44	6.26	6.0	4.3
10	6.0	4.4	2.34	1.70	10.78	7.96	7.0	5.1
11	7.1	5.2	2.67	1.98	13.45	9.94	8.0	5.9
12	8.2	6.0	2.99	2.18	16.44	12.12	9.0	6.8
13	9.3	6.8	3.20	2.44	19.64	14.56	9.9	7.6
14	10.5	7.5	3.47	2.69	23.11	17.25	10.8	8.4
15	11.5	8.3	3.73	2.81	26.84	20.06	11.6	9.0
16	12.6	8.9	3.97	3.00	30.81	23.06	12.3	9.6
17	13.5	9.6	4.08	3.14	34.89	26.20	13.1	10.1
18	14.4	10.2	4.30	3.18	39.19	29.38	13.8	10.5
19	15.2	10.9	4.52	3.31	43.71	32.69	14.5	10.9
20	16.1	11.5	4.74	3.40	48.45	36.09	15.2	11.2
21	17.0	[a]	4.81	[a]	53.26	[a]	15.9	
22	17.9	[a]	5.00	[a]	58.26	[a]	16.5	[a]
23	18.6	[a]	5.15	[a]	63.41	[a]	17.1	[a]
24	19.4	[a]	5.28	[a]	68.69	[a]	17.4	[a]

[a]No data given because females are usually not marketed after 20 weeks of age.

Table Poultry-7
BODY WEIGHTS AND FEED CONSUMPTION OF LARGE-TYPE TURKEYS DURING THE HOLDING AND BREEDING PERIODS

| Age (weeks) | Females | | | Males | |
	Weight (kg)	Egg Production (%)	Feed per Turkey Daily (g)	Weight (kg)	Feed per Turkey Daily (g)
20	8.4	0	260	14.3	500
25	9.8	0	320	16.4	570
30	11.1	0[a]	310	19.1	630
35	11.1	68	280	20.7	620
40	10.8	64	280	21.8	570
45	10.5	58	280	22.5	550
50	10.5	52	290	23.2	560
55	10.5	45	290	23.9	570
60	10.6	38	290	24.5	580

Note: These values are based on experimental data involving "in-season" egg production (that is, November through July) of commercial stock. It is estimated that summer breeders would produce 70% to 90% as many eggs and consume 60% to 80% as much feed as in-season breeders.

[a]Light stimulation is begun at this point.

Table Swine-1

DIETARY AMINO ACID REQUIREMENTS OF GROWING PIGS ALLOWED FEED AD LIBITUM (90% DRY MATTER)[a]

	Body Weight (kg)					
	3–5	*5–10*	*10–20*	*20–50*	*50–80*	*80–120*
Average weight in range (kg)	4	7.5	15	35	65	100
DE content of diet (kcal/kg)	3,400	3,400	3,400	3,400	3,400	3,400
ME content of diet (kcal/kg)[b]	3,265	3,265	3,265	3,265	3,265	3,265
Estimated DE intake (kcal/day)	855	1,690	3,400	6,305	8,760	10,450
Estimated ME intake (kcal/day)[b]	820	1,620	3,265	6,050	8,410	10,030
Estimated feed intake (g/day)	250	500	1,000	1,855	2,575	3,075
Crude protein (%)[c]	26.0	23.7	20.9	18.0	15.5	13.2
Amino Acid Requirements[d]						
True Ileal Digestible Basis (%)						
Arginine	0.54	0.49	0.42	0.33	0.24	0.16
Histidine	0.43	0.38	0.32	0.26	0.21	0.16
Isoleucine	0.73	0.65	0.55	0.45	0.37	0.29
Leucine	1.35	1.20	1.02	0.83	0.67	0.51
Lysine	1.34	1.19	1.01	0.83	0.66	0.52
Methionine	0.36	0.32	0.27	0.22	0.18	0.14
Methionine + cystine	0.76	0.68	0.58	0.47	0.39	0.31
Phenylalanine	0.80	0.71	0.61	0.49	0.40	0.31
Phenylalanine + tyrosine	1.26	1.12	0.95	0.78	0.63	0.49
Threonine	0.84	0.74	0.63	0.52	0.43	0.34
Tryptophan	0.24	0.22	0.18	0.15	0.12	0.10
Valine	0.91	0.81	0.69	0.56	0.45	0.35
Apparent Ileal Digestible Basis (%)						
Arginine	0.51	0.46	0.39	0.31	0.22	0.14
Histidine	0.40	0.36	0.31	0.25	0.20	0.16
Isoleucine	0.69	0.61	0.52	0.42	0.34	0.26
Leucine	1.29	1.15	0.98	0.80	0.64	0.50
Lysine	1.26	1.11	0.94	0.77	0.61	0.47

Table Swine-1 (Continued)

	Body Weight (kg)					
	3–5	*5–10*	*10–20*	*20–50*	*50–80*	*80–120*
Methionine	0.34	0.30	0.26	0.21	0.17	0.13
Methionine + cystine	0.71	0.63	0.53	0.44	0.36	0.29
Phenylalanine	0.75	0.66	0.56	0.46	0.37	0.28
Phenylalanine + tyrosine	1.18	1.05	0.89	0.72	0.58	0.45
Threonine	0.75	0.66	0.56	0.46	0.37	0.30
Tryptophan	0.22	0.19	0.16	0.13	0.10	0.08
Valine	0.84	0.74	0.63	0.51	0.41	0.32
Total Basis (%)[e]						
Arginine	0.59	0.54	0.46	0.37	0.27	0.19
Histidine	0.48	0.43	0.36	0.30	0.24	0.19
Isoleucine	0.83	0.73	0.63	0.51	0.42	0.33
Leucine	1.50	1.32	1.12	0.90	0.71	0.54
Lysine	1.50	1.35	1.15	0.95	0.75	0.60
Methionine	0.40	0.35	0.30	0.25	0.20	0.16
Methionine + cystine	0.86	0.76	0.65	0.54	0.44	0.35
Phenylalanine	0.90	0.80	0.68	0.55	0.44	0.34
Phenylalanine + tyrosine	1.41	1.25	1.06	0.87	0.70	0.55
Threonine	0.98	0.86	0.74	0.61	0.51	0.41
Tryptophan	0.27	0.24	0.21	0.17	0.14	0.11
Valine	1.04	0.92	0.79	0.64	0.52	0.40

[a]Mixed gender (1:1 ratio of barrows to gilts) of pigs with high-medium lean growth rate (325 g/day of carcass fat-free lean) from 20 to 120 kg body weight.

[b]Assumes that ME is 96% of DE. In corn–soybean meal diets of these crude protein levels, ME is 94% to 96% of DE.

[c]Crude protein levels apply to corn–soybean meal diets. In 3–10 kg pigs fed diets with dried plasma and/or dried milk products, protein levels will be 2% to 3% less than shown.

[d]Total amino acid requirements are based on the following types od diets: 3–5 kg pigs, corn–soybean meal diet that includes 5% dried plasma and 25% to 50% dried milk products; 5–10 kg pigs, corn–soybean meal diet that includes 5% to 25% dried milk products; 10–120 kg pigs, corn–soybean meal diet.

[e]The total lysine percentages for 3–20 kg pigs are estimated from empirical data. The other amino acids for 3–20 kg pigs are based on the ratios of amino acids to lysine (true digestible basis); however, there are very few empirical data to support these ratios. The requirements for 20–120 kg pigs are estimated from the growth model.

Table Swine-2
DIETARY MINERAL, VITAMIN, AND FATTY ACID REQUIREMENTS OF GROWING PIGS ALLOWED FEED AD LIBITUM (90% DRY MATTER)[a]

Requirements (% or Amount/kg of Diet)

	Body Weight (kg)					
	3–5	*5–10*	*10–20*	*20–50*	*50–80*	*80–120*
Average weight in range (kg)	4	7.5	15	35	65	100
DE content of diet (kcal/kg)	3,400	3,400	3,400	3,400	3,400	3,400
ME content of diet (kcal/kg)[b]	3,265	3,265	3,265	3,265	3,265	3,265
Estimated DE intake (kcal/day)	855	1,690	3,400	6,305	8,760	10,450
Estimated ME intake (kcal/day)[b]	820	1,620	3,265	6,050	8,410	10,030
Estimated feed intake (g/day)	250	500	1,000	1,855	2,575	3,075
Mineral elements						
Calcium (%)[c]	0.90	0.80	0.70	0.60	0.50	0.45
Phosphorus, total (%)[c]	0.70	0.65	0.60	0.50	0.45	0.40
Phosphorus, available (%)[c]	0.55	0.40	0.32	0.23	0.19	0.15
Sodium (%)	0.25	0.20	0.15	0.10	0.10	0.10
Chlorine (%)	0.25	0.20	0.15	0.08	0.08	0.08
Magnesium (%)	0.04	0.04	0.04	0.04	0.04	0.04
Potassium (%)	0.30	0.28	0.26	0.23	0.19	0.17
Copper (mg)	6.00	6.00	5.00	4.00	3.50	3.00
Iodine (mg)	0.14	0.14	0.14	0.14	0.14	0.14
Iron (mg)	100	100	80	60	50	40
Manganese (mg)	4.00	4.00	3.00	2.00	2.00	2.00
Selenium (mg)	0.30	0.30	0.25	0.15	0.15	0.15
Zinc (mg)	100	100	80	60	50	50
Vitamins						
Vitamin A (IU)[d]	2,200	2,200	1,750	1,300	1,300	1,300
Vitamin D₃ (IU)[d]	220	220	200	150	150	150
Vitamin E (IU)[d]	16	16	11	11	11	11
Vitamin K (menadione) (mg)	0.50	0.50	0.50	0.50	0.50	0.50
Biotin (mg)	0.08	0.05	0.05	0.05	0.05	0.05

Table Swine-2 (Continued)

	Body Weight (kg)					
	3–5	5–10	10–20	20–50	50–80	80–120
Choline (g)	0.60	0.50	0.40	0.30	0.30	0.30
Folacin (mg)	0.30	0.30	0.30	0.30	0.30	0.30
Niacin, available (mg)[e]	20.00	15.00	12.50	10.00	7.00	7.00
Pantothenic acid (mg)	12.00	10.00	9.00	8.00	7.00	7.00
Riboflavin (mg)	4.00	3.50	3.00	2.50	2.00	2.00
Thiamin (mg)	1.50	1.00	1.00	1.00	1.00	1.00
Vitamin B_6 (mg)	2.00	1.50	1.50	1.00	1.00	1.00
Vitamin B_{12} (µg)	20.00	17.50	15.00	10.00	5.00	5.00
Linoleic acid (%)	0.10	0.10	0.10	0.10	0.10	0.10

[a]Pigs of mixed gender (1:1 ratio of barrows to gilts). The requirements of certain minerals and vitamins may be slightly higher for pigs having high lean growth rates (>325 g/day of carcass fat-free lean), but no distinction is made.

[b]Assumes that ME is 96% of DE. In corn–soybean meal diets, ME is 94–96% of DE, depending on crude protein level of the diet.

[c]The percentages of calcium, phosphorus, and available phosphorus should be increased by 0.05 to 0.1 percentage points for developing boars and replacement gilts from 50 to 120 kg body weight.

[d]Conversions: 1 IU vitamin A – 0.344 µg (g retinyl acetate; 1 IU vitamin D_3 = 0.025 µg cholecalciferol; 1 IU vitamin E = 0.67 mg of D α-tocopherol or 1 mg of DL–α-tocopheryl acetate.

[e]The niacin in corn, grain sorghum, wheat, and barley is unavailable. Similarly, the niacin in by-products made from these cereal grains is poorly available unless the by-products have undergone a fermentation or wet milling process.

Table Swine-3
DIETARY AMINO ACID REQUIREMENTS OF GESTATING SOWS (90% DRY MATTER)[a]

	Body Weight at Breeding (kg)					
	125	150	175	200	200	200
	Gestation Weight Gain (kg)[b]					
	55	45	40	35	30	35
	Anticipated Pigs in Litter					
	11	12	12	12	12	14
DE content of diet (kcal/kg)	3,400	3,400	3,400	3,400	3,400	3,400
ME content of diet (kcal/kg)[c]	3,265	3,265	3,265	3,265	3,265	3,265
Estimated DE intake (kcal/day)	6,660	6,265	6,405	6,535	6,115	6,275
Estimated ME intake (kcal/day)[c]	6,395	6,015	6,150	6,275	5,870	6,025
Estimated feed intake (kg/day)	1.96	1.84	1.88	1.92	1.80	1.85
Crude protein (%)[d]	12.9	12.8	12.4	12.0	12.1	12.4
Amino Acid Requirements						
True Ileal Digestible Basis (%)						
Arginine	0.04	0.00	0.00	0.00	0.00	0.00
Histidine	0.16	0.16	0.15	0.14	0.14	0.15
Isoleucine	0.29	0.28	0.27	0.26	0.26	0.27
Leucine	0.48	0.47	0.44	0.41	0.41	0.44
Lysine	0.50	0.49	0.46	0.44	0.44	0.46
Methionine	0.14	0.13	0.13	0.12	0.12	0.13
Methionine + cystine	0.33	0.33	0.32	0.31	0.32	0.33
Phenylalanine	0.29	0.28	0.27	0.25	0.25	0.27
Phenylalanine + tyrosine	0.48	0.48	0.46	0.44	0.44	0.46
Threonine	0.37	0.38	0.37	0.36	0.37	0.38
Tryptophan	0.10	0.10	0.09	0.09	0.09	0.09
Valine	0.34	0.33	0.31	0.30	0.30	0.31
Apparent Ileal Digestible Basis (%)						
Arginine	0.03	0.00	0.00	0.00	0.00	0.00
Histidine	0.15	0.15	0.14	0.13	0.13	0.14

Table Swine-3 (Continued)

Isoleucine	0.26	0.26	0.25	0.24	0.24	0.25
Leucine	0.47	0.46	0.43	0.40	0.40	0.43
Lysine	0.45	0.45	0.42	0.40	0.40	0.42
Methionine	0.13	0.13	0.12	0.11	0.12	0.12
Methionine + cystine	0.30	0.31	0.30	0.29	0.30	0.31
Phenylalanine	0.27	0.26	0.24	0.23	0.23	0.24
Phenylalanine + tyrosine	0.45	0.44	0.42	0.40	0.41	0.43
Threonine	0.32	0.33	0.32	0.31	0.32	0.33
Tryptophan	0.08	0.08	0.08	0.07	0.07	0.08
Valine	0.31	0.30	0.28	0.27	0.27	0.28

Total Basis (%)[d]

Arginine	0.06	0.03	0.00	0.00	0.00	0.00
Histidine	0.19	0.18	0.17	0.16	0.17	0.17
Isoleucine	0.33	0.32	0.31	0.30	0.30	0.31
Leucine	0.50	0.49	0.46	0.42	0.43	0.45
Lysine	0.58	0.57	0.54	0.52	0.52	0.54
Methionine	0.15	0.15	0.14	0.13	0.13	0.14
Methionine + cystine	0.37	0.38	0.37	0.36	0.36	0.37
Phenylalanine	0.32	0.32	0.30	0.28	0.28	0.30
Phenylalanine + tyrosine	0.54	0.54	0.51	0.49	0.49	0.51
Threonine	0.44	0.45	0.44	0.43	0.44	0.45
Tryptophan	0.11	0.11	0.11	0.10	0.10	0.11
Valine	0.39	0.38	0.36	0.34	0.34	0.36

[a]Daily intakes of DE and feed and the amino acid requirements are estimated by the gestation model.
[b]Weight gain includes maternal tissue and products of conception.
[c]Assumes that ME is 96% of DE.
[d]Crude protein and total amino acid requirements are based on a corn–soybean meal diet.

Table Swine-4

DIETARY MINERAL, VITAMIN, AND FATTY ACID REQUIREMENTS OF GESTATING AND LACTATING SOWS (90% DRY MATTER)[a]

	Gestation	*Lactation*
DE content of diet (kcal/kg)	3,400	3,400
ME content of diet (kcal/kg)[b]	3,265	3,265
DE intake (kcal/day)	6,290	17,850
ME intake (kcal/day)[b]	6,040	17,135
Feed intake (kg/day)	1.85	5.25

Requirements (% or Amount/kg of Diet)

	Gestation	*Lactation*
Mineral elements		
Calcium (%)	0.75	0.75
Phosphorus, total (%)	0.60	0.60
Phosphorus, available (%)	0.35	0.35
Sodium (%)	0.15	0.20
Chlorine (%)	0.12	0.16
Magnesium (%)	0.04	0.04
Potassium (%)	0.20	0.20
Copper (mg)	5.00	5.00
Iodine (mg)	0.14	0.14
Iron (mg)	80	80
Manganese (mg)	20	20
Selenium (mg)	0.15	0.15
Zinc (mg)	50	50
Vitamins		
Vitamin A (IU)[c]	4,000	2,000
Vitamin D_3 (IU)[c]	200	200
Vitamin E (IU)[c]	44	44
Vitamin K (menadione) (mg)	0.50	0.50
Biotin (mg)	0.20	0.20
Choline (g)	1.25	1.00
Folacin (mg)	1.30	1.30
Niacin, available (mg)[d]	10	10
Pantothenic acid (mg)	12	12
Riboflavin (mg)	3.75	3.75
Thiamin (mg)	1.00	1.00
Vitamin B_6 (mg)	1.00	1.00
Vitamin B_{12} (μg)	15	15
Linoleic acid (%)	0.10	0.10

[a]The requirements are based on the daily consumption of 1.85 and 5.25 kg of feed, respectively. If lower amounts of feed are consumed, the dietary percentage may need to be increased.

[b]Assumes that ME is 96% of DE.

[c]Conversions: 1 IU vitamin A = 0.344 μg retinyl acetate; 1 IU vitamin D_3 = 0.025 μg cholecalciferol; 1 IU vitamin E = 0.67 mg of D-α-tocopherol or 1 mg of DL-α-tocopheryl acetate.

[d]The niacin in corn, grain sorghum, wheat, and barley is unavailable. Similarly, the niacin in by-products made from these cereal grains is poorly available unless the by-products have undergone a fermentation or wet-milling process.

Table Swine-5

DIETARY AND DAILY AMINO ACID, MINERAL, VITAMIN, AND FATTY ACID REQUIREMENTS OF SEXUALLY ACTIVE BOARS (90% DRY MATTER)[a]

DE content of diet (kcal/kg)	3,400
ME content of diet (kcal/kg)	3,265
DE intake (kcal/day)	6,800
ME intake (kcal/day)	6,530
Feed intake (kg/day)	2.00
Crude protein (%)[b]	13.0

Requirements

	% or amount/kg of diet	amount/day
Amino acids (total basis)[b]		
Arginine	—	—
Histidine	0.19%	3.8 g
Isoleucine	0.35%	7.0 g
Leucine	0.51%	10.2 g
Lysine	0.60%	12.0 g
Methionine	0.16%	3.2 g
Methionine + cystine	0.42%	8.4 g
Phenylalanine	0.33%	6.6 g
Phenylalanine + tyrosine	0.57%	11.4 g
Threonine	0.50%	10.0 g
Tryptophan	0.12%	2.4 g
Valine	0.40%	8.0 g
Mineral elements		
Calcium	0.75%	15.0 g
Phosphorus, total	0.60%	12.0 g
Phosphorus, available	0.35%	7.0 g
Sodium	0.15%	3.0 g
Chlorine	0.12%	2.4 g
Magnesium	0.04%	0.8 g
Potassium	0.20%	4.0 g
Copper	5 mg	10 mg

continued

Table Swine-5 (Continued)

Nutrient				
Iodine	0.14	mg	0.28	mg
Iron	80	mg	160	mg
Manganese	20	mg	40	mg
Selenium	0.15	mg	0.3	mg
Zinc	50	mg	100	mg
Vitamins				
Vitamin A [c]	4,000	IU	8,000	IU
Vitamin D$_3$ [c]	200	IU	400	IU
Vitamin E [c]	44	IU	88	IU
Vitamin K (menadione)	0.50	mg	1.0	mg
Biotin	0.20	mg	0.4	mg
Choline	1.25	g	2.5	g
Folacin	1.30	mg	2.6	mg
Niacin, available [d]	10	mg	20	mg
Pantothenic acid	12	mg	24	mg
Riboflavin	3.75	mg	7.5	mg
Thiamin	1.0	mg	2.0	mg
Vitamin B$_6$	1.0	mg	2.0	mg
Vitamin B$_{12}$	15	μg	30	μg
Linoleic acid	0.1	%	2.0	g

[a]The requirements are based on the daily consumption of 2.0 kg of feed. Feed intake may need to be adjusted, depending on the weight of the boar and the amount of weight gain desired.

[b]Assumes a corn–soybean meal diet. The lysine requirement was set as 0.60% (12.0 g/day). Other amino acids were calculated using ratios (total basis) similar to those for gestating sows.

[c]Conversions: 1 IU vitamin A = 0.344 μg retinyl acetate; 1 IU vitamin D$_3$ = 0.025 μg cholecalciferol; 1 IU vitamin E = 0.67 mg of D-α-tocopherol or 1 mg of DL-α-tocopheryl acetate.

[d]The niacin in corn, grain sorghum, wheat, and barley is unavailable. Similarly, the niacin in by-products made from these cereal grains is poorly available unless the by-products have undergone a fermentation or wet-milling process.

Table-Feeds-Ruminant-1
COMPOSITION OF FEEDS COMMONLY USED IN RUMINANT DIETS ON A 100% DRY MATTER BASIS

Entry No.	Feed Name Description	International Feed Number[a]	Dry Matter (%)	Values as Determined at Maintenance Intake TDN (%)	DE (Mcal/kg)	ME (Mcal/kg)	Production Growing Cattle NEM (Mcal/kg)	NEG (Mcal/kg)	Lactating Cows NEL (Mcal/kg)	Crude Protein (%)
	Alfalfa. *Medicago sativa*									
001	—meal dehydrated, 15% protein	1-00-022	90	59	2.60	2.18	1.27	0.70	1.33	17.3
002	—meal dehydrated, 17% protein	1-00-023	92	61	2.69	2.27	1.34	0.77	1.38	18.9
003	—meal dehydrated, 20% protein	1-00-024	92	62	2.73	2.31	1.38	0.80	1.40	22.0
004	—meal dehydrated, 22% protein	1-07-851	93	67	2.95	2.53	1.54	0.94	1.52	23.9
005	—hay, sun-cured, early vegetative	1-00-050	90	66	2.91	2.49	1.51	0.92	1.50	23.0
006	—hay, sun-cured, late vegetative	1-00-054	90	63	2.78	2.36	1.41	0.83	1.42	20.0
007	—hay, sun-cured, early bloom	1-00-059	90	60	2.65	2.22	1.31	0.74	1.35	18.0
008	—hay, sun-cured, midbloom	1-00-063	90	58	2.56	2.13	1.24	0.68	1.30	17.0
009	—hay, sun-cured, full bloom	1-00-068	90	55	2.43	2.00	1.14	0.58	1.23	15.0
010	—silage, wilted, 25–45% dry matter (see similar maturity descriptions of hays)		—	—	—	—	—	—	—	—
	Almond. *Prunus amygdalus*									
011	—hulls	4-00-	90	59	2.60	2.18	1.27	0.70	1.33	2.1
012	—hulls, 15% crude fiber	4-00-359	90	52	2.28	1.85	1.04	0.49	1.15	2.1
013	—hulls and shells	1-27-475	90	45	1.98	1.55	0.78	0.25	0.98	1.7
	Apples. *Melus* spp.									
014	—pomace, oat hulls added, dehydrated	4-28-096	89	68	3.00	2.58	1.57	0.97	1.55	5.1
015	—pomace, no hulls added, dehydrated	4-00-423	89	69	3.04	2.62	1.60	1.00	1.57	4.9
	Bahiagrass. *Paspalum notatum*									
016	—fresh	2-00-464	30	54	2.38	1.95	1.11	0.55	1.20	8.9
017	—hay, sun cured, late vegetative	1-20-787	91	44	1.94	1.51	0.75	0.22	0.96	9.5
	Bakery Waste									
018	—dehydrated (dried bakery product)	4-00-466	92	89	3.92	3.51	2.20	1.52	2.06	10.7

continued

	Barley. *Hordeum vulgare*								
019	—grain	4-00-549	88	84	3.70	3.29	2.06	1.40	1.94	13.5
020	—grain, Pacific Coast	4-07-939	89	86	3.79	3.38	2.12	1.45	1.99	10.8
021	—grain screenings	4-00-542	89	80	3.53	3.11	1.94	1.30	1.84	13.1
022	—hay, sun-cured	1-00-495	87	56	2.47	2.04	1.18	0.62	1.25	8.7
023	—malt sprouts, dehydrated	5-00-545	94	71	3.13	2.71	1.67	1.06	1.62	28.1
024	—straw	1-00-498	91	49	2.16	1.73	0.93	0.53	1.08	4.3
	Bean, Navy. *Phaseolus vulgaris*								
025	—seeds	5-00-623	89	84	3.70	3.29	2.06	1.40	1.94	25.3
	Beet, Mangels. *Beta vulgaris macrorhiza*								
026	—roots, fresh	4-00-637	11	80	3.53	3.12	1.94	1.30	1.84	11.8
	Beet, Sugar. *Beta vulgaris altissima*								
027	—aerial part w crowns, silage	3-00-660	22	51	2.25	1.82	1.00	0.45	1.13	13.4
028	—pulp, dehydrated	4-00-669	91	78	3.44	3.02	1.88	1.24	1.79	9.7
029	—pulp, wet	4-00-671	11	78	3.44	3.02	1.88	1.24	1.79	9.9
030	—pulp w molasses, dehydrated	4-00-672	92	78	3.44	3.02	1.88	1.24	1.79	10.1
	Bermudagrass, Coastal. *Cynodon dactylon*								
031	—hay, sun-cured, early vegetative	1-00-713	94	61	2.69	2.27	1.35	0.77	1.38	16.0
032	—hay, sun-cured, late vegetative	1-20-900	91	54	2.38	1.96	1.11	0.55	1.20	16.5
033	—hay, sun-cured, 15–28 days' growth	1-09-207	92	55	2.43	2.00	1.14	0.58	1.23	16.0
034	—hay, sun-cured, 29–42 days' growth	1-09-209	93	50	2.21	1.78	0.97	0.42	1.11	12.0
035	—hay, sun-cured, 43–56 days' growth	1-09-210	93	43	1.90	1.47	0.72	0.19	0.93	8.0
	Blood								
036	—meal	5-00-380	92	66	2.91	2.49	1.51	0.92	1.50	87.2
	Bluegrass, Canada. *Poa compressa*								
037	—fresh, early vegetative	2-00-763	26	71	3.13	2.71	1.67	1.06	1.62	18.7
038	—hay, sun-cured, late vegetative	1-20-889	97	71	3.13	2.71	1.67	1.06	1.62	—
	Bluegrass, Kentucky. *Poa pratensis*								
039	—fresh, early vegetative	2-00-777	31	72	3.17	2.76	1.69	1.08	1.64	17.4
040	—fresh, early bloom	2-00-779	35	69	3.04	2.62	1.60	1.00	1.57	16.6
041	—fresh, milk stage	2-00-782	42	63	2.78	2.36	1.41	0.83	1.42	11.6
042	—fresh, mature	2-00-784	42	56	2.47	2.04	1.18	0.62	1.25	9.5

Table-Feeds-Ruminant-1 (Continued)

Entry No.	Ether Extract (%)	Total Ash (%)	Crude Fiber (%)	Neutral Detergent Fiber (%)	Acid Detergent Fiber (%)	Cellulose (%)	Lignin (%)	Calcium	Chlorine	Magnesium	Phosphorus	Potassium	Sodium	Sulfur
										Macrominerals (%)				
001	2.5	10.0	29.4	51	41	29	12	1.37	0.48	0.31	0.24	2.48	0.08	0.24
002	3.0	10.6	26.2	45	35	24	11	1.52	0.52	0.32	0.25	2.60	0.11	0.24
003	3.7	11.3	22.5	42	31	22	8	1.74	0.51	0.36	0.30	2.73	0.14	0.29
004	4.4	11.0	19.8	39	28	20	8	1.82	0.56	0.33	0.33	2.58	0.13	0.32
005	4.0	10.2	20.5	38	28	22	5	1.80	0.34	0.26	0.35	2.21	0.22	0.33
006	3.8	9.2	22.0	40	29	23	7	1.54	0.34	0.24	0.29	2.56	0.15	0.31
007	3.0	9.6	23.0	42	31	24	8	1.41	0.38	0.33	0.22	2.52	0.14	0.28
008	2.6	9.1	26.0	46	35	26	9	1.41	0.38	0.31	0.24	1.71	0.12	0.28
009	2.0	8.9	29.0	50	37	28	10	1.25	0.35	0.31	0.22	1.53	0.11	0.27
010	—	—	—	—	—	—	—	—	—	—	—	—	—	—
011	3.6	7.6	11.0	25	20	14	6	0.23	—	0.13	0.11	0.53	0.02	0.11
012	3.0	6.5	15.0	32	28	19	9	0.23	—	0.13	0.11	0.53	0.02	0.11
013	2.2	8.8	19.0	40	36	25	11	0.23	—	0.13	0.11	0.53	0.02	0.11
014	5.2	3.5	20.0	—	32	23	9	0.13	—	0.07	0.12	0.49	0.14	0.02
015	5.1	2.2	17.0	—	26	—	—	0.13	—	0.07	0.11	0.46	0.12	0.02
016	1.6	11.1	30.4	68	38	31	7	0.46	—	0.25	0.22	1.45	—	—
017	1.7	9.6	33.0	73	38	31	6	0.28	—	0.27	0.21	1.80	—	—

continued

018	12.7	4.4	1.3	18	13	12	1	0.14	1.61	0.26	0.26	0.53	1.24	0.02
019	2.1	2.6	5.7	19	7	5	2	0.05	0.18	0.15	0.38	0.47	0.03	0.17
020	2.0	3.1	7.1	21	9	—	—	0.06	0.17	0.14	0.39	0.58	0.02	0.16
021	2.6	3.4	9.6	—	—	—	—	0.34	—	0.14	0.33	0.75	0.02	0.15
022	2.1	7.6	27.5	47	18	14	3	0.23	0.39	0.18	0.26	1.18	0.14	0.17
023	1.4	7.0	16.0	80	59	37	11	0.23	0.67	0.20	0.75	0.23	1.26	0.85
024	1.9	7.1	42.0	—	—	—	—	0.30	0.06	0.23	0.07	2.37	0.14	0.17
025	1.5	5.2	5.0	—	—	—	—	0.18	1.41	0.15	0.59	1.47	0.05	0.26
026	0.7	9.6	7.4	54	33	31	2	0.18	—	0.20	0.22	2.30	0.63	0.20
027	2.8	32.5	13.7	54	33	31	2	1.56	0.04	1.07	0.29	5.74	0.54	0.57
028	0.6	4.4	19.8	44	25	22	3	0.69	—	0.27	0.10	0.20	0.21	0.22
029	0.6	4.7	18.1	66	30	26	4	0.87	—	0.22	0.10	0.19	0.19	0.22
030	0.6	6.1	16.5	70	32	28	4	0.61	—	0.16	0.10	1.78	0.53	0.42
031	2.5	6.1	26.8	74	33	28	4	—	—	—	—	—	—	—
032	1.8	7.7	27.3	76	38	30	6	—	—	—	—	—	—	—
033	2.8	11.0	27.0	78	43	33	7	0.40	—	0.21	0.27	2.20	—	—
034	2.1	10.0	33.0	—	—	—	—	0.32	—	0.16	0.20	1.70	—	—
035	1.4	9.0	36.0	—	—	—	—	0.26	0.30	0.13	0.18	1.30	0.35	0.37
036	1.4	5.8	1.1	—	—	—	—	0.32	—	0.24	0.26	0.10	0.14	0.17
037	3.7	9.1	25.5	—	—	—	—	0.39	—	0.16	0.39	2.04	0.11	0.13
038	—	—	—	55	29	26	3	0.30	—	0.33	0.29	1.59	0.14	0.17
039	3.6	9.4	25.3	65	32	28	4	0.50	—	0.18	0.44	2.27	—	—
040	3.9	7.1	27.4	68	38	33	5	0.46	—	0.17	0.39	2.01	0.13	0.16
041	3.6	7.3	30.3	69	40	34	6	0.33	—	0.15	0.33	1.77	0.13	0.15
042	3.1	6.2	32.2	—	—	—	—	0.26	—	0.16	0.27	1.52	—	—

Table-Feeds-Ruminant-1 (Continued)

| Entry No. | Feed Name Description | International Feed Number[a] | Dry Matter (%) | Values as Determined at Maintenance Intake | | | Production | | | Crude Protein (%) |
| | | | | TDN (%) | DE (Mcal/kg) | ME (Mcal/kg) | Growing Cattle | | Lactating Cows | |
							NEM (Mcal/kg)	NEG (Mcal/kg)	NEL (Mcal/kg)	
	Brewers Grains									
043	—dehydrated	5-02-141	92	66	2.91	2.49	1.51	0.91	1.50	25.4
044	—wet	5-02-142	21	66	2.91	2.49	1.51	0.91	1.50	25.4
	Brewers Dried Yeast—see Yeast, brewers									
	Brome. *Bromus* spp.									
045	—fresh, early vegetative	2-00-892	34	74	3.26	2.85	1.75	1.13	1.69	18.0
046	—fresh, mature	2-00-898	57	57	2.51	2.09	1.21	0.64	1.28	6.4
047	—hay, sun-cured, late vegetative	1-00-887	88	68	3.00	2.58	1.57	0.97	1.55	16.0
048	—hay, sun-cured, late bloom	1-00-888	89	59	2.60	2.18	1.27	0.70	1.33	10.0
	Buckwheat, Common.									
	Fagopyrum sagittatum									
049	—grain	4-00-994	88	72	3.17	2.76	1.69	1.08	1.64	12.5
050	—middlings	5-00-991	89	84	3.70	3.29	2.06	1.40	1.94	33.5
	Canola Meal—see Rape, *Brassica* spp.									
	Carrot. *Daucus* spp.									
051	—roots, fresh	4-01-145	12	84	3.70	3.29	2.06	1.40	1.94	9.9
	Casein									
052	—dehydrated (cattle)	5-01-162	91	89	3.92	3.51	2.20	1.52	2.06	92.7
	Citrus. *Citrus* spp.									
053	—pulp, silage	3-01-234	21	78	3.44	3.02	1.88	1.24	1.79	7.3
054	—pulp wo fines, dehydrated (dried citrus pulp)	4-01-237	91	77	3.40	2.98	1.86	1.22	1.77	6.7
	Clover, Alsike. *Trifolium hybridum*									
055	—fresh, early vegetative	2-01-314	19	66	2.91	2.49	1.51	0.92	1.50	24.1
056	—hay, sun-cured	1-01-313	88	58	2.56	2.13	1.25	0.68	1.30	14.9
	Clover, Crimson. *Trifolium incarnatum*									
057	—fresh, early vegetative	2-20-890	18	63	2.78	2.36	1.41	0.83	1.42	17.0
058	—hay, sun-cured	1-01-328	87	57	2.51	2.09	1.21	0.64	1.28	16.0

continued

	Clover, Ladino. Trifolium repens									
059	—fresh, early vegetative	2-01-380	19	68	3.00	2.58	1.57	0.97	1.55	24.7
060	—hay, sun-cured	1-01-378	90	65	2.87	2.45	1.47	0.88	1.47	22.0
	Clover, Red. Trifolium pratense									
061	—fresh, early bloom	2-01-428	20	69	3.04	2.62	1.60	1.00	1.57	19.4
062	—fresh, full bloom	2-01-429	26	64	2.82	2.40	1.44	0.85	1.45	14.6
063	—hay, sun-cured	1-01-415	89	55	2.43	2.00	1.14	0.58	1.23	16.0
	Coconut. Cocos nucifera									
064	—meats, meal mech-extd (copra meal)	5-01-572	92	82	3.62	3.20	2.01	1.35	1.89	22.4
065	—meats, meal solv-extd (copra meal)	5-01-573	91	75	3.31	2.89	1.79	1.16	1.72	23.4
	Corn, Dent yellow. Zea mays indentata									
066	—aerial part wo ears, wo husks, sun-cured (stover, straw)	1-28-233	85	50	2.21	1.78	0.97	0.42	1.11	5.9
067	—cobs, ground	1-28-234	90	50	2.21	1.78	0.97	0.42	1.11	3.2
068	—distillers grains, dehydrated	5-28-235	94	86	3.79	3.38	2.12	1.45	1.99	23.0
069	—distillers grains w solubles, dehydrated	5-28-236	92	88	3.88	3.47	2.18	1.50	2.04	25.0
070	—distillers solubles, dehydrated	5-28-237	93	88	3.88	3.47	2.18	1.50	2.04	29.7
071	—ears, ground (corn and cob meal)	4-28-238	87	83	3.66	3.25	2.03	1.37	1.91	9.0
072	—ears w husks, silage	3-28-239	44	74	3.26	2.85	1.75	1.13	1.69	8.9
073	—germ, meal wet-milled, solv-extd	5-28-240	91	74	3.26	2.85	1.75	1.13	1.69	22.3
074	—gluten, meal	5-28-241	91	86	3.79	3.38	2.12	1.45	1.99	46.8
075	—gluten, meal, 60% protein	5-28-242	90	89	3.92	3.51	2.20	1.52	2.06	67.2
076	—gluten w bran (corn gluten feed)	5-28-243	90	83	3.66	3.25	2.03	1.37	1.91	25.6
077	—grain, cracked	4-20-698	89	80	3.53	3.12	1.94	1.30	1.84	10.0
078	—grain, flaked	4-28-244	89	88	3.88	3.47	2.18	1.50	2.04	10.0
079	—grain, ground	4-26-023	88	85	3.75	3.34	2.10	1.43	1.96	10.0
080	—grain, high-moisture	4-20-770	77	88	3.88	3.47	2.18	1.50	2.04	10.0
081	—grain, opaque 2 (high-lysine)	4-28-253	90	89	3.92	3.51	2.20	1.52	2.06	11.3
082	—grits by-product (hominy feed)	4-03-001	90	87	3.84	3.42	2.16	1.48	2.01	11.5

Table-Feeds-Ruminant-1 (Continued)

Entry No.	Ether Extract (%)	Total Ash (%)	Crude Fiber (%)	Neutral Detergent Fiber (%)	Acid Detergent Fiber (%)	Cellulose (%)	Lignin (%)	Calcium	Chlorine	Macrominerals (%) Magnesium	Phosphorus	Potassium	Sodium	Sulfur
043	6.5	4.8	14.9	46	24	18	6	0.33	0.17	0.16	0.55	0.09	0.23	0.32
044	6.5	4.8	14.9	42	23	18	5	0.33	0.17	0.16	0.55	0.09	0.23	0.32
045	3.7	10.7	24.0	56	31	27	3	0.50	—	0.18	0.30	2.30	0.02	0.20
046	2.2	—	38.0	72	44	35	9	0.20	—	0.18	0.26	1.25	0.02	0.20
047	2.6	9.4	30.0	65	35	32	4	0.32	—	0.18	0.37	2.32	0.02	0.20
048	2.3	8.4	37.0	68	43	36	8	0.30	—	0.18	0.35	2.32	0.02	0.20
049	2.8	2.3	11.8	—	—	—	—	0.11	0.05	0.12	0.37	0.51	0.06	0.16
050	8.2	5.5	8.3	—	—	—	—	—	—	—	—	—	—	—
051	1.4	8.2	9.7	9	8	7	0	0.40	0.50	0.20	0.35	2.80	1.04	0.17
052	0.7	2.4	0.2	0	0	0	0	0.67	—	0.01	0.90	0.01	0.01	—
053	9.7	5.5	15.6	—	—	—	—	2.04	—	0.16	0.15	0.62	0.09	0.02
054	3.7	6.6	12.7	23	22	18	3	1.84	—	0.17	0.12	0.79	0.09	0.08
055	3.2	12.8	17.5	—	—	—	—	1.29	0.78	0.41	0.26	2.46	0.46	0.19
056	3.0	8.7	30.1	—	—	—	—	1.29	0.78	0.41	0.26	2.46	0.46	0.19
057	—	—	28.0	—	—	—	—	1.40	0.63	0.28	0.22	2.40	0.39	0.28
058	2.4	11.0	30.1	—	—	—	—	1.40	0.63	0.28	0.22	2.40	0.39	0.28
059	2.5	13.5	14.0	—	—	—	—	1.35	0.30	0.48	0.31	2.62	0.13	0.21
060	2.7	10.1	21.2	36	32	25	7	1.35	0.30	0.48	0.31	2.62	0.13	0.21

continued

061	5.0	10.2	23.2	40	31	27	4	2.26	0.32	0.51	0.38	2.49	0.20	0.17
062	2.9	7.8	26.1	43	35	28	7	1.31	0.32	0.51	0.27	1.96	0.20	0.17
063	2.8	8.5	28.8	46	36	28	8	1.53	0.32	0.43	0.25	1.62	0.19	0.17
064	6.9	7.3	12.8	—	19	—	—	0.22	0.03	0.33	0.66	1.62	0.04	0.36
065	2.7	7.4	16.0	—	24	—	—	0.19	0.03	0.36	0.66	1.63	0.04	0.37
066	1.3	7.2	34.4	67	39	25	11	0.57	—	0.40	0.10	1.45	0.07	0.17
067	0.7	1.7	36.2	89	35	28	7	0.12	—	0.07	0.04	0.87	0.47	0.47
068	9.8	2.4	12.1	43	17	12	5	0.11	0.08	0.07	0.43	0.18	0.10	0.46
069	10.3	4.8	9.9	44	18	14	4	0.15	0.18	0.18	0.71	0.44	0.57	0.33
070	9.2	7.8	5.0	23	7	6	1	0.35	0.28	0.65	1.37	1.80	0.25	0.40
071	3.7	1.9	9.4	28	11	9	2	0.07	0.05	0.14	0.27	0.53	0.02	0.16
072	3.8	2.8	11.6	—	14	—	—	0.10	—	0.12	0.29	0.49	0.01	0.13
073	4.1	4.2	13.1	—	—	—	—	0.04	0.04	0.34	0.47	0.31	0.08	0.33
074	2.4	3.4	4.8	37	9	8	1	0.16	0.07	0.06	0.50	0.03	0.10	0.39
075	2.4	1.8	2.2	14	5	4	1	0.08	0.10	0.09	0.54	0.21	0.06	0.72
076	2.4	7.5	9.7	45	12	—	—	0.36	0.25	0.36	0.82	0.64	0.15	0.23
077	4.3	1.6	2.6	9	3	2	1	0.03	0.05	0.14	0.29	0.37	0.03	0.12
078	4.3	1.6	2.6	9	3	2	1	0.03	0.05	0.14	0.29	0.37	0.03	0.12
079	4.3	1.6	2.6	9	3	2	1	0.03	0.05	0.14	0.29	0.37	0.03	0.12
080	4.3	1.6	2.6	9	3	2	1	0.02	0.05	0.14	0.32	0.35	0.01	0.14
081	4.8	1.8	3.7	—	—	—	—	0.03	—	0.14	0.22	0.39	—	0.11
082	7.7	3.1	6.7	55	13	10	2	0.05	0.06	0.26	0.57	0.65	0.09	0.03

Table-Feeds-Ruminant-1 (Continued)

Entry No.	Feed Name Description	International Feed Number[a]	Dry Matter (%)	Values as Determined at Maintenance Intake TDN (%)	DE (Mcal/kg)	ME (Mcal/kg)	Production Growing Cattle NEM (Mcal/kg)	NEG (Mcal/kg)	Lactating Cows NEL (Mcal/kg)	Crude Protein (%)
083	—silage, aerial part wo ears, wo husks (stalkage, stover)	3-28-251	31	55	2.43	2.00	1.14	0.58	1.23	5.9
084	—silage, few ears	3-28-245	29	62	2.73	2.31	1.38	0.80	1.40	8.4
085	—silage, well-eared	3-28-250	33	70	3.09	2.67	1.63	1.03	1.60	8.1
	Corn, Sweet *Zea mays saccharata*									
086	—process residue, silage	3-07-955	32	67	2.95	2.53	1.54	0.91	1.52	7.7
	Cotton. *Gossypium* spp.									
087	—hulls	1-01-599	91	45	1.98	1.55	0.78	0.25	0.98	4.1
088	—seeds, w lint	5-01-614	92	96	4.23	3.83	2.41	1.69	2.23	23.0
089	—seeds, wo lint	5-01-	90	96	4.23	3.82	2.41	1.69	2.23	25.0
090	—seeds, meal mech-extd, 36% protein	5-01-625	92	73	3.22	2.80	1.73	1.11	1.67	41.9
091	—seeds, meal mech-extd, 41% protein	5-01-617	93	78	3.44	3.02	1.88	1.24	1.79	44.3
092	—seeds, meal prepressed, solv-extd, 41% protein	5-07-872	91	76	3.35	2.93	1.82	1.19	1.74	45.6
093	—seeds, meal prepressed, solv-extd, 44% protein	5-07-873	91	75	3.31	2.89	1.79	1.16	1.72	48.9
094	—seeds, meal solv-extd, low[a] gossypol	5-01-633	93	71	3.13	2.71	1.67	1.06	1.62	44.8
095	—seeds wo hulls, meal prepressed, solv-extd, 50% protein	5-07-874	93	75	3.31	2.89	1.79	1.16	1.72	54.0
	Cowpea, Common. *Vigna sinensis*									
096	—hay, sun-cured	1-01-645	90	59	2.60	2.18	1.27	0.70	1.33	19.4
	Distillers Grains—see Corn; Sorghum									
	Fats and Oils (not exceeding 3% of diet)									
097	—fat, animal, hydrolyzed	4-00-376	99	177	7.30	7.30	5.84	5.84	5.84	—
098	—fat, swine (lard)	4-04-790	99	177	7.30	7.30	5.84	5.84	5.84	—
099	—oil, soybean	4-07-983	100	177	7.30	7.30	5.84	5.84	5.84	—

continued

		IFN								
	Fesgue. *Festuca* spp.									
100	—hay, sun-cured, early vegetative (South)	1-06-132	91	61	2.69	2.27	1.35	0.77	1.38	12.4
101	—hay, sun-cured, late vegetative (South)	1-13-582	91	58	2.56	2.13	1.25	0.68	1.30	10.5
102	—hay, sun-cured, early bloom (South)	1-01-871	92	48	2.12	1.69	0.90	0.36	1.06	9.5
	Fescue, Kentucky 31. *Festuca arundinacea*									
103	—fresh, vegetative	2-01-902	29	67	2.91	2.49	1.51	0.92	1.50	14.5
104	—hay, sun-cured, early bloom	1-09-186	91	64	2.82	2.40	1.44	0.85	1.45	20.2
105	—hay, sun-cured, midbloom	1-09-187	92	60	2.65	2.22	1.31	0.74	1.35	16.4
106	—hay, sun-cured, full bloom	1-09-188	92	58	2.56	2.13	1.25	0.68	1.30	12.1
107	—hay, sun-cured, mature	1-09-189	90	56	2.47	2.04	1.18	0.62	1.25	9.2
	Fish, Anchovy. *Engraulis ringen*									
108	—meal mech-extd	5-01-985	92	79	3.48	3.07	1.91	1.27	1.82	71.2
	Fish, Menhaden. *Brevoortia tyrannus*									
109	—meal mech-extd	5-02-009	92	73	3.22	2.80	1.73	1.11	1.67	66.7
	Fish, White. Family *Gadidae*; family *Lophiidae*									
110	—meal mech-extd	5-02-025	91	77	3.40	2.98	1.86	1.22	1.77	68.2
	Flax. *Linum usitatissimum*									
111	—seed screenings	4-02-056	91	64	2.82	2.40	1.44	0.85	1.45	18.2
112	—seeds, meal mech-extd (linseed meal)	5-02-045	91	82	3.62	3.20	2.01	1.35	1.89	37.9
113	—seeds, meal solv-extd (linseed meal)	5-02-048	90	78	3.44	3.02	1.88	1.24	1.79	38.3
	Grain Screenings—see Barley, Wheat									
	Grape. *Vitis* spp.									
114	—mare, dehydrated (pomace)	1-02-208	91	33	1.46	1.02	0.34	0.0	0.69	13.0
	Hemicellulose Extract									
115	—molasses, wood	4-08-030	76	60	2.65	2.22	1.31	0.74	1.35	0.7

Table-Feeds-Ruminant-1 (Continued)

Entry No.	Ether Extract (%)	Total Ash (%)	Crude Fiber (%)	Neutral Detergent Fiber (%)	Acid Detergent Fiber (%)	Cellulose (%)	Lignin (%)	Calcium	Chlorine	Macrominerals (%) Magnesium	Phosphorus	Potassium	Sodium	Sulfur
083	2.1	11.6	31.3	67	43	25	8	0.38	—	0.31	0.31	1.54	0.03	0.11
084	3.0	7.2	32.3	53	30	23	5	0.34	—	0.23	0.19	1.41	—	0.08
085	3.1	4.5	23.7	51	28	24	4	0.23	—	0.19	0.22	0.96	0.01	0.15
086	5.2	4.9	27.0	—	34	—	—	0.30	—	0.24	0.90	1.15	0.03	0.11
087	1.7	2.8	47.8	90	73	59	24	0.15	0.02	0.14	0.09	0.87	0.02	0.09
088	20.0	4.8	24.0	44	34	24	10	0.21	—	0.46	0.64	1.00	0.01	0.26
089	23.8	4.5	17.2	37	26	12	14	0.12	—	0.41	0.54	1.18	0.01	—
090	4.6	7.3	15.5	—	—	—	—	0.20	—	0.58	1.04	1.46	0.05	0.28
091	5.0	6.6	12.8	28	20	13	6	0.21	0.05	0.58	1.16	1.45	0.05	0.43
092	1.3	7.0	14.1	26	19	12	6	0.22	0.04	0.55	1.21	1.39	0.04	0.34
093	1.7	6.7	12.1	28	21	13	7	0.17	0.04	0.55	1.00	1.39	0.04	0.34
094	1.3	6.3	13.7	—	—	—	—	—	—	—	—	—	—	—
095	1.4	7.1	8.8	—	—	—	—	0.19	0.05	0.50	1.24	1.56	0.06	0.56
096	3.1	11.3	26.7	—	—	—	—	1.40	0.17	0.45	0.35	2.26	0.27	0.35
097	99.5	—	—	—	—	—	—	—	—	—	—	—	—	—
098	100.0	—	—	—	—	—	—	—	—	—	—	—	—	—
099	99.9	—	—	—	—	—	—	—	—	—	—	—	—	—

100	3.4	12.0	26.0	57	32	28	3	0.51	—	0.22	0.36	2.30	—	—
101	3.0	10.5	33.0	64	36	31	4	0.40	—	0.20	0.34	2.00	—	—
102	2.0	10.0	37.0	72	39	33	5	0.30	—	0.19	0.26	1.70	—	—
103	5.5	9.9	24.6	—	—	—	—	0.51	—	—	0.37	—	—	—
104	6.6	9.8	23.6	59	32	29	3	—	—	—	—	—	—	—
105	6.1	9.1	25.5	63	35	30	4	—	—	—	—	—	—	—
106	5.3	7.9	27.4	67	39	32	5	—	—	—	—	—	—	—
107	4.3	6.4	32.6	70	42	34	7	—	—	—	—	—	—	—
108	4.5	16.1	1.1	—	—	—	—	4.08	1.08	0.27	2.70	0.78	0.95	0.84
109	10.5	20.8	1.0	—	—	—	—	5.65	0.60	0.16	3.16	0.76	0.43	0.49
110	5.1	25.4	0.8	—	—	—	—	8.02	0.55	0.20	4.17	0.91	0.85	0.53
111	10.2	6.8	13.2	25	17	10	7	0.37	—	0.43	0.47	0.84	—	0.25
112	6.0	6.3	9.6	25	19	13	6	0.45	0.04	0.64	0.96	1.34	0.12	0.41
113	1.5	6.5	10.1	55	54	—	35	0.43	0.04	0.66	0.89	1.53	0.15	0.43
114	7.9	10.3	31.9	—	—	—	—	0.61	0.01	—	0.06	0.62	0.09	—
115	0.4	4.1	1.0	—	2	—	—	1.03	—	—	0.09	—	—	—

Table-Feeds-Ruminant-1 (Continued)

Entry No.	Feed Name Description	International Feed Number[a]	Dry Matter (%)	Values as Determined at Maintenance Intake			Production			Crude Protein (%)
							Growing Cattle		Lactating Cows	
				TDN (%)	DE (Mcal/kg)	ME (Mcal/kg)	NEM (Mcal/kg)	NEG (Mcal/kg)	NEL (Mcal/kg)	
	Hominy Feed—see Corn, grits by-product									
	Johnsongrass—see Sorghum; Johnsongrass									
	Kentucky Bluegrass—see Bluegrass, Kentucky									
	Lespedeza. Common, *Lespedeza striata*, or Korean, *Lespedeza stipulacea*									
116	—hay, sun-cured, late vegetative	1-26-024	92	59	2.60	2.18	1.27	0.70	1.33	17.8
117	—hay, sun-cured, early bloom	1-26-025	93	55	2.43	2.00	1.14	0.58	1.23	15.5
118	—hay, sun-cured, midbloom	1-26-026	93	50	2.21	1.78	0.97	0.42	1.11	14.5
119	—hay, sun-cured, full bloom	1-26-027	93	47	2.07	1.64	0.86	0.32	1.03	13.4
	Meat									
120	—meal rendered	5-00-385	94	71	3.13	2.71	1.67	1.06	1.62	54.8
121	—w blood, meal rendered (tankage)	5-00-386	92	72	3.17	2.76	1.69	1.08	1.64	64.5
122	—w blood and bone, meal rendered (tankage)	5-00-387	93	68	3.00	2.58	1.57	0.97	1.55	50.2
123	—w bone, meal rendered	5-00-388	93	71	3.13	2.71	1.67	1.06	1.62	54.1
	Milk									
124	—dehydrated (cattle)	5-01-167	96	119	5.25	4.85	3.06	2.22	2.80	26.5
125	—fresh (cattle)	5-01-168	12	129	5.69	5.29	3.34	2.16	3.04	26.7
126	—skimmed dehydrated (cattle)	5-01-175	94	85	3.75	3.34	2.10	1.43	1.96	35.8
127	—skimmed fresh (cattle)	5-01-170	10	92	4.06	3.65	2.30	1.60	2.13	31.2
	Millet, Foxtail. *Setaria italica*									
128	—fresh	2-03-101	28	63	2.78	2.36	1.41	0.83	1.42	9.5
129	—grain	4-03-102	89	85	3.75	3.34	2.10	1.43	1.96	13.5
130	—hay, sun-cured	1-03-099	87	59	2.60	2.18	1.27	0.70	1.33	8.6

continued

#	Feedstuff	Ref. No.								
131	Millet, Proso. *Panicum miliaceum* —grain	4-03-120	90	84	3.70	3.29	2.06	1.40	1.94	12.9
132	Molasses and Syrup —beet, sugar, molasses, more than 48% invert sugar, more than 79.5 degrees brix	4-00-668	78	75	3.31	2.89	1.79	1.16	1.72	8.5
133	—citrus, syrup (citrus molasses)	4-01-241	68	75	3.31	2.89	1.79	1.16	1.72	8.2
134	—sugarcane, molasses, dehydrated	4-04-695	94	70	3.09	2.67	1.63	1.03	1.60	10.3
135	—sugarcane, molasses, more than 46% invert sugar, more than 79.5 degrees brix (Black strap)	4-04-696	75	72	3.17	2.76	1.69	1.08	1.64	5.8
136	Napiergrass. *Pennisetum purpureum* —fresh, late vegetative	2-03-158	20	55	2.43	2.00	1.14	0.58	1.23	8.7
137	—fresh, late bloom	2-03-162	23	53	2.34	1.91	1.08	0.52	1.18	7.8
138	Oats. *Avena sativa* —cereal by-product, less than 4% fiber (feeding oat meal, oat middlings)	4-03-303	91	95	4.19	3.78	2.39	1.67	2.21	16.4
139	—grain	4-03-309	89	77	3.40	2.98	1.86	1.22	1.77	13.3
140	—grain, light, less than 34.7 kg/hl (less than 27 lb/bu)	4-03-318	91	66	2.91	2.49	1.51	0.92	1.50	13.1
141	—grain, Pacific Coast	4-07-999	91	78	3.44	3.02	1.88	1.24	1.79	10.0
142	—groats	4-03-331	90	94	4.14	3.74	2.35	1.64	2.18	17.7
143	—hay, boot stage	1-03-	90	72	3.17	2.76	1.69	1.08	1.64	17.5
144	—hay, head emerging	1-03-	90	60	2.65	2.73	1.31	0.74	1.35	14.0
145	—hay, dough stage	1-03-	90	53	2.34	1.91	1.08	0.52	1.18	11.5
146	—hulls	1-03-281	92	35	1.54	1.11	0.41	0.00	0.74	3.9
147	—straw	1-03-283	92	50	2.21	1.78	0.97	0.42	1.11	4.4
148	—silage, wilted, 25–40% dry matter (see similar maturity description of hays)		—	—	—	—	—	—	—	—
149	Orange. *Citrus sinensis* —pulp wo fines, dehydrated (orange pulp)	4-01-254	88	78	3.44	3.02	1.88	1.24	1.79	8.5
150	—fresh, whole	4-01-	13	78	3.44	3.02	1.88	1.24	1.79	7.5

Table-Feeds-Ruminant-1 (Continued)

Entry No.	Ether Extract (%)	Total Ash (%)	Crude Fiber (%)	Neutral Detergent Fiber (%)	Acid Detergent Fiber (%)	Cellulose (%)	Lignin (%)	Calcium	Chlorine	Macrominerals (%) Magnesium	Phosphorus	Potassium	Sodium	Sulfur
116	—	—	24.0	—	—	—	—	1.12	—	—	0.28	1.28	—	—
117	—	—	28.0	—	—	—	—	1.23	—	0.26	0.25	1.00	—	—
118	—	—	30.0	—	—	—	—	—	—	0.25	—	—	—	—
119	—	—	32.0	—	—	—	—	—	—	0.24	—	—	—	—
120	9.7	28.8	2.8	—	—	—	—	9.44	1.27	0.29	4.74	0.61	1.37	0.50
121	9.7	23.4	2.2	—	—	—	—	6.37	1.88	0.39	3.33	0.60	1.81	0.76
122	13.7	30.4	2.4	—	—	—	—	12.01	—	—	5.82	—	—	0.28
123	10.4	31.5	2.4	—	—	—	—	11.06	0.80	1.09	5.48	1.43	0.77	0.27
124	27.8	5.7	0.2	—	—	—	—	0.95	0.92	0.10	0.74	1.08	0.38	0.32
125	29.5	6.3	—	—	—	—	—	0.95	0.92	0.10	0.76	1.12	0.38	0.32
126	0.9	8.4	0.2	—	—	—	—	1.36	0.96	0.13	1.09	1.70	0.49	0.34
127	1.0	6.9	—	—	—	—	—	1.31	0.96	0.12	1.04	1.90	0.47	0.32
128	3.1	8.7	31.6	—	—	—	—	0.32	—	—	0.19	1.94	—	—
129	4.6	4.0	9.3	—	—	—	—	—	—	—	0.22	0.35	—	—
130	2.9	8.6	29.6	—	—	—	—	0.33	0.13	0.23	0.19	1.94	0.10	0.16

continued

131	3.9	2.9	6.8	—	17	—	4	0.03	—	0.18	0.34	0.48	—	—
132	0.2	11.3	—	—	—	—	—	0.17	1.64	0.29	0.03	6.07	1.48	0.60
133	0.3	7.9	6.7	—	—	—	—	1.72	0.11	0.21	0.13	0.14	0.41	0.23
134	0.9	13.3	—	—	—	—	—	1.10	—	0.47	0.15	3.60	0.20	0.46
135	0.1	13.1	—	—	—	—	—	1.00	3.10	0.43	0.11	3.84	0.22	0.47
136	3.0	8.6	33.0	70	45	33	10	0.60	—	0.26	0.41	1.31	0.01	0.10
137	1.4	5.3	39.0	75	47	35	14	0.35	—	0.26	0.30	1.31	0.01	0.10
138	7.0	2.5	3.9	32	—	—	—	0.08	0.06	0.16	0.49	0.55	0.10	0.24
139	5.4	3.4	12.1	—	16	11	3	0.07	0.11	0.14	0.38	0.44	0.08	0.23
140	4.9	4.6	15.9	—	—	—	—	—	—	—	—	—	—	—
141	5.5	4.2	12.3	—	—	—	—	0.11	0.13	0.19	0.34	0.42	0.07	0.22
142	6.9	2.4	2.8	—	—	—	4	0.08	0.09	0.13	0.48	0.39	0.06	0.22
143	2.6	6.5	29.0	58	35	31	6	—	—	—	—	—	—	—
144	3.3	8.3	32.0	62	39	33	9	—	—	—	—	—	—	—
145	4.2	6.9	27.0	56	34	24	8	0.15	0.08	0.09	0.15	0.62	0.04	0.15
146	1.8	6.6	33.4	78	42	30	14	0.24	0.78	0.18	0.06	2.57	0.42	0.23
147	2.2	7.8	40.5	70	47	40	—	—	—	—	—	—	—	—
148	—	—	—	—	—	—	—	—	—	—	—	—	—	—
149	1.7	3.8	9.6	21	16	—	—	0.71	—	0.16	0.11	0.62	0.09	0.02
150	1.9	4.4	11.3	—	14	—	—	—	—	—	—	—	—	—

Table-Feeds-Ruminant-1 (Continued)

Entry No.	Feed Name Description	International Feed Number[a]	Dry Matter (%)	Values as Determined at Maintenance Intake TDN (%)	DE (Mcal/kg)	ME (Mcal/kg)	Production Growing Cattle NEM (Mcal/kg)	NEG (Mcal/kg)	Lactating Cows NEL (Mcal/kg)	Crude Protein (%)
	Orchardgrass. *Dactylis glomerata*									
151	—fresh, early vegetative	2-03-439	23	72	3.17	2.76	1.69	1.08	1.64	18.4
152	—hay, sun-cured, early bloom	1-03-425	89	65	2.87	2.45	1.47	0.88	1.47	15.0
153	—hay, sun-cured, late bloom	1-03-428	91	54	2.38	1.96	1.11	0.55	1.20	8.4
	Pangolagrass. *Digitaria decumbens*									
154	—hay, sun-cured, 15–28 days' growth	1-10-638	91	51	2.25	1.82	1.01	0.46	1.13	11.5
155	—hay, sun-cured, 29–42 days' growth	1-26-214	91	45	1.98	1.55	0.78	0.25	0.98	7.1
156	—hay, sun-cured, 43–56 days' growth	1-29-573	91	40	1.76	1.33	0.60	0.07	0.86	5.5
	Paper									
157	—corrugated	1-28-257	93	69	3.04	2.62	1.60	1.00	1.57	—
	Pea. *Pisum spp.*									
158	—seeds	5-03-600	89	87	3.84	3.42	2.16	1.48	2.01	25.3
159	—vines wo seeds, silage	3-03-596	25	57	2.51	2.09	1.21	0.64	1.28	13.1
	Peanut. *Arachis hypogaea*									
160	—hay, sun-cured	1 03 619	91	55	2.43	2.00	1.14	0.58	1.23	10.8
161	—hulls	1-08-028	91	22	0.97	0.53	—	—	0.42	7.8
162	—kernels, meal mech-extd (peanut meal)	5-03-649	93	83	3.66	3.25	2.03	1.37	1.91	52.0
163	—kernels, meal solv-extd (peanut meal)	5-03-650	92	77	3.40	2.98	1.86	1.22	1.77	52.3
	Pearlmillet. *Pennisetum glaucum*									
164	—fresh	2-03-115	21	61	2.69	2.27	1.35	0.77	1.38	8.5
165	—silage	3-20-903	30	59	2.60	2.18	1.27	0.70	1.33	9.2
	Pineapple. *Ananas comosus*									
166	—process residue, dehydrated (pineapple bran)	4-03-722	87	68	3.00	2.58	1.57	0.97	1.55	4.6
	Potato. *Solanum tuberosum*									
167	—process residue, dehydrated	4-03-775	89	90	3.97	3.56	2.24	1.55	2.09	8.4

continued

No.	Feed	Ref. No.								
168	—tubers, dehydrated	4-07-850	91	81	3.57	3.16	1.97	1.32	1.87	8.9
169	—tubers, fresh	4-03-787	23	81	3.57	3.16	1.97	1.32	1.87	9.5
170	—tubers, silage	4-03-768	25	82	3.62	3.20	2.01	1.35	1.89	7.6
	Rape. *Brassica spp.*									
171	—fresh, early vegetative	2-03-865	18	81	3.57	3.16	1.97	1.32	1.87	16.4
172	—seeds, meal mech-extd	5-03-870	92	76	3.35	2.93	1.82	1.19	1.74	38.7
173	—seeds, meal solv-extd	5-03-871	91	69	3.04	2.62	1.60	1.00	1.57	40.6
	Rape, Summer. *Brassica napus*									
174	—seeds, meal mech-extd (canola meal)	5-08-136	94	74	3.26	2.85	1.75	1.13	1.69	37.4
175	—seeds, meal prepressed, solv-extd (canola meal)	5-08-135	92	75	3.31	2.89	1.79	1.16	1.72	44.0
	Redtop. *Agrostis alba*									
176	—fresh	2-03-897	29	63	2.78	2.36	1.41	0.83	1.42	11.6
177	—hay, sun-cured, midbloom	1-03-886	94	57	2.51	2.09	1.21	0.64	1.28	11.7
	Rice. *Oryza sativa*									
178	—bran w germ (rice bran)	4-03-928	91	70	3.09	2.67	1.63	1.03	1.60	14.1
179	—grain, ground (ground rough rice, ground paddyrice)	4-03-938	89	79	3.48	3.07	1.91	1.27	1.82	8.9
180	—grain, polished and broken (brewers rice)	4-03-932	89	89	3.92	3.51	2.20	1.52	2.06	8.6
181	—groats, polished (rice, polished)	4-03-942	89	88	3.88	3.47	2.18	1.50	2.04	8.2
182	—hulls	1-08-075	92	12	0.53	0.08	0.00	0.00	0.17	3.3
	Rye. *Secale cereale*									
183	—distillers grains, dehydrated	5-04-023	92	61	2.69	2.27	1.35	0.77	1.38	23.5
184	—grain	4-04-047	88	84	3.70	3.29	2.06	1.40	1.94	13.8
185	—silage	3-04-020	32	53	2.34	1.91	1.08	0.52	1.18	12.8
	Ryegrass, Italian. *Lolium multiflorum*									
186	—hay, sun-cured, early vegetative	1-04-064	89	68	3.00	2.58	1.57	0.97	1.55	15.2
187	—hay, sun-cured, late vegetative	1-04-065	86	62	2.73	2.31	1.38	0.80	1.40	10.3
188	—hay, sun-cured, early bloom	1-04-066	83	54	2.38	1.96	1.11	0.55	1.20	5.5
	Ryegrass, Perennial. *Lolium perenne*									
189	—hay, sun-cured	1-04-077	86	64	2.82	2.40	1.44	0.85	1.45	8.6

Table-Feeds-Ruminant-1 (Continued)

Entry No.	Ether Extract (%)	Total Ash (%)	Crude Fiber (%)	Neutral Detergent Fiber (%)	Acid Detergent Fiber (%)	Cellulose (%)	Lignin (%)	Macrominerals (%)						
								Calcium	Chlorine	Magnesium	Phosphorus	Potassium	Sodium	Sulfur
151	4.9	11.3	24.7	55	31	25	3	0.58	0.08	0.31	0.54	3.58	0.04	0.21
152	2.8	8.7	31.0	61	34	29	5	0.27	—	0.11	0.34	2.91	0.01	—
153	3.4	10.1	37.1	72	45	39	9	0.26	—	0.11	0.30	2.67	0.01	—
154	2.2	8.5	34.0	70	41	33	6	0.58	—	0.20	0.21	1.70	—	—
155	2.0	8.0	36.0	73	43	35	6	0.46	—	0.15	0.23	1.40	—	—
156	2.0	7.6	38.0	77	46	37	7	0.38	—	0.14	0.18	1.10	—	—
157	—	—	—	92	82	70	12	—	—	—	—	—	—	—
158	1.4	3.3	6.9	—	—	—	—	0.15	0.06	0.14	0.44	1.13	0.05	—
159	3.3	9.0	29.8	59	49	34	9	1.31	—	0.39	0.24	1.40	0.01	0.25
160	3.4	8.6	33.2	—	—	—	—	1.23	—	0.49	0.15	1.38	—	0.23
161	2.0	4.2	62.9	74	65	40	23	0.26	—	0.17	0.07	0.95	0.13	0.10
162	6.3	5.5	7.5	14	6	5	—	0.20	0.03	0.31	0.61	1.25	0.23	0.29
163	1.4	6.3	10.8	—	—	—	—	0.29	0.03	0.17	0.68	1.23	0.08	0.33
164	2.2	10.0	31.5	—	—	—	—	—	—	—	—	—	—	—
165	—	—	38.0	—	—	—	—	—	—	—	—	—	—	—
166	1.5	3.5	20.9	73	37	—	7	0.23	—	—	0.13	—	—	—

continued

167	0.4	3.4	7.3	—	—	—	—	0.16	—	—	0.25	—	—	—
168	0.5	8.7	2.3	—	—	—	—	0.08	0.40	0.12	0.22	2.15	0.01	0.09
169	0.4	4.8	2.4	—	—	—	—	0.04	0.28	0.14	0.24	2.17	0.09	0.09
170	0.4	5.5	4.0	—	—	—	—	0.04	—	0.14	0.23	2.13	0.09	0.23
171	4.0	11.4	13.0	—	—	—	—	—	—	—	—	—	—	—
172	7.9	7.5	13.1	—	—	—	—	0.72	—	0.54	1.14	0.90	0.50	—
173	1.8	7.5	13.2	34	16	18	8	0.67	0.11	0.60	1.04	1.36	0.10	1.25
174	7.4	7.2	16.5	36	18	18	7	0.76	—	0.59	1.15	—	—	—
175	1.2	7.8	10.1	64	—	—	8	0.73	—	0.58	1.13	—	—	—
176	3.9	8.1	26.7	—	18	11	—	0.46	0.09	0.23	0.29	2.35	0.05	0.19
177	2.6	6.5	30.7	33	—	—	—	0.63	—	—	0.35	1.69	—	—
178	15.1	12.8	12.8	—	18	18	—	0.08	0.08	1.04	1.70	1.92	0.04	0.20
179	1.9	5.3	10.0	16	—	—	—	0.07	0.09	0.15	0.32	0.36	0.06	0.05
180	0.8	0.8	0.7	—	1	1	—	0.03	0.09	0.12	0.30	0.15	0.08	0.05
181	0.5	0.6	0.4	—	—	—	—	0.03	0.04	0.02	0.13	0.12	0.02	0.09
182	0.8	20.6	42.9	82	72	33	16	0.10	0.08	0.83	0.08	0.57	0.12	0.09
183	7.8	2.5	13.4	—	—	—	—	0.16	0.05	0.18	0.52	0.08	0.18	0.48
184	1.7	1.9	2.5	—	—	—	—	0.07	0.03	0.14	0.37	0.52	0.03	0.17
185	3.3	7.9	34.0	—	—	—	—	0.39	—	—	0.32	—	—	—
186	3.2	13.0	19.7	61	38	33	3	—	—	—	—	—	—	—
187	2.4	11.0	23.8	64	42	36	6	0.62	—	—	0.34	1.56	—	—
188	0.9	8.4	36.3	69	45	36	9	—	—	—	—	—	—	—
189	2.2	11.5	24.6	41	30	28	2	0.65	—	—	0.32	1.67	—	—

Table-Feeds-Ruminant-1 (Continued)

Entry No.	Feed Name Description	International Feed Number[a]	Dry Matter (%)	Values as Determined at Maintenance Intake TDN (%)	DE (Mcal/kg)	ME (Mcal/kg)	Production Growing Cattle NEM (Mcal/kg)	NEG (Mcal/kg)	Lactating Cows NEL (Mcal/kg)	Crude Protein (%)
	Safflower. *Carthamus tinctorius*									
190	—seeds	4-07-958	94	89	3.92	3.51	2.20	1.52	2.06	17.4
191	—seeds, meal mech-extd	5-04-109	91	60	2.65	2.22	1.31	0.74	1.35	22.1
192	—seeds, meal solv-extd	5-04-110	92	57	2.51	2.09	1.21	0.64	1.28	25.4
193	—seeds wo hulls, meal solv-extd	5-07-959	92	73	3.22	2.80	1.73	1.11	1.67	46.9
	Screenings—see Barley; Wheat									
	Sesame. *Sesamum indicum*									
194	—seeds, meal mech-extd	5-04-220	93	77	3.40	2.98	1.86	1.22	1.77	49.1
	Sorghum. *Sorghum bicolor*									
195	—aerial part w heads, sun-cured (fodder)	1-07-960	89	58	2.56	2.13	1.25	0.68	1.30	7.5
196	—aerial part wo heads, sun-cured (stover)	1-04-302	88	54	2.38	1.96	1.11	0.55	1.20	5.2
197	—distillers grains, dehydrated	5-04-374	94	83	3.66	3.25	2.03	1.37	1.91	34.4
198	—grain, less than 8% protein	4-20-892	88	81	3.57	3.16	1.97	1.32	1.87	7.9
199	—grain, 8–10% protein	4-20-893	87	80	3.53	3.12	1.94	1.30	1.84	9.7
200	—grain, more than 10% protein	4-20-894	88	79	3.48	3.06	1.91	1.27	1.82	13.0
201	—silage	3-04-323	30	60	2.65	2.22	1.31	0.74	1.35	7.5
202	—silage, dough stage	3-04-321	28	55	2.43	2.00	1.14	0.58	1.23	6.0
	Sorghum, Johnsongrass. *Sorghum halepense*									
203	—hay, sun-cured	1-04-407	89	53	2.34	1.91	1.08	0.52	1.18	9.5
	Sorghum, Sorgo. *Sorghum bicolor saccharatum*									
204	—silage	3-04-468	27	58	2.56	2.14	1.25	0.68	1.30	6.2
	Sorghum, Sudangrass. *Sorghum bicolor sudanense*									
205	—fresh, early vegetative	2-04-484	18	70	3.09	2.67	1.63	1.03	1.60	16.8
206	—fresh, midbloom	2-04-485	23	63	2.78	2.36	1.41	0.83	1.42	8.8
207	—hay, sun-cured, full bloom	1-04-480	91	56	2.47	2.04	1.18	0.62	1.25	8.0
208	—silage	3-04-499	28	55	2.43	2.00	1.14	0.58	1.23	10.8

continued

209	Soybean. *Glycine max*								
209	—hay, sun-cured, midbloom	1-04-538	94	53	2.34	1.91	1.08	0.52	17.8
210	—hay, sun-cured, dough stage	1-04-542	88	61	2.69	2.27	1.35	0.77	16.8
211	—hulls	1-04-560	91	77	3.40	2.98	1.86	1.22	12.1
212	—seeds	5-04-610	92	91	4.01	3.60	2.27	1.57	42.8
213	—seeds, heat-processed	5-04-597	90	94	4.14	3.74	2.35	1.64	42.2
214	—seeds, meal mech-extd	5-04-600	90	85	3.75	3.34	2.10	1.43	47.7
215	—seeds, meal solv-extd, 44% protein	5-20-637	89	84	3.70	3.29	2.06	1.40	49.9
216	—seeds wo hulls, meal solv-extd	5-04-612	90	87	3.84	3.42	2.16	1.48	55.1
217	—silage	3-04-581	27	55	2.43	2.00	1.14	0.58	17.3
	Sudangrass—see Sorghum Sugarcane. *Saccharum officinarum*								
218	—bagasse, dehydrated	1-04-686	91	44	1.94	1.51	0.75	0.22	1.5
	Sunflower, Common. *Helianthus annuus*								
219	—seeds, meal solv-extd	5-09-340	90	44	1.94	1.51	0.75	0.22	25.9
220	—seeds, wo hulls, meal mech-extd	5-04-738	93	74	3.26	2.85	1.75	1.13	44.6
221	—seeds wo hulls, meal solv-extd	5-04-739	93	65	2.87	2.45	1.47	0.88	49.8
222	—silage, low-oil	5-04-	30	61	2.69	2.27	1.35	0.77	11.1
222	—silage, high-oil	5-04-	30	66	2.91	2.49	1.51	0.92	12.5
	Sweetclover, Yellow. *Melilotus officinalis*								
223	—hay, sun-cured	1-04-754	87	54	2.38	1.96	1.11	0.55	15.7
	Timothy. *Phleum pratense*								
224	—hay, sun-cured, late vegetative	1-04-881	89	66	2.91	2.49	1.51	0.92	17.0
225	—hay, sun-cured, early bloom	1-04-882	90	61	2.69	2.27	1.35	0.77	15.0
226	—hay, sun-cured midbloom	1-04-883	89	58	2.56	2.13	1.25	0.68	9.1
227	—hay, sun-cured, full bloom	1-04-884	89	56	2.47	2.04	1.18	0.62	8.1
228	—hay, sun-cured, late bloom	1-04-885	88	54	2.38	1.95	1.11	0.55	7.8
229	—hay, sun-cured, milk stage	1-04-886	92	52	2.29	1.87	1.04	0.49	7.0
230	—silage, wilted, 25–45% dry matter (see similar maturity descriptions of hays)	—	—	—	—	—	—	—	—

Table-Feeds-Ruminant-1 (Continued)

Entry No.	Ether Extract (%)	Total Ash (%)	Crude Fiber (%)	Neutral Detergent Fiber (%)	Acid Detergent Fiber (%)	Cellulose (%)	Lignin (%)	Calcium	Chlorine	Macrominerals (%)				
										Magnesium	Phosphorus	Potassium	Sodium	Sulfur
190	35.1	3.1	28.6	—	—	—	—	0.26	—	0.36	0.67	0.79	0.06	—
191	6.7	4.1	35.4	59	41	—	—	0.27	—	0.36	0.78	0.79	0.05	—
192	1.5	5.9	32.5	58	41	27	14	0.37	—	0.37	0.81	0.82	0.05	0.14
193	1.4	8.2	14.7	17	17	15	2	0.38	0.18	1.11	1.40	1.19	0.05	0.22
194	7.5	12.1	6.1	17	—	—	—	2.17	0.07	0.50	1.46	1.35	0.04	0.35
195	2.4	9.4	26.9	—	—	—	—	0.40	—	0.29	0.21	1.47	0.02	—
196	1.7	11.0	33.5	—	—	—	—	0.52	—	0.28	0.13	1.20	0.02	—
197	9.5	3.8	12.7	—	—	—	—	0.16	—	0.19	0.74	0.38	0.05	0.18
198	3.5	2.1	2.0	—	—	—	1	0.03	0.10	0.19	0.32	0.38	0.05	0.18
199	3.4	2.1	2.0	18	9	8	—	0.04	0.10	0.18	0.34	0.40	0.01	0.09
200	3.3	2.1	2.0	—	—	—	—	0.04	0.10	0.14	0.36	0.38	0.01	0.13
201	3.0	8.7	27.9	—	38	—	6	0.35	0.13	0.29	0.21	1.37	0.02	0.11
202	3.3	9.3	28.5	—	—	—	—	0.29	0.11	0.27	0.26	1.02	0.03	0.14
203	2.4	8.2	33.5	—	—	—	—	0.84	—	0.35	0.28	1.35	0.01	0.10
204	2.6	6.4	28.3	—	—	—	—	0.34	0.06	0.27	0.17	1.12	0.15	0.10
205	3.9	9.0	23.0	55	29	26	3	0.43	—	0.35	0.41	2.14	0.01	0.11
206	1.8	10.5	30.0	65	40	34	5	0.43	—	0.35	0.36	2.14	0.01	0.11
207	1.8	9.6	36.0	68	42	35	6	0.55	—	0.51	0.30	1.87	0.02	0.06
208	2.8	9.8	33.1	68	42	38	5	0.46	—	0.44	0.21	2.25	0.02	0.06
209	5.4	8.8	29.8	—	40	—	—	1.26	—	0.79	0.27	0.97	0.12	0.26

continued

210	4.1	6.8	28.5	—	39	—	—	1.29	—	0.79	0.33	0.97	0.12	0.26
211	2.1	5.1	40.1	67	50	46	2	0.49	—	—	0.21	1.27	0.01	0.09
212	18.8	5.5	5.8	—	10	—	—	0.27	0.03	0.29	0.65	1.82	0.02	0.24
213	20.0	5.1	5.6	—	11	—	—	0.28	—	0.23	0.66	1.89	0.03	0.24
214	5.3	6.7	6.6	—	—	—	—	0.29	0.08	0.28	0.68	1.98	0.03	0.37
215	1.5	7.3	7.0	8.0	10	5.0	—	0.30	0.08	0.30	0.68	1.98	0.03	0.37
216	1.0	6.5	3.7	—	6.0	—	—	0.29	0.05	0.32	0.70	2.30	0.03	0.48
217	2.7	9.7	28.4	40	40	—	12	1.36	—	0.38	0.47	0.93	0.09	0.30
218	0.4	5.5	49.0	—	61	—	—	0.90	—	0.10	0.29	0.50	0.20	0.10
219	1.2	6.3	35.1	—	33	21	—	0.23	—	0.75	1.03	1.06	—	0.33
220	8.7	7.1	13.1	—	—	—	16	0.42	0.20	0.78	1.14	1.14	0.24	—
221	3.1	8.1	12.2	45	—	—	12	0.44	0.11	0.77	0.98	1.14	0.24	—
222	7.1	—	33.5	42	42	26	—	0.80	—	—	0.30	—	—	—
223	10.7	8.8	31.0	—	39	27	3	1.50	0.37	0.49	0.25	1.60	0.09	0.47
224	2.0	7.1	33.4	55	29	28	4	1.27	—	0.14	0.34	1.68	0.18	—
225	2.8	5.7	27.0	61	32	31	5	0.66	—	0.14	0.25	1.62	0.18	—
226	2.9	6.3	28.0	67	36	33	6	0.53	0.62	0.16	0.22	1.59	0.18	—
227	2.6	5.2	31.0	68	38	34	7	0.48	—	0.14	0.20	1.64	0.18	—
228	3.1	5.4	32.0	70	40	34	8	0.43	—	0.13	0.18	1.61	0.07	—
229	2.8	6.3	32.5	71	41	—	—	0.38	—	0.12	0.18	1.00	0.01	—
230	2.3	—	33.9	—	—	—	—	0.28	—	—	—	—	—	—

Table-Feeds-Ruminant-1 (Continued)

| Entry No. | Feed Name Description | International Feed Number[a] | Dry Matter (%) | Values as Determined at Maintenance Intake | | | Production | | | Crude Protein (%) |
| | | | | TDN (%) | DE (Mcal/kg) | ME (Mcal/kg) | Growing Cattle | | Lactating Cows | |
							NEM (Mcal/kg)	NEG (Mcal/kg)	NEL (Mcal/kg)	
	Tomato. *Lycopersicon esculentum*									
231	—pomace, dehydrated	5-05-041	92	58	2.56	2.13	1.25	0.68	1.30	23.5
	Torula Dried Yeast—see Yeast, Torula									
	Trefoil, Birdsfoot. *Lotus corniculatus*									
232	—fresh	2-20-786	24	66	2.91	2.49	1.51	0.92	1.50	21.0
233	—hay, sun-cured	1-05-044	92	59	2.60	2.18	1.27	0.70	1.33	16.3
	Triticale. *Triticale hexaploide*									
234	—grain	4-20-362	90	84	3.70	3.29	2.06	1.40	1.94	17.6
235	—silage, head emerging	4-20-	35	55	2.43	2.00	1.14	0.58	1.23	10.0
	Turnip. *Brassica rapa*									
236	—roots, fresh	4-05-067	90	85	3.75	3.34	2.10	1.43	1.96	11.8
	Urea									
237	—45% nitrogen, 281% protein equivalent	5-05-070	99	0	0.0	0.0	0.0	0.0	0.0	281.0
	Vetch. *Vicia spp.*									
238	—hay, sun-cured	1-05-106	89	57	2.51	2.09	1.21	0.64	1.28	20.8
	Wheat. *Triticum aestivum*									
239	—bran	4-05-190	89	70	3.09	2.67	1.63	1.03	1.60	17.1
240	—flour by-product, less than 7% fiber (wheat shorts)	4-05-201	88	73	3.22	2.80	1.73	1.11	1.67	18.6

continued

241	—flour by-product, less than 9.5% fiber (wheat middlings)	4-05-205	89	69	3.04	2.62	1.60	1.00	1.57	18.4
242	—fresh, early vegetative	2-05-176	22	73	3.22	2.80	1.73	1.11	1.67	28.6
243	—germ, ground	5-05-218	88	94	4.14	3.74	2.35	1.64	2.18	28.1
244	—grain	4-05-211	89	88	3.88	3.47	2.18	1.50	2.04	16.0
245	—grain, hard red winter	4-05-268	88	88	3.88	3.47	2.18	1.50	2.04	14.4
246	—grain, soft red winter	4-05-294	88	89	3.92	3.51	2.20	1.52	2.06	13.0
247	—grain, soft white winter	4-05-337	89	89	3.92	3.51	2.20	1.52	2.06	11.3
248	—grain screenings	4-05-216	89	71	3.13	2.71	1.67	1.06	1.61	15.8
249	—hay, sun-cured	1-05-172	88	58	2.56	2.13	1.25	0.68	1.30	8.5
250	—mill run, less than 9.5% fiber	4-05-206	90	79	3.48	3.07	1.91	1.27	1.82	17.2
251	—silage, early vegetative	3-05-184	30	57	2.51	2.09	1.21	0.64	1.28	11.9
252	—straw	1-05-175	89	44	1.94	1.51	0.75	0.22	0.96	3.6
	Whey									
253	—fresh (cattle)	4-08-134	7	81	3.57	3.16	1.97	1.32	1.87	14.2
254	—dehydrated (cattle)	4-01-182	93	81	3.57	3.16	1.97	1.32	1.87	14.2
255	—low-lactose, dehydrated (dried whey product, cattle)	4-01-186	93	79	3.48	3.07	1.91	1.27	1.82	17.9
	Yeast. *Saccharomyces cerevisiae*									
256	—brewers, dehydrated	7-05-527	93	79	3.48	3.07	1.91	1.27	1.82	46.9
	Yeast, Torula. *Torulopsis utilis*									
257	—torula, dehydrated	7-05-534	93	78	3.44	3.02	1.88	1.24	1.79	52.7

[a]Some specific numbers have not been assigned by the USDA Feed Composition Data Bank.

Table-Feeds-Ruminant-1 (Continued)

| Entry No. | Ether Extract (%) | Total Ash (%) | Crude Fiber (%) | Neutral Detergent Fiber (%) | Acid Detergent Fiber (%) | Cellulose (%) | Lignin (%) | Macrominerals (%) | | | | | | |
								Calcium	Chlorine	Magnesium	Phosphorus	Potassium	Sodium	Sulfur
231	10.3	7.5	26.4	55	50	—	11	0.43	—	0.20	0.60	3.63	—	—
232	2.7	9.0	24.7	—	—	—	—	1.91	—	0.28	0.22	1.99	0.07	0.25
233	2.5	7.0	30.7	47	36	24	9	1.70	—	0.51	0.27	1.92	0.07	0.25
234	1.7	2.0	4.4	—	8	—	—	0.06	—	—	0.33	0.40	—	0.17
235	—	—	—	—	38	—	—	—	—	—	—	—	—	—
236	1.9	8.9	11.5	44	34	—	10	0.59	0.65	0.22	0.26	2.99	1.05	0.43
237	0.0	—	0.0	0	0	0	0	—	—	—	—	—	—	—
238	3.0	9.1	30.6	48	33	—	8	1.18	—	0.25	0.32	2.32	0.52	0.15
239	4.4	6.9	11.3	51	15	11	3	0.13	0.05	0.60	1.38	1.56	0.04	0.25
240	5.2	4.9	7.7	37	10	—	—	0.10	0.08	0.28	0.91	1.06	0.03	0.22
241	4.9	5.2	8.2	52	30	24	4	0.13	0.04	0.40	0.99	1.13	0.19	0.20
242	4.4	13.3	17.4	—	—	—	—	0.42	—	0.21	0.40	3.50	0.18	0.22
243	9.5	4.7	3.5	—	—	—	—	0.06	0.08	0.28	1.05	1.09	0.03	0.28
244	2.0	1.9	2.9	—	8	8	—	0.04	0.08	0.16	0.42	0.42	0.05	0.18
245	1.8	1.9	2.8	—	4	—	—	0.05	0.06	0.13	0.43	0.49	0.02	0.15
246	1.8	2.1	2.4	14	—	—	—	0.05	0.08	0.11	0.43	0.46	0.01	0.12
247	1.9	1.8	2.6	—	4	—	—	0.07	0.09	0.13	0.36	0.46	0.04	0.16
248	3.9	6.1	7.7	68	—	6	8	0.15	—	0.18	0.39	0.58	0.10	0.22
249	2.2	7.1	28.1	—	41	—	7	0.15	—	0.12	0.20	1.00	0.21	0.22
250	4.6	5.9	9.2	—	—	—	—	0.11	—	0.52	1.13	1.33	0.24	0.34
251	2.5	7.5	26.9	85	—	—	—	0.27	0.07	0.62	0.27	1.39	0.07	0.24
252	1.8	7.8	41.6	0	54	39	14	0.18	0.32	0.12	0.05	1.42	0.14	0.19
253	0.7	9.8	0.2	0	0	0	0	0.73	—	—	0.65	2.75	—	—
254	0.7	9.8	0.2	0	0	0	0	0.92	0.08	0.14	0.82	1.23	0.70	1.12
255	1.1	16.5	0.2	0	0	0	0	1.71	1.10	0.23	1.12	3.16	1.54	1.15
256	0.9	7.1	3.1	—	—	—	—	0.13	0.08	0.27	1.49	1.79	0.08	0.45
257	1.7	8.3	2.4	—	—	—	—	0.54	0.02	0.18	1.71	2.04	0.04	0.59

Table Beef-Dairy-1
COMPOSITION OF MINERAL SUPPLEMENTS
FOR CATTLE ON A 100% DRY MATTER BASIS

Entry No.	Feed Name Description	International Feed Number	Dry Matter (%)	Protein Equivalent— N × 6.25 (%)	Macrominerals (%)						
					Calcium	Chlorine	Magnesium	Phosphorus	Potassium	Sodium	Sulfur
	Ammonium										
01	—phosphate, monobasic, $(NH_4)H_2PO_4$	6-09-338	97	70.9	0.28	—	0.46	24.74	0.01	0.06	1.46
02	—phosphate, dibasic, $(NH_4)_2HPO_4$	6-00-370	97	115.9	0.52	—	0.46	20.60	0.01	0.05	2.16
03	—sulfate	6-09-339	100	134.1	—	—	—	—	—	—	24.10
	Bone										
04	—charcoal (bone black, bone char)	6-00-402	90	9.4	30.11	—	0.59	14.14	0.16	—	—
05	—meal, steamed	6-00-400	97	13.2	30.71	—	0.33	12.86	0.19	5.69	2.51
	Calcium										
06	—carbonate, $CaCO_3$	6-01-069	100	—	39.39	—	0.05	0.04	0.06	0.06	—
07	—phosphate, monobasic, from defluorinated phosphoric acid	6-01-082	97	—	16.40	—	0.61	21.60	0.08	0.06	1.22
08	—phosphate, dibasic, from defluorinated phosphoric acid (dicalcium phosphate)	6-01-080	97	—	22.00	—	0.59	19.30	0.07	0.05	1.14
09	—sulfate, dihydrate, $CaSO_4 \cdot 2H_2O$, cp[a]	6-01-089	97	—	23.28	—	—	—	—	—	18.62
	Colloidal Clay										
10	—clay (soft rock phosphate); see also PHOSPHATE	6-03-947	100[b]	—	17.00	—	0.38	9.00	—	0.10	—
	Cobalt										
11	—carbonate, $CoCO_3$	6-01-566	99[b]	—	—	—	—	—	—	—	0.20
	Copper (Cupric)										
12	—sulfate, pentahydrate, $CuSO_4 \cdot 5H_2O$, cp[a]	6-01-720	100	—	—	—	—	—	—	—	12.84
	Curacao										
13	—phosphate	6-05-586	99[b]	—	34.34	—	0.81	14.14	—	0.20	—
	Ethylenediamine										
14	—dihydroiodide	6-01-842	98[b]	—	—	—	—	—	—	—	—
	Iron (Ferrous)										
15	—sulfate, heptahydrate	6-20-734	98[b]	—	—	—	—	—	—	—	12.35
	Limestone										
16	—limestone, ground	6-02-632	100	—	34.00	0.03	2.06	0.02	0.12	0.06	0.04
17	—magnesium (dolomitic)	6-02-633	99[b]	—	22.30	0.12	9.99	0.04	0.36	—	—
	Magnesium										
18	—carbonate, $MgCO_3Mg(OH)_2$	6-02-754	98[b]	—	0.02	0.00	30.81	—	—	—	—
19	—oxide, MgO	6-02-756	98	—	3.07	—	56.20	—	—	—	—

continued

			Microminerals (mg/kg)				
Cobalt	Copper	Fluorine	Iodine	Iron	Manganese	Selenium	Zinc
10	10	2,500	—	17,400	400	—	100
—	10	2,100	—	12,400	400	—	100
—	1	—	—	10	1	—	—
—	—	—	—	—	—	—	—
—	—	—	—	26,700	—	—	100
—	—	—	—	300	300	—	—
10	10	2,100	—	15,800	360	—	90
10	10	1,800	—	14,400	300	—	100
—	—	—	—	—	—	—	—
—	—	15,000	—	19,000	1,000	—	—
460,000	—	—	—	500	—	—	—
—	254,500	—	—	—	—	—	—
—	—	5,500	—	3,500	—	—	—
—	—	—	803,400	—	—	—	—
—	—	—	—	218,400	—	—	—
—	—	—	—	3,500	—	—	—
—	—	—	—	770	—	—	—
—	—	—	—	220	—	—	—
—	—	200	—	—	100	—	—

continued

Table Beef-Dairy-1 (Continued)

Entry No.	Feed Name Description	International Feed Number	Dry Matter (%)	Protein Equivalent— N × 6.25 (%)	Macrominerals (%)						
					Calcium	Chlorine	Magnesium	Phosphorus	Potassium	Sodium	Sulfur
	Manganese (Manganous)										
20	—oxide, MnO, cp[a]	6-03-056	99[b]	—	—	—	—	—	—	—	—
21	—carbonate, MnCO$_3$	6-03-036	97	—	—	—	—	—	—	—	—
	Oystershell										
22	—ground (flour)	6-03-481	99	—	38.00	0.01	0.30	0.07	0.10	0.21	—
	Phosphate										
23	—defluorinated	6-01-780	100	—	32.00	—	0.42	18.00	0.08	4.90	—
24	—rock	6-03-945	100	—	35.00	—	0.41	13.00	0.06	0.03	—
25	—rock, low-fluorine	6-03-946	100	—	36.00	—	—	14.00	—	—	—
26	—rock, soft; see also Calcium	6-03-947	100	—	17.00	—	0.38	9.00	—	0.10	—
27	—phosphate, monobasic, monohydrate, NaH$_2$PO$_4$ · H$_2$O; see also Sodium	6-04-288	97	—	—	—	—	22.50	—	16.68	—
	Phosphoric Acid										
28	—H$_3$PO$_4$	6-03-707	75	—	0.05	—	0.51	31.60	0.02	0.04	1.55
	Potassium										
29	—bicarbonate, KHCO$_3$, cp[a]	6-29-493	99[b]	—	—	—	—	—	39.05	—	—
30	—chloride, KCl	6-03-755	100	—	0.05	47.30	0.34	—	50.00	1.00	0.45
31	—iodide, KI	6-03-759	100[b]	—	—	—	—	—	21.00	—	—
32	—sulfate, K$_2$SO$_4$	6-06-098	98[b]	—	0.15	1.55	0.61	—	41.84	0.09	17.35
	Sodium										
33	—bicarbonate, NaHCO$_3$	6-04-272	100	—	—	—	—	—	—	27.00	—
34	—chloride, NaCl	6-04-152	100	—	—	60.66	—	—	—	39.34	—
35	—phosphate, monobasic, monohydrate, NaH$_2$PO$_4$ · H$_2$O; see also Phosphate	6-04-288	97	—	—	—	—	22.50	—	16.68	—
36	—selenite, Na$_2$SeO$_3$	6-26-013	98[b]	—	—	—	—	—	—	26.60	—
37	—sulfate, decahydrate, Na$_2$SO$_4$ · 10H$_2$O, cp[a]	6-04-292	97[b]	—	—	—	—	—	—	14.27	9.95
38	—tripolyphosphate, Na$_5$P$_3$O$_{10}$	6-08-076	96	—	—	—	—	25.00	—	31.00	—
	Zinc										
39	—oxide, ZnO	6-05-533	100	—	—	—	—	—	—	—	—
40	—sulfate, monohydrate, ZnSO$_4$ · H$_2$O	6-05-555	99[b]	—	0.02	0.015	—	—	—	—	17.68

NOTE: The compositions of hydrated mineral ingredients (e.g., CaSO$_4$ · 2H$_2$O) are shown including the waters of hydration. Mineral compositions of feed-grade mineral supplements vary by source, mining site, and manufacturer. The manufacturer's analysis should be used when it is available.

[a]cp = Chemically pure.
[b]Dry matter values have been estimated for these minerals.

continued

			Microminerals (mg/kg)				
Cobalt	Copper	Fluorine	Iodine	Iron	Manganese	Selenium	Zinc
—	—	—	—	—	774,500	—	—
—	—	—	—	—	478,000	—	—
—	—	—	—	2,870	100	—	—
10	20	1,800	—	6,700	200	—	60
10	10	35,000	—	16,800	200	—	100
—	—	—	—	—	—	—	—
—	—	15,000	—	1,000	19,000		—
—	—	—	—	—	—	—	—
10	10	3,100	—	17,500	500	—	130
—	—	—	—	—	—	—	—
—	—	—	—	600	—	—	—
—	—	—	681,700	—	—	—	—
—	—	—	—	710	10	—	—
—	—	—	—	—	—	—	—
—	—	—	—	—	—	—	—
—	—	—	—	—	—	—	—
—	—	—	—	—	—	456,000	—
—	—	—	—	—	—	—	—
—	—	—	—	40	—	—	—
—	—	—	—	—	—	—	780,000
—	—	—	—	10	10	—	363,600

Table Beef-Dairy-2
RUMINAL UNDEGRADABILITY OF PROTEIN IN SELECTED FEEDS

Feed	Number of Determinations	Undegradability		
		Mean	*S.D.*	*C.V.*
Alfalfa, dehydrated	8	0.59	0.17	29
Alfalfa hay	12	0.28	0.07	25
Alfalfa silage	6	0.23	0.08	36
Alfalfa-bromegrass	1	0.21		
Barley	16	0.27	0.10	37
Barley, flaked	1	0.67		
Barley, micronized	1	0.47		
Barley silage	1	0.27		
Bean meal, field	1	0.46		
Beans	2	0.16	0.02	14
Beet pulp	4	0.45	0.14	30
Beet pulp molasses	2	0.35	0.03	8
Beets	3	0.20	0.03	16
Blood meal	2	0.82	0.01	1
Brewers dried grains	9	0.49	0.13	27
Bromegrass	1	0.44		
Casein	3	0.19	0.06	32
Casein, HCHO	2	0.72	0.08	11
Clover, red	3	0.31	0.04	12
Clover, red, silage	1	0.38		
Clover, white	1	0.33		
Clover-grass	2	0.54	0.11	21
Clover-grass silage	7	0.28	0.06	22
Coconut	1	0.57		
Coconut meal	5	0.63	0.07	11
Corn	11	0.52	0.18	34
Corn, 0% cottonseed hulls	1	0.46		
Corn, 7% cottonseed hulls	1	0.43		
Corn, 14% cottonseed hulls	1	0.59		
Corn, 21% cottonseed hulls	1	0.48		
Corn, 10.5% protein, 0% $NaHCO_3$	1	0.36		
Corn, 10.5% protein, 3.5% $NaHCO_3$	1	0.30		
Corn, 12% protein, 0% $NaHCO_3$	1	0.29		
Corn, 12% protein, 3.5% $NaHCO_3$	1	0.24		
Corn, dry rolled	6	0.60	0.07	12
Corn, dry rolled, 0% roughage	1	0.54		
Corn, dry rolled, 21% roughage	1	0.49		
Corn, flaked	1	0.58		
Corn, flakes	1	0.65		
Corn, high-moisture acid	1	0.56		
Corn, high-moisture ground	1	0.80		
Corn, micronized	1	0.29		

continued

Table Beef-Dairy-2 (Continued)

Feed	Number of Determinations	Undegradability		
		Mean	S.D.	C.V.
Corn, steam flaked	1	0.68		
Corn, steam flaked, 0% roughage	1	0.51		
Corn, steam flaked, 21% roughage	1	0.47		
Corn gluten feed	1	0.25		
Corn gluten feed dry	2	0.22	0.11	51
Corn gluten feed wet	1	0.26		
Corn gluten meal	3	0.55	0.08	14
Corn silage	3	0.31	0.06	20
Cottonseed meal	21	0.43	0.11	25
Cottonseed meal, HCHO	2	0.64	0.15	23
Cottonseed meal, prepressed	2	0.36	0.02	6
Cottonseed meal, screwpressed	2	0.50	0.10	20
Cottonseed meal, solvent	6	0.41	0.13	32
Distillers dried grain with solubles	4	0.47	0.18	39
Distillers dried grains	1	0.54		
Distillers wet grains	1	0.47		
Feather meal, hydrolyzed	1	0.71		
Fish meal	26	0.60	0.16	26
Fish meal, stale	1	0.48		
Fish meal, well-preserved	1	0.78		
Grapeseed meal	1	0.45		
Grass	4	0.40	0.10	26
Grass pellets	2	0.46	0.05	11
Grass silage	20	0.29	0.06	20
Guar meal	1	0.34		
Linseed	1	0.18		
Linseed meal	5	0.35	0.10	27
Lupin meal	1	0.35		
Manoic meal	1	0.36		
Meat and bonemeal	5	0.49	0.18	37
Meat meal	1	0.76		
Oats	4	0.17	0.03	15
Palm cakes	6	0.66	0.06	9
Peanut meal	8	0.25	0.11	45
Peas	4	0.22	0.03	15
Rapeseed meal	10	0.28	0.09	31
Rapeseed meal, protected	1	0.70		
Rye	1	0.19		
Ryegrass, dehydrated	4	0.22	0.14	66
Ryegrass, dried artificially	1	0.71		
Ryegrass, dried artificially, chopped	1	0.30		
Ryegrass, dried artificially, ground	1	0.73		
Ryegrass, dried artificially, pelleted	1	0.54		

continued

Table Beef-Dairy-2 (Continued)

Feed	Number of Determinations	Undegradability		
		Mean	S.D.	C.V.
Ryegrass, fresh	1	0.48		
Ryegrass, fresh or frozen	3	0.41	0.18	44
Ryegrass, frozen	1	0.52		
Ryegrass silage, HCHO	1	0.93		
Ryegrass silage, HCHO dried	1	0.83		
Ryegrass silage, unwilted	1	0.22		
Sanfoin	1	0.81		
Sorghum grain	2	0.54	0.02	4
Sorghum grain, dry ground	1	0.49		
Sorghum grain, dry rolled	2	0.64	0.08	12
Sorghum grain, micronized	1	0.64		
Sorghum grain, reconstituted	2	0.42	0.32	75
Sorghum grain, steam flaked	2	0.47	0.07	15
Soybean meal	39	0.35	0.12	33
Soybean meal, dried 120 C	1	0.59		
Soybean meal, dried 130 C	1	0.71		
Soybean meal, dried 140 C	1	0.82		
Soybean meal, 35% concentrate	1	0.18		
Soybean meal, 65% concentrate	1	0.46		
Soybean meal, HCHO	3	0.80	0.11	14
Soybean meal, unheated	1	0.14		
Soybean-rapeseed meal, HCHO	2	0.78	0.02	3
Soybeans	2	0.26	0.11	40
Subterranean clover	2	0.40	0.18	45
Sunflower meal	9	0.26	0.05	20
Timothy, dried artificially, chopped	1	0.32		
Timothy, dried artificially, pelleted	1	0.53		
Wheat	4	0.22	0.06	27
Wheat bran	4	0.29	0.10	34
Wheat gluten	1	0.17		
Wheat middlings	3	0.21	0.02	11
Yeast	1	0.42		
Zein	1	0.60		

Table Feed-1

COMPOSITION (EXCLUDING AMINO ACIDS) OF SOME FEEDS COMMONLY USED FOR POULTRY AND SWINE (DATA ON AS-FED BASIS)

Entry No.	Feed Name Description	International Feed Number[a]	Dry Matter (%)	MF_n (kcal/kg)	TMF_n (kcal/kg)	Protein (%)	Ether Extract (%)	Linoleic Acid (%)	Crude Fiber (%)	Calcium (%)	Total Phosphorus (%)	Nonphytate Phosphorus (%)	Potassium (%)	Chlorine (%)
	Alfalfa *Medicago sativa*													
01	meal dehydrated, 17% protein	1-00-023	92	1,200	1,011	17.5	2.5	0.47	24.1	1.44	0.22	0.22	2.15	0.47
02	meal dehydrated, 20% protein	1-00-024	92	1,630	—	20.0	3.6	0.58	20.2	1.67	0.28	—	2.15	0.47
	Bakery													
03	waste, dehydrated (dried bakery product)	4-00-466	92	3,862	3,696	10.5	11.7	—	1.2	0.13	0.24	—	0.35	1.23
	Barley *Hordeum vulgare*													
04	grain	4-00-549	89	2,640	2,900	11.0	1.8	0.83	5.5	0.03	0.36	0.17	0.48	0.15
05	grain, Pacific coast	4-07-939	89	2,620	—	9.2	2.0	0.85	6.4	0.05	0.32	—	0.53	0.15
	Broadbean *Vicia faba*													
06	seeds	5-09-262	87	2,431	2,339	24.0	1.4	—	7.0	0.11	0.54	—	1.2	—
	Blood													
07	meal, vat dried	5-00-380	94	2,830	—	81.1	1.6	—	0.5	0.55	0.42	—	0.18	0.27
08	meal, spray or ring dried	5-00-381	93	3,420	3,625	88.9	1.0	0.10	0.6	0.41	0.30	—	0.18	0.27
	Brewer's Grains													
09	dehydrated	5-02-141	92	2,080	—	25.3	6.2	2.94	15.3	0.29	0.52	—	0.08	0.12
	Buckwheat, common *Fagopyrum saxitatum*													
10	grain	4-00-994	88	2,660	2,755	10.8	2.5	—	10.5	0.09	0.32	0.12	0.10	0.04
	Cane Molasses—see Molasses Canola *Brassica napus-Brassica campestris*													
11	seeds, meal prepressed solvent extracted, low erucic acid, low glucosinolates	5-06-145	93	2,000	2,070	38.0	3.8	—	12.0	0.68	1.17	0.30	1.29	—
	Casein													
12	dehydrated	5-01-162	93	4,130	4,134	87.2	0.8	—	0.2	0.61	1.00	1.00	0.01	—
13	precipitated dehydrated	5-20-837	92	4,118	—	85.0	0.06	—	0.2	0.68	0.82	0.82	0.01	—
	Cattle													
14	skim milk, dehydrated	5-01-175	93	2,537	—	36.1	1.0	—	0.2	1.28	1.02	1.02	1.60	0.90
	Coconut *Cocos nucifera*													
15	kernels with coats, meal solvent extracted (copra meal)	5-01-573	92	1,525	—	19.2	2.1	—	14.4	0.17	0.65	—	1.41	0.03
	Corn, Dent Yellow *Zea mays indentata*													
16	distillers' grains, dehydrated	5-28-235	94	1,972	—	27.8	9.2	—	12.0	0.10	0.40	0.39	0.17	0.07

continued

Table Feed-1 (continued)

Entry No.	Feed Name Description	International Feed Number[a]	Dry Matter (%)	MF$_n$ (kcal/kg)	TMF$_n$ (kcal/kg)	Protein (%)	Ether Extract (%)	Linoleic Acid (%)	Crude Fiber (%)	Calcium (%)	Total Phosphorus (%)	Nonphytate Phosphorus (%)	Potassium (%)	Chlorine (%)
17	distillers' grains with solubles, dehydrated	5-28-236	93	2,480	3,097	27.4	9.0	4.55	9.1	0.17	0.72	0.39	0.65	0.17
18	distillers' solubles, dehydrated	5-28-237	92	2,930	—	28.5	9.0	4.55	4.0	0.35	1.27	1.17	1.75	0.26
19	gluten, meal, 60% protein	5-28-242	90	3,720	3,811	62.0	2.5	—	1.3	—	0.50	0.14	0.35	0.05
20	gluten with bran (corn gluten feed)	5-28-243	90	1,750	2,228	21.0	2.5	—	8.0	0.40	0.80	—	0.57	0.22
21	grain	4-02-935	89	3,350	3,470	8.5	3.8	2.20	2.2	0.02	0.28	0.08	0.30	0.04
22	grits by-product (hominy feed)	4-03-011	90	2,896	3,269	10.4	8.0	3.28	5.0	0.05	0.52	—	0.59	0.05
	Cotton *Gossypium* spp.													
23	seeds, meal mechanically extracted, 41% protein (expeller)	5-01-617	93	2,320	—	40.9	3.9	2.47	12.0	0.20	1.05	—	1.19	0.04
24	seeds, meal prepress solvent extracted, 41% protein	5-07-872	90	2,400	—	41.4	0.5	—	13.6	0.15	0.97	0.22	1.22	0.03
25	seeds, meal prepressed solvent extracted, 44% protein	5-07-873	91	1,857	2,135	44.7	1.6	—	11.1	0.15	1.25	0.37	—	—
	Feathers—see Poultry													
	Fish													
26	solubles, condensed	5-01-969	51	1,460	—	31.5	7.8	—	0.2	0.30	0.76	—	1.74	2.65
27	solubles, dehydrated	5-01-971	92	2,830	—	63.6	9.3	0.12	0.5	1.23	1.63	—	0.37	—
28	Fish, Anchovy *Engraulis ringen* meal mechanically extracted	5-01-985	92	2,580	—	64.2	5.0	0.20	1.0	3.73	2.43	—	0.69	0.60
29	Fish, Herring *Clupea harengus* meal mechanically extracted	5-02-000	93	3,190	—	72.3	10.0	0.15	0.7	2.29	1.70	—	1.09	0.90
30	Fish, Menhaden *Brevoortia tyrannus* meal mechanically extracted	5-02-009	92	2,820	2,977	60.05	9.4	0.12	0.7	5.11	2.88	—	0.65	0.60
31	Fish, White Gadidae (family)-Lophiidae (family)-Rajidae (family) meal mechanically extracted	5-02-025	91	2,593	—	62.6	4.6	0.08	0.7	7.31	3.81	—	0.83	0.50
	Gelatin													
32	process residue (gelatin by-products)	5-14-503	91	2,360	3,029	88.0	0.0	—	—	0.50	Trace	—	—	—
	Hominy Feed—see Corn													
33	Livers meal	5-00-389	92	2,860	—	65.6	15.0	—	1.4	0.56	1.25	—	—	—
	Meat													
34	meal rendered	5-00-385	92	2,195	—	54.4	7.1	0.28	2.7	8.27	4.10	—	0.60	continued 0.91
35	with bone, meal rendered	5-00-388	93	2,150	2,495	50.4	10.0	0.36	2.8	10.30	5.10	—	1.45	0.69
	Millet Pearl *Pennisetum glaucum*													
36	grain	4-03-118	91	2,675	3,367	14.0	4.3	0.84	3.0	0.05	0.32	0.12	0.43	0.14
	Millet, Proso *Panicum miliaceum*													

Table Feed-1 (Continued)

Entry No.	Iron (mg/kg)	Magnesium (%)	Manganese (mg/kg)	Sodium (%)	Sulfur (%)	Copper (mg/kg)	Selenium (mg/kg)	Zinc (mg/kg)	Biotin (mg/kg)	Choline (mg/kg)	Folacin (mg/kg)	Niacin (mg/kg)	Pantothenic Acid (mg/kg)	Pyridoxine (mg/kg)	Riboflavin (mg/kg)	Thiamin (mg/kg)	Vitamin B_{12} (µg/kg)	Vitamin E (mg/kg)
01	480	0.36	30	0.09	0.17	10	0.34	24	0.30	1,401	4.2	38	25.0	6.5	13.6	3.4	4	125
02	390	0.36	42	0.09	0.43	11	0.29	25	0.33	1,419	3.3	40	34.0	8.0	15.2	5.8	4	144
03	28	0.24	65	1.14	0.02	5	—	15	0.07	923	0.2	26	8.3	4.3	1.4	2.9	—	41
04	78	0.14	18	0.04	0.15	10	0.10	30	0.15	990	0.07	55	8.0	3.0	1.8	1.9	—	20
05	110	0.12	16	0.02	0.15	8	0.10	15	0.15	1,034	0.05	48	7.0	2.9	1.6	4.0	—	20
06	70	0.13	8	0.08	—	4	—	42	0.09	1.7	—	22	3.0		1.6	5.5	—	1
07	2,020	0.16	5	0.32	0.32	10	0.01	4	0.08	695	0.1	29	3.0	4.4	2.6	0.4	44	—
08	3,000	0.40	6	0.33	0.32	8	—	306	0.20	280	0.4	13	5.0	4.1	1.3	0.5	44	—
09	250	0.16	38	0.26	0.31	21	0.70	98	0.96	1,723	7.1	29	8.0	0.7	1.4	0.5	—	25
10	44	0.09	34	0.05	0.14	10	—	9	—	440	—	19	12.0		5.5	4.0	—	—
11	159	0.64	54	—	—	—	1.00	71	0.90	6,700	2.3	160	9.5		3.7	5.2	—	—
12	18	0.01	4	0.01	—	4	—	33	0.05	205	0.5	1	3.0	0.4	1.5	1.5	—	—
13	17	0.01	4	0.01	—	4	—	32	0.04	208	0.5	1	2.7	0.4	1.5	0.5	—	—
14	8	0.12	2	0.51	0.32	12	0.12	39	0.33	1,393	0.62	11.5	36.4	4.1	19.1	3.7	51	9
15	—	0.31	54	0.04	—	—	—	—	—	1,089	0.30	23.8	6.5	4.4	3.5		—	—
16	300	0.25	22	0.09	0.43	25	0.45	55	0.49	1,180	0.9	37	11.7	4.4	5.2	1.7	—	40
17	280	0.19	24	0.48	0.30	57	0.39	80	0.78	2,637	0.9	71	11.0	2.2	8.6	2.9	—	55
18	560	0.64	74	0.26	0.37	83	0.33	85	1.10	4,842	1.1	116	21.0	10.0	17.0	6.9	3	24
19	400	0.15	4	0.02	0.43	26	1.00	33	0.15	330	0.2	55	3.0	6.2	2.2	0.3	—	15
20	460	0.29	24	0.15	0.22	48	0.10	70	0.33	1,518	0.3	66	17.0	15.0	2.4	2.0	—	22
21	45	0.12	7	0.02	0.08	3	0.03	18	0.06	620	0.4	24	4.0	7.0	1.0	3.5	—	—
22	67	0.24	15	0.08	0.03	13	0.10	3	0.13	1,155	0.3	47	8.2	11.0	2.1	8.1	—	—
23	160	0.52	73	0.04	0.40	19	0.25	64	0.60	2,753	1.0	38	10.0	5.3	5.1	6.4	—	39
24	110	0.40	20	0.04	0.31	18	—	70	0.55	2,933	2.7	40	7.0	3.0	4.0	3.3	—	15
25	—	—	—	—	—	—	—	—	—	2,685	0.9	46	14.5		4.7		—	—
26	160	0.02	14	2.62	0.12	45	2.00	38	0.18	3,519	0.02	169	35.0	12.2	14.6	5.5	347	—
27	300	0.30	50	0.3	0.40	—	—	76	0.26	5,507	0.06	271	55.0	23.8	7.7	7.4	401	—
28	220	0.24	10	0.65	0.54	9	1.36	103	0.23	4,408	0.2	100	15.0	4.0	7.1	0.1	352	4
29	140	0.15	5	0.61	0.69	6	1.93	132	0.31	5,306	0.3	93	17.0	4.0	9.9	0.1	403	22
30	440	0.16	33	0.65	0.45	11	2.10	147	0.20	3,056	0.3	55	9.0	4.0	4.9	0.5	104	7
31	181	0.18	12	0.78	0.48	6	1.62	90	0.08	3,099	0.3	59	9.9	5.9	9.1	1.7	90	9
32		0.05																
33	630	—	9	1.15	0.49	89	—	—	0.02	11,311	5.5	204	29.0		46.3	0.2	498	—
34	440	0.58	10	0.70	0.50	10	0.42	103	0.17	2,077	0.3	57	5.0	3.0	5.5	0.2	68	1
35	490	1.12	14			2	0.25	93	0.14	1,996	0.3	46	4.1	12.8	4.4	0.8	70	1
36	25	0.16	31	0.04	0.13	22	—	13	—	793	—	53	7.8		1.6	6.7	—	—

continued

Table Feed-1 (Continued)

Entry No.	Feed Name Description	International Feed Number[a]	Dry Matter (%)	MF_n (kcal/kg)	TMF_n (kcal/kg)	Protein (%)	Ether Extract (%)	Linoleic Acid (%)	Crude Fiber (%)	Calcium (%)	Total Phosphorus (%)	Nonphytate Phosphorus (%)	Potassium (%)	Chlorine (%)
37	grain	4-03-120	90	2,898	—	11.6	3.5	—	6.1	0.03	0.30	0.14	0.43	—
	Oats *Avena sativa*													
38	grain	4-03-309	89	2,550	2,625	11.4	4.2	1.47	10.8	0.06	0.27	0.05	0.45	0.11
39	grain, Pacific coast	4-07-999	91	2,610	—	9.0	5.0	—	11.0	0.08	0.30	—	0.37	0.12
40	hulls	1-03-281	92	400	—	4.6	1.4	—	28.7	0.13	0.10	—	0.53	0.10
	Pea *Pisum* spp.													
41	seeds	5-03-600	90	2,570	2,654	23.8	1.3	—	5.5	0.11	0.42	—	1.02	0.06
	Peanut *Arachis hypogaea*													
42	kernels, meal mechanically extracted (peanut meal) (expeller)	5-03-649	90	2,500	—	42.0	7.3	1.43	12.0	0.16	0.56	—	1.15	0.03
43	kernels, meal solvent extracted (peanut meal)	5-03-650	92	2,200	2,462	50.7	1.2	0.24	10.0	0.20	0.63	0.13	1.15	0.03
	Poultry													
44	by-product, meal rendered (viscera with feet and heads)	5-03-798	93	2,950	3,120	60.0	13.0	2.54	1.5	3.00	1.70	—	0.55	0.54
45	feathers, meal hydrolyzed	5-03-795	93	2,360	3,276	81.0	7.0	—	1.0	0.33	0.55	—	0.30	0.28
	Rice *Oryza sativa*													
46	bran with germ (rice bran)	4-03-928	91	2,980	3,085	12.9	13.0	3.57	11.4	0.07	1.50	0.22	1.73	0.07
47	grain, polished and broken (brewer's rice)	4-03-932	89	2,990	3,536	8.7	0.7	—	9.8	0.08	0.08	0.03	0.13	0.08
48	polishings	4-03-943	90	3,090	—	12.2	11.0	3.58	4.1	0.05	1.31	0.14	1.06	0.11
	Rye *Secale cereale*													
49	grain	4-04-047	88	2,626	2,931	12.1	1.5	—	2.2	0.06	0.32	0.06	0.46	0.03
	Safflower *Carthamus tinctorius*													
50	seeds, meal solvent extracted	5-04-110	92	1,193	—	23.4	1.4	1.13	30.0	0.34	0.75	—	0.76	—
51	seeds without hulls, meal solvent extracted	5-07-959	92	1,921	—	43.0	1.3	0.82	13.5	0.35	1.29	0.39	1.10	0.16
	Sesame *Sesamum indicum*													
52	seeds, meal mechanically extracted (expeller)	5-04-220	93	2,210	1,978	43.8	6.5	1.90	7.0	1.99	1.37	0.34	1.20	0.06
	Sorghum *Sorghum bicolor*													
53	grain, 8–10% protein	4-20-893	87	3,288	3,376	8.8	2.9	1.13	2.3	0.04	0.30	—	0.35	0.09
54	grain, more than 10% protein	4-20-894	88	3,212	—	11.0	2.6	0.82	2.3	0.04	0.32	—	0.33	0.09
	Soybean *Glycine max*													
55	flour by-product (soybean mill feed)	4-04-594	89	720	—	13.3	1.6	—	33.0	0.37	0.19	—	1.50	0.02

Table Feed-1 (Continued)

Entry No.	Feed Name Description	International Feed Number[a]	Dry Matter (%)	MF$_n$ (kcal/kg)	TMF$_n$ (kcal/kg)	Protein (%)	Ether Extract (%)	Linoleic Acid (%)	Crude Fiber (%)	Calcium (%)	Total Phosphorus (%)	Nonphytate Phosphorus (%)	Potassium (%)	Chlorine (%)
56	protein concentrate, more than 70% protein	5-08-038	93	3,500	—	84.1	0.4	—	0.2	0.02	0.80	0.32	0.18	0.02
57	seeds, heat processed	5-04-597	90	3,300	2,990	37.0	18.0	8.46	5.5	0.25	0.58	—	1.61	0.03
58	seeds, meal solvent extracted	5-04-604	89	2,230	—	44.0	0.8	0.40	7.0	0.29	0.65	0.27	2.00	0.05
59	seeds without hulls, meal solvent extracted	5-04-612	90	2,440	2,485	48.5	1.0	0.40	3.9	0.27	0.62	0.22	1.98	0.05
	Sunflower, common *Helianthus annuus*													
60	seeds, meal solvent extracted	5-09-340	90	1,543	—	32.0	1.1	0.60	24.0	0.21	0.93	0.14	0.96	—
61	seeds without hulls, meal solvent extracted	5-04-739	93	2,320	2,060	45.4	2.9	1.59	12.2	0.37	1.00	0.16	1.00	0.10
	Triticale *Triticale hexaploide*													
62	grain	4-20-362	90	3,163	3,144	14.0	1.5	—	4.0	0.05	0.30	0.10	0.36	—
	Wheat *Triticum aestivum*													
63	bran	4-05-190	89	1,300	1,725	15.7	3.0	1.70	11.0	0.14	1.15	0.20	1.19	0.06
64	flour by-product, less than 4% fiber (wheat red dog)	4-05-203	88	2,568	—	15.3	3.3	—	2.6	0.04	0.49	0.14	0.51	0.14
65	flour by-product, less than 9.5% fiber (wheat middlings)	4-05-205	88	2,000	2,708	15.0	3.0	1.87	7.5	0.12	0.85	0.30	0.99	0.03
66	flour by-product, less than 7% fiber (wheat shorts)	4-05-201	88	2,162	2,061	16.5	4.6	—	6.8	0.09	0.81	—	0.93	0.07
67	grain, hard red winter	4-05-268	87	2,900	3,167	14.1	2.5	0.59	3.0	0.05	0.37	0.13	0.45	0.05
68	grain, soft white winter	4-05-337	89	3,120	—	11.5	2.5	—	3.0	0.05	0.31	—	0.42	0.05
	Whey *Bos taurus*													
69	dehydrated	4-01-182	93	1,900	693	13.0	0.8	0.01	0.2	0.97	0.76	—	1.05	1.5
70	low lactose, dehydrated (dried whey product)	4-01-186	91	2,090	—	16.0	1.0	0.01	0.3	1.95	0.98	—	3.0	1.03
	Yeast, Brewer's *Saccharomyces cerevisiae*													
71	dehydrated	7-05-527	93	1,990	2,634	44.4	1.0	—	2.7	0.12	1.40	—	1.70	0.12
	Yeast, Torula *torulopsis utilis*													
72	dehydrated	7-05-534	93	2,160	—	47.2	2.5	0.05	2.4	0.58	1.67	—	1.70	0.12

NOTE: Dash indicates that no data were available.

[a]First digit is class of feed: 1, dry forages and roughages; 2, pasture, range plants, and forages fed green; 3, silages; 4, energy feeds; 5, protein supplements; 6, minerals; 7, vitamins; 8, additives; the other five digits are the International Feed Number.

continued

Table Feed-1 (Continued)

Entry No.	Iron (mg/kg)	Magnesium (%)	Manganese (mg/kg)	Sodium (%)	Sulfur (%)	Copper (mg/kg)	Selenium (mg/kg)	Zinc (mg/kg)	Biotin (mg/kg)	Choline (mg/kg)	Folacin (mg/kg)	Niacin (mg/kg)	Pantothenic Acid (mg/kg)	Pyridoxine (mg/kg)	Riboflavin (mg/kg)	Thiamin (mg/kg)	Vitamin B$_{12}$ (µg/kg)	Vitamin E (mg/kg)
37	71	0.16	—	—	—	—	—	—	—	440	—	23	11.0	—	3.8	7.3	—	—
38	85	0.16	43	0.08	0.21	8	0.30	38	0.27	946	0.3	12	7.8	1.0	1.1	6.0	—	20
39	73	0.17	38	0.06	0.20	—	0.07	—	0.22	959	0.3	14	13.0	1.3	1.1	0.6	—	—
40	100	0.08	14	0.04	0.14	3	—	0.1	—	284	1.0	7	3.0	2.2	1.5	0.6	—	—
41	50	0.13	—	—	—	—	—	30	0.18	642	0.4	34	10.0	1.0	2.3	4.6	—	3
42	156	0.33	25	0.06	0.29	15	0.28	30	0.33	1,655	0.4	166	47.0	10.0	5.2	7.1	—	3
43	142	0.04	29	0.07	0.30	15	—	20	0.39	2,396	0.4	170	53.0	10.0	11.0	5.7	—	3
44	440	0.22	11	0.40	0.51	14	0.75	120	0.30	5,952	1.0	40	12.3	4.4	11.0	1.0	—	2
45	76	0.20	10	0.69	1.50	7	0.84	54	0.04	891	0.2	27	10.0	3.0	2.1	0.1	310	—
46	190	0.95	250	0.07	0.18	13	0.40	30	0.42	1,135	2.2	293	23.0	14.0	2.5	22.5	78	60
47	—	0.11	18	0.07	0.06	—	0.27	17	0.08	800	0.2	30	8.0	28.0	0.7	1.4	—	14
48	160	0.65	12	0.10	0.17	3	—	26	0.61	1,237	0.2	520	47.0	—	1.8	19.8	—	90
49	60	0.12	58	0.02	0.15	7	0.38	31	0.06	419	0.6	19	8.0	2.6	1.6	3.6	—	15
50	495	0.35	18	0.05	0.13	10	—	41	1.43	820	0.5	11	33.9	—	2.3	—	—	1
51	484	1.02	39	0.04	0.20	9	—	33	1.67	3,248	1.6	22	39.1	11.3	2.4	4.5	—	1
52	93	0.77	48	0.04	0.43	—	—	100	0.34	1,536	—	30	6.0	12.5	3.6	2.8	—	—
53	45	0.15	15	0.01	0.08	10	0.20	15	0.26	668	0.2	41	12.4	5.2	1.3	3.0	—	7
54	—	0.12	—	0.01	0.11	—	—	—	—	—	—	—	—	—	1.1	—	—	—
55	—	0.12	29	0.25	0.06	—	—	—	0.22	640	0.3	24	13.0	2.2	3.5	2.2	—	—
56	130	0.01	1	0.07	0.71	7	0.10	23	0.3	2	2.5	6	4.2	5.4	1.2	0.2	—	—
57	80	0.28	30	0.03	0.22	16	0.11	25	0.27	2,860	4.2	22	11.0	10.8	2.6	11.0	—	40
58	120	0.27	29	0.01	0.43	22	0.10	40	0.32	2,794	1.3	29	16.0	6.0	2.9	4.5	—	2
59	170	0.30	43	0.02	0.44	15	0.10	55	0.32	2,731	1.3	22	15.0	5.0	2.9	3.2	—	3
60	140	0.68	34	0.2	0.30	35	—	100	1.45	3,791	—	264	29.9	11.1	3.0	3.0	—	—
61	30	0.75	23	0.2	—	4	—	98	—	2,894	—	220	24.0	16.0	4.7	3.1	—	—
62	44	—	43	—	0.15	8	—	32	—	462	—	—	—	—	0.4	—	—	—
63	170	0.52	113	0.05	0.22	14	0.85	100	0.48	1,232	1.2	186	31.0	7.0	4.6	8.0	—	14
64	46	0.16	55	0.04	0.24	6	0.30	65	0.11	1,534	0.8	42	13.3	4.6	2.2	22.8	—	33
65	50	0.16	118	0.12	0.26	18	0.80	100	0.37	1,439	0.8	98	13.0	9.0	2.2	16.5	—	40
66	73	0.25	117	0.02	0.20	12	0.43	109	—	1,813	1.7	107	22.3	7.2	4.2	19.1	—	54
67	60	0.17	32	0.04	0.12	6	0.20	34	0.11	1,090	0.4	48	9.9	3.4	1.4	4.5	—	13
68	40	0.10	24	0.06	0.12	7	0.06	28	0.11	1,002	0.4	57	11.0	4.0	1.2	4.3	—	13
69	130	0.13	6	1.3	1.04	46	0.08	3	0.34	1,369	0.08	10	44.0	4.0	27.1	4.1	23	13
70	238	0.25	8	1.50	1.05	7	0.10	7	0.64	4,392	1.4	19	69.0	4.0	45.8	5.7	23	0.2
71	120	0.23	5	0.07	0.38	33	1.00	39	1.05	3,984	9.9	448	109.0	42.8	37.0	91.8	1	2
72	90	0.13	13	0.07	0.34	14	1.00	99	1.39	2,881	22.4	500	73.0	36.3	47.7	6.2	4	—

Table Feed-2

AMINO ACID COMPOSITION OF SOME FEEDS COMMONLY USED FOR POULTRY AND SWINE (DATA ON AS-FED BASIS)

Entry No.	Feed Name Description	International Feed Number[a]	Dry Matter (%)	Protein (%)	Arginine (%)	Glycine (%)	Serine (%)	Histidine (%)	Isoleucine (%)	Leucine (%)	Lysine (%)	Methionine (%)	Cystine (%)	Phenylalanine (%)	Tyrosine (%)	Threonine (%)	Tryptophan (%)	Valine (%)
01	Alfalfa *Medicago sativa* meal dehydrated, 17% protein	1-00-023	88.0	17.0	0.69	0.82	0.72	0.57	0.67	1.19	0.73	0.24	0.19	0.81	0.81	0.69	0.23	0.84
02	meal dehydrated, 20% protein	1-00-024	92.0	20.0	0.92	0.97	0.89	0.34	0.88	1.30	0.87	0.31	0.25	0.85	0.59	0.76	0.33	0.97
03	Bakery waste dehydrated (dried bakery product)	4-00-466	92.0	9.8	0.47	0.82	0.65	0.13	0.45	0.73	0.31	0.17	0.17	0.40	0.41	0.49	0.10	0.42
04	Barley *Hordeum vulgare* grain	4-00 549	89.0	11.0	0.52	0.44	0.46	0.27	0.37	0.76	0.40	0.18	0.24	0.56	0.35	0.37	0.14	0.52
05	grain, Pacific coast	4-07-939	89.0	9.0	0.48	0.36	0.32	0.21	0.40	0.60	0.29	0.13	0.18	0.48	0.31	0.30	0.12	0.46
06	Broadbean *Vicia faba* seeds	5-09-262	87.0	23.6	2.12	1.02	1.15	0.82	0.95	1.76	1.50	0.18	0.28	1.00	0.80	0.85	0.70	1.07
07	Blood meal, vat dried	5-00-380	94.0	81.1	3.63	4.59	3.14	3.52	0.95	10.53	7.05	0.55	0.52	5.66	2.07	3.15	1.29	7.28
08	meal, spray or ring dried	5-00-381	93.0	88.9	3.62	3.95	4.25	5.33	0.98	11.32	7.88	1.09	1.03	5.85	2.63	3.92	1.35	7.53
09	Brewer's Grains dehydrated	5-02-141	92.0	25.3	1.28	1.09	0.80	0.57	1.44	2.48	0.90	0.57	0.39	1.45	1.19	0.98	0.34	1.66
10	Buckwheat, Common *Fagopyrum sagittatum* grain	4-00-994	88.0	10.8	1.02	0.71	0.41	0.26	0.37	0.56	0.61	0.20	0.20	0.44	0.21	0.46	0.19	0.54
11	Canola *Brassica napus-Brassica campestris* seeds, meal prepressed solvent extracted, low erucic acid, low glucosinolates	5-06-145	88.0	34.8	2.08	1.82	1.53	0.93	1.37	2.47	1.94	0.71	0.87	1.44	1.09	1.53	0.44	1.76
12	Casein dehydrated	5-01-162	93.0	87.2	3.61	1.79	5.81	2.78	4.82	9.00	7.99	2.65	0.21	4.96	5.37	4.29	1.05	6.46
13	precipitated dehydrated	5-20-837	92.0	85.0	3.42	1.81	5.52	2.52	4.77	8.62	7.31	2.80	0.15	4.81	5.17	4.00	0.98	5.82
14	Cattle skim milk, dehydrated	5-01-175	93.0	36.1	1.21	0.73	2.05	1.03	1.83	3.59	2.80	0.90	0.29	1.75	1.83	1.59	0.50	2.28
15	Coconut *Cocos nucifera* kernels with coats, meal solvent extracted (copra meal)	5-01-573	92.6	19.2	1.97	0.82	0.79	0.36	0.63	1.18	0.50	0.28	0.28	0.88	0.44	0.58	0.12	0.91
	Corn, Dent Yellow *Zea mays indentata*																	

continued

Table Feed-2 (Continued)

Entry No.	Feed Name Description	International Feed Number[a]	Dry Matter (%)	Protein (%)	Arginine (%)	Glycine (%)	Serine (%)	Histidine (%)	Isoleucine (%)	Leucine (%)	Lysine (%)	Methionine (%)	Cystine (%)	Phenylalanine (%)	Tyrosine (%)	Threonine (%)	Tryptophan (%)	Valine (%)
16	distillers' grains, dehydrated	5-28-235	94.0	27.9	0.97	0.49	0.70	0.62	0.99	3.01	0.78	0.40	0.24	0.94	0.84	0.49	0.20	1.18
17	distillers' grains with solubles, dehydrated	5-28-236	93.0	27.2	0.98	0.57	1.61	0.66	1.00	2.20	0.75	0.60	0.40	1.20	0.74	0.92	0.19	1.30
18	distillers' solubles, dehydrated	5-28-237	92.0	28.5	1.05	1.10	1.30	0.70	1.25	2.11	0.90	0.50	0.40	1.30	0.95	1.00	0.30	1.39
19	gluten, meal, 60% protein	5-28-242	88.0	60.2	1.82	1.67	2.96	1.20	2.45	10.04	1.03	1.49	1.10	3.56	3.07	2.00	0.36	2.78
20	gluten with bran (corn gluten feed)	5-28-243	90.0	22.0	1.01	0.99	0.80	0.71	0.65	1.89	0.63	0.45	0.51	0.77	0.58	0.89	0.10	0.05
21	grain	4-02-935	88.0	8.5	0.38	0.33	0.37	0.23	0.29	1.00	0.26	0.18	0.18	0.38	0.30	0.29	0.06	0.40
22	grits by-product (hominy feed)	4-03-011	90.0	10.0	0.47	0.40	0.50	0.20	0.40	0.84	0.40	0.13	0.13	0.35	0.49	0.40	0.10	0.49
	Cotton *Gossypium* spp.																	
23	seeds, meal mechanically extracted, 41% protein (expeller)	5-01-617	91.4	41.0	4.35	1.69	1.68	1.07	1.31	2.23	1.59	0.55	0.59	2.20	1.09	1.30	0.50	1.84
24	seeds, meal direct solvent extracted, 41% protein	5-07-872	90.4	41.4	4.66	1.69	1.78	1.10	1.33	2.41	1.76	0.51	0.62	2.23	1.14	1.34	0.52	1.82
25	seeds, meal prepressed solvent extracted, 41% protein	5-07-873	89.9	41.4	4.59	1.70	1.74	1.10	1.33	2.43	1.71	0.52	0.62	2.22	1.13	1.32	0.47	1.88
	Fish																	
26	solubles, condensed	5-01-969	51.0	31.5	1.61	3.41	0.83	1.56	1.06	1.86	1.73	0.50	0.30	0.93	0.40	0.86	0.31	1.16
27	solubles, dehydrated	5-01-971	92.0	63.6	2.78	5.89	2.02	2.18	1.95	3.16	3.28	1.00	0.66	1.48	0.78	1.35	0.51	2.22
	Fish, Anchovy *Engraulis ringen*																	
28	meal mechanically extracted	5-01-985	90.0	65.0	3.81	3.68	2.51	1.59	3.06	4.98	5.07	1.95	0.65	2.75	2.22	2.82	0.78	3.46
	Fish, Herring *Clupea harengus*																	
29	meal mechanically extracted	5-02-000	92.0	72.0	4.21	4.30	2.75	1.74	3.23	5.46	5.47	2.16	0.72	2.82	2.25	3.07	0.83	3.90
	Fish, Menhaden *Brevoortia tyrannus*																	
30	meal mechanically extracted	5-02-009	92.1	61.3	3.68	4.46	2.37	1.42	2.28	4.16	4.51	1.63	0.57	2.21	1.80	2.46	0.49	2.77
	Fish, White *Gadidae* (family)-*Lophiidae* (family)-*Rajidae* (family)																	
31	meal mechanically extracted	5-02-025	91.0	62.2	4.02	4.42	3.06	1.34	2.72	4.36	4.53	1.68	0.75	2.28	1.83	2.57	0.67	3.02
	Gelatin																	
32	process residue (gelatin by-products)	5-14-503	91.0	88.0	7.40	20.00	2.80	0.85	1.40	3.10	3.70	0.68	0.09	1.70	0.26	1.30	0.09	1.80

continued

Table Feed-2 (Continued)

Entry No.	Feed Name Description	International Feed Number[a]	Dry Matter (%)	Protein (%)	Arginine (%)	Glycine (%)	Serine (%)	Histidine (%)	Isoleucine (%)	Leucine (%)	Lysine (%)	Methionine (%)	Cystine (%)	Phenylalanine (%)	Tyrosine (%)	Threonine (%)	Tryptophan (%)	Valine (%)
	Hominy Feed—see Corn																	
	Livers																	
33	meal	5-00-389	92.0	65.6	4.14	5.57	2.49	1.47	3.09	5.28	4.80	1.22	0.89	2.89	1.69	2.48	0.59	4.13
	Meat																	
34	meal rendered	5-00-385	92.0	54.4	3.73	6.30	1.60	1.30	1.60	3.32	3.00	0.75	0.66	1.70	0.84	1.74	0.36	2.30
35	with bone, meal rendered	5-00-388	93.4	51.6	3.28	6.65	2.20	0.96	1.54	3.28	2.61	0.69	0.69	1.81	1.20	1.74	0.27	2.36
	Millet, Pearl *Peninsum glaucum*																	
36	grain	4-03-118	90.0	15.7	0.74	0.47	0.74	0.31	0.37	1.14	0.45	0.25	0.24	0.56	0.35	0.48	0.08	0.49
	Millet, Proso *Panicum miliaceum*																	
37	grain	4-03-120	87.5	9.1	0.35	0.31	0.40	0.22	0.35	1.14	0.21	0.16	0.17	0.47	0.34	0.29	0.08	0.44
	Oats *Avena sativa*																	
38	grain	4-03-309	89.0	11.4	0.79	0.50	0.40	0.24	0.52	0.89	0.50	0.18	0.22	0.59	0.53	0.43	0.16	0.68
39	grain, Pacific coast	4-07-999	91.0	9.0	0.60	0.40	0.30	0.10	0.40	0.30	0.40	0.13	0.17	0.44	0.20	0.20	0.12	0.51
40	hulls	1-03-281	92.0	4.6	0.14	0.14	0.14	0.07	0.14	0.25	0.14	0.07	0.06	0.13	0.14	0.13	0.07	0.20
	Pea *Pisum* spp.																	
41	seeds	5-03-600	88.8	23.8	2.23	1.00	1.08	0.59	0.97	1.65	1.68	0.74	0.33	1.10	0.73	0.84	0.18	1.10
	Peanut *Arachis hypogaea*																	
42	kernels, meal mechanically extracted (peanut meal) (expeller)	5-03-649	90.0	40.0	4.35	2.18	1.83	0.87	1.27	2.42	1.26	0.45	0.52	1.97	1.47	1.01	0.39	1.53
43	kernels, meal solvent extracted (peanut meal)	5-03-650	91.9	49.0	5.33	2.67	2.25	1.07	1.55	2.97	1.51	0.54	0.64	2.41	1.80	1.24	0.48	1.87
	Poultry																	
44	by-product, meal rendered (viscera with feet and heads)	5-03-798	94.2	59.5	3.94	6.17	2.71	1.07	2.16	3.99	3.10	0.99	0.98	2.29	1.68	2.17	0.37	2.87
45	feathers, meal hydrolyzed	5-03-795	91.0	82.9	5.57	6.13	8.52	0.95	3.91	6.94	2.28	0.57	4.34	3.94	2.48	3.81	0.55	5.93
	Rice *Oryza sativa*																	
46	bran with germ (rice bran)	4-03-928	89.1	13.7	0.96	0.70	0.59	0.35	0.45	0.91	0.59	0.26	0.27	0.60	0.42	0.48	0.12	0.68
47	grain, polished and broken (brewer's rice)	4-03-932	89.2	10.0	0.74	0.50	0.44	0.26	0.37	0.74	0.43	0.22	0.21	0.48	0.33	0.36	0.10	0.54
48	polishings	4-03-943	90.0	12.2	0.78	0.71	1.36	0.24	0.41	0.80	0.57	0.22	0.10	0.46	0.63	0.40	0.13	0.76
	Rye *Secale cereale*																	
49	grain	4-04-047	88.0	12.1	0.53	0.49	0.52	0.26	0.47	0.70	0.42	0.17	0.19	0.56	0.26	0.36	0.11	0.56
	Safflower *Carthamus tinctorius*																	
50	seeds, meal solvent extracted	5-04-110	92.0	27.0	2.21	1.53	0.99	0.61	1.02	1.74	0.90	0.42	0.45	1.10	0.71	0.85	0.37	1.42
51	seeds without hulls, meal solvent extracted	5-07-959	92.0	43.0	3.65	2.32	—	1.07	1.56	2.46	1.27	0.68	0.70	1.75	1.07	1.30	0.59	2.33
	Sesame *Sesamum indicum*																	

continued

Table Feed-2 (Continued)

#	Feed	Intl Feed No.[a]																
52	seeds, meal mechanically extracted	5-04-220	90.0	41.0	4.68	2.04	1.72	0.99	1.51	2.68	0.91	1.22	0.72	1.93	1.48	1.40	0.62	1.91
	Sorghum *Sorghum bicolor*																	
53	grain, 8–10% protein	4-20-893	87.5	9.1	0.35	0.31	0.40	0.22	0.35	1.14	0.21	0.16	0.17	0.47	0.34	0.29	0.08	0.44
54	grain, more than 10% protein	4-20-894	88.0	10.0	0.35	0.32	0.45	0.23	0.43	1.37	0.22	0.15	0.11	0.52	0.17	0.33	0.09	0.54
	Soybean *Glycine max5*																	
55	flour by-product (Soybean mill feed)	4-04-594	89.0	13.3	0.94	0.40	—	0.18	0.40	0.57	0.48	0.10	0.21	0.37	0.23	0.30	0.10	0.37
56	protein concentrate, more than 70% protein	5-08-038	93.0	84.1	6.70	3.30	5.30	2.10	4.60	6.60	5.50	0.81	0.49	4.30	3.10	3.30	0.81	4.40
57	seeds, heat processed	5-04-597	88.0	35.5	2.59	1.55	1.87	0.99	1.56	2.75	2.25	0.53	0.54	1.78	1.34	1.41	0.51	1.65
58	seeds, meal solvent extracted	5-04-604	88.2	44.0	3.14	1.90	2.29	1.17	1.96	3.39	2.69	0.62	0.66	2.16	1.91	1.72	0.74	2.07
59	seeds without hulls, meal solvent extracted	5-04-612	88.4	47.5	3.48	2.05	2.48	1.28	2.12	3.74	2.96	0.67	0.72	2.34	1.95	1.87	0.74	2.22
	Sunflower, common *Helianthus annuus*																	
60	seeds, meal solvent extracted	5-09-340	90.0	23.3	2.30	—	1.00	0.55	1.00	1.60	1.00	0.50	0.50	1.15	—	1.05	0.45	1.60
61	seeds without hulls, meal solvent extracted	5-04-739	89.8	36.8	2.85	2.03	1.49	0.87	1.43	2.22	1.24	0.80	0.64	1.66	0.91	1.29	0.41	1.74
	Triticale *Triticale hexaploide*																	
62	grain	4-20-362	88.0	11.8	0.57	0.48	0.52	0.26	0.39	0.76	0.39	0.26	0.26	0.49	0.32	0.36	0.14	0.51
	Wheat *Triticum aestivum*																	
63	bran	4-05-190	88.0	15.4	1.02	0.81	0.67	0.46	0.47	0.96	0.61	0.23	0.32	0.61	0.46	0.50	0.23	0.70
64	flour by-product, less than 4% fiber (wheat red dog)	4-05-203	88.0	15.3	0.96	0.74	0.75	0.41	0.55	1.06	0.59	0.23	0.37	0.66	0.46	0.50	0.10	0.72
65	flour by-product, less than 9.5% fiber (wheat middlings)	4-05-205	88.0	16.0	1.15	0.63	0.75	0.37	0.58	1.07	0.69	0.21	0.32	0.64	0.45	0.49	0.20	0.71
66	flour by-product, less than 7% fiber (wheat shorts)	4-05-201	88.0	16.5	1.18	0.96	0.77	0.45	0.58	1.09	0.79	0.27	0.36	0.67	0.47	0.60	0.21	0.83
67	grain, hard red winter	4-05-268	88.1	13.3	0.60	0.59	0.59	0.31	0.44	0.89	0.37	0.21	0.30	0.60	0.43	0.39	0.16	0.57
68	grain, soft white winter	4-05-337	89.0	10.2	0.40	0.49	0.55	0.20	0.42	0.59	0.31	0.15	0.22	0.45	0.39	0.32	0.12	0.44
	Whey *Bos taurus*																	
69	dehydrated	4-01-182	93.0	12.0	0.34	0.30	0.32	0.18	0.82	1.19	0.97	0.19	0.30	0.33	0.25	0.89	0.19	0.68
70	low lactose, dehydrated (dried whey product)	4-01-186	91.0	15.5	0.67	1.04	0.76	0.25	0.90	1.35	1.47	0.57	0.57	0.50	0.35	0.85	0.23	0.83
	Yeast, Brewer's *Saccharomyces cerevisiae*																	
71	dehydrated	7-05-527	93.0	44.4	2.19	2.09	—	1.07	2.14	3.19	3.23	0.70	0.50	1.81	1.49	2.06	0.49	2.32
	Yeast, Torula *Torulopsis utilis*																	
72	dehydrated	7-05-534	93.0	47.2	2.60	2.60	2.76	1.40	2.90	3.50	3.80	0.80	0.60	3.00	2.10	2.60	0.50	2.90

Note: Dash indicates that no data were available.

[a]First digit is class of feed: 1, dry forages and roughages; 2, pasture, range plants, and forages fed green; 3, silages; 4, energy feeds; 5, protein supplements; 6, minerals; 7, vitamins; 8, additives; the other five digits are the International Feed Number.

Table Feed-3
MINERAL CONCENTRATIONS IN MACRO MINERAL SOURCES (DATA ON AS-FED BASIS)

Entry No.	Description	International Feed Number	Calcium[a] (%)	Phosphorus (%)	Phosphorus Bioavailability[b] (%)	Sodium (%)	Chlorine (%)	Potassium (%)	Magnesium (%)	Sulfur (%)	Iron (%)	Manganese (%)
01	Bonemeal, steamed	6-00-400	29.80	12.50	80 to 90	0.04	—	0.20	0.30	2.40	—	0.03
02	Calcium carbonate	6-01-069	38.50	0.02	—	0.08	0.02	0.08	1.61	0.08	0.06	0.02
03	Calcium phosphate (dicalcium)	6-01-080	20 to 24	18.50	95 to 100	0.18	0.47	0.15	0.80	0.80	0.79	0.14
04	Calcium phosphate (monocalcium)	6-26-334	17.00	21.10	100	0.20	—	0.16	0.90	0.80	0.75	0.01
05	Calcium sulfate, dihydrate	6-01-090	21.85	—	—	—	—	—	0.48	16.19	—	—
06	Limestone, ground[c]	6-02-632	35.84	0.01	—	0.06	0.02	0.11	2.06	0.04	0.35	0.02
07	Magnesium carbonate	6-02-754	0.02	—	—	—	—	—	30.20	—	—	0.01
08	Magnesium oxide	6-02-756	1.69	—	—	—	—	0.02	55.00	0.10	1.06	—
09	Magnesium sulfate, heptahydrate	6-02-758	0.02	—	—	—	0.01	—	9.60	13.04	—	—
10	Phosphate, defluorinated	6-01-780	32.00	18.00	85 to 95	3.27	—	0.10	0.29	0.13	0.84[d]	0.05
11	Phosphate, monoammonium	6-09-338	0.35	24.20	100	0.20	—	0.16	0.75	1.50	0.41	0.01
12	Phosphate, rock curaçao, ground	6-05-586	35.09	14.23	40 to 60	0.20	—	—	0.80	—	0.35	—
13	Phosphate, rock, soft	6-03-947	16.09	9.05	30 to 50	0.10	—	—	0.38	—	1.92	0.10
14	Potassium chloride	6-03-755	0.05	—	—	1.00	46.93	51.37	0.23	0.32	0.06	0.001
15	Potassium and magnesium sulfate	6-06-177	0.06	—	—	0.76	1.25	18.45	11.58	21.97	0.01	0.002
16	Potassium sulfate	6-08-098	0.15	—	—	0.09	1.50	43.04	0.60	17.64	0.07	0.001
17	Sodium carbonate	6-12-316	—	—	—	43.30	—	—	—	—	—	—
18	Sodium bicarbonate	6-04-272	0.01	—	—	27.00	—	0.01	—	—	—	—
19	Sodium chloride	6-04-152	0.30	—	—	39.50	59.00	—	0.005	0.20	0.01	—
20	Sodium phosphate, dibasic	6-04-286	—	21.15	100	31.04	—	—	—	—	—	—
21	Sodium phosphate, monobasic	6-04-288	0.09	24.94	100	18.65	0.02	0.01	0.01	—	—	—
22	Sodium sulfate, decahydrate	6-04-291	—	—	—	13.80	—	—	—	9.70	—	—

NOTE: The mineral supplements used as feed supplements are not chemically pure compounds, and the composition may vary substantially among sources. The supplier's analysis should be used if it is available. For example, feed-grade dicalcium phosphate contains some monocalcium phosphate and feed-grade monocalcium phosphate contains some dicalcium phosphate. Dashes indicate that no data were available.

[a]Estimates suggest 90% to 100% bioavailability of calcium in most sources of monocalcium phosphate, dicalcium phosphate, tricalcium phosphate, defluorinated phosphate, calcium carbonate, calcium sulfate, and calcitic limestone. The calcium in high-magnesium limestone or dolomitic limestone is less bioavailable (50% to 80%).

[b]Bioavailability estimates are generally expressed as a percentage of monosodium phosphate or monocalcium phosphate.

[c]Most calcitic limestones will contain 38% or more calcium and less magnesium than shown.

[d]Iron in defluorinated phosphate is about 65% as available as the iron in ferrous sulfate.

Table Feed-4
INORGANIC SOURCES AND ESTIMATED
BIOAVAILABILITIES OF TRACE MINERALS[a]

Mineral Element and Source[b]	Chemical Formula	Mineral Content (%)	Relative Bioavailability (%)
Copper			
Cupric sulfate (pentahydrate)	$CuSO_4 \cdot 5H_2O$	25.2	100
Cupric chloride, tribasic	$Cu_2(OH)_3Cl$	58.0	100
Cupric oxide	CuO	75.0	0 to 10
Cupric carbonate (monohydrate)	*$CuCO_3 \cdot Cu(OH)_2 \cdot H_2O$*	*50 to 55*	*60 to 100*
Cupric sulfate (anhydrous)	*$CuSO_4$*	*39.9*	*100*
Iron			
Ferrous sulfate (monohydrate)	$FeSO_4 \cdot H_2O$	30.0	100
Ferrous sulfate (heptahydrate)	$FeSO_4 \cdot 7H_2O$	20.0	100
Ferrous carbonate	$FeCO_3$	38.0	15 to 80
Ferric oxide	*Fe_2O_3*	*69.9*	*0*
Ferric chloride (hexahydrate)	*$FeCl_3 \cdot 6H_2O$*	*20.7*	*40 to 100*
Ferrous oxide	*FeO*	*77.8*	*—[c]*
Iodine			
Ethylenediamine dihydroiodide (EDDI)	$C_2H_8N_2 2HI$	79.5	100
Calcium iodate	$Ca(IO_3)_2$	63.5	100
Potassium iodide	KI	68.8	100
Potassium iodate	*KIO_3*	*59.3*	*—[c]*
Cupric iodide	*CuI*	*66.6*	*100*
Manganese			
Manganous sulfate (monohydrate)	$MnSO_4 \cdot H_2O$	29.5	100
Manganous oxide	MnO	60.0	70
Manganous dioxide	*MnO_2*	*63.1*	*35 to 95*
Manganous carbonate	*$MnCO_3$*	*46.4*	*30 to 100*
Manganous chloride (tetrahydrate)	*$MnCl_2 \cdot 4H_2O$*	*27.5*	*100*
Selenium			
Sodium selenite	Na_2SeO_3	45.0	100
Sodium selenate (decahydrate)	*$Na_2SeO_4 \cdot 10H_2O$*	*21.4*	*100*
Zinc			
Zinc sulfate (monohydrate)	$ZnSO_4 \cdot H_2O$	35.5	100
Zinc oxide	ZnO	72.0	50 to 80
Zinc sulfate (heptahydrate)	*$ZnSO_4 \cdot 7H_2O$*	*22.3*	*100*
Zinc carbonate	*$Zn \cdot CO_3$*	*56.0*	*100*
Zinc chloride	*$ZnCl_2$*	*48.0*	*100*

[a]The mineral source listed first under each mineral element was generally the standard with which the other sources were compared to establish relative bioavailability.

[b]Less commonly used sources in italic.

[c]—Indicates no data available.

71

Glossary of Terms Frequently Used in Discussing Matters Related to Feeds and Feeding

ABOMASUM The fourth compartment of a ruminant's stomach. Sometimes called the *true stomach*.

ABORTION The expulsion of a nonviable, immature fetus.

ABSCESS A collection of pus in any part of the body.

ABSORPTION The movement of nutrients or other substances from the digestive tract or through the skin into the blood and/or lymph system.

ACETIC ACID One of the volatile fatty acids with the formula CH_3COOH. Commonly found in silage, rumen contents, and vinegar.

ADDITIVE An ingredient or a combination of ingredients added, usually in small quantities, to a basic feed mix for the purpose of fortifying the basic mix with certain essential nutrients and/or medicines.

ADIPOSE Of a fatty nature.

AD LIBITUM As desired by the animal.

ADRENAL Near the kidney.

AERIAL PART The above-ground part of a plant.

AEROBIC Living or functioning in the presence of air or molecular oxygen.

AFTERBIRTH   The membranes expelled from the uterus following delivery of a fetus.

ALANINE   One of the nonessential amino acids.

ALIMENTARY   Having to do with feed or food.

ALIMENTARY TRACT   Same as *digestive tract.*

AMINO ACID   Any one of a class of organic compounds that contain both the amino (NH$_2$) group and the carboxyl (COOH) group.

AMMONIATED   Combined or impregnated with ammonia or an ammonium compound.

AMYLASE   Any one of several enzymes that effect a hydrolysis of starch to maltose. Examples are pancreatic amylase (amylopsin) and salivary amylase (ptyalin).

ANABOLISM   The conversion of simple substances into more complex substances by living cells. Constructive metabolism.

ANAEROBIC   Living or functioning in the absence of air or molecular oxygen.

ANEMIC   Lacking in size and/or number of red blood cells.

ANIMAL PROTEIN FACTOR   What was once an unidentified growth factor essential for poultry and swine and present in protein feeds of animal origin. It is now known to be the same as vitamin B$_{12}$.

ANTACID   A substance that counteracts acidity.

ANTIBIOTIC   A substance produced by one microorganism that has an inhibiting effect on the growth of another.

ANTIBODY   Substance produced in the body that acts against disease.

ANTIOXIDANT   A material capable of chemically protecting other substances from oxidation.

ANUS   The posterior end and opening of the digestive tract.

ARACHIDONIC ACID   A 20-carbon unsaturated fatty acid having 4 double bonds.

ARGININE   One of the essential amino acids.

ARTIFICIALLY DRIED   Dried by other than natural means. Dehydrated.

ASCORBIC ACID   Same as *vitamin C,* the antiscorbutic vitamin.

AS FED   As consumed by the animal.

ASH   The incombustible residue remaining after incineration at 600°C for several hours.

ASPARTIC ACID   One of the nonessential amino acids.

ASPHYXIA   Suffocation or the suspension of animation as the result of suffocation.

ASPIRATED   Removal of light materials from heavier material by use of air.

ATROPHY   A wasting away of a part of the body.

AVIDIN   A protein in egg albumen that can combine with biotin to render the latter unavailable to the animal.

BACTERIA   Very small, unicellular plant organisms.

BALANCED   Containing essential nutrients in the proper proportions.

BALANCED DAILY RATION   A combination of feeds that will provide the essential nutrients in the proper amounts to nourish a given animal for a 24-hour period.

BALANCED RATION   A combination of feeds that will provide the essential nutrients in the proper proportions.

BASAL METABOLISM   The heat production of an animal during physical, digestive, and emotional rest.

BILE   A greenish-yellow fluid formed in the liver, stored in the gall bladder (except in the horse, which has no gall bladder), and secreted via the bile duct into the upper small intestine. It functions in digestion.

BIOCHEMISTRY   The chemistry of living things.

BIOLOGICAL   Pertaining to the science of life.

BIOLOGICAL FUNCTION   The role played by a chemical compound in living organisms.

BIOLOGICAL VALUE   The efficiency with which a protein furnishes the proper proportions of the essential amino acids. A protein that has a high biological value is said to be of *good quality.*

BIOSYNTHESIS   The formation of chemical substances from other chemical substances in a living organism.

BIOTIN   One of the B vitamins.

BOILING POINT   The temperature at which the vapor pressure of a liquid equals the atmospheric pressure.

BOLTED   Separated from parent material by means of a bolting cloth.

BOMB CALORIMETER   An instrument used for determining the gross energy content of a material.

BRAN   The pericarp or seed coat of grain removed during processing.

B.T.U. (BRITISH THERMAL UNIT)   The amount of heat energy required to raise the temperature of 1 lb of water 1°F. Equal to 252 calories.

BUFFER   Any substance that can counteract changes in free acid or alkali concentration.

BUSHEL   A certain volume equal to 2150.42 cubic inches (approximately 1.25 cubic feet).

BUTYRIC ACID   One of the volatile fatty acids with the formula $CH_3CH_2CH_2COOH$. Commonly found in rumen contents and poor-quality silage.

CAECUM   Same as *cecum.*

CALCIFICATION   Process by which organic tissue becomes hardened by a deposit of calcium salts.

CALORIC   Pertaining to heat or energy.

CALORIE   The amount of energy as heat required to raise the temperature of 1 gram of water 1°C (precisely from 14.5°C to 15.5°C).

CALORIMETER   An instrument for measuring heat.

CALORIMETRY   The science of measuring heat.

CARBOHYDRATE   Organic substances that contain carbon, hydrogen, and oxygen, with the hydrogen and oxygen present in the same proportions as in water.

CARCASS   The body of an animal less the viscera and usually the head, skin, and lower leg.

CARCINOGEN   Any cancer-producing substance.

CARCINOGENIC   Cancer-producing.

CARDIOVASCULAR   Pertaining to the heart and blood vessels.

CARIES   Areas of tooth decay.

CAROTENE   A yellow organic compound that is a precursor of vitamin A.

CARRIER   An edible material that is used to facilitate the addition of micronutrients to a ration.

CARTILAGE   The gristle or connective tissue attached to the ends of bones.

CASEIN   The protein precipitated from milk by acid and/or rennin.

CATABOLISM   The conversion of complex substances into more simple compounds by living cells. Destructive metabolism.

CATALYST   A substance that speeds up the rate of a chemical reaction but is not itself used up in the reaction.

CECUM   An intestinal pouch located at the junction of the large and small intestine. Also *caecum.*

CELL   The structural and functional microscopic unit of plant and animal organisms.

CELL PLATELET   A small, colorless, disk-shaped cell in the blood concerned with blood coagulation.

CELLULOSE   A polysaccharide having the formula $(C_6H_{10}O_5)_n$. Found in the fibrous portion of plants. Low in digestibility.

CELSIUS   Same as *Centigrade.*

CENTIGRADE   A thermometer scale in which water freezes at 0° and boils at 100°. Same as *Celsius.*

CHLOROPHYLL   The green coloring matter present in growing plants.

CHOLESTEROL   The most common member of the sterol group.

CHOLINE   One of the B vitamins.

CHOPPED   Reduced in particle size by cutting.

CHROMATOGRAPHY   A technique for separating complex mixtures of chemical substances.

CITRULLINE   One of the nonessential amino acids.

CLIPPED   With oat grain, the more fibrous end has been removed.

COAGULATED   Curdled, clotted, or congealed.

COAGULATION   The change from a fluid state to a thickened jelly, curd, or clot.

COENZYME   A partner required by some enzymes to produce enzymatic activity.

COLLAGEN   The main supportive protein of connective tissue.

COLOSTRUM MILK   The milk secreted during the first few days of lactation.

COMBUSTION   The combination of substances with oxygen accompanied by the liberation of heat.

COMMERCIAL FEED   Any material produced by a commercial company and distributed for use as a feed or feed component.

COMPLETE RATION   A single feed mixture into which has been included all of the dietary essentials, except water, of a given class of livestock.

CONCENTRATE   Any feed low (less than about 20%) in crude fiber and high (more than about 60%) in TDN on an air-dry basis. Opposite of roughage. Also, a concentrated source of one or more nutrients used to enhance the nutritional adequacy of a supplement mix.

CONGENITAL   Existing at birth.

CONGESTION   Excessive accumulation of blood in a part of the body.

CONVULSION   A violent involuntary contraction or series of contractions of the voluntary muscles.

CORONARY   Of or relating to the heart.

CREATININE   A nitrogenous compound arising from protein metabolism and secreted in the urine.

CRIMPED   Having been passed between rollers with corrugated surfaces.

CRUDE FAT   That part of a feed that is soluble in ether. Also referred to as *ether extract.*

CRUDE FIBER   The more fibrous, less digestible, portion of a feed. Consists primarily of cellulose and lignin.

CRUDE PROTEIN   Total ammoniacal nitrogen × 6.25, based on the fact that feed protein on the average contains 16% nitrogen.

CUD   A bolus of previously eaten food that has been regurgitated by a ruminant animal for further chewing.

CURD   The semi-solid mass that is formed when milk comes in contact with an acid or the enzyme rennin. It consists mainly of the protein casein.

CYANOCOBALAMIN   Same as *vitamin B_{12}*.

CYSTINE   One of the nonessential amino acids. It is sulfur containing and may be used to meet, in part, the need for methionine.

CYSTITIS   Inflammation of the bladder.

DEFICIENCY DISEASE   A disease resulting from an inadequate dietary intake of some nutrient.

DEFLUORINATED   Having had the fluorine content reduced to a level that is nontoxic under normal use.

DEHYDRATED   Having had most of the moisture removed through artificial drying.

DERMATITIS   Inflammation of the skin.

DESICCATE   To dry completely.

DEXTRIN   An intermediate polysaccharide product obtained during starch hydrolysis.

DIGESTIBLE ENERGY   The part of the gross energy of a feed that does not appear in the feces.

DIGESTION   The processes involved in the conversion of feed into absorbable forms.

DIGESTIVE TRACT   The passage from the mouth to the anus through which feed passes following consumption as it is subjected to various digestive processes. Primarily the gullet, stomach, and intestines.

DIP   Degraded intake protein (broken down in the rumen).

DISACCHARIDE   Any one of several so-called compound sugars that yield two monosaccharide molecules upon hydrolysis. Sucrose, maltose, and lactose are the most common.

DISPENSABLE AMINO ACID   Basically the same as *nonessential amino acid.*

DRY MATTER   The part of feed that is not water. Sometimes referred to as *dry substance* or *total solids.* Is the sum of the crude protein, crude fat, crude fiber, nitrogen-free extract, and ash.

DRY-RENDERED   Having been heat processed for the removal of fat without the addition of water or steam.

DUODENUM   The upper portion of the small intestine, which extends from the stomach to the jejunum.

DYSTOCIA   Difficult parturition.

EDEMA   Swelling of part of, or of the entire body due to an accumulation of excess water.

ELEMENT   Any one of the fundamental atoms of which all matter is composed.

EMACIATED   An excessively thin condition of the body.

EMULSIFY   To disperse small drops of one liquid into another liquid.

ENDEMIC   Occurring in low incidence but more or less constantly in a given population.

ENDOCRINE   Pertaining to internal secretions.

ENDOGENOUS   Originating from within the organism.

ENDOMETRIUM   The mucous membrane that lines the uterus.

ENERGY   The capacity to perform work.

ENSILAGE   The same as *silage*.

ENSILED   Having been subjected to anaerobic fermentation to form silage.

ENTERITIS   Inflammation of the intestines.

ENVIRONMENTAL   Pertaining to surrounding influences.

ENZYMATIC   Related to an enzyme.

ENZYME   One of a class of organic compounds, formed by living cells, capable of producing or accelerating specific organic reactions. An organic catalyst.

EPIDEMIC   When many individuals in a given region are attacked by some disease at the same time.

EPITHELIAL   Refers to the cells that form the outer layer of the skin and other membranes.

ERGOSTEROL   One of the sterols that, upon exposure to ultraviolet light, is converted to vitamin D_2. It is of plant origin.

ESOPHAGUS   The passageway leading from the mouth to the stomach. Sometimes called the *gullet*.

ESSENTIAL AMINO ACID   Any one of several amino acids that are needed by animals and cannot be synthesized by them in the amount needed and so must be present in the protein of the feed as such.

ESTROGENS   Estrus-producing hormones secreted by the ovaries.

ESTRUS   The recurring periods of sexual receptivity in female mammals. The period of *heat*.

ETIOLOGY   The causes of a disease or disorder.

EXCRETA   The products of excretion—primarily feces and urine.

EXOGENOUS.   Originating from outside of the organism.

EXPANDED   As applied to feed—having been increased in volume as the result of a sudden reduction in surrounding pressure.

EXPELLER PROCESS   A process for the mechanical extraction of oil from seeds, involving the use of a screw press.

EXTRINSIC FACTOR   A factor coming from or originating from outside an organism.

EXTRUDED   As applied to feed—having been forced through a die under pressure.

FACTOR   In nutrition, any chemical substance found in feed.

FAHRENHEIT   A thermometer scale in which water freezes at 32° and boils at 212°.

FAT   The product formed when 3 fatty acids combine with 1 glycerol. The glyceryl ester of a fatty acid. Stearin, palmitin, and olein are examples.

FAT SOLUBLE   Soluble in fats and fat solvents but generally not soluble in water.

FATTENING   This is the deposition of unused energy in the form of fat within the body tissues.

FATTY ACID   Any one of several organic compounds containing carbon, hydrogen, and oxygen, which combine with glycerol to form fat.

FAUNA   The animal life present. Frequently used to refer to the overall protozoal population present.

FECES   The excreta discharged from the digestive tract through the anus.

FEED   Any material eaten by an animal as a part of its daily ration.

FEED GRADE   Suitable for animal but not for human consumption.

FERMENTATION   Chemical changes brought about by enzymes produced by various microorganisms.

FETUS   The unborn young of animals.

FIBROUS   High in content of cellulose and/or lignin.

FINISH   To fatten a slaughter animal. Also, the degree of fatness of such an animal.

FISTULA   An abnormal tube-like passage from some part of the body to another part or to the exterior—sometimes surgically inserted.

FLAKED   Rolled or cut into flat pieces.

FLORA   The plant life present. In nutrition it generally refers to the bacteria present in the digestive tract.

FODDER   The entire above-ground part of nearly mature corn or sorghum in the fresh or cured form.

FOLACIN   Same as *folic acid.* One of the B vitamins.

FOLIC ACID   Same as *folacin,* which is one of the B vitamins.

FORAGE   Crops used as pasture, hay, haylage, silage, or green chop for feeding purposes.

FORMULA FEED   A feed consisting of 2 or more ingredients mixed in specified proportions.

FORTIFY   Nutritionally, to add 1 or more nutrients to a feed.

FRACTIONATION   The laboratory separation of natural materials into their component parts.

FREE CHOICE   Free to eat 2 or more feeds at will.

FRESH   Usually denotes the green or wet form of a feed material.

FRUCTOSE   A hexose monosaccharide found especially in ripe fruits and honey. Obtained along with glucose from sucrose hydrolysis. Commonly known as *fruit sugar.*

GALACTOSE   A hexose monosaccharide obtained along with glucose from lactose hydrolysis.

GALL BLADDER   A membranous sac lying next to the liver of all farm livestock, except the horse, in which bile is stored.

GASTRIC   Pertaining to the stomach.

GASTRIC JUICE   A clear liquid secreted by the wall of the stomach. It contains hydrochloric acid and the enzymes rennin, pepsin, and gastric lipase.

GASTRITIS   Inflammation of the stomach.

GASTROENTERITIS   Inflammation of the stomach and intestines.

GASTROINTESTINAL   Pertaining to the stomach and intestines.

GENETIC   Pertaining to heredity.

GENITOURINARY   Refers to the organs of reproduction and urine excretion.

GERM   Embryo of a seed.

GESTATION   The period of pregnancy.

GINGIVITIS   Inflammation of the gums.

GLAND   An organ that produces and secretes a chemical substance in the body.

GLUCOSE   A hexose monosaccharide obtained upon the hydrolysis of starch and certain other carbohydrates. Also called *dextrose.* The basic product of photosynthesis.

GLUTAMIC ACID   One of the nonessential amino acids.

GLYCEROL   An alcohol containing 3 carbons and 3 hydroxy groups.

GLYCINE   One of the nonessential amino acids.

GLYCOGEN   A polysaccharide with the formula $(C_6H_{10}O_5)_n$, which is formed in the liver and depolymerized to glucose to serve as a ready source of energy when needed by the animal. Known also as *animal starch.*

GOITER   An enlargement of the thyroid gland located in the neck. Sometimes caused by an iodine deficiency.

GOSSYPOL   A substance present in cottonseed and cottonseed meal that is toxic to swine and certain other nonruminant animals.

GRAVID   Pregnant.

GREEN CHOP   Forage harvested and fed in the green, chopped form.

GROAT   Grain from which the hull has been removed.

GROSS ENERGY   The total heat of combustion of a material as determined by the use of a bomb calorimeter.

GROUND   Reduced in particle size by impact, shearing, or attrition.

GROWTH   An increase in muscle, bone, vital organs, and connective tissue as contrasted to fattening or fat deposition.

HAY   The aerial part of finer-stemmed forage crops stored in the dry form for animal feeding.

HEAT INCREMENT   The heat that is unavoidably produced by an animal incidental with nutrient digestion and utilization. Was originally called *work of digestion.*

HEAT LABILE   Unstable to heat.

HEMOGLOBIN   The oxygen-carrying, red-pigmented protein of the red corpuscles.

HEMORRHAGE   Copious loss of blood through bleeding.

HEPATITIS   Inflammation of the liver.

HEXOSAN   A hexose-based polysaccharide having the general formula $(C_6H_{10}O_5)_n$. Cellulose, starch, and glycogen are the most common.

HEXOSE   A 6-carbon monosaccharide having the formula $C_6H_{12}O_6$. Glucose, fructose, and galactose are common examples.

HISTIDINE   One of the essential amino acids.

HOMOGENIZED   The fat within a fluid having been reduced to globules so small they remain in suspension for an extended period of time.

HORMONE   A chemical substance secreted into the body fluids by an endocrine gland that has a specific effect on other tissues.

HULLS   The outer protective covering of seeds.

HUSKS   Usually refers to the fibrous covering of an ear of corn.

HYDRAULIC PROCESS   A process for the mechanical extraction of oil from seeds, involving the use of a hydraulic press. Sometimes referred to as the *old process.*

HYDROGENATION   The chemical addition of hydrogen to any unsaturated compound.

HYDROLYSIS   The splitting of a substance into the smaller units by its chemical reaction with water.

HYDROXYPROLINE   One of the nonessential amino acids.

HYPER   A prefix meaning in excess of the normal.

HYPEREMIA   An excess of blood in any part of the body.

HYPERTENSION   An abnormally high tension—usually associated with high blood pressure.

HYPERTHYROIDISM   Overactivity of the thyroid gland.

HYPERTROPHIED   Having increased in size independent of natural growth.

HYPERVITAMINOSIS   An abnormal condition resulting from the intake of an excess of one or more vitamins.

HYPO   A prefix denoting less than the normal amount.

HYPOMAGNESEMIA   An abnormally low level of magnesium in the blood.

HYSTERITIS   Inflammation of the uterus.

ILEUM   The lower portion of the small intestine extending from the jejunum to the cecum.

IMPERMEABLE   Not capable of being penetrated.

INACTIVATE   To render a substance inactive.

INCIDENCE   The frequency of occurrence of a situation or a condition.

INDISPENSABLE AMINO ACID   Basically the same as *essential amino acid.*

INERT   Relatively inactive.

INGEST   To eat or take in through the mouth.

INORGANIC   Denotes chemical compounds that do not contain carbon in chain structure.

INOSITOL   One of the B vitamins.

INSULIN   A hormone secreted by the pancreas into the blood. It regulates sugar metabolism.

INTESTINAL JUICE   A clear liquid secreted by glands in the wall of the small intestine. It contains the enzymes intestinal lactase, maltase, and sucrase, and several peptidases.

INTESTINAL TRACT   The small and large intestine.

INTESTINE, LARGE   The tube-like part of the digestive tract lying between the small intestine and the anus. Larger in diameter but shorter in length than the small intestine.

INTESTINE, SMALL   The long, tortuous, tube-like part of the digestive tract leading from the stomach to the cecum and large intestine. Smaller in diameter but longer than the large intestine.

INTRINSIC FACTOR   A chemical substance in normal stomach juice necessary for the absorption of vitamin B_{12}.

INULIN   A polysaccharide found especially in Jerusalem artichokes that yields fructose upon hydrolysis.

IODINE NUMBER   A number that denotes the degree of unsaturation of a fat or fatty acid. It is the amount of iodine in grams that can be taken up by 100 g of the fat or fatty acid.

IRRADIATION   The act of treating with ultraviolet light.

ISOLEUCINE   One of the essential amino acids.

JEJUNUM   The middle portion of the small intestine that extends from the duodenum to the ileum.

KERATIN   A sulfur-containing protein that is the primary component of epidermis, hair, wool, hoof, horn, and the organic matrix of the teeth.

KERNEL   A dehulled seed.

KILOCALORIE   1,000 calories.

LABILE   Unstable. Easily destroyed.

LACTASE   An enzyme present in intestinal juice that acts on lactose to produce glucose and galactose.

LACTATION   The secretion of milk.

LACTIC ACID   An organic acid, one form ($CH_3CHOH \cdot COOH$) of which is commonly found in sour milk, sauerkraut, and silage. Other forms enter into body metabolism.

LACTOSE   A disaccharide found in milk having the formula $C_{12}H_{22}O_{11}$. It hydrolyzes to glucose and galactose. Commonly known as *milk sugar.*

LD$_{50}$ A dose that is lethal for 50% of the test animals.

LESION   Any unhealthy change in the structure of a part of the body.

LEUCINE   One of the essential amino acids.

LIGNIN   An indigestible compound that, along with cellulose, is a major component of the cell wall of certain plant materials such as wood, hulls, straws, and overripe hays.

LINOLEIC ACID   An 18-carbon unsaturated fatty acid having 2 double bonds. It reacts with glycerol to form linolein.

LINOLEIN   An unsaturated fat formed from the reaction of linoleic acid with glycerol.

LINOLENIC ACID   An 18-carbon unsaturated fatty acid having 3 double bonds.

LIPASE   A fat-splitting enzyme. Gastric lipase is present in gastric juice and pancreatic lipase is present in pancreatic juice. Both act on fats to produce fatty acids and glycerol.

LIPIDS   A broad term for fats and fat-like substances.

LYMPH   The slightly yellow, transparent fluid occupying the lymphatic channels of the body.

LYSINE   One of the essential amino acids.

MALFORMATION   Any abnormal development of a part of the body.

MALIGNANT   Virulent or destructive as applied to cancer.

MALTASE   An enzyme that acts on maltose to produce glucose. Salivary maltase is in saliva, and intestinal maltase is in intestinal juice.

MALTOSE   A disaccharide having the formula $C_{12}H_{22}O_{11}$. Obtained from the partial hydrolysis of starch. It hydrolyzes to glucose.

MAMMARY GLANDS   The milk-secreting glands.

MANURE   The refuse from animal quarters consisting of excreta with or without litter or bedding.

MATRIX   The intercellular framework of a tissue.

MEAL   A feed ingredient having a particle size somewhat larger than flour.

MECHANICALLY EXTRACTED   Having had its fat content removed by the application of heat and mechanical pressure. The hydraulic and expeller processes are both methods of mechanical extraction.

MEDIUM   A nutrient substrate used for supporting the growth of microorganisms.

MEGACALORIE   1,000 kilocalories or 1,000,000 calories.

METABOLISM   The sum of all the physical and chemical processes taking place in a living organism.

METABOLITE   Any substance produced by metabolism.

METABOLIZABLE ENERGY   Digestible energy minus the energy of the urine and fermentation gases.

METHIONINE   One of the essential amino acids. It is sulfur containing and may be replaced in part by cystine.

METRITIS   Inflammation of the uterus.

MICROBE   Same as *microorganism.*

MICROBIOLOGICAL   Pertaining to microorganisms.

MICROFLORA   The gross overall bacterial population present. Is sometimes used to include the protozoa as well as the bacteria.

MICROGRAM   One millionth of a gram or one thousandth of a milligram.

MICRO-INGREDIENT   Any ration component normally measured in milligrams or micrograms per kilogram or in parts per million.

MICROORGANISM   A very small living organism—usually microscopic in size.

MIDDLINGS   A by-product of flour milling consisting of varying proportions of small particles of bran, endosperm, and germ.

MILLIGRAM   One-thousandth of a gram.

MILL RUN   A product as it comes from the mill, having no definite specifications.

MISCIBLE   Capable of being mixed easily with another substance.

MOLASSES   A thick, viscous, usually dark colored, liquid product containing a high concentration of soluble carbohydrates, minerals, and certain other materials.

MOLECULE   A chemical combination of 2 or more atoms.

MONOSACCHARIDE   Any one of several simple, nonhydrolyzable sugars. Glucose, fructose, galactose, arabinose, xylose, and ribose are examples.

MORBIDITY   A state of sickness.

MORIBUND   In a dying state—near death.

MUCOSA   The membrane that lines the passages and cavities of the body.

MUCOUS MEMBRANE   A membrane lining the cavities and canals of the body, kept moist by mucus.

MUCUS   A slimy liquid secreted by the mucous glands and membranes.

MYCOTOXIN   A fungus or bacterial toxin. Sometimes present in feed material.

NECROSIS   Death of a part of the cells making up a living tissue.

NEONATE   A newly born animal.

NEPHRITIS   Inflammation of the kidneys.

NET ENERGY   The part of metabolizable energy that the animal has complete control over the use of. It is metabolizable energy minus the heat increment.

NEURITIC   Pertaining to the nerves.

NEW PROCESS   Pertains to the extraction of oil from seeds. Same as expeller process.

NIACIN   Same as *nicotinic acid.* One of the B vitamins. Nicotinamide also has niacin activity.

NICOTINAMIDE   The amide of nicotinic acid. It has niacin activity.

NICOTINIC ACID   Same as *niacin.* One of the B vitamins.

NITROGEN-FREE EXTRACT   The part of feed dry matter that is not crude protein, crude fat, crude fiber, or ash. It consists mostly of sugars and starches. Sometimes referred to as *NFE.*

NONESSENTIAL AMINO ACID   Any one of several amino acids that are required by animals but can be synthesized in adequate amounts by an animal in its tissues from other amino acids.

NONPROTEIN   Any one of a group of ammoniacal nitrogen containing compounds that are not true proteins. Urea is a common example.

NONRUMINANT   A simple-stomached animal that does not ruminate. Examples are swine, horses, dogs, and humans.

NUTRIENT   Any chemical compound having specific functions in the nutritive support of animal life.

NUTRITURE   Nutritional status.

OBESE   Being overweight due to a surplus of body fat.

OIL   Usually a mixture of pure fats that is liquid at room temperature.

OLD PROCESS   Pertains to the extraction of oil from seeds. Same as *hydraulic process.*

OLEIC ACID   An 18-carbon unsaturated fatty acid (one double bond) that reacts with glycerol to form olein.

OLEIN   The fat formed from the reaction of oleic acid with glycerol.

OMASUM   The third compartment of a ruminant's stomach. Sometimes called the *manyplies.*

ORGANIC   Refers to chemical compounds that contain carbon in chain structure.

ORGANIC ACID   Any organic compound that contains a carboxyl group (COOH).

ORTS   The portion of an animal's feed that it refuses to eat.

OSMOSIS   The passage of a solute or a solution through a semipermeable membrane toward effecting an equalization of the concentration of the fluids on opposite sides of the membrane.

OSMOTIC PRESSURE   The pressure exerted by the movement of a solvent through a semipermeable membrane toward equalizing solution concentration on opposite sides of the membrane.

OSSIFICATION   The process of bone formation.

OSTEITIS   Inflammation of a bone.

OSTEOMALACIA   A weakening of the bones due to a calcium, phosphorus, and/or vitamin D deficiency. A negative calcium and phosphorus balance.

OSTEOPOROSIS   An abnormal porousness of bone as the result of a calcium, phosphorus, and/or vitamin D deficiency.

OVULATION   The discharge of the ovum or egg from the graafian follicle of the ovary.

PABA   Para-aminobenzoic acid.

PALMITIC ACID   A 16-carbon saturated fatty acid.

PALMITIN   The fat formed from the reaction of palmitic acid with glycerol.

PANCREAS   A large, elongated gland located near the stomach. It produces pancreatic juice, which is secreted into the upper small intestine via the pancreatic duct.

PANCREATIC JUICE   A thick, transparent liquid secreted by the pancreas into the upper small intestine. It contains the enzymes pancreatic amylase, pancreatic lipase, and trypsin, and the hormone insulin.

PANDEMIC   Widely spread throughout several countries.

PANTOTHENIC ACID   One of the B vitamins.

PAPILLAE   Small nipple-shaped projections located on the interior of the rumen wall.

PARA-AMINOBENZOIC ACID   One of the B vitamins. Often abbreviated *PABA.*

PARAKERATOSIS   Any abnormality of the outermost or horny layer of the skin.

PARALYSIS   Loss of power of voluntary motion.

PARATHYROID   Any 1 of 4 small glands situated beside the thyroid gland, concerned chiefly with calcium and phosphorus metabolism.

PARTURITION   The act of giving birth to young.

PASTURE   Forages that are harvested by grazing animals.

PATHOGEN   Any disease-producing microorganism or material.

PATHOLOGY   The branch of medicine that deals with the special nature of disease.

PELLETS   Compacted particles of feed formed by forcing ground material through die openings.

PENTOSAN   A pentose-based polysaccharide having the general formula $(C_5H_8O_4)_n$. Araban and xylan are examples. Not nearly as abundant as the hexosans.

PENTOSE   A 5-carbon monosaccharide having the formula $C_5H_{10}O_5$. Arabinose, xylose, and ribose are examples. Not abundant in the free form in nature.

PEPSIN   The proteolytic enzyme present in the gastric juice. It acts on protein to form proteoses, peptones, and peptides.

PERMEABLE   Capable of being penetrated.

PERSPIRATION   Sweat or the act of sweating.

pH   A measure of hydrogen ion concentration or the degree of acidity.

PHAGOCYTE   Any cell that can ingest particles or cells that are foreign or harmful to the body.

PHENYLALANINE   One of the essential amino acids.

PHOSPHOLIPIDS   Fat-like substances containing phosphorus and nitrogen, along with fatty acids and cholesterol.

PHYSIOLOGICAL   Pertaining to the science that deals with the functions of living organisms or their parts.

PITUITARY   A gland in the lower part of the brain that produces a number of hormones.

PLASMA   The colorless fluid portion of the blood in which the corpuscles are suspended.

POLYSACCHARIDE   Any one of a group of carbohydrates consisting of a combination of a large but undetermined number of monosaccharide molecules, such as starch, dextrin, glycogen, cellulose, inulin, etc.

POSTPARTUM   Following the birth of young.

POTENT   Strong, powerful, concentrated.

POULTRY LITTER   The fibrous material used on the floor of poultry houses along with the excreta that accumulates therein.

PRECONCEPTIONAL   Before pregnancy.

PRECURSOR   A compound that can be used by the body to form another compound.

PREGNANT   The state of having a developing embryo in the body. Gravid.

PRE-MIX   A uniform mixture of 1 or more microingredients and a carrier, used in the introduction of microingredients into a larger mixture.

PRESSURE COOKER   An airtight container for the cooking of feed at high temperature under steam pressure.

PROGESTERONE   A sex hormone produced by the corpus lutea of the ovary.

PROLINE   One of the nonessential amino acids.

PROPIONIC ACID   One of the volatile fatty acids with the formula CH_3CH_2COOH, commonly found in rumen contents but not in silage.

PROTEIN   Any one of many complex organic nitrogenous compounds formed from various combinations of different amino acids.

PROTOPLASM   The essential protein substance of living cells.

PROTOZOA   Very small, unicellular animal organisms.

PROVITAMIN A   Carotene.

PUBERTY   The age at which the reproductive organs become functionally active.

PULP   The solid residue that remains following the removal of the juices from plant materials.

PUTREFACTION   The decomposition of proteins by microorganisms under anaerobic conditions.

PYREXIA   A feverish condition.

PYRIDOXINE   The same as *vitamin B6*.

RADIOACTIVE   Giving off atomic energy in the form of alpha, beta, or gamma rays.

RADIOISOTOPE   A radioactive form of an element.

RANCID   A term used to describe fats that have undergone partial decomposition.

RANGE CUBES   Large pellets produced for feeding in the pasture on the ground.

RENNIN   The milk-curdling enzyme present in the gastric juice of milk-consuming animals.

RESIDUE   That which remains of a particular substance.

RESORPTION   A return of the nutritive components of a partially formed fetus and fetal membrane to the system of the mother.

RESPIRATION   The act of breathing.

RETICULUM   The second compartment of a ruminant's stomach. Also called the *honeycomb* or *waterbag*.

RIBOFLAVIN   Same as *vitamin B2*. Formerly known as vitamin G.

ROLLED   Compressed into flat particles by having been passed between rollers.

ROUGHAGE   Any feed high (more than about 20%) in crude fiber and low (less than about 60%) in TDN, on an air-dry basis. Opposite of concentrate.

RUMEN   The first compartment of a ruminant's stomach. Also called the *paunch*.

RUMINANT   Any of a group of hoofed mammals that have a 4-compartment stomach and that ruminate or chew a cud. Examples are cattle, sheep, goats, and deer.

RUMINATE   To regurgitate previously eaten feed for further chewing. To chew a cud.

SALIVA   A clear, somewhat viscid solution secreted by the salivary glands into the mouth. It contains the enzymes salivary amylase and salivary maltase.

SALMONELLA   A pathogenic, diarrhea-producing organism sometimes present in contaminated feeds.

SAPONIFIABLE   Having the capacity to react with alkali to form soap.

SAPONIFICATION   The formation of soap and glycerol from the reaction of fat with alkali.

SARCOMA   A tumor of fleshy consistency—often highly malignant.

SATURATED FAT   A fat formed from the reaction of glycerol with any one of several saturated fatty acids. Stearin and palmitin are examples.

SATURATED FATTY ACID   Any one of several fatty acids containing no double bonds. Stearic and palmitic acids are examples.

SEDENTARY   Sitting most of the time.

SELF-FED   Provided with a part or all of the ration on a continuous basis, thus permitting the animal to eat at will.

SEMI-DISPENSABLE AMINO ACID   An amino acid that is essential only under certain circumstances or that may replace in part one of the essential amino acids. Arginine, cystine, and tyrosine fall into this group.

SEPTICEMIA   A diseased condition resulting from the presence of pathogenic bacteria and their associated poisons in the blood.

SERINE   One of the nonessential amino acids.

SERUM   The colorless fluid portion of blood remaining after clotting and removal of corpuscles. It differs from plasma in that the fibrinogen has been removed.

SHORTS   A by-product of flour milling consisting of a mixture of small particles of bran and germ, the aleurone layer, and coarse flour.

SILAGE   The feed resulting from the storage and fermentation of green or wet crops under anaerobic conditions.

SILO   A semi-airtight to airtight structure designed for use in the production and storage of silage.

SOAP   A compound formed along with glycerol from the reaction of fat with alkali.

SOLID   A substance that does not perceptibly flow.

SOLUTION   A uniform liquid mixture of 2 or more substances molecularly dispersed within one another.

SOLVENT PROCESS   A process for the extraction of oil from seeds involving the use of an organic solvent.

SPECIFIC GRAVITY   The ratio of the weight of a body to the weight of an equal volume of water.

SPECIFIC HEAT   The heat-absorbing capacity of a substance in relation to that of water.

SPORE   An inactive reproductive form of certain microorganisms.

STABILIZED   Made more resistant to chemical change by the addition of a particular substance.

STARCH   A polysaccharide having the formula $(C_6H_{10}O_5)_n$. An important source of energy for livestock. Yields glucose upon complete hydrolysis.

STEARIC ACID   An 18-carbon saturated fatty acid that reacts with glycerol to form stearin.

STEARIN   The fat formed from the reaction of stearic acid with glycerol.

STERILE   Free from living microorganisms. Also, not capable of producing young.

STERILITY   An inability to produce young.

STEROL   One of a class of complex, fat-like substances widely distributed in nature.

STOMACH   The part of the digestive tract lying between the esophagus and the small intestine. A 4-compartment organ in ruminants; a single compartment organ in nonruminants.

STRAW   The part of the mature plant remaining after the removal of the seed by threshing or combining.

STRESS   Any circumstance that tends to disrupt the normal, steady functioning of the body and its parts.

SUBSTRATE   A substance upon which an enzyme acts. Same as *zymolyte*.

SUCRASE   An enzyme present in intestinal juice that acts on sucrose to produce glucose and fructose.

SUCROSE   A disaccharide having the formula $C_{12}H_{22}O_{11}$. It hydrolyzes to glucose and fructose. Commonly known as *cane, beet,* or *table sugar.*

SUN-CURED   Dried by exposure to the sun.

SUPPLEMENT   A semi-concentrated source of 1 or more nutrients used to enhance the nutritional adequacy of a daily ration or a complete ration mixture.

SYNDROME   A medical term meaning a set of symptoms that occur together.

SYNTHESIS   The bringing together of 2 or more substances to form a new material.

TDN   Total digestible nutrients.

TETANY   A syndrome involving sharp flexion of the wrist and ankle joints, muscle twitching, cramps, and convulsions.

THERAPEUTIC   Pertaining to the medical treatment of disease.

THERAPY   The medical treatment of disease.

THERMAL Refers to heat.

THIAMINE The same as *thiamin, thiamine hydrochloride,* or *vitamin B₁*.

THREONINE One of the essential amino acids.

THROMBOSIS The obstruction of a blood vessel by the formation of a blood clot.

THYROID The gland in the neck that secretes the hormone thyroxin.

TOCOPHEROL Any of 4 different forms of an alcohol that is also known as *vitamin E*.

TOTAL DIGESTIBLE NUTRIENTS A figure that indicates the relative energy value of a feed to an animal. It is the sum of the digestible protein, digestible nitrogen-free extract, digestible crude fiber, and ($2.25 \times$ the digestible fat).

TOXIC Of a poisonous nature.

TRACE MINERAL Any one of several mineral elements that are required by animals in very minute amounts. Same as *micromineral*.

TRACER ELEMENT A radioactive element used in biological and other research to trace the fate of a substance.

TRAUMA A wound or injury.

TRUE PROTEIN A nitrogenous compound that will hydrolyze completely to amino acids.

TRYPTOPHAN One of the essential amino acids.

TYROSINE One of the nonessential amino acids.

UIP Undegraded intake protein (not broken down in the rumen).

UNSATURATED FAT A fat formed from the reaction of glycerol with any one of several unsaturated fatty acids. Olein and linolein are examples.

UNSATURATED FATTY ACIDS Any one of several fatty acids containing one or more double bonds. Oleic, linoleic, linolenic, and arachidonic acids are examples.

UREA A white, crystalline, water-soluble substance with the formula $CO(NH_2)_2$. It is the most extensively used source of nonprotein nitrogen for animal feeding.

UREASE An enzyme that acts on urea to produce carbon dioxide and ammonia. It is found in the jackbean and the soybean and is produced by certain microorganisms in the rumen.

UREMIA A toxic accumulation of urinary constituents in the blood.

VALINE One of the essential amino acids.

VASCULAR Pertaining to the blood vessels of the body.

VERTEBRATES Animals with backbones.

VFA Volatile fatty acid(s).

VILLI Small thread-like projections attached to the interior side of the wall of the small intestine.

VISCERA   The organs of the great cavities of the body that are normally removed at slaughter.

VISCOSITY   The freedom of flow of liquids.

VITAMIN   One of a group of organic substances that in relatively small amounts are essential for life.

VOLATILE FATTY ACID   Any one of several volatile organic acids found especially in rumen contents and/or silage. Acetic, propionic, and butyric acids are ordinarily the most prevalent.

WET-RENDERED   Cooked with steam under pressure in closed tanks.

WHEY   The watery portion of milk remaining after the removal of the fat and curd.

WORK   Movement of matter through space.

Index

G

W

Y

Z